Forest Management and Planning

Forest Management and Planning

Second Edition

Pete Bettinger

Warnell School of Forestry and Natural Resources,
University of Georgia,
Athens, GA, United States

Kevin Boston

Department of Forestry and Wildland Sciences,
Humboldt State University,
Arcata, CA, United States

Jacek P. Siry

Warnell School of Forestry and Natural Resources,
University of Georgia,
Athens, GA, United States

Donald L. Grebner

Department of Forestry,
Mississippi State University,
Mississippi State, MS, United States

ACADEMIC PRESS

An imprint of Elsevier
elsevier.com

Academic Press is an imprint of Elsevier
125 London Wall, London EC2Y 5AS, United Kingdom
525 B Street, Suite 1800, San Diego, CA 92101-4495, United States
50 Hampshire Street, 5th Floor, Cambridge, MA 02139, United States
The Boulevard, Langford Lane, Kidlington, Oxford OX5 1GB, United Kingdom

Notices
Knowledge and best practice in this field are constantly changing. As new research and experience broaden our understanding, changes in research methods, professional practices, or medical treatment may become necessary.

Practitioners and researchers must always rely on their own experience and knowledge in evaluating and using any information, methods, compounds, or experiments described herein. In using such information or methods they should be mindful of their own safety and the safety of others, including parties for whom they have a professional responsibility.

To the fullest extent of the law, neither the Publisher nor the authors, contributors, or editors, assume any liability for any injury and/or damage to persons or property as a matter of products liability, negligence or otherwise, or from any use or operation of any methods, products, instructions, or ideas contained in the material herein.

British Library Cataloguing-in-Publication Data
A catalogue record for this book is available from the British Library

Library of Congress Cataloging-in-Publication Data
A catalog record for this book is available from the Library of Congress

ISBN: 978-0-12-809476-1

For Information on all Academic Press publications
visit our website at https://www.elsevier.com

  Working together to grow libraries in developing countries

www.elsevier.com • www.bookaid.org

Publisher: Candice Janco
Acquisition Editor: Mary Preap
Editorial Project Manager: Pat Gonzalez
Production Project Manager: Lucía Pérez
Designer: Greg Harris

Typeset by MPS Limited, Chennai, India

Dedication

This book is dedicated to our wives, children, families, and students

Contents

Preface   xi

1. Management of Forests and Other Natural Resources

I. Introduction   1
II. Forest Management   2
III. The Need for Forest Management Plans   3
    A. The Necessity of Plans, Planners, and Planning Processes   3
    B. Information Necessary to Develop a Forest Management Plan   4
    C. Plan Development Challenges   6
IV. General Emphasis of Forest Management Plans   7
    A. Organization-Specific Plans   8
    B. Landscape Plans   8
    C. Community or Cooperative Forest Plans   8
    D. Ecological Approaches to Plans   9
    E. Urban Forest Plans   9
V. Example Forest Plans   10
    A. North American Small Private Landowner Plan   10
    B. North American National Forest Plan   10
    C. Australian State Forest Plan   10
    D. European Estate Plan   11
    E. Asian Private and Communal Forest Area Management Plan   11
    F. South American Community Forest Plan   12
    G. African Participatory Management Plan   12
    H. North American Urban Forest Plan   13
    I. North American Industrial Forest Plan   13
VI. Characterizing the Decision-Making Process   13
    A. A View From the Management Sciences   13
    B. A Broad View on Planning Within Natural Resource Management Organizations   15
    C. A Hierarchy of Planning Within Natural Resource Management Organizations   17
VII. Summary   18
Questions   18
References   19

2. Valuing and Characterizing Forest Conditions

I. Introduction   21
II. Structural Evaluation of Natural Resources   21
    A. Trees per Unit Area   21
    B. Average Diameter of Trees   22
    C. Diameter Distribution of Trees   23
    D. Basal Area   23
    E. Quadratic Mean Diameter   24
    F. Average Height   24
    G. Timber Volume   24
    H. Mean Annual Increment, Periodic Annual Increment   25
    I. Snags per Unit Area   27
    J. Down Woody Debris   27
    K. Crown or Canopy Cover   28
    L. Tree, Stand, or Forest Age   28
    M. Biomass and Carbon   30
    N. Pine Straw   30
    O. Other Nontimber Forest Products   30
    P. Site Quality   31
    Q. Stocking and Density   33
III. Economic Evaluation of Natural Resources   34
    A. Basic Concepts: Present and Future Values   35
    B. Prices and Costs   43
    C. Net Present Value   46
    D. Internal Rate of Return   47
    E. Benefit/Cost Ratio   48
    F. Equal Annual Equivalent   48
    G. Soil Expectation Value   49
    H. Other Mixed-Method Economic Assessments   50
    I. Selecting Discount Rates   51
    J. Forest Taxation   52

IV. Environmental and Social Evaluation
of Natural Resources   53
  A. Habitat Suitability   53
  B. Recreation Values   54
  C. Water Resources   55
  D. Aquatic Habitat Values   56
  E. Air Quality   56
  F. Income and Employment   58
V. Summary   58
Questions   59
References   61

3. **Geographic Information and Land Classification in Support of Forest Planning**

I. Introduction   65
II. Geographic Information Systems   65
  A. Geographic Data Collection
    Processes   66
  B. Geographic Data Structures   68
  C. Geographic Data Used in This
    Book   70
  D. Geographic Information Processes   71
III. Land Classification   76
  A. Strata-based Land Classifications   80
  B. Land Classification Based on Units
    of Land   82
  C. Land Classification Based on Spatial
    Position   82
IV. Summary   84
Questions   84
References   84

4. **Estimation and Projection of Stand and Forest Conditions**

I. Introduction   87
II. The Growth of Forests   87
  A. Growth of Even-Aged Stands   89
  B. Growth of Uneven-Aged Forests   91
  C. Growth of Two-Aged Forests   94
  D. Growth Transition Through Time   94
III. Projecting Stand Conditions   101
  A. Growth and Yield Tables   101
  B. Growth and Yield Simulators   103
  C. Brief Summary of Some Growth
    and Yield Simulators   106
IV. Output From Growth and Yield
  Models   108
V. Model Evaluation   108
VI. Summary   109
Questions   109
References   110

5. **Optimization of Tree- and Stand-Level Objectives**

I. Introduction   113
II. Tree-level Optimization   114
III. Stand-level Optimization   116
  A. Optimum Timber Rotation   116
  B. Optimum Thinning Timing   120
  C. Optimum Stand Density or Stocking 121
  D. Recent Developments in the
    Scientific Literature   122
IV. Decision Tree Analysis   122
V. Mathematical Models for Optimizing
  Stand-level Management Regimes   124
VI. Dynamic Programming   124
  A. Recursive Relationships   125
  B. Caveats of Dynamic Programming   126
  C. Disadvantages of Dynamic
    Programming   126
  D. Dynamic Programming Example—
    An Evening Out   126
  E. Dynamic Programming Example—
    Western Stand Thinning, Fixed
    Rotation Length   129
  F. Dynamic Programming
    Example—Southern Stand Thinning,
    Varying Rotation Lengths   132
VII. Summary   135
Questions   136
References   137

6. **Graphical Solution Techniques for Two-Variable Linear Problems**

I. Introduction   139
II. Example Problems in Natural Resource
  Management   140
  A. The Road Construction Plan   140
  B. The Plan for Developing Snags
    to Enhance Wildlife Habitat   144
  C. The Plan for Fish Habitat Development 146
  D. The Hurricane Clean-up Plan   147
III. Optimality, Feasibility, and Efficiency 148
IV. Summary   150
Questions   151
References   152

7. **Linear Programming**

I. Introduction   153
II. Four Assumptions Inherent in
  Standard Linear Programming Models   153
  A. The Assumption of Proportionality   154
  B. The Assumption of Additivity   154
  C. The Assumption of Divisibility   154
  D. The Assumption of Certainty   154

III. Objective Functions for Linear
Programming Problems   154
IV. Accounting Rows for Linear
Programming Problems   156
   A. Accounting Rows Related to Land
      Areas Scheduled for Treatment   156
   B. Wood Flow-Related Accounting
      Rows   157
   C. Habitat-Related Accounting Rows   158
V. Constraints for Linear Programming
Problems   160
   A. Resource Constraints   160
   B. Policy Constraints   160
VI. Detached Coefficient Matrix   162
VII. Model I, II, and III Linear
Programming Problems   163
VIII. Interpretation of Results Generated
from Linear Programming Problems   164
   A. Objective Function Value, Variable
      Values, and Reduced Costs   165
   B. Slack and Dual Prices   166
IX. Assessing Alternative Management
Scenarios   167
X. Case Study: Western United States
Forest   168
XI. Case Study: Northern United States
Hardwood Forest   170
XII. Summary   174
Questions   175
References   176

8. Advanced Planning Techniques

I. Introduction   177
II. Extensions to Linear Programming   177
   A. Mixed Integer Programming   178
   B. Integer Programming   179
   C. Goal Programming   182
III. Binary Search   184
IV. Heuristic Methods   189
   A. Monte Carlo Simulation   190
   B. Simulated Annealing   190
   C. Threshold Accepting   191
   D. Tabu Search   192
   E. Genetic Algorithms   192
   F. Other Heuristics   194
V. Forest Planning Software   194
   A. Spectrum   194
   B. Habplan   195
   C. Magis   195
   D. Remsoft Spatial Planning System
      (Woodstock/Stanley)   195
   E. Tigermoth   196
VI. Summary   196
Questions   197
References   198

9. Forest and Natural Resource
Sustainability

I. Introduction   201
II. Sustainability of Production   202
III. Sustainability of Multiple Uses   205
IV. Sustainability of Ecosystems and
Social Values   206
V. Incorporating Measures of Sustainability
Into Forest Plans   208
VI. Sustainability Beyond the Immediate
Forest   211
VII. Summary   211
Questions   212
References   212

10. Models of Desired Forest Structure

I. Introduction   215
II. The Normal Forest   215
III. The Regulated Forest   221
IV. Irregular Forest Structures   222
V. Structures Guided by a Historical
Range of Variability   224
VI. Structures Not Easily Classified   225
VII. Summary   226
Questions   227
References   227

11. Control Techniques for Commodity
Production and Wildlife Objectives

I. Controlling the Area Scheduled   229
II. Controlling the Volume Scheduled   231
   A. The Hanzlik Formula for Volume
      Control   233
   B. The Von Mantel Formula for Volume
      Control   235
   C. The Austrian Formula for Volume
      Control   236
   D. The Hundeshagen Formula for
      Volume Control   238
   E. The Meyer Amortization Method for
      Volume Control   238
   F. The Heyer Method for Volume
      Control   239
   G. Structural Methods for Volume
      Control   240
III. Application of Area and Volume
Control to the Putnam Tract   240
   A. Area Control   241
   B. Volume Control—Hanzlik Formula   241
   C. Volume Control—Von Mantel
      Formula   241

D. Volume Control—Austrian Formula   241
E. Volume Control—Hundeshagen
Formula   241
F. Volume Control—Meyer Formula   242
IV. Area—Volume Check   242
V. Wildlife Habitat Control   242
VI. The Allowable Cut Effect   243
VII. Summary   245
Questions   246
References   246

12. Spatial Restrictions and Considerations in Forest Planning

I. Adjacency and Green-up Rules as
They Relate to Clearcut Harvesting   249
II. Adjacency and Green-up of Group
Selection Patch Harvests   254
III. Habitat Quality Considerations   255
A. Case 1: Elk Habitat Quality   255
B. Case 2: Bird Species Habitat
Considerations   255
C. Case 3: Red-Cockaded Woodpecker
Habitat Considerations   256
D. Case 4: Spotted Owl Habitat Quality   258
IV. Road and Trail Maintenance
and Construction   260
A. Case 1: Road Management Problem   261
B. Case 2: Trail Development Problem   262
V. Summary   264
Questions   265
References   266

13. Hierarchical System for Planning and Scheduling Management Activities

I. Strategic Planning   269
II. Tactical Planning   270
III. Operational Planning   271
IV. Vertical Integration of Planning
Processes   272
V. Blended, Combined, and Adaptive
Approaches   273
VI. Your Involvement in Forest Planning
Processes   275
VII. Summary   275
Questions   276
References   276

14. Forest Supply Chain Management

I. Introduction   279
II. Components of a Forestry Supply Chain   281
III. Association With the Hierarchy
of Forest Planning   282

IV. Mathematical Formulations Associated
With Forestry Supply Chain Components   286
V. Sources of Variation in the Forestry
Supply Chain   288
VI. Summary   289
Questions   289
References   290

15. Forest Certification and Carbon Sequestration

I. Introduction   291
II. Forest Certification Programs   294
A. Sustainable Forestry Initiative   295
B. Forest Stewardship Council   296
C. American Tree Farm System   297
D. Green Tag Forestry System   298
E. Canadian Standards Association   298
F. International Organization for
Standardization, Standard 14001   299
G. Programme for the Endorsement
of Forest Certification   299
III. Cost and Benefits of Forest Certification   299
IV. Forest Carbon Sequestration   300
V. Opportunities and Challenges in
Increasing Forest Carbon Storage   301
VI. Emissions Trading   301
VII. Selected US Carbon Reporting and
Trading Schemes   302
VIII. Forest Carbon Implications for
Forest Management   303
IX. Summary   304
Questions   304
References   304

16. Scenario Analysis in Support of Strategic Planning

I. Introduction   307
II. An Overview of the Role
of Scenario Analysis   308
III. Developing Scenarios   310
IV. Applying Scenario Analysis
to Forest Planning   312
V. Summary   313
Questions   313
References   313

Appendix A: Databases Used Throughout *Forest
Management and Planning*   315
Appendix B: The Simplex Method for Solving
Linear Planning Problems   331
Appendix C: Writing a Memorandum or Report   339
Index   343

Preface

Forest Management and Planning arose from our desire to provide for students in forestry and natural resource management programs a focused treatment of the topics that are important for upper-level forest management courses. This book presents an extensive overview of the methodology one might use to develop forest and natural resource management plans, and to analyze a number of resource issues that are encountered by managers. A portion of the book is devoted to the development of information to support stand-level and forest-level management planning processes. In this regard, we discuss commonly used economic and ecological criteria for assessing the value and relative differences between plans of action at both the stand- and forest-level. At the forest-level, we emphasize the development of traditional commodity production forest plans as well as the development of forest plans containing both wildlife goals and other ecosystem services. We also present alternative methods for developing forest-level plans, such as those that involve discrete *yes* or *no* management decisions.

Many of the topics included in upper-level university natural resource management courses have remained stable over the past 40 years. These topics generally include economic and physiological assessments of forest structure to determine whether proposed courses of action can meet a landowner's needs. However, quantitative forest planning has broadened and now includes complex wildlife goals, spatial restrictions on forest management plans, and other advanced issues. In addition, forest sustainability and forest certification are central issues for land management organizations, and wood supply chain-of-custody and carbon certification issues are now becoming important in forest management planning. Therefore, although this book begins with a discussion of methods for assessing and valuing fine-scale decisions (e.g., a single project), it concludes with discussions of how we might use them to address broader-scale issues for the management of natural resources.

Our various experiences in forest management over the last 40 years have helped us to craft this book. While each of the authors has taken and taught forest management courses, we have also acquired valuable practical experience throughout North and Central America,

Oceania, Asia, and Europe. Although we currently work in academia, we have worked for the forest industry and forestry consultants, as well as state, federal, and international organizations. In addition, our extensive travels have allowed us to gain experience and understanding of forest management challenges in other parts of the world. Our goal was to develop a book that avoided taking an advocacy position on important topics such as sustainability and forest certification, since many of these alternative management paradigms are used in today's natural resource management environment. We attempt to provide impartial treatment of these types of topics, since many are value-laden. As a result, the book provides an overview of the issues and discusses many of the challenges and opportunities related to managing forests under alternative philosophies.

The first part of *Forest Management and Planning* describes the management planning process (see Chapter 1: Management of Forests and Other Natural Resources) and the development of information necessary for valuing and characterizing forest conditions (see Chapter 2: Valuing and Characterizing Forest Conditions). Included in Chapter 2, are physical, economic, and ecological methods for valuing and characterizing forest conditions. The first part of the book also provides an overview of geographic databases (see Chapter 3: Geographic Information and Land Classification in Support of Forest Planning) and the methods used to estimate and project conditions into the future (see Chapter 4: Estimation and Projection of Stand and Forest Conditions). We then turn our attention to tree- and stand-level optimization techniques (see Chapter 5: Optimization of Tree- and Stand-Level Objectives), graphical techniques for envisioning linear planning problems (see Chapter 6: Graphical Solution Techniques for Two-Variable Linear Problems), and linear programming (see Chapter 7: Linear Programming), a commonly used mathematical problem-solving technique. Chapter 8, Advanced Planning Techniques, focuses on advanced forest planning techniques such as mixed-integer programming, goal programming, binary search, and heuristics. Forest-level planning generally utilizes linear programming or these advanced techniques, thus an understanding of their similarities and differences is important for natural resource managers. Starting with Chapter 9, Forest and Natural Resource Sustainability), we begin to associate the planning techniques with broader issues prevalent within the field of natural resource management. Chapter 10, Models of Desired Forest Structure, describes a number of models of desired forest structure, and Chapter 11, Control Techniques for Commodity Production and Wildlife Objectives, discusses a number of control techniques that one might use to move forests to a desired structure. Here one will find the classical concepts of area and volume control. Spatial restrictions are increasingly being incorporated into forest plans, therefore we provide a discussion of several of these in Chapter 12, Spatial Restrictions and Considerations in Forest. The remaining chapters of the book cover other issues of importance in forest management and planning, including the hierarchy of planning processes typically found in land management organizations (see Chapter 13: Hierarchical System for Planning and Scheduling Management Activities), the wood supply chain and its management (see Chapter 14: Forest Supply Chain Management), and forest certification and carbon sequestration (see Chapter 15: Forest Certification and Carbon Sequestration). New to the second edition, Chapter 16, Scenario Analysis in Support of Strategic Planning, provides a discussion of how scenario analysis might be used to further explore the trade-offs among alterative strategic forest plans.

Three appendices are provided in this book to enhance the learning process. Appendix A, Databases Used Throughout *Forest Management and Planning*, provides data that is used in a number of examples throughout the book. One set of data involves a 100-year projection of a single western North American conifer stand, using 5-year time period increments. The development of the stand in each time period is illustrated with a stand table and several summary statistics. Two forests, composed of 80 or more stands, are described in the

Appendix as well. The geographic information system databases related to these forests can be acquired from the authors. Appendix B, The Simplex Method for Solving Linear Planning Problems, provides a description of the Simplex Method, which is a process used within linear programming to locate optimal solutions to linear planning problems. Appendix C, Writing a Memorandum or Report, provides a discussion and helpful hints for writing memorandums and reports.

Although the book contains a number of graphics to help students visualize management problems, we incorporated several photographs as well to associate the concepts described to the management of the land. Most of the photographs provided in the book were captured by Kelly A. Bettinger, a wildlife biologist, through her extensive travels. The exception is the photograph of Hurricane Katrina storm damage in Chapter 6, which was taken by Andrew J. Londo, assistant director for agriculture and natural resources extension at Ohio State University. The photograph on the cover of the book is from Durango, Mexico, courtesy of Donald L. Grebner. Finally, we are grateful for the review of our uneven-aged forest linear programming model contained in Chapter 7, by Dr. John Wagner, Professor of Forest Resource Economics at the State University of New York, College of Environmental Sciences and Forestry (SUNY-ESF).

We hope that readers of this book will find it to be a useful learning tool and a valuable reference in their future careers in natural resource management. Our goal is to provide readers with descriptions and examples of forest management and planning tools, so that they may become confident and competent natural resource managers.

Pete Bettinger
Kevin Boston
Jacek P. Siry
Donald L. Grebner

Chapter 1

Management of Forests and Other Natural Resources

Objectives

As we progress through the 21st century, and as the human population continues to expand, the management of natural resources is becoming one of maintaining the consumptive needs of society while also caring for the integrity and function of ecological systems. A large number of natural resource managers today continue to manage for wood production objectives, which in itself is a noble endeavor. A large number of natural resource managers also research and advise on the management of forests as it relates to wildlife, fisheries, recreational, and other environmental and social services. On many lands a balance must be struck between commodity production and ecosystem goals. This balance is explored through planning processes performed at the national, regional, and local levels. This introductory chapter covers issues related to forest management and planning and the decision-making environment within which we must operate. To be successful as land and resource managers, we must understand the system within which we work, as well as the social system within which we live and participate as professionals. Upon completion of this introductory chapter, you should be able to:

1. Understand the basic forms of decision-making processes as viewed by the management sciences.
2. Understand the steps in a general planning process, and how they might vary from one natural resource management organization to the next.
3. Understand the hierarchy of planning common to natural resource management organizations.
4. Understand the challenges related to natural resource planning.
5. Understand how information related to planning efforts flows within an organization.

I. INTRODUCTION

The management of forested lands is an important endeavor. As a society, we expect that forest land managers will meet our current needs for forest-related services and sustain forest resources so that future generations of people will be able to enjoy the various outcomes from forests that we enjoy today. The ability to meet this expectation is often expressed through a plan, which might include statements that reflect our beliefs of what the management of the land may provide. For example, various actions involved in the management of forests may lead to the generation of revenue or supply of forest products. A plan might then describe how these actions maintain, improve, or otherwise affect aesthetic values, biodiversity, the water producing value of a forest, or the productive capacity of the land. Usually a plan describes desired forest conditions and illustrates land use allocations, along with a description of lands suitable for various management activities. A plan is informed by the management practices appropriate for the land, and the objectives and constraints of the landowner. Often, the goals of the landowner are addressed through actions, which may or may not be financially beneficial, yet which address their perspective on sustainability. These ideas are not new; the thoughts provided in this paragraph were drawn from both a recent United States Forest Service management plan (US Department of Agriculture, Forest Service, 2014) and a proposed plan for forests in upstate New York that was developed over a century ago (Hosmer and Bruce, 1901). Both reflect what was noted in a review by Olson (2010), that the tension between use of the landscape and the need to prevent overuse of the landscape is the heart of the problem for landowners and land managers. For many reasons, the more recent of the two plans is more extensive in its evaluation of resources, yet the themes of the plans are essentially the same even though the perspectives on sustainability may differ.

The need for management and planning of forests perhaps becomes stronger every year as human populations continue to increase, as societal values evolve, and as immediate expression of thoughts and ideas are facilitated by the Internet. Often, forest planning situations are unique with regard to the problem setting, the character of forests, the risks involved, the long-term vision of the land manager or landowner, and the desires of the populace (Korjus,

Forest Management and Planning. DOI: http://dx.doi.org/10.1016/B978-0-12-809476-1.00001-1

2014). This book therefore presents concepts, new and old, that help landowners and land managers develop and evaluate plans of action for forests.

II. FOREST MANAGEMENT

Forest management involves the integration of silvicultural practices and business concepts (e.g., analyzing economic alternatives) in such a way as to best achieve a landowner's objectives. Management of forests requires a plan (however developed), and an assessment of the activities necessary to meet the objectives. In addition, a recognition of the

important ecological and social concerns associated with a forest may influence the character and depth of a plan. In a more general way, forest management can involve the collective application of silvicultural practices so that an entire forest remains healthy and vigorous by imposing treatments on the various stands (Heiligmann, 2002). The range of forest management activities (Table 1.1) can include those focused on the economics of forest businesses, or on the ecology of the ecosystem. Activities can include tree planting, herbaceous weed control, fertilization, precommercial thinning, commercial thinning, final harvests, harvests for habitat improvement, preservation, road construction, road

TABLE 1.1 Types of Management Activities a Land Manager Might Consider

Activity	For Even-Aged Forests	For Uneven-Aged Forests
Site Preparation		
Burn	✓	
Chop	✓	
Rake	✓	
Plow	✓	
Bed	✓	
Herbicide application	✓	
Tree Establishment		
Plant	✓	
Coppice		✓
Seed	✓	✓
Early Tending		
Release	✓	✓
Weed	✓	
Fuel reduction	✓	✓
Prune	✓	
Prescribed burn	✓	✓
Fertilization	✓	
Tree Cutting Activities		
Precommercial thin	✓	
Commercial thin	✓	
Shelterwood	✓	
Seed tree harvest	✓	
Single tree selection harvest		✓
Variable retention harvest		✓
Group selection harvest		✓
Clearcut	✓	

Source: Grebner, D.L., Bettinger, P., Siry, J.P., 2013. Introduction to Forestry and Natural Resources. Academic Press, New York, NY. 508 p.

obliteration, and prescribed fire, among others. Each may have a cost and a benefit, depending on the objectives of the landowner. Choosing the timing and placement of activities is the main task of forest planning.

From a forest manager's perspective, activities implemented within a forest may affect the natural succession of forest growth. One way for a forest manager to view the development of a forest is to visualize the orderly change in character of a vegetative community over time, or the succession of vegetation. Forest succession is thus the sequential change in tree species, character, and structure of trees within a given area (Grebner et al., 2013), either naturally or through human intervention. *Primary succession* is one of two types of ecological succession of plant life, and in our case relates to a forest becoming established in a barren area with no substrate (soil), such as land surfaces wiped clean by landslides or overtaken by sand, rock or lava. When forests become established in areas where substrate is available (e.g., after fires, after harvests), and which supported vegetation previously, the process is called *secondary succession*. Tree planting activities are one form of establishment of forests through secondary succession. Afforestation, the planting of trees on former agricultural or developed lands, is another form of secondary succession that has been used widely in Europe, China, and elsewhere over the last century (Krawczyk, 2014). Natural succession on these types of lands can also occur through seed distributed by wind, water and animals. It should be no surprise that management activities vary in their use from one region to another, and vary depending on the tree species desired. For example, after a final harvest in the southern United States, a land manager may use various site preparation practices (i.e., raking, herbaceous weed control) to develop a site suitable for planting a loblolly pine (*Pinus taeda*) forest. However, if the desire of the land manager was to develop a deciduous forest on this site, they may consider other practices to assure that the desired trees become established through growth from coppice (stumps or roots) or seed. The management of uneven-aged or multiaged forests may require other approaches that match desired conditions with natural disturbance regimes (O'Hara, 2009) and other functions of forests (e.g., hunting opportunities) that are desired by landowners.

Later in this book we discuss concepts related to forest and natural resource sustainability. In Chapter 9, Forest and Natural Resource Sustainability, we discuss the sustainability of timber production, multiple uses, and ecological systems. The term *sustainable forest management* tends to favor the latter two approaches, because those who use it suggest that it involves management actions that are ecologically sound, economically viable, and socially acceptable. This approach to forest management is similar to, if not consistent with, ecosystem-based forest management approaches, where management plans are developed within a larger framework, take a big-picture perspective, and involve a number of values derived in and around the area being managed (Palmer, 2000). We attempt to stay neutral when it comes to favoring any approach, since each form of sustainability is used today, depending on the landowner and the landowner's objectives. Thus our goal is to describe the approaches used in practice, and to provide some guidance for young professionals on the methods that might be used within each for developing a forest plan.

III. THE NEED FOR FOREST MANAGEMENT PLANS

Forest plans are descriptions of the activities that should be used to best meet the objectives a landowner has for their property. Managing a forest without a plan in mind may be guided by short-term operational considerations, but this may in turn have long-term undesirable or unforeseen consequences for the landowner (Demers et al., 2001). As a result, the planning process is an important aspect of forest management. If a forest plan is not carefully and thoughtfully prepared, the activities that are implemented in the near future may not yield the result that is desired by the landowner over a longer period of time. Most of the larger natural resource management organizations in North America have developed a plan of action for the land that they manage. However, many small forest landowners do not (Joshi et al., 2015; Butler et al., 2004). More broadly speaking, it has been estimated that management plans have been developed for 52% of the world's forests (Food and Agriculture Organization of the United Nations, 2010). Whether planning occurs through a traditional process that uses mathematical tools such as linear programming to allocate activities to forest strata, a more elaborate process that uses heuristic methods to develop a spatially explicit harvest schedule, or a seat-of-the-pants (back of the envelope, scratch of the head) method to determine what to do next, some form of planning is generally used. In many cases, quantitative relationships are employed to separate the better plans from the mediocre or poor plans.

A. The Necessity of Plans, Planners, and Planning Processes

Why do people develop natural resource management plans? Organizations that undergo forest planning generally are interested in plans that will provide them guidance for (1) implementing activities, (2) predicting future harvest levels, (3) optimizing the use of limited resources, and (4) maintaining or developing habitat areas, perhaps while simultaneously balancing several other concerns (budgets, personnel, etc.). Today's natural resource management environment in many areas of the world places as much, if not more, emphasis on ecological and social concerns as it does on economic or commodity production interests. It is imperative that natural resource managers efficiently use the resources at their disposal to meet the goals they consider important. To the displeasure of many college students, quantitative methods typically are used to justify or support decisions. These techniques include economic, biometric, and operations research analysis tools. To be an effective natural resource manager, and to be able to consider multiple objectives and constraints simultaneously, it may be necessary to use contemporary simulation and optimization techniques to assist in developing forest plans. Therefore, although students may not become an expert in these fields, they must understand how these tools can be applied, and how the outcomes can facilitate the development of a plan.

Periodically, we see natural resource management issues making headlines in the news media, which underscores one important responsibility entrusted to us as natural resource managers. That is, if we claim to manage land scientifically, and if our intent is to meet our landowner's objectives, then we need to be able to confidently and competently assess the conditions and outcomes of current and future forests, range, and wildlife habitat. If this is not possible, and if we cannot communicate well the trade-offs, then it will be difficult for us to convince our clients (the landowner, supervisor, stockholder, or the general public) that their goals are (or will be) met. It will also be difficult to convince the general public that we (natural resource managers) know what we are doing. To develop trust amongst various groups interested in the management of natural resources, land managers need to demonstrate that economic, ecological, and social goals are all being considered in the development of management plans. Planning processes that proceed in a systematic, organized, and quantitative fashion may help to ensure that the resulting plans can withstand rigorous scrutiny. The content of this book should help you develop some of these tools, or at the very least understand the concepts that you might encounter in your career as natural resource managers.

B. Information Necessary to Develop a Forest Management Plan

A forest plan begins with a statement of the goals and objectives of the landowner. These must be ascertained through an understanding of the landowner's desires. Effective communication with a landowner is essential. Small, private landowners may require one-on-one meetings and tours of their property. Other larger landowners may require numerous meetings with stakeholders and managers to effectively gage the goals and objectives. This information is important in that it guides the development of sampling efforts (e.g., timber cruises) and the development of management alternatives. Without knowledge of the goals and objectives of the landowner, a planner may be developing a plan that could be of no interest to the landowner.

Next, maps, tables, and photographs of the property should be compiled to provide context and data for the management plan. This information is used to help landowners and other readers of the plan understand where the property is located in relation to other geographic features, and provides a descriptive tool showing where actions will occur. People can then evaluate the impacts of management on places that are important to them. Maps and tables that demonstrate how ecological, economic, and social goals will be achieved over time help people understand that these goals are being taken into consideration. An understanding of the most current state of the resources being managed is essential for building a plan of action. If maps or photographs are several years old, then they may need to be updated prior to the development of a plan, especially if activities have been implemented since their development (in the case of maps) or capture (in the case of photographs).

Inventories of the resources that are under the control of the landowner, and that may be affected by the actions described in a management plan, must then be collected or compiled. These inventories might include forest conditions, water conditions, soil conditions, wildlife populations or habitat conditions, and recreational area and trail conditions. This information is necessary to understand the current conditions of the resources of interest located within the property. The inventories may be given to a growth and yield computer model to allow a planner to summarize the current condition of a forest in terms of tree density or volume per unit area. In addition to understanding the current condition, projections for all alternatives to be considered are needed to understand where the forest resources are headed under different management regimes. The appropriate forest growth and yield models can assist in these determinations as well. Economic, ecological, and social outcomes, where appropriate, then need to be assessed to determine the value associated with each alternative management regime. In addition, natural resources may be functionally connected, and actions that are applied to one resource (e.g., the trees) may affect another (e.g., wildlife habitat). Understanding these functional relationships is essential in assessing alternative plans of action.

Ultimately, a forest plan will provide a management recommendation that describes how a plan of action (a set of activities over time) will contribute to the goals and objectives of the landowner, and how these activities may affect other natural resources of interest. In addition, the forest plan should provide a comparison of how the management recommendation differs from some set of alternative management scenarios. This comparison allows landowners to understand the "what if" questions that they might have contemplated. The different management recommendations may be developed by incorporating the information derived both from the assessments of current stand conditions and the projections of future stand conditions into a decision model to help landowners and land managers understand the efficiency and effectiveness of the alternatives. The most efficient alternative is often described as the one that best uses the resources available; sometimes this is represented in financial terms (maximizing returns on an investment) and other times this is represented in more abstract terms (minimizing deviations from goals). The most effective alternative can be described as the one that seems to adhere best to the constraints of the management situation while contributing well to the objectives.

Once a management alternative has been selected for a property, a timeline describing the implementation of the activities should be provided, suggesting how the management activities proposed will interact economically, ecologically, and socially, and how they will contribute to the overall goals and objectives of the landowner (Table 1.2). Timelines are helpful to landowners, particularly for budgeting purposes. Notice in Table 1.2, for example, that there are costs associated with planned activities in 2019

and 2021, yet no revenues are generated in those years. Management plans should be designed to help landowners understand the options available, and although they provide guidance, it is ultimately up to the landowner to determine the course of action to take.

During a typical planning cycle of a medium-sized natural resource management organization, field-level managers are implementing natural resource management plans and collecting data about the resources to the best of their ability. Within this period of time, numerous treatments may be prescribed, markets may change, natural disasters may occur, and land may change owners. Near the end of the cycle, data related to changes in the resources are compiled by the field managers and sent to a central office, where the official "corporate" databases are updated and new plans are designed and selected (Fig. 1.1). The cycle occurs on a yearly basis in some industrial forestry organizations, and occurs over a longer period of time in some public land management agencies. However, what should be of interest to young natural resource professionals beginning their careers as field managers are three thoughts: (1) the quality of the resulting management plan depends on the data provided to the planners by yourself and your colleagues, (2) the plan itself is developed through a process that you should understand, because you will be implementing the plan, and you should know how it was developed (the general quantitative methods used to generate outcomes for each alternative) and how the plan alternative was selected (the type of planning process that was used), and (3) the

TABLE 1.2 A Summary of Activities Related to the Management of a Small Forest (Several Stands)

Year	Activity	Revenue ($)	Cost ($)
2018	Final harvest - even-aged stand	100,000	
	Commercial thinning	20,000	
	Fertilization		15,000
	Road maintenance		4,000
2019	Site preparation		15,000
	Planting		5,000
2020	Commercial thinning	15,000	
	Prescribed burning		2,000
2021	Herbaceous weed control		5,000
	Habitat improvement		3,000
2022	Partial harvest - uneven-aged stand	75,000	
	Road maintenance		4,000
2023	Commercial thinning	18,000	

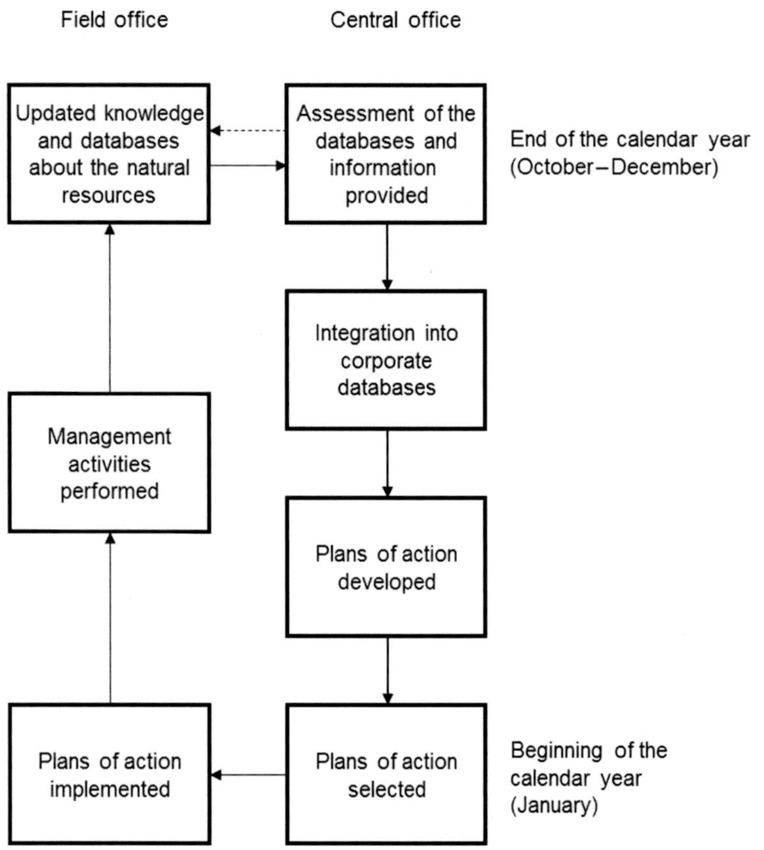

Field office   Central office

FIGURE 1.1   Movement of information during a planning cycle.

```
┌─────────────────┐      ┌─────────────────┐
│ Updated knowledge│ ◄ - -│ Assessment of the│
│ and databases   │      │ databases and   │     End of the calendar year
│ about the natural│ ────►│ information     │     (October–December)
│ resources       │      │ provided        │
└─────────────────┘      └─────────────────┘
        ▲                         │
        │                         ▼
        │                ┌─────────────────┐
        │                │ Integration into│
        │                │ corporate       │
┌─────────────────┐      │ databases       │
│ Management      │      └─────────────────┘
│ activities      │               │
│ performed       │               ▼
└─────────────────┘      ┌─────────────────┐
        ▲                │ Plans of action │
        │                │ developed       │
        │                └─────────────────┘
        │                         │
        │                         ▼
┌─────────────────┐      ┌─────────────────┐
│ Plans of action │ ◄────│ Plans of action │     Beginning of the
│ implemented     │      │ selected        │     calendar year
└─────────────────┘      └─────────────────┘     (January)
```

operational details of your daily activities are related to both the tactical and the strategic goals of the organization as described in the plan.

With the movement to field-level use of geographic information systems (GISs) and the notion that recent graduates should be more computer literate than their predecessors, more responsibility on data quality and data development is being placed on field-level land managers. This information (e.g., timber cruises, other samples of resources, maps of resources, adjustments to maps) now often directly replaces older, dated information, and improves upon the organization's knowledge of the land and vegetation that they own or manage. Although central offices may still monitor and control the data standards, young professionals are being asked to enter jobs with these skills already in hand. Hopefully, you will gain some of these important skills as you work through this book.

C. Plan Development Challenges

A forest plan is developed through careful assessment of past and current conditions of the land, and thoughtful consideration of desired future conditions. In general, it begins with a person or team defining the goals and

measures necessary to assess attainment of goals. As we suggested, this process involves understanding a landowner's objectives and constraints. The past and current states of the land and trees are also assessed. An analysis of how the forest might develop, given potential activities employed, would then be performed. Here, the costs and benefits of actions (including doing nothing) would be determined. If a distinct objective was noted, the efficiency of alternatives would be compared. Forest management is a rewarding experience for those who are drawn to the profession, yet it faces challenges from a number of areas. As an example, given the property and stakeholders involved, the entire process may require a significant amount of time, depending on the size of the forest.

As you may also expect, there are numerous economic challenges. For example, there may be the need to make a profit, the need to break even financially, the need to operate within a budget (perhaps at the activity level), the need to generate income, or the need to generate competitive financial returns when compared to other investments. These economic challenges usually are expressed in financial terms, and may involve discounting or compounding monetary values if the need arises. There are a number of environmental challenges as well, including

FIGURE 1.2 Management of natural resources may involve a balance between commodity production goals and goals related to wildlife habitat maintenance and development. *Photo courtesy of Kelly A. Bettinger.*

those related to wildlife habitat maintenance and development (Fig. 1.2), water quality, soil quality, air quality, biological diversity, and fish habitat conditions. A number of these concerns are embedded in statutes and regulations, others are simply the desire of landowners to protect or maintain certain values. There are also a number of social challenges facing forest management. For example, the use of prescribed fire is becoming a severe social challenge, because as people move out into the rural landscape, air quality around their homes becomes more of an issue. However, prescribed fire may be needed to restore and maintain native forest and range ecosystems, which is an important social and environmental concern.

Convincing the public that land is being managed responsibly is another social issue that we address in Chapter 15, Forest Certification and Carbon Sequestration, with a discussion of forest certification. Policy instruments (laws and regulations) guide the management of public lands and influence the management of private lands. The development of additional policies to guide the management of private forests is a contentious issue. Janota and Broussard (2008) found that absentee landowners and landowners who view their forests as long-term investments are more supportive of policies that encourage sustainable management, whereas landowners who view the effects of their management actions as isolated from the broader landscape are less favorable toward these types of policies. In addition to these challenges to the management of forests, there is also the social need to provide jobs to local communities, and the need to pay these employees a reasonable wage.

There are a number of technological challenges related to forest management as well, and we will allude to some of these as we discuss the various planning processes. Other forest management challenges, such as those related to silvicultural systems or operational methods (harvesting, fuel reduction, etc.), are perhaps best left to be described in

other texts. The long production period associated with the growing of forests sets this type of management apart from that incurred in agricultural operations, and as a result the outcomes of management are subject to many more potential environmental and human-caused risks. However, the development of management plans for forested areas must be accomplished in light of these uncertainties, which can be numerous for plans of action that cover large areas and long periods of time.

Planning and decision-making processes often are hampered by a number of challenges internal to an organization. These include technological limitations (obsolete computer systems, inadequate software programs, and so on), personnel issues, lack of data, and limited support from an organization's management team. For example, the state of the technology used within natural resource organizations comes as a mildly disappointing surprise, sometimes, to newly hired young professionals. Technology may be so obsolete that it becomes the bottleneck in the planning process (e.g., an alternative may take hours of computing time to generate and report). Overcoming this challenge to forest planning may require planning itself. To correct this situation, we may need to develop an estimate of the budget that would be required to purchase new equipment (i.e., gather information), then assess the alternatives (purchase system X or system Y), and finally, make a decision.

In many forest planning processes, the development of data can account for nearly half (or more) of the time spent in the planning process. What we are referring to here include GIS databases, growth and yield data for each management prescription, prices, costs, measures of potential habitat quality, and levels of constraints that will be applied. Collecting, managing, correcting, and formatting this data generally is performed by several people in a natural resource organization, and is, unfortunately, one of the most underappreciated tasks by upper-level management. People's motivation to assist with the planning process may also be a challenge; one of the frequent reasons for this attitude among people is the perception that the success of an organization does not depend on the timely development of a new plan. We have mentioned here only a few of the challenges, but the suite of setbacks that could occur is broad, and few planning processes can avoid them entirely. However, many of the challenges to planning that are internal to a natural resource management organization can be overcome, if they are recognized and acknowledged.

IV. GENERAL EMPHASIS OF FOREST MANAGEMENT PLANS

A forest management plan acts as a guide for a landowner or land manager; it is a description of the activities to perform within a given time frame (a planning horizon) to best achieve the management goals. In the next few subsections we describe the general emphasis of different types of forest management plans, beginning with the more common organization-specific plans and progressing to other types that have different scopes or levels of public participation.

A. Organization-Specific Plans

Most forest plans developed for a specific property ignore the resources that may exist, and the activities that may be implemented, in lands surrounding that property. However, there are instances where nearby activities (e.g., development of roads) or features (e.g., homes) can influence the alternatives that are considered for all or part of the property. Additionally, those organizations that are required to address cumulative effects (through certification programs or regulatory processes) must describe the impact of their proposed activities within the context of likely activities on neighboring lands. These instances aside, an organization-specific plan would only emphasize the objectives and constraints facing the landowner and the regulatory issues that affect the landowner. The land managers would develop information (maps, inventories) specifically for the property. An analyst would then compile, summarize, and present this information in a way that communicates a message to the decision-maker (the landowner or land manager). Future states of the property, as projected through the alternatives, would also be communicated through summaries of projected forest characteristics and maps to help the decision-maker understand issues such as forest sustainability. The plan developed would then describe the actions that would seem necessary to best meet the goals and objectives of the landowner in an efficient and effective manner. A number of examples of these types of plans can be found in the book *Forest Plans of North America* (Siry et al., 2015). Organization-specific plans are not limited to private landowners; they represent the types of plans commonly developed by forest companies, states, counties, towns, and federal agencies.

B. Landscape Plans

Some large resource management organizations now develop plans that recognize the resources that may exist nearby, and the activities that have been implemented on lands surrounding the property. This effort can be used to satisfy the cumulative effects analysis that is required for public planning projects, and can be viewed as part of the compliance process for voluntary forest certification programs. Landscape plans attempt to associate the guidance suggested for an organization with the larger social and environmental issues within the region where the property is located. These plans are more difficult to develop, since information concerning resources owned or managed by others seems necessary. Further, an understanding of the management perspectives of other landowners would also seem necessary. Therefore a landscape plan would not only emphasize the objectives and constraints facing the landowner, but place these in context with the broader social and environmental context. As with organization-specific plans, the land managers would develop information (maps, inventories) specifically for the property, but also information for other properties. An analyst would then compile, summarize, and present this information in a way that communicates a message to the decision-maker (the landowner or land manager). Future states of the property (and perhaps other nearby properties), as projected through the alternatives, would also be communicated through summaries and maps to help the decision-maker understand sustainability. The plan developed would then describe the actions that would seem necessary to best meet the goals and objectives of the landowner in an efficient and effective manner. A few examples of these types of plans can also be found in *Forest Plans of North America* (Siry et al., 2015), particularly the plans developed for federal or provincial government (Crown) land in Canada.

C. Community or Cooperative Forest Plans

Collaborative forest management—or community forestry— is a system where communities and governmental agencies work together to collectively develop a plan for managing natural resources, and each share responsibilities associated with the plan. The idea of a community-driven forest management and planning process is not new. Brown (1938) discussed the concept 80 years ago, and noted some requirements for community forests in North America:

> *To initiate a community forest, one would require cheap land, large areas of forests near towns or cities, and markets that are nearby.*

The emphasis of these types of plans is thus on meeting the needs (economic, social, and environmental) of the community that lives in the area within or surrounding the forest.

Improvements in forest protection and ecological values often are noted as some of the benefits of these types of forest management programs. However, in developing countries, community interest in these programs generally is based on basic needs for fuel, timber, food, water, and other nontimber forest products, and when these are marginally available the interest in collaborative planning and management may wane (Matta and Kerr,

2006). Aspects of successful collaborative planning programs include measurable benefits (financial and others) which the community can gain from, local organizational control over the natural resources, and an absence of governmental control (Crook and Decker, 2006). These types of management and planning systems require that groups reach consensus on contentious forest-related issues, and find agreement on the use of communal forest resources. The planning process may be lengthy and challenging, particularly when environmental and economic objectives are both important (Konstant et al., 1999).

Admittedly, much of the discussion and analysis within this book assumes that planning processes occur within a single property and involve a single landowner (organization-specific plans). However, crossownership planning, or cooperative management, has been suggested as a way in which the effects of forest fragmentation can be mitigated, and as a way to improve the economics associated with small-scale decisions. Stevens et al. (1999) suggested from a survey of nonindustrial landowners in the northeastern United States that over half would either be interested in sharing the costs associated with recreation projects, or would be interested in adjusting the timing of management activities such that they are concurrent with those of other landowners. There may be a spatial context associated with this form of collaborative planning, since it may be feasible only for landowners whose properties are within some proximity to others. In addition, some landowners may require observation of such collaboration before choosing to enter into agreements with their neighbors (Brunson et al., 1996).

D. Ecological Approaches to Plans

In some management environments, the goals of forest management are adapted to newly emerging circumstances (Bončina and Čavlović, 2009). This *adaptive management* and associated planning process involves many of the same planning processes as we have described in this chapter, with one exception. When utilizing this approach, a monitoring program is employed to provide feedback at various stages of plan implementation, which could allow the managers of a property to better recognize some of the uncertainties related to management activities. The emphasis of this type of plan, whether developed as an organization-specific, community, or landscape plan, is to allow the plan to evolve during the period in which the plan is in place (the time horizon of the plan). With this approach, the success or failure of management actions to produce the desired effects is evaluated both quantitatively and qualitatively. The conditions under which management activities fail to produce the desired outcomes are considered, and revised management prescriptions, constraints, or objectives

are developed. An updated plan is then created using the adjusted, and perhaps improved, management prescriptions, goals, and objectives. Grumbine (1994) suggests that adaptive management is a learning process, where the outcomes from previous management experiences are evaluated and future decisions are adjusted with this new information. This allows land managers to adapt to uncertain situations. Adaptive management and planning has been closely associated with *ecosystem management* on some public lands in North America, in an effort to manage for various environmental goals. However, we could extend the notion of adaptive management to the short-term tactical plans developed by many timber companies as well. Further, it may be alluded to in the guidelines for forest certification programs. For example, private landowners who acquire certification through the American Tree Farm System follow guidelines that suggest that the management plans developed should be adaptive. In these cases, updated information may be collected annually, and plans adjusted given the changing circumstances of the landscape, markets, and landowner objectives.

E. Urban Forest Plans

Urban forest plans are developed for the trees contained within cities and municipalities. They are similar to organization-specific plans in that trees located outside of these areas are generally of no interest to the planners. However, they differ from organization-specific plans in that the owners of the trees could be the cities themselves or the thousands of private landowners within the cities. The trees considered within these plans could be street trees, trees located within the properties of private landowners, or trees located in parks and other areas. An urban forest plan would emphasize the objectives and constraints of a city, with regard to tree structure. The percent canopy cover, the size and state (health) of trees, and the species distribution are common metrics that are used to evaluate current and future conditions of an urban forest. Urban foresters would then develop information (maps, inventories) specifically for the city. An analyst would then compile, summarize, and present this information in a way that communicates a message to the decision-maker (the city managers or the public). Future states of the city, as projected through the alternatives, would also be communicated through summaries and maps to help the city managers or the public understand issues such as forest sustainability. The plan developed would then describe the actions that would seem necessary to best meet the goals and objectives of the city in an efficient and effective manner. A number of cities have urban forest plans, but not all. A good example is the City of San Francisco plan (Swae, 2015) that we describe in more detail in Section V.

V. EXAMPLE FOREST PLANS

Forest plans come in all shapes and sizes, from the extensive, voluminous plans developed for United States National Forests, to the shorter, briefer plans developed by consultants for private landowners. Often, especially for smaller properties, the plan describes very explicitly the timing and placement of management activities. However, for larger properties such as United States National Forests, actual management activities are not prescribed, and a separate project-level planning process is employed for each individual proposed action (Bettinger et al., 2015). The contents and subject matter of a forest plan will therefore vary depending on the objectives of the forest landowner, the constraints, opportunities, and risks facing the landowner, and the area of the world in which the forest is situated. As examples, in the next set of subsections we briefly describe nine different forest plans from around the world.

A. North American Small Private Landowner Plan

The McPhail Tree Farm is located in the Piedmont region of South Carolina. The tree farm is enrolled in the American Tree Farm System, which requires the development of a forest plan. The property is an important element of the family's investment portfolio and retirement plan (Straka and Cushing, 2015). The forests (1,005 acres, 407 hectares) contain both southern coniferous and deciduous tree species. The planning process began with a forester meeting with the landowner to understand the goals and objectives related to ownership of the property. The management objectives included a desire to practice sustainable forest management, emphasizing commercial timber production and income generation, a desire to achieve stable levels of future cash flows, a desire to promote the development of wildlife habitat, and a desire to maintain or enhance the aesthetic and recreational qualities of the property. The landowners indicated that hunting and camping is common on the property. In the American Tree Farm System plan for this property, special sites are delineated, issues with adjacent management activities and ownerships are described (e.g., aesthetics, wildfire concerns, privacy, etc.), and access and boundary line issues are addressed. A broad treatment of general management issues that affect the property is provided in the plan, and stand-specific information is presented to address management activities. Information regarding the resources contained in the tree farm was collected through forest inventories and an analysis of aerial photographs. Alternative management scenarios were developed, and the decision (the choice of management alternative) was made by the landowner. Finally, forest plan documents and an implementation strategy were developed for the landowner. The plan included maps, charts, tables, schedules of proposed activities, descriptions of expected future resource conditions, and projected cash flows. An equal focus of the forest plan was placed on financial results and forest sustainability. Given the size of the property and the various goals and constraints of the landowner, the planning process required quantitative forest planning methods to develop an efficient course of action to achieve the landowner's goals related to future cash flow stability and the desire to sustain forest values.

B. North American National Forest Plan

The Chattahoochee-Oconee National Forest in Georgia is comprised of two national forest tracts of land administered as a single unit. The forest is 866,000 acres (nearly 350,500 hectares) in size, and over 90% of it was accumulated through land acquisitions or exchanges in association with a law enacted in 1911, the Weeks Act (36 Stat. 961). In the northern part of the state, the Chattahoochee portion is mainly composed of deciduous forests and land with significant topographic relief. Management in the Chattahoochee portion is often guided by recommendations for development and maintenance of recreational opportunities. In the central part of the state, the Oconee portion is mainly composed of coniferous forests situated on flatter lands, and management is often guided by recommendations for development and maintenance of red-cockaded woodpecker (*Picoides borealis*) habitat. Therefore the two main guiding issues facing the management of the forest are ecosystem restoration and the provision of recreational opportunities. The most recent strategic forest plan (US Department of Agriculture, Forest Service, 2004) required 8 years to develop, involved a significant amount of public interaction, and resulted in a complex array of management issues. In terms of the management regimes considered by the planning team, final harvests, shelterwood harvests, and uneven-aged management harvest entries were assumed as some of the possible actions that might be used to meet the goals of the plan. The overall objective of the plan was an economic one (maximize the net present value), and constraints were related to concerns about early successional habitat development, wood-flow (nondeclining yields), and other various forest management requirements. As the plan notes, it is strategic in nature, and does not prescribe actual management activities. A project-level planning process is employed for design and implementation of individual forest management activities.

C. Australian State Forest Plan

State-owned forests in Victoria, Australia, are managed by VicForests, a state-owned business. Authority for VicForests to manage the forest resources arises from an Australian law, the Sustainable Forests (timber) Act of 2004. VicForests manages about 4.5 million acres (1.82 million hectares) of forests, with the over-arching goal of managing and maintaining the resource without compromising future social amenities and environmental concerns (VicForests, 2015a). Only about 1.2 million acres (490,000 hectares) of the land is available for implementation of forest management activities, and for these lands a sustainable timber supply was assessed for the near-term (10 years) and longer-term (100 years). The forests are mainly of two types: ash forests, generally composed of a single overstory species of eucalypts (*Eucalyptus* spp.), and mixed forests, generally composed of various eucalypt species. A variety of management activities are employed to meet the goals noted in the plan, including final harvests, seed tree harvests, uneven-aged partial harvests, salvage harvests, thinnings, and variable retention partial harvests. Regeneration of forests is accomplished mainly through natural means (seed or coppice). Silvicultural decisions are guided by forest type and regeneration concerns, but also by the need to protect areas of high conservation value, areas with sensitive soils, and areas with significant social or cultural value.

The forest plan that VicForests developed is both strategic and tactical in nature. It was designed to provide a long-range vision for the forests, and to indicate in general where the appropriate forest management activities can take place during various periods of the planning horizon. Associated with the forest plan are a set of forest coupe (harvest area) plans that are operational in nature. The planning process allows annual public input on proposed harvest activities. This in turn allows VicForests to change or adapt the plans; therefore it is considered an adaptive management plan. The planning process also involved field assessments and evaluations of digital information (maps and inventories) to identify forest management issues of concern. After a preliminary plan was developed, public input was sought, and the plan was revised before the final plan was published (VicForests, 2015b).

D. European Estate Plan

The Windsor Estate, southwest of London, is a 15,567 acre (6,300 hectare) property containing forests, parks, gardens and other landscape features typical of southern England. About 7,670 acres (3,104 hectares) is managed by the forestry department through a plan (The Crown Estate Forestry Department, 2015) that has been approved by the United Kingdom Woodland Assurance Scheme. The plan was devised to manage, conserve and protect important resources while assuring the development of sustainable timber yields. The plan is both strategic (20 years) and tactical (10-year action plan) and begins with a commendable vision statement: *the resurrection of the*

traditional landscape of the Royal Forest, a broadleaved woodland dominated by oak and beech. The use of silvicultural practices is central to the management of a good portion of the forest. Natural regeneration and shelterwood or selective harvests can be used to meet the goals for each woodland area, parkland, healthland, and sites of special scientific interest. In the southern portion of the forest, the sustained yield of conifers is emphasized, and the even-aged management of Douglas-fir (*Pseudotsuga menziesii*) and other conifer species is practiced. Management objectives for each forest and woodland group range from the production of wood to the protection of significant forest features. Strategies were developed for each woodland group to achieve the management objectives. These strategies suggest the forest management practices (such as thinning, selective felling activities, final harvest, shelterwood treatments, and regeneration activities, among others) needed to meet the objectives. Pests and diseases, invasive species, significant biological threats are considered in the plan, along with the need to accommodate recreational activities. Finally, a monitoring plan was developed to associate broad objectives (economic sustainability, biological sustainability, forest resilience, public access) with specific indicators of success. The monitoring plan therefore helps forest managers assess the progress and success of the plan, and supports adaptive management.

E. Asian Private and Communal Forest Area Management Plan

The forests of northern India are often located in somewhat rugged and steep areas. In 2007, a management plan was developed for forests in the Dasuya region (Lal, 2007). The plan covered 68,417 acres (27,688 hectares) of private lands, common lands, and communal (panchayat) forests; all are considered *private forests*, not owned by the Punjab State government. The forests are managed by the state government in some respects, yet the responsibility for protection of the forest areas rests with the individuals, communities and panchayats that own the land. In essence, the state government can regulate, restrict or prohibit such things as the harvest of trees or grazing within the area covered by the plan, but individuals and villages are empowered to implement the management activities. Due to strict forest laws, the forest plan developers found it important to enlist the support of local villagers to ensure long-term conservation objectives are achieved, and therefore measures were suggested to promote increased local acceptance of the plan. The management plan also includes a 5-year harvesting program listing the villages allowed to harvest timber each year. Permits are required for the extraction of certain tree species, and general guidance is provided for silvicultural activities. Natural regeneration is mainly achieved through coppice, some artificial regeneration is employed, and specific tree marking guidelines and tree felling processes are provided.

Overall, it was suggested that the forest management plan should be considered a *conservation management plan*. The most important goal, according to national forest policy, is to maintain ecological balance. This might be achieved through subgoals that promote activities to reverse degradation of ecosystems, to conserve soil resources, to encourage public participation in the management of the land, to promote ecotourism, to manage invasive plant species to improve biodiversity, and to enhance productivity of forested areas. The management of nontimber forest products, ecotourism, and biodiversity are all considered in the plan, in conjunction with other potential forest management activities. From a practical point of view, given all of the stakeholders involved, the authors note that the plan may be revised based on feedback and progress associated with the first 5 years of plan implementation.

F. South American Community Forest Plan

In 2011, a forest management plan developed for residents and producers associated with the Chico Mendes Extractive Reserve near the town of Xapuri, in the State of Acre, in western Brazil (Agapejev de Andrade and Thaines, 2011). The area represents a 46,388 acre (18,773 hectare) community forest comprised of 62 property owners. This area is formally considered public lands where people have the right to use them through a concession provided by the state. The plan, while very detailed in the current condition of the area and in the potential outcomes, is strategic in nature. Each resident decides upon the timing and location of activities, and due to community dynamics it was very difficult to explicitly describe an implementation plan. One purpose of the plan was to suggest an alternative stream of income, from the harvest of forest products and utilization of waste materials, that could complement local uses of the land that include agricultural activities (rice, maize, banana, sugar cane, pineapple, etc.), Brazil nut (*Bertholletia excelsa*) harvests, rubber (*Hevea brasiliensis*) extraction, and cattle farming. Therefore, the plan was developed to provide forest management guidance to the community forest, to strengthen the community with respect to economic, ecological, and social sustainability. The plan is quite specific in that it describes forest inventory methods, tree felling techniques, road design principles, safety guidelines, and chain of custody protocols. The plan also provides guidelines for avoiding negative environmental impacts, through best management practices for the planned management activities.

The forests covered by the plan were considered "open" with respect to the canopy, containing species such as chestnut, mahogany, and cherry. In addition, palm and bamboo vegetation grow throughout the forests. The typical management activity described by the plan involves cutting entries on 25-year intervals, extracting about 214 ft^3 per acre (15 m^3 per hectare) of the growing stock. The plan also provides estimates of projected financial values per unit of wood, per unit area, and per year for the property as a whole. Although the plan seems to have a 25-year time horizon, it appears that it will be revisited every 5 years.

G. African Participatory Management Plan

The Kakamega Forest Ecosystem management plan (Kenya Wildlife Service and Kenya Forest Service, 2012) covers a set of forests, forest reserves, and nature reserves in Kenya. The 54,400 acre (22,014 hectare) area is a watershed for some of the rivers that flow into Lake Kenya, and is generally composed of tropical rainforests, pastures, and developed sites, and plantations of cypress, pines and eucalypts. Human activity within and around the forest is an important issue. Impacts associated with human activity include encroachment upon the forest boundaries, overgrazing of the land, over-exploitation of valuable plant species, and deforestation. The planning process used a participatory approach involving stakeholders associated with the area. This provided people an opportunity to engage in the development of the management plan for the area. The planning process included a scoping meeting to identify management issues and stakeholders, inventories of resources, initial stakeholder consultations, mid-process stakeholder reviews, meetings with management experts, and draft plan stakeholder reviews. A land classification (zonation scheme) was developed to provide a framework for the conservation of natural resources, regulation and promotion of visitor use, and sustainable use of forest resources. The classification divided the property into four zones: protection, core forest (for forest restoration, rehabilitation and connectivity purposes), potential utilization, and livelihood support.

The forest management plan describes the general management approach for the Kakamega Forest Ecosystem and the goals to be sought over a 10-year time horizon. Five management programs are incorporated into the plan. These programs were aimed at tourism development, community outreach and education, forest operations and security, forest resource management, and ecological management of the land. Each of these programs has objectives, actions, a 3-year activity plan, and a monitoring program to assess impacts from the implementation of management activities. For example, one objective of the forest resource management program is to maintain and enhance the productivity of plantations and increase the efficiency of wood utilization to meet the needs of the growing human population in the area. An action item for this objective is to provide a sustainable wood supply, mainly through tree planting activities.

H. North American Urban Forest Plan

Given its location near the Pacific Ocean and San Francisco Bay, the City of San Francisco is famous for its scenic views. About 669,000 trees are located within the city, providing about 14% tree canopy cover (Swae, 2015). The Department of Public Works has jurisdiction over trees in the public right-of-way, but maintenance of some of these is the responsibility of private landowners. The Recreation and Park Department is responsible for trees in city parks, natural areas, and public golf courses. Other trees are managed by a variety of state and federal agencies and a local nonprofit organization carries out the majority of street tree planting. The City of San Francisco represents a highly altered natural environment, as most of the natural landscape has been transformed through development and urbanization activities.

The city's urban forest plan arose out of the need to ensure the health and sustainability of the city's trees. The plan has three phases that address three different sets of urban trees: street trees, trees in parks and open spaces, and trees on private property. The first phase, the plan for street trees, has been completed. This plan phase was informed by a series of public meetings, workshops, and other events coordinated with city residents, agencies, landscape professionals, and urban forestry specialists. The plan was also shaped by the results of a census and a financing study. The street tree plan has four sections that contain a vision for the urban forest, a policy framework, recommendations, and an implementation strategy. The goals are to grow the extent of the urban forest, to protect the existing trees, to manage the design of the urban forest with sustainability in mind, to provide stable financing, and to engage the public in urban forest development and maintenance efforts. The plan was developed with the goal of advancing the sustainability policies and programs of the city, which include targets for reducing waste and greenhouse gases. The plan also encourages the planting of trees and other vegetation to support local wildlife of interest, and promotes social equity by emphasizing tree planting in neighborhoods where there is a disproportionate lack of trees.

I. North American Industrial Forest Plan

Rayonier, Inc. manages nearly 1.9 million acres (about 769,000 hectares) of land in the southern United States. About 42% of the land contains loblolly pine (*Pinus taeda*) plantations, about 23% contains slash pine (*Pinus elliottii*) plantations, and about 35% contains natural pine and deciduous forests that reside on soils too wet to intensively manage. The lands are owned and managed with the objective of maximizing the net present value for their shareholders. One of the main questions addressed through the planning process (McTague and Oppenheimer, 2015) is the time required for the forest estate to progress to a fully regulated condition (approximately equal areas in each forest age class). Long-range forest planning is therefore conducted. The company also adheres to the standards for forest land management that were developed by the Sustainable Forestry Initiative, a voluntary certification process (discussed in Chapter 15: Forest Certification and Carbon Sequestration), which includes the determination and use of long-term harvest levels that are sustainable and the development of harvest plans address the size, shape, and placement of final harvests for visual quality purposes. These issues are examined through the forest planning process, as are wood flow constraints and land sales opportunities. Although the planning horizon is 30 years long, the Rayonier southwide harvest schedule plan is generated once every 2 years for nearly 28,000 stands. Outcomes from the planning process include projections of harvest volumes, by product grade, from different management activities such as final harvests and thinnings, and lists of harvest activities for each stand. As a means of monitoring and adapting to changes, annual inspections of activities are used to identify possible departures from forest sustainability goals and from the path desired to a regulated forest.

VI. CHARACTERIZING THE DECISION-MAKING PROCESS

Decisions regarding management plans are made in natural resource management organizations usually by a team of people with various educational and cultural backgrounds, and various lengths of experience in professional settings. One main characteristic of planning efforts is that the time frame for the tasks performed by the team members usually is limited. In addition, the tasks the team members must perform may require a high degree of knowledge, judgment, and expertise (Cohen and Bailey, 1997). More often than not, people on these teams have developed individualized sets of behaviors and decision-making styles based on previous experiences, which makes group decision-making an interesting and sometimes controversial event.

A. A View From the Management Sciences

The work that has been performed to explore how groups make decisions is vast, and a number of theories regarding how and why decisions are made have been put forward (Bettenhausen, 1991; Salas, 1995). Generally speaking, in the management sciences, there are three types of decision-making processes: rational, irrational, and something in-between called the "garbage can" process. These models are more thoroughly discussed in the management sciences literature, and our objective here is simply to provide a brief description of each. In the *rational model*, a decision-making team gathers all the data needed, analyzes all the possible scenarios, and reaches the best solution based on this complete set of information. Of course, this process is used only when there is a sufficient amount of time and resources (Smith, 1998), and may involve decisions that are easily resolved by means of mathematical formulas (Mian and Dal, 1999). However, this is rarely the case in natural resource management. In fact, some may argue that there never are enough resources available (such as time, funding, or people) for this model to be used in forest or natural resource planning. Further, the rational model assumes that the planning team is sufficiently involved to provide the appropriate amount of attention to the attributes of the plan for which they have expertise. Given the multiple demands on a natural resource manager's time, this assumption may not hold true. And it will eventually become obvious that decisions concerning the development of a plan are inherently value-laden, even though we may believe that we are objectively assessing the management of a landscape. It is for these and other reasons that the *best solution* to a problem may not be the plan chosen by the land manager or the landowner.

The *irrational model* of decision-making is the opposite of the rational model: decisions are made based on limited (or no) data, and few (or no) alternatives are assessed. In this model of decision-making, decisions are based on limited information. Although we would hope that important natural resource management decisions are made using a more conscientious effort, we acknowledge that these types of decisions often do occur. More commonly, a decision model similar to this is used, one called the *semirational model* (or bounded rationality) (Simon, 1972). With this model, decisions are based on the best available information that can be collected during a limited time period, thus planners recognize the uncertainties and shortcomings of the databases and models. When using this decision-making model, we assume that incomplete information is the *status quo*, that a subset of alternatives are considered due to a lack of information or time, and that decision-makers will select a management alternative that is *good enough*.

A third alternative model often used (but rarely recognized) in decision-making efforts is known as the *garbage can model*, which was coined by Cohen et al. (1972). This model differs from the others in at least one of these aspects: (1) the goals and objectives are unclear, they may be problematic, or may be a loose collection of ideas; (2) the technology for achieving the goals and objectives is unclear, or the processes required to develop results may be misunderstood by the team members; or (3) team member involvement in the decision-making effort varies, depending on the amount of time and effort each member can devote to the tasks in the decision-making process. Cohen et al. (1972) noted that these conditions are particularly conspicuous in public and educational group decision-making efforts. This alternative model was designed to explain situations where teams are confronted with unclear criteria for decision-making, and where goals are subjective and conflicting (Mian and Dal, 1999). Without being formally introduced or recognized, this model may be more prevalent in natural resource management decision-making situations than the rational or semirational approaches.

Decision-making is the process of creating and selecting management alternatives, and is based on the values and preferences of the decision-makers. In making a decision, we usually assume that several alternatives were considered, and the one selected best fits our goals and objectives. However, this is not universally the case. Risk is inherent in almost every decision we make, and very few decisions are made with absolute certainty about the outcomes and impacts, because a complete understanding of all the alternatives is almost impossible to obtain. In situations where time constraints pressure the planning process, the alternatives assessed may be limited due to the effort necessary to gather information. Plan developers must also guard against the use of selective information. That is, in some cases planners choose to use a set of information containing only those facts that support their preconceived position or their notion of a desired outcome. Consideration of alternative management scenarios or management pathways may help reduce the risk of making poor decisions.

Throughout this book we emphasize the need to optimize the use of a set of resources. Optimization involves strategies for choosing the best possible solution to the problem given a limit on one or more resources or given limits imposed by policies. Along the way, the optimization process hopefully evaluates as many alternatives as possible and suggests the choice of the very best option given the problem at hand. Many natural resource managers cringe at the thought of implementing an optimal plan because the human element largely has been ignored, and a number of economic, ecological, and social concerns may have not been incorporated into the problem-solving process. One of the main features of decisions related to the management of natural resources is that they may have politically relevant side effects, and as a result decisions made using strict optimality criteria might be viewed by some as inadequate (Gezelius and Refsgaard, 2007). In reality, as plans are implemented, other acceptable options may arise that are satisfactory with respect to the objectives of the plan.

Plans are then adjusted marginally to take into account those factors that were not recognized in the initial plan development process. However, throughout this book we suggest the need to develop optimal decisions for managing natural resources. Beginning with the most efficient decision related to the management of resources allows you to understand the trade-offs involved when satisficing is necessary.

B. A Broad View on Planning Within Natural Resource Management Organizations

Our description of a planning model is very general in nature; actual processes used within each specific natural resource management organization may deviate from this model. Most decision-making processes, particularly those that involve the public or public land, include the following 10 steps:

1. Allow public participation and comment on the management of an area.
2. Determine the goals for a management area.
3. Inventory the conditions necessary to evaluate the goals.
4. Analyze trends in land use changes and vegetative growth.
5. Formulate alternatives for the area.
6. Assess the alternatives for the area.
7. Select an alternative and develop a management plan.
8. Implement the management plan.
9. Monitor the management plan.
10. Update the management plan.

The steps may be rearranged, depending on the planning model used by a specific natural resource management organization. For example, the public participation step may occur later in the process, as alternatives are being formulated for the landscape. Alternatively, some steps may be omitted from planning models. In many cases, planning processes associated with private landowners may forgo or minimize the use of the public participation step. However, many elements in the decision-making process are consistent among natural resource management organizations, such the statement of goals, the assessment of alternatives, and the selection and implementation of the plan.

One major difference in the planning processes for public and private land is that planning processes may be mandated for public land, and only suggested for private land. For example, United States National Forest planning efforts are required by the Forest and Rangeland Renewable Resources Planning Act of 1974 and the National Forest Management Act of 1976, yet there is no similar national law pertaining to private lands. Several themes permeate the National Forest planning process and

differentiate it from private land planning processes. First, it should take an interdisciplinary approach, and a team composed of professionals from several disciplines is used to integrate their knowledge and experience into the planning process. Second, the public is encouraged to participate throughout the planning process. Third, the plan being developed must be coordinated with other planning efforts of other federal, state, or local governments as well as Indian tribes. And finally, the public has the ability to appeal the decision made regarding the final forest plan. These themes make the National Forest planning process distinctly different than, say, the process used by a timber company, where public participation, coordination, and appeals may be limited. As overarching guidelines for United States National Forest planning processes, the National Forest Management Act (United States Congress, 1990), Part 219.1(a) states that:

The resulting plans shall provide for multiple use and sustained yield of goods and services from the National Forest System in a way that maximizes long term net public benefits in an environmentally sound manner.

The importance of planning is emphasized as well, as Part 219.1(b) states that:

Plans guide all natural resource management activities and establish management standards and guidelines for the National Forest System. They determine resource management practices, levels of resource production and management, and the availability and suitability of lands for resource management.

Example

As an example of a specific United States National Forest planning process, the Humboldt-Toiyabe National Forest (Nevada) embarked on a planning process for a portion of the forest (Middle Kyle Canyon) in 2005. The process began with the development of data from which all future work would be based. A number of maps were generated, and presented at various scales, to help people understand the issues that affect the analysis area. The National Forest then held meetings with community and government representatives in an effort to understand their needs, their expectations, and any other relevant information regarding the planning effort. The information obtained from the meetings was then synthesized, and a set of goals for the analysis area was developed. Three management options for the analysis area were proposed, each in an effort to address, in different ways, the goals. They included a *no action alternative*, a *market-supported alternative*, and a *high development alternative*. The options represented different approaches to public use, facility development, vegetation management, and other interests. The intent of this effort was to span the range of possible alternatives, and to demonstrate the potential for the national forest to address economic, ecological, and social objectives. The options then were analyzed to determine the impacts on the objectives, and subsequently a second round of public participation was employed. With some modifications, the market-supported alternative was selected (US Department of Agriculture, Forest Service, 2009; Shapins Associates, 2005).

State forest planning processes are similar to federal forest planning processes. For example, in developing the 2011 Elliott State Forest plan (Oregon), a core team of interdisciplinary professionals was organized, and while guided by a steering committee, they were directly responsible for managing all technical elements of the planning process (Oregon Department of State Lands and Oregon Department of Forestry, 2011). The technical elements included developing current and future descriptions of the resources, developing the goals of the plan, developing strategies for reaching each goal, and finding a way to balance the competing goals through a modeling process that examined multiple alternatives. The public was involved in the process as well, through meetings, field tours, and newsletters.

Example

The managers of the Brule River State Forest (Wisconsin) developed broad goals for the forest with an emphasis on restoring, enhancing, or maintaining ecosystems. In addition, the managers of the forest constructed objectives for providing angling, hunting, canoeing, kayaking, camping, and cross-country skiing opportunities (Van Horn et al., 2003). The steps that the forest used in the planning process included:

- Conduct research and gather data on the property (Step 3 earlier)
- Identify key issues (Step 2 earlier)
- Draft vision statement and property goals (Step 2 earlier)
- Develop and evaluate a range of reasonable alternatives (Steps 5 and 6 earlier)
- Develop and evaluate a preferred alternative (Step 7 earlier)
- Develop the draft plan and environmental impact statement (EIS)
- Distribute the draft plan and EIS for public and governing body review (Step 1 earlier)
- Receive written comment
- Hold public hearings (Step 1 earlier)
- Submit the draft plan, EIS, and comments to the Natural Resources Board for review
- Receive decision from Natural Resources Board
- Implement the plan (Step 8 earlier)

In addition to broad vision and goal statements, the Brule River State Forest plan includes specific forest-wide goals for recreation use (in the form of visitor days), watersheds (protect and maintain stream conditions), and land

management (annual targets for thinning, clearcutting, prescribed burning), as well as specific objectives for areas within the forest.

Some counties and cities in the United States also have developed plans for the management of their natural resources. For example, Erie County (New York) developed a plan for its lands that would create educational and economic opportunities, utilize an educational center, conduct research, reduce taxes through timber sales, provide clean water, enhance wildlife habitat, and encourage recreational use (Grassia and Miklasz, 2003). The county developed "guiding principles" to ensure that the forest management practices suggested will build public confidence and ensure acceptance of the plan. Their strategy for achieving success revolved around frequent communication of the benefits of the plan to the residents of the county.

What distinguishes public land management from private land management is that usually Step 1 is limited when developing a plan for private land, and used extensively when developing a plan for public land. In addition, whereas the goals for private landowners may focus on economic values or commodity production, the goals on public land are generally broader (recreation, wildlife, water, timber, etc.). Finally, the planning process, particularly when performed by industrial landowners, is repeated every year or two, whereas on public land the process may be repeated at much longer intervals (5 or 10 years).

Example

Molpus Timberlands Management, LLC, based in Hattiesburg, Mississippi, is a private timberland investment organization that is active in acquiring and managing forested properties. For each of their properties they implement a planning process to determine the management approach given the goals and objectives of their investors. The steps that they use in their planning process include:

- Collect preplanning data about the forested property (Step 3 earlier)
- Develop the forest planning team
- Assess local conditions, markets, and other limitations (Step 3 earlier)
- Get field foresters to take ownership in developing the management plan (Step 1 earlier)
- Identify the main objective and all relevant constraints for the forested property (Step 2 earlier)
- Conduct stratification of inventory (Step 4 earlier)
- Develop management regimes (Step 5 earlier)
- Calibrate and test growth and yield models and expected silvicultural responses to allow for the development and evaluation of alternatives
- Select harvest scheduling tools and methods
- Formulate a plan (Step 5 earlier)

- Initialize and solve unconstrained planning model (Step 5 earlier)
- Review and provide feedback of the forest plan by the forest planning team (Step 6 earlier)
- Improve models and conduct subsequent opportunities for review and feedback as deemed necessary (Step 6 earlier)
- Select final planning model (Step 7 earlier)
- Report results to the forest planning team for evaluation of strategic and tactical concerns
- Construct "what if" scenarios and track results (Step 6 earlier)
- Implement the plan (Step 8 earlier)
- Update and improve the plan over time (Steps 9 and 10 earlier)

One distinct feature of this process is that it incorporates constant feedback and exchange between the field staff and the planning office. In general, Timber Investment Management Organizations (TIMOs) commonly try to maximize the net present value of their clients' timberland investments through commodity production activities. Some common constraints that they face involve the state of the ending inventory (standing volume at the end of the time horizon associated with the plan), and the product and harvest volume stipulations contained within wood supply agreements.

Example

The American Tree Farm System (2016) is the largest forest certification program for private landowners in the United States. As we described in Section V, the American Tree Farm System encourages the practice of sustainable forestry on family-owned private forests, and their forest certification system was designed to include components that one would normally find in a family forest management plan that focuses on sustainable forest management. Forest management plans developed through this system generally follow a process that involves the following steps:

- Describe the property in terms of owner and legal description
- Describe the history of the property (Step 4 earlier)
- Determine the forest management goals (Step 2 earlier)
- Develop maps of the property (Step 3 earlier)
- Determine the appropriate management actions for different sites
- Develop objectives for each stand in the forest (Step 2 earlier)
- Understand current stand conditions (Step 3 earlier)
- Develop desired future stand conditions (Step 2 earlier)
- Select management activities for each stand (Step 6 earlier)
- Schedule and track management activities (Steps 7, 8, and 9 earlier)

These types of forest plans are often developed for a landowner by a service forester (state or county agent) or a consulting forester. A significant amount of personal contact between the two may be necessary to best meet the needs of the landowner. Objectives and constraints will vary considerably from one landowner to the next, and although the time frame of the plan may often be long, interaction between the landowner and plan developer may continue throughout the life of the plan.

C. A Hierarchy of Planning Within Natural Resource Management Organizations

Planning, at a small or large scale, can be viewed as a hierarchy (Fig. 1.3). At the highest level in the hierarchy are strategic planning processes, which focus on the long-term achievement of management goals. Here, goals such as the development of wildlife habitat or the production of timber harvest volume usually are modeled over long time frames and large areas and are general in nature. Spatial aspects of management plans generally are ignored here, although with recent advances in computer technology and software, there are fewer reasons to avoid these issues in strategic planning. At lower levels of the planning hierarchy spatial relationships usually are recognized. For example, in tactical planning processes, issues such as the location of management activities over space and time are acknowledged. Plans that involve spatial habitat models are tactical plans, because the locational relationships between habitat units (usually timber stands) are recognized. This level of planning identifies site-specific actions that contribute to the larger purpose of the plan, but the technical details of implementing the actions are limited.

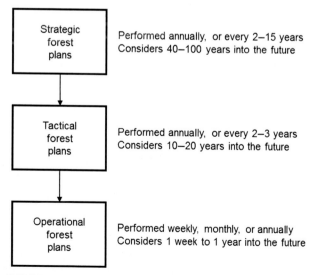

FIGURE 1.3 A hierarchy of natural resource planning processes.

At the lowest level in the hierarchy is operational planning. This is the day-to-day, weekly, monthly, or annual planning that is required to actually implement a management action. Some examples of this type of planning include scheduling seedlings for the planting season, loggers for harvest areas, equipment for stream improvement projects, or fire crews for prescribed burning efforts. Operational plans (weekly, monthly, annually) are guided by tactical plans (annually, biannually), which are guided by strategic plans (longer term). The level of detail increases as we move from strategic to operational planning. Conversely, the number of people involved increases from operational to strategic planning. Although many natural resource management organizations develop and use management plans, they may not use all three types. Most, in fact, have developed a strategic plan and use various forms of operational plans. Each level of planning has been enhanced with the expanded use of GIS, which give us the ability to view resource conditions and management scenarios quickly, and let us recognize spatial relationships among resources at lower levels of planning.

As a recreation or range manager, forester, wildlife biologist, soils scientist, or hydrologist, sometime in your career (perhaps immediately) you will be involved in decision-making and planning processes. At a minimum, you may be placed in a position to manage summer students or interns, and subsequently manage the budget required to pay their salaries. It is not uncommon, however, for an entry-level forester to be placed in charge of a planting or site preparation program, or for a biologist to manage a budget related to habitat improvements. How you decide to allocate the budget to the alternatives at your disposal requires quantitative analysis and decision-making techniques. Further, at some point in your career, you will likely be asked to provide input to one or more of the three general types of planning processes. This description of the different types of planning processes was admittedly brief, however Chapter 13, Hierarchical System for Planning and Scheduling Management Activities, is devoted to a more extensive treatment of the hierarchical system.

VII. SUMMARY

Quantitative and qualitative planning methods are meant to assist the human mind in determining objectively rational courses of action. Planning methods are employed to help us sort through and understand the complexities inherent in our management alternatives. As economic and ecological conditions change, and as society's impression of how the landscape should be managed change, we need to address how our management of natural resources should change. This requires a planning process, which is facilitated by information, such as field data, potential management prescriptions, and forest plan alternatives. To be able to use quantitative methods, we may make simplifying assumptions so that problems are tractable (useable). Therefore, the most we should expect from the results is "guidance" for how natural resources should be managed. As a natural resource manager, you will also need to rely on your judgment in making decisions.

This book covers some concepts that will be important to your careers in natural resource management. These concepts include an overview of measures of forest structure, forest growth dynamics, economic evaluation methods, and planning techniques. Although these subjects may seem daunting or displeasurable, rest assured that there are few positions in natural resource management that avoid them entirely. Economics commonly is used to help us objectively sort through the various management choices available. Planning helps us organize the alternatives for the land we manage, and provides a framework for comparing and choosing among these alternatives. Thus at some point in your career you will be involved, for better or worse, in forest and natural resource planning. The concepts we cover in this book should not only be of value in your career, but should also be of value in your personal lives, particularly the subject of the "time value of money."

QUESTIONS

1. *Assessment of a forest plan.* Either through a search of the Internet, through information provided in a book (Siry et al., 2015), or through an investigation of the forest plans contained in your college's library, locate relevant information concerning a federal, state, or county forest plan. From the official documentation of the plan, report the following two features:
   a. What goals or objectives guided the development of the plan?
   b. What were the steps used in the planning process?

2. *Forest planning process.* Assume you are employed by a small natural resource consulting firm (three people), and you needed to develop a management plan for a private landowner in central Pennsylvania. What types of internal (to your consulting firm) organizational challenges related to the development of the management plan should you consider?

3. *Types of forest planning processes.* Assume you are employed by a small forest products company in northern Minnesota, and the owner of the company wants your team (several foresters, a biologist, an engineer and a few technical staff managing the inventory and GIS) to develop a strategic forest plan for the property that you manage. The owner has suggested that they want a rational plan to be developed, one that explores several alternatives. Develop a one-page memorandum to the landowner describing the three

general types of planning processes, and the advantages and disadvantages of each with respect to the landowner's suggestion. You might emphasize the time required and the extent of management alternatives that might be assessed.

4. *Cooperative planning and adaptive management.* Assume that you are a natural resource management consultant in a small town in central New York. As part of your nonprofessional life, you serve on your town's land planning committee. The committee is actively involved in the management of a small public forest within the town's limits, yet none of the other committee members have your natural resource background. They have mentioned at various points in time over the last year the need for adaptive management and cooperative planning. Develop a short memorandum for the committee that describes the two approaches.

5. *Public and private forest planning.* Assume that you are having dinner with some of your friends and during the various conversations that arise, you learn that one of them has a very negative opinion of how management plans are developed for public lands. Further, they dislike how private landowners seem to not do any planning at all for the management of natural resources. These are generalities, of course, so to help clarify the matter, describe briefly the similarities and differences between management plans developed for public land and private land.

6. *American Tree Farm System.* One of the performance measures associated with a forest management plan developed for the American Tree Farm System notes that landowners shall have a written plan that is consistent with the size of their forest and the scale and intensity of the management activities implemented on their forest. What actions can a landowner take to indicate that they have met this standard?

7. *Idaho Panhandle Forest Plan.* In 2015, the Idaho Panhandle National Forests in the United States revised their land management plan (US Department of Agriculture, Forest Service, 2015). In a short summary, please address the following questions:
   a. About how large is this property?
   b. For about how many years does the plan provide strategic guidance to the national forest managers?
   c. What three wildlife species (or species groups) are of great interest to the national forest?

REFERENCES

Agapejev de Andrade, I., Thaines, F., 2011. Plano de manejo florestal sustentável, AMOPREX—Assoc. dos moradores e produtores da Resex Chico Mendes em Xapuri. Tecnologia e Manejo Florestal, Ria Branco, 93 p.

American Tree Farm System, 2016. American Tree Farm System. Washington, DC. Available from: https://www.treefarmsystem.org/ (Accessed 1/23/2016).

Bettenhausen, K., 1991. Five years of group research: what we have learned and what needs to be addressed. J. Manage. 17 (2), 345–381.

Bettinger, P., Merry, K., Mavity, E., Rightmyer, D., Stevens, R., 2015. Chattahoochee-Oconee National Forest, Georgia, United States of America. In: Siry, J.P., Bettinger, P., Merry, K., Grebner, D.L., Boston, K., Cieszewski, C. (Eds.), Forest Plans of North America. Academic Press, New York, NY, pp. 277–284.

Bončina, A., Čavlović, J., 2009. Perspectives of forest management planning: Slovenian and Croatian experience. Croat. J. For. Eng. 30 (1), 77–87.

Brown, N.C., 1938. Community forestry: a neglected phase of the American forestry system. J. For. 36 (7), 687–694.

Brunson, M.W., Yarrow, D.T., Roberts, S.D., Guynn Jr., D.C., Kuhns, M.R., 1996. Nonindustrial private forest owners and ecosystem management: can they work together? J. For. 94 (6), 14–21.

Butler, B.J., Leatherberry, E.C., Best, C., Kilgore, M.A., Sampson, R.N., Larson, K., 2004. America's family forest owners. J. For. 102 (7), 4–14.

Cohen, M.D., March, J.G., Olsen, J.P., 1972. A garbage can model of organizational choice. Adm. Sci. Q. 17 (1), 1–25.

Cohen, S.G., Bailey, D.E., 1997. What makes teams work: group effectiveness research from the shop floor to the executive suite. J. Manage. 23 (3), 239–290.

Crook, B.J., Decker, E., 2006. Factors affecting community-based natural resource use programs in southern Africa. J. Sustainable For. 22 (3/4), 111–133.

Demers, C., Long, A., Clausen, R., 2001. What is in a Natural Resource Management Plan? School of Forest Resources and Conservation, Florida Cooperative Extension Service, Institute of Food and Agricultural Sciences University of Florida, Gainesville, FL, Extension Report SS-FOR-14.

Food and Agriculture Organization of the United Nations, 2010. Global Forest Resources Assessment 2010. Food and Agriculture Organization of the United Nations, Rome, FAO Forestry Paper 163.

Gezelius, S.S., Refsgaard, K., 2007. Barriers to rational decision-making in environmental planning. Land Use Policy. 24 (2), 338–348.

Grassia, B., Miklasz, C., 2003. Erie County Forest Management Plan (Draft): Creating Sustainable Forests in Erie County for the 21st Century. Erie County Department of Parks, Recreation and Forestry, Buffalo, NY, 277 p. Available from: http://www2.erie.gov/parks/sites/www2.erie.gov.parks/files/uploads/pdfs/ECFMP_draft_plan_1-04.pdf (Accessed 1/16/2016).

Grebner, D.L., Bettinger, P., Siry, J.P., 2013. Introduction to Forestry and Natural Resources. Academic Press, New York, NY, 508 p.

Grumbine, R.E., 1994. What is ecosystem management? Conserv. Biol. 8 (1), 27–38.

Heiligmann, R.B., 2002. Forest Management, Developing a Plan to Care for Your Forest. School of Natural Resources, Ohio State University, Columbus, OH, Extension Fact Sheet F-34-02.

Hosmer, R.S., Bruce, E.S., 1901. A Working Plan for Township 40, Totten and Crossfield Purchase. Hamilton County, New York State Forest Preserve. US Department of Agriculture, Division of Forestry, Washington, DC., Bulletin No. 30. 64 p.

Janota, J.J., Broussard, S.R., 2008. Examining private forest policy preferences. For. Policy Econ. 10 (3), 89–97.

Joshi, O., Grebner, D.L., Munn, I.A., Grala, R.K., 2015. Issues concerning landowner management plan adoption decisions: a recursive bivariate probit approach. Inter. J. For. Res. 2015, Article ID 926303.

Kenya Wildlife Service, Kenya Forest Service, 2012. Kakamega Forest Ecosystem Management Plan 2012-2022. Kenya Wildlife Service and Kenya Forest Service, Nairobi, 152 p.

Konstant, T.L., Newton, A.C., Taylor, J.H., Tipper, R., 1999. The potential for community-based forest management in Chiapas, Mexico: a comparison of two case studies. J. Sustainable For. 9 (3/4), 169—191.

Korjus, H., 2014. Challenges in forest management planning. For. Res. Open Access. 3 (3), 1 p.

Krawczyk, R., 2014. Afforestation and secondary succession. Leśne Prace Badawcze. 75 (4), 423—427.

Lal, K., 2007. Management Plan of Private Forest Areas Closed Under Sec. 4 & 5 of Punjab Land Preservation Act—1900 (PLPA-1900), Dasuya Forest Division (2007-2008 to 2016-2016). Dasuya Forest Division, Dasuya, India, 113 p.

Matta, J., Kerr, J., 2006. Can environmental services payments sustain collaborative forest management? J. Sustainable For. 23 (2), 63—79.

McTague, J.P., Oppenheimer, M.J., 2015. Rayonier, Inc., Southern United States of America. In: Siry, J.P., Bettinger, P., Merry, K., Grebner, D.L., Boston, K., Cieszewski, C. (Eds.), Forest Plans of North America. Academic Press, New York, NY, pp. 395—402.

Mian, S.A., Dal, C.X., 1999. Decision-making over the project life cycle: an analytical hierarchy approach. Project Manage. J. 30 (1), 40—52.

O'Hara, K.L., 2009. Multiaged silviculture in North America. J. For. Sci. 55 (9), 432—436.

Olson, B.A., 2010. Paper trails: The Outdoor Recreation Resource Review Commission and the rationalization of recreational resources. Geoforum. 41, 447—456.

Oregon Department of State Lands, Oregon Department of Forestry, 2011. Elliott State Forest Management Plan. Oregon Department of State Lands and Oregon Department of Forestry, Salem, OR.

Palmer, B., 2000. Forest Management for Missouri Landowners. Missouri Department of Conservation, Jefferson City, MO, 108 p.

Salas, E., 1995. Military team research: ten years of progress. Mil. Psychol. 7 (2), 55—75.

Shapins Associates, 2005. Middle Kyle Canyon Framework Plan. US Department of Agriculture, Forest Service, Forest Service, Intermountain Region, Humboldt-Toiyabe National Forest, Las Vegas, NV, 117 p.

Simon, H., 1972. Theories on bounded rationality. In: Radnor, C., Radnor, R. (Eds.), Method and Appraisal. North-Holland, Cambridge, pp. 161—176.

Siry, J.P., Bettinger, P., Merry, K., Grebner, D.L., Boston, K., Cieszewski, C. (Eds.), 2015. Forest Plans of North America. Academic Press, New York, NY, 458 p.

Smith Sr., C.L., 1998. Computer-Supported Decision-Making: Meeting the Decision Demands of Modern Organizations. Ablex Publishing Corp., Greenwich, CT, 172 p.

Stevens, T.H., Dennis, D., Kittredge, D., Richenbach, M., 1999. Attitudes and preferences toward co-operative agreements for management of private forestlands in the North-eastern United States. J. Environ. Manage. 55 (2), 81—90.

Straka, T.J., Cushing, T.L., 2015. McPhail Tree Farm, South Carolina, United States of America. In: Siry, J.P., Bettinger, P., Merry, K., Grebner, D.L., Boston, K., Cieszewski, C. (Eds.), Forest Plans of North America. Academic Press, New York, NY, pp. 87—96.

Swae, J., 2015. City of San Francisco, California, United States of America. In: Siry, J.P., Bettinger, P., Merry, K., Grebner, D.L., Boston, K., Cieszewski, C. (Eds.), Forest Plans of North America. Academic Press, New York, NY, pp. 285—292.

The Crown Estate Forestry Department, 2015. The Crown Estate Windsor Forest Management Plan (Draft) 2015-2034. The Crown Estate Office, Windsor, United Kingdom, 67 p.

United States Congress, 1990. National forest management act regulations. Title 36—Parks, Forests, and Public Property, Chapter II—Forest Service, Department of Agriculture, Part 219—Planning, Subpart A—National Forest System Land and Resource Management Planning.

US Department of Agriculture, Forest Service, 2004. Land and Resource Management Plan, Chattahoochee-Oconee National Forests. US Department of Agriculture, Forest Service, Southern Region, Atlanta, GA, Management Bulletin R8-MB 113 A.

US Department of Agriculture, Forest Service, 2009. Record of Decision, Middle Kyle Complex. US Department of Agriculture, Forest Service, Forest Service, Intermountain Region, Humboldt-Toiyabe National Forest, Las Vegas, NV.

US Department of Agriculture, Forest Service, 2014. Revised Land and Resource Management Plan. George Washington National Forest. US Department of Agriculture, Forest Service, Roanoke, VA, Management Bulletin R8-MB 143 A.

US Department of Agriculture, Forest Service, 2015. Land Management Plan, 2015 Revisions, Idaho Panhandle National Forests. US Department of Agriculture, Forest Service, Northern Region, Missoula, MT, 187 p.

Van Horn, K., Brokaw, K., Petersen, S., 2003. Brule River State Forest Master Plan and Environmental Impact Statement. Wisconsin Department of Natural Resources, Madison, WI, 261PUB-FR-225.

VicForests, 2015a. Ecologically Sustainable Forest Management Plan, Working Plan Version 1.0. VicForests, Melbourne, VIC.

VicForests, 2015b. Planning Process. VicForests, Melbourne, VIC. Available from: http://www.vicforests.com.au/planning-1/planning-process/stage-one (Accessed 1/12/2016).

Chapter 2

Valuing and Characterizing Forest Conditions

Objectives

The need to evaluate the current and future state of forests is a necessary step in the assessment of alternatives for the management of these resources. When we indicate that a resource needs to be valued, we are suggesting that quantitative measures or qualitative labels are applied to the current and potential conditions of the area. These help you, as a natural resource manager, to understand the potential outcomes of your decisions. These values may relate to the structural condition of the resources, such as the basal area, wood volume, or tree density. They may also relate to economic outcomes associated with the revenues and costs of management activities. In this chapter, we describe a number of structural, economic, and ecological values and conditions that may be assessed to further put into context the impact of management on natural resources. After completing this chapter, you should be able to:

1. Understand the plethora of biological measures available for evaluating the structural conditions of a forest before and after planned management activities.
2. Understand the basic concepts of estimating future and present values.
3. Understand the common financial criteria used in forest resource management for making decisions.
4. Develop an initial understanding of contemporary societal issues that forest managers face when managing either private or public forest lands.

I. INTRODUCTION

An evaluation of the resources that are located within and around the forest that you manage is an important first step in understanding the management framework within which decisions can be made. Further, an assessment of the outcomes, either economic, ecological, or social, that arise as a result of implementing management activities is necessary to determine whether the course of action being suggested will meet the expectations of the landowner. In 1905, Gifford Pinchot (1905) stated in his book *A Primer of Forestry*, that:

A forest working plan is intended to give all the information needed to decide upon and carry out the best business policy in handling and perpetuating a forest. It gives this information in the form of a written statement . . . The working plan also predicts the future yield of the forest . . . Finally, it estimates the future return in money, taking into account taxes, interest . . . In order to make this estimate entirely safe, it is usually based on the present price of stumpage, although its future value will certainly be much higher.

Although modern forest and natural resource management plans may have expanded their goals and objectives beyond timber harvest levels (depending on the organization), the basic point remains: to develop a plan of action, some assessment of the current and future state of the resource is necessary. In addition, it is necessary to quantify as much of the assessment as possible in monetary or economic terms. This chapter therefore provides coverage of economic, ecological, and social measures commonly used in natural resource management for assessing the current and future value of resources.

II. STRUCTURAL EVALUATION OF NATURAL RESOURCES

The structural evaluation of a property and surrounding area involves understanding the current state of the resources that can be managed, and involves evaluating the future conditions of those resources after management activities have been applied. This section of the chapter provides a brief overview of many of the structural metrics used in natural resource management to describe forested conditions.

A. Trees per Unit Area

Perhaps the most basic structural evaluation conducted within each stand of a forest is a determination of the number of stems or trees per unit area. In the United States this is most commonly referred to as trees per acre (TPA), whereas in other parts of the world it is often referred to

Forest Management and Planning. DOI: http://dx.doi.org/10.1016/B978-0-12-809476-1.00002-3

in terms of trees per hectare (TPH). A *stand table* is simply a description of the number of trees per acre by diameter class. It can be presented either in tabular form (Table 2.1) or in graphical form (Fig. 2.1). The range of trees per acre that are common with even-aged stands is 0 (for a recently site prepared area) to about 1,500. Common planting densities in the southern United States are around 600 trees per acre, if trees are planted using a 12-foot by 6-foot (3.7 m by 1.8 m) spacing. The density of trees could reach 10,000–20,000 per acre (25,000–50,000 per hectare) if a significant amount of natural regeneration occurs within gaps of even-aged or uneven-aged stands, which could lead to a stand density

management issue if competition-related mortality does not reduce the density sufficiently.

B. Average Diameter of Trees

One of the most common measures of the size of live and dead trees is the diameter at breast height (DBH). Breast height is considered to be 4.5 ft (1.37 m) above ground level. Since trees are not necessarily uniform in size or shape, there are a number of standards that you should consider when measuring the DBH of trees. Although covered in more detail in forest measurements courses, these standards for measuring the DBH of trees include the following:

1. DBH should always be measured on the uphill side of a tree.
2. DBH should not be measured where it could include limbs, vines, or other objects that are not part of the main tree bole.
3. If a tree leans, DBH should be measured perpendicular to the lean.
4. If a tree forks below 4.5 ft, each fork is considered a separate tree.
5. If a tree forks below, but near 4.5 ft, each fork is considered a separate tree, and DBH should be measured a foot or so above the fork.
6. If a tree forks above 4.5 ft, it is considered one tree.
7. If a tree has an unusual bulge around 4.5 ft, the DBH measurement should be made a foot or so above the bulge.
8. If a tree has a bottleneck near 4.5 ft, such as what you might find in a baldcypress tree (*Taxodium distichum*), the DBH measurement should be made a foot or so above the bottleneck.

TABLE 2.1 A Stand Table for an Even-Aged Stand of Loblolly Pine in Georgia

DBH Class (in.)	Trees per Acre	Trees per Hectare
6	1	2.5
7	7	17.3
8	17	42.0
9	42	103.8
10	58	143.3
11	61	150.7
12	22	54.4
13	9	22.2
14	2	4.9
Total	219	541.1

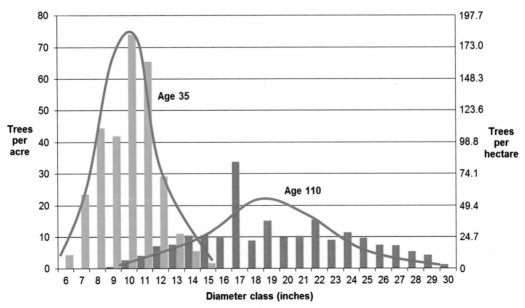

FIGURE 2.1 Even-aged stand diameter distributions (stand tables).

The devices that you could use to measure a tree's DBH include a DBH tape, calipers, relaskop, Biltmore stick, or a Bitterlich sector fork. The average DBH of a stand provides the relative size of the samples that were obtained through field measurements. This, in conjunction with other measures of structure, can help you visualize the quality of the forest. The average diameter is occasionally a requirement for habitat suitability models as well.

Example

The average DBH of the initial condition of a coastal Douglas-fir (*Pseudotsuga menziesii*) forest that we describe in Appendix A is 4.2 inches at a stand age of 15 years. Once the stand has projected to an age of 35, we find that the average DBH has risen to 9.9 inches.

Through time, the average DBH rises as we would expect in an even-aged stand, even though the trees per acre may decline as a result of competition-related mortality or thinning operations that remove trees from the lower end of the diameter distribution. In uneven-aged stands, the average DBH may remain relatively constant, when partial cutting activities target trees of all diameter classes.

C. Diameter Distribution of Trees

When developing a diameter distribution we first group all the individual trees (or tree records) into diameter classes. Assuming we are using the English system of measurement, the typical diameter classes are 1 or 2 inches in size. If using the metric system, we may group the trees into 1- or 2-cm (or greater) diameter classes. A graph would

then be constructed to illustrate the number of trees per unit area by DBH class. For an even-aged stand, the distribution should be approximately normal (Fig. 2.1). Uneven-aged stands have more than one distinct age class, and usually consist of numerous small trees that fill in the gaps in the canopy. As a result, when developing a diameter distribution for uneven-aged stands, the distribution should approximate a reverse J-shape (Fig. 2.2). We will explore further the intricacies of uneven-aged diameter distributions in Chapter 4, Estimation and Projection of Stand and Forest Conditions.

D. Basal Area

The basal area of a stand of trees is the sum of the cross-sectional surface areas of each live tree, measured at DBH, and reported on a per unit area basis. Basal area is a measure of tree density, and widely used in forestry, wildlife, and other natural resource management professions. To calculate basal area, assume that a tree is cut off at 4.5 ft above ground (DBH). Since the area of a circle is πr^2, and since we commonly measure the diameter of a tree rather than the radius, we can substitute (DBH/2) into the equation, which then becomes:

$$\text{Basal area (units}^2) = \pi \left(\frac{DBH}{2}\right)^2 \text{ or } \pi \left(\frac{DBH^2}{4}\right)$$

Since DBH commonly is measured in inches in the United States, and since we desire basal area expressed in square feet per acre in the United States, we simply

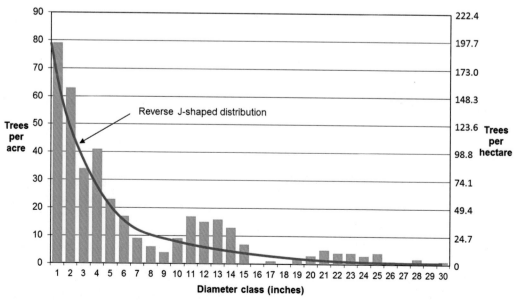

FIGURE 2.2 Uneven-aged stand reverse-J shaped diameter distribution.

divide everything by 144 (the number of square inches in a square foot).

$$\text{Basal area (feet}^2) = \left(\frac{\pi \left(\dfrac{DBH^2}{4} \right)}{144} \right)$$

When the DBH is measured in inches, and the outcome is reported in ft^2, the condensed version of the basal area equation then becomes:

$$\text{Basal area (feet}^2) = 0.005454 \times DBH^2$$

We need to estimate the basal area in *square feet per acre* because these are the units that commonly are communicated among natural resource professionals in the United States. In addition, they are the units commonly used in a number of wildlife habitat suitability models. In almost every other part of the world, including Canada, basal area is expressed in *square meters per hectare*, and uses diameters of trees commonly measured in centimeters. Therefore, when the DBH is measured in centimeters, and the outcome is reported in m^2, the basal area equation one might use in these cases is:

$$\text{Basal area (m}^2) = 0.00007854 \times DBH^2$$

Example

The initial basal area of the western forest described in Appendix A is 50.1 ft^2 per acre. This converts to 11.5 m^2 per hectare ((50.1 ft^2 per acre/10.765 ft^2 per m^2)*2.471 acres per hectare). When the stand is projected to age 35, the basal area is estimated to be 166.1 ft^2 per acre, which converts to 38.1 m^2 per hectare. Reasonable ranges of basal area are 0−250 ft^2 per acre in the eastern part of North America, and 0−500 ft^2 in the western part of North America.

E. Quadratic Mean Diameter

The quadratic mean diameter (QMD) of trees is the diameter of the tree represented by the average tree basal area of the stand. For example, if the basal area per acre of a stand of trees were 150 ft^2 per acre, and there were 217 trees per acre in the stand, the average tree basal area would be (150 ft^2 per acre/217 trees per acre), or 0.69 ft^2 per tree. The QMD is then the diameter of a tree that would provide a basal area of 0.69 ft^2. To arrive at this, you would use the following equation:

$$QMD \text{ (inches)} = \sqrt{\frac{\text{Average basal area per tree}}{0.005454}}$$

Using the earlier example (150 ft^2 per acre, 217 trees per acre), the QMD for this stand would be 11.26 inches. As a check on this work, the basal area of a 11.26 inch tree is 0.69 ft^2 (0.005454 × 11.26^2). If there are 217 of these trees per acre, the stand's basal area is (217 × 0.69 ft^2), or 150 ft^2 per acre.

Example

The QMD of the western stand described in Appendix A can be determined by first computing the average basal area per tree. This can be accomplished by dividing the average basal area per acre (50.1 ft^2) by the number of trees per acre (465.5). Using the equation provided earlier, the QMD at age 15 is:

$$QMD \text{ (inches)} = \sqrt{\frac{(50.1 \text{ ft}^2/465.5 \text{ } TPA)}{0.005454}} = 4.4 \text{ inches}$$

F. Average Height

The average height of trees in a stand lets us visualize the size of the trees relative to other stands in the nearby vicinity. The average height is also closely related to the site index of the stand (described in Section II, Part P). To arrive at the average height, estimates (either from field measurements or through height calculations) of all sizes of trees are necessary. Field measurements of tree heights are one of the most expensive tasks in field inventories. As a result, sometimes tree heights are estimated using equations that are based on the diameter of each tree. Site index computations utilize only the heights of the dominant and codominant trees; therefore, the intermediate and suppressed tree heights would need to be removed or ignored to facilitate site index estimation. Field-measured heights provide a static metric of the height of a stand. Projected heights for even-aged stands should increase with age (Fig. 2.3), even as trees per unit area decline. Projected heights for uneven-aged stands should be relatively constant once the uneven-aged stand has matured.

G. Timber Volume

Timber volumes are common measures of inventory and of output used in a forest management plan. They are often a function of the DBH and height of trees. Timber volumes are directly related to revenue, and thus are intimately tied to the economic evaluation of activities. Timber volumes can also be generated through habitat improvement activities, particularly those that involve reducing the density of a stand of trees so that it is more suitable for certain species of wildlife. For example,

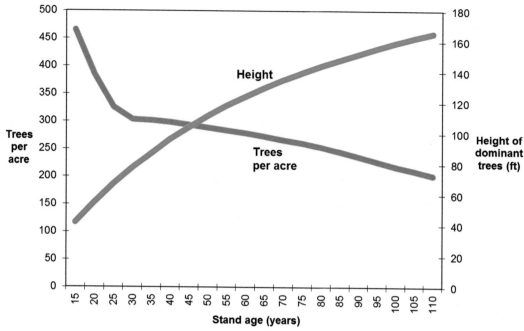

FIGURE 2.3   Height development over time in an even-aged coastal Douglas-fir stand.

maximum basal area for good quality red-cockaded woodpecker (*Picoides borealis*) habitat, for example, is suggested to be 80 ft^2 per acre in pine stands (preferably less), according to the recovery standard developed by the US Department of the Interior, Fish and Wildlife Service (2003). Since stands of trees tend to grow in size and density over time, maintaining good habitat for the woodpecker may require periodic removals of trees, which could generate timber volume.

Timber volume can be expressed as a solid wood unit (m^3, ft^3, cunit, cord); as a manufactured wood unit (board feet, thousand board feet (MBF)) expressed by either a Doyle, Scribner, or International log rule; or as a weight (ton, or metric ton). In today's global economy, it would be advantageous for you to understand how to convert between metric and English units. A cord is a solid area of wood that is 4 ft wide, 4 ft tall, and 8 ft long, or 128 ft^3. Since air pockets are present in cut, stacked wood, a cord generally is considered to contain only about 90 ft^3 of wood even though it might require 128 ft^3 of space to occupy. A board foot is a 12 inch square of wood that is 1 inch thick. Theoretically, you could extract 12 board feet from a solid cubic foot of wood. However, given the sawdust (kerf) that is generated by cutting and separating the boards, in general you should expect to generate only 4−6 board feet from each cubic foot of solid wood. These conversion factors will vary depending on the species of tree being processed, and the equipment used to perform the processing.

H. Mean Annual Increment, Periodic Annual Increment

The mean annual increment (MAI) is the average yearly growth computed for volume, weight, or other measure, up to the time of measurement or projection. MAI can be calculated for a tree or a stand of trees, and if for the latter, it represents the average growth rate per unit area per year. The MAI will change over the life of a tree or stand of trees, with slow growth rates initially, higher rates of growth in the mid-life of a tree or stand, and decreasing growth rates with older ages. The point at which MAI peaks commonly is referred to as *biological maturity*, and sometimes used as a guide for harvesting decisions.

$$MAI = \left(\frac{\text{Volume or weight per acre}}{\text{Age of stand}} \right)$$

MAI can be expressed as a function of site index, which is described in more detail in Section II, Part P. A number of MAI equations for the western United States were developed based on the site index of a stand of trees (Hanson et al., 2002). For example, the equation:

$$MAI = 0.00473 \times SI^{2.04}$$

possibly could be used to express yield in cubic feet per acre per year for Douglas-fir stands in eastern Oregon and Washington (Cochran, 1979). These relationships between site index and MAI should however be used with care, since as we have noted, MAI changes over the life of a tree. The MAI of an even-aged stand will

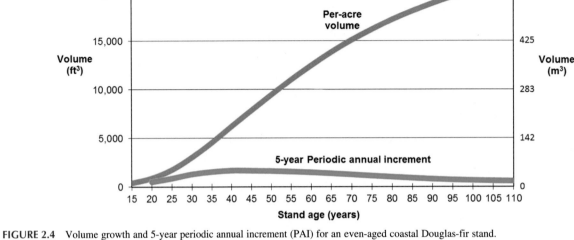

FIGURE 2.4 Volume growth and 5-year periodic annual increment (PAI) for an even-aged coastal Douglas-fir stand.

also change over its life; however, the MAI of an uneven-aged stand may or may not change over time, based on the condition of the uneven-aged stand and the intensity of periodic cuttings. The MAI equations summarized in (Hanson et al., 2002) produce a single estimate of MAI that represents the average increment over the time period ranging from stand establishment to the age at which MAI culminates (reaches the maximum value).

The periodic annual increment (PAI) is the growth rate of a tree or stand of trees over some period of time, whether that period is 1 year, 5 years, a decade, or longer (Fig. 2.4). For example, some government agencies develop growth projections for private and public forests using permanently installed inventory plots. With periodic measurements of these, analysts can describe the change in forest conditions over a period of time for measures such as merchantable volumes, using commonly collected tree measurements (e.g., DBH and total tree height). The US Forest Service regularly produces forest status reports for each state, as do some Canadian Provinces, such as the 10-year PAI reports developed by the Nova Scotia Department of Natural Resources (2000).

PAI can be expressed in terms of an annual growth rate, and under this condition, would be the same as both the periodic MAI and the current annual increment (CAI) that are used in natural resource management. When graphed, the point at which the PAI curve (or CAI curve, if expressed on an annual basis) and the MAI curve meet is also the point at which MAI culminates (Fig. 2.5), and is considered by many to be representative of the

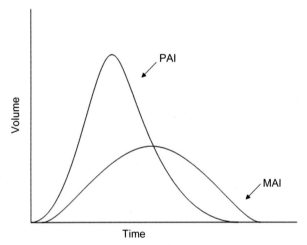

FIGURE 2.5 Theoretical relationship between mean annual increment (MAI) and periodic annual increment (PAI).

biological rotation age for an even-aged stand. The CAI computation in equation form can be expressed as:

$$CAI = \text{(Volume at the end of a year}$$
$$- \text{Volume at the beginning of the year)}$$

If we were interested in the PAI, and the "periods" were longer than 1 year, the PAI equation would be a modified version of the CAI equation:

$$PAI = \text{(Volume at the end of a period}$$
$$- \text{Volume at the beginning of the period)}/$$
$$\text{Length of the period}$$

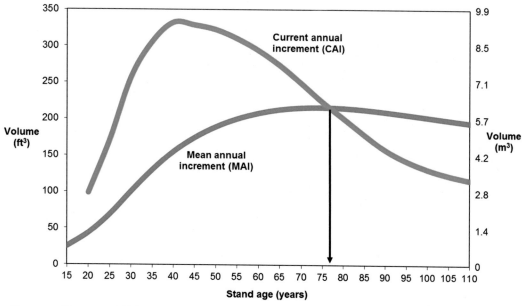

FIGURE 2.6 Mean annual increment (MAI) and current annual increment (CAI) for an even-aged coastal Douglas-fir stand.

Example

Using the western forest stand data provided in Appendix A, if we plotted the cubic foot volume for each 5-year time period, we would find a somewhat sigmoid curve that represented the volume per acre (Fig. 2.6), and if we computed the change from one 5-year time period to the next, then converted this to an annual rate of change, the CAI would range from several hundred cubic feet of growth per 5-year time period to almost 350 ft^3 when the stand was 40 years old. If we divided the cubic feet per acre by the age of the stand, then we would arrive at the MAI (Fig. 2.6), which culminates somewhere between age 75 and 80 for this example stand at about 216 ft^3 of growth per acre per year. As we suggested, the point at which the CAI curve and the MAI curve meet is also the point at which MAI culminates, and is considered by many to be representative of the optimal biological rotation age for the stand. Although the specific annual growth rate from 1 year to the next (CAI) peaks earlier, around age 40, the average annual growth rate peaks at the point where MAI and CAI cross.

I. Snags per Unit Area

Trees that recently have died and remain standing are considered *snags*. They can be considered a hazard for logging and fire control purposes, yet they are of value for a number of wildlife habitat purposes. For example, the habitat suitability model for the downy woodpecker (*Picoides pubescens*) includes two variables, one that is a function of the basal area per unit area in a stand, and the other that is a function of the number of snags per unit area that are greater than 6 inches DBH (Schroeder, 1983).

Snags per unit area can be estimated using the same field sampling techniques that are used to determine live trees per unit area. Projecting snag availability through time (into the future) involves a more complex procedure. The number of new snags in each projected time period can be obtained by assessing the difference between live trees per unit area from one time period to the next. For example, using the western stand described in Appendix A, 81.1 trees per acre died between the ages of 15 and 20. However, determining how large each tree was when it died is difficult with this data, since a simple comparison of the trees per unit area by diameter class is not a straightforward computation. In other words, some of the trees in each diameter class may have grown to the next higher class prior to their expiration. However, if an estimate of the trees per unit area by diameter class that have died could be ascertained, the question then becomes how long they will continue to stand and function as wildlife habitat. Decay rates have been proposed to estimate the length of time a dead tree will continue to stand, and degree of breakage over time that will occur. A method for estimating this was proposed for coniferous forests in the western coastal United States (Mellen and Ager, 1998).

J. Down Woody Debris

Down wood are former standing trees that are now lying near, or on, the ground surface. For stream surveys, down wood is considered as logs that are lying within the stream reach or suspended above the channel (Frazier et al., 2005). Idol et al. (1999) discuss measurement techniques

TABLE 2.2 Down Wood Decay Classes

	Decay Class				
	I	II	III	IV	V
Bark	Intact	Mostly intact	Mostly intact	Absent	Absent
Integrity	Sound	Sapwood rotting	Heartwood sound	Heartwood rotten	None
Branches	All present	Larger twigs present	Larger branches present	Branch stubs present	Absent

Source: Thomas, J.W., 1979. Wildlife habitat in managed forests in the Blue Mountains of Oregon and Washington. US Department of Agriculture, Forest Service, Washington, DC. Agricultural Handbook 553.

for down woody debris in oak-hickory (*Quercus* spp.– *Carya* spp.) forests. To measure the volume of a piece of down wood, two measurements are required: the length of the wood, and the mid-point diameter. If they are more accessible, the small-end diameter and large-end diameter can be averaged to obtain the mid-point diameter. The volume can be estimated using the following equation, as long as the diameter and length are expressed in the same units.

$$\text{Volume} = \pi \left(\frac{\text{Mid-point diameter (feet)}}{2}\right)^2 \text{Length (feet)}$$

Other assumptions are needed when assessing down wood volume, including the decay class (Table 2.2) above which, and including, the logs that need to be measured. This relates to the soundness of the logs. Thomas (1979) describes one type of down woody debris classification system for the western United States. In addition, the minimum diameter after which the down wood does not "count" (since it is too small to be of value) is important.

K. Crown or Canopy Cover

A number of measurements of the crowns of trees may be necessary for growth and yield modeling as well as habitat quality assessments (e.g., the habitat model for the red-spotted newt (*Notophthalmus viridescens viridescens*) (Sousa, 1985)). Two of the basic measurements include the length of the crown from the tip of the tree to the base of the live branches, and the crown ratio (total tree height/crown length). Crown diameters can be measured in the field by projecting vertical lines up the sides of a tree and measuring the distance from one side of the crown to another. Crown diameter measurements can also be made from aerial photographs, although some portions

of the crown may be obscured in a photograph by other nearby trees. Crown closure, or canopy cover, is a measure of the amount of ground area that is covered by the canopy of trees in a stand. In some cases, crown closure is used as a proxy for stand density. Crown radius and crown closure could be estimated from variables (e.g., DBH) commonly measured during forest inventories, since ground measurement of crown cover is a time-consuming process (Gill et al., 2000). The crown radius of ponderosa pine (*Pinus ponderosa*), for example, could be estimated using the following equation:

$$\text{Crown radius} = 0.9488 + 0.0356 \times DBH$$

If crown cover for individual trees could be estimated from aerial photographs or other remotely sensed imagery, the DBH of individual trees could be estimated using relationships such as:

Shortleaf pine (*Pinus echinata*) DBH (inches)
$$= 0.6733 + 0.5287 \times (\text{Crown diameter in feet})$$

from Gering and May (1995), which are developed locally or regionally to reduce the time required to capture field measurements of individual stands.

L. Tree, Stand, or Forest Age

Age is a useful measurement for describing a condition of an even-aged forest, and is helpful in predicting future growth and yield of trees. Age of trees, stands, or forests can be described in a number of ways, including:

- Elapsed time since germination of seed
- Elapsed time since budding or sprouting of seedlings
- Elapsed time since planting of trees or seeding of a site
- Elapsed time since trees were 4.5 ft tall, otherwise called *breast height age*

In temperate climates, most trees record their growth history in annual growth rings. Each annual ring is made up of earlywood and latewood. Earlywood is the light colored, fast growing wood that is developed in the spring and summer. Latewood is the dark colored, slow growing wood developed in the fall. Annual rings are distinctive in conifers, such as pines and Douglas-fir, and some deciduous species such as oaks. However, they are not as easy to see in other hardwood species, such as birches (*Betula* spp.) and maples (*Acer* spp.). Increment borers can be used to estimate stand age, as can branch whorls for some trees such as eastern white pine (*Pinus strobus*). Management history (if maintained by natural resource management organizations) is another source of stand age values.

Age classifications are another age-related characteristic that can be assigned to individual stands. The two basic age classifications that are used frequently in natural resource management are even-aged and uneven-aged. Even-aged stands are those where the ages of the trees are generally within about 20% of the average stand age. If we were to develop a tree age distribution for an even-aged stand, we would find that it often resembles a very tight bell-shaped, or normal distribution (Fig. 2.7). Uneven-aged stands are those where there are two or more distinct age ranges of trees within a stand. If you were to develop an age distribution for an uneven-aged stand, you will likely see the age cohorts stand out, such

as the group of 30−40-year-old trees and the smaller group of 55−65-year-old trees in Fig. 2.7.

Age classes are different from a single stand age value in that they represent the distribution of stand ages across an ownership or landscape. Age class distributions are similar to diameter distributions in that both are histograms reflecting conditions of forests in an area. However, age class distributions group all similar stands together into classes, to illustrate to landowners or land managers the ranges of ages within an ownership (Fig. 2.8). Typically, these are grouped into 1-, 5-, or 10-year classes.

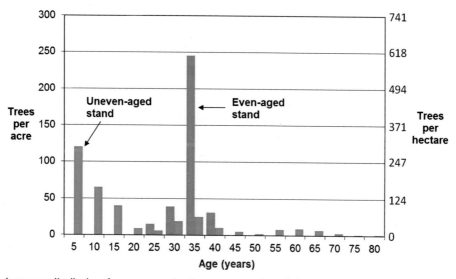

FIGURE 2.7 Example tree age distributions for an even-aged and an uneven-aged stand of trees.

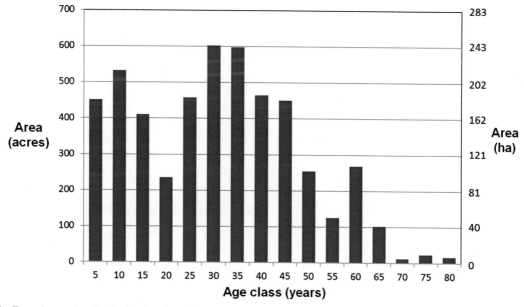

FIGURE 2.8 Example age class distribution for a large forest.

M. Biomass and Carbon

Carbon sequestration and carbon accounting are hot topics in forest management, and we provide more detail on these issues in Chapter 15, Forest Certification and Carbon Sequestration. Standard forest inventory data (DBH, tree heights, and basal area) have been shown to be strongly correlated with tree biomass (Bartelink, 1996; Mikšys et al., 2007). The measurement of carbon in a tree or stand of trees can either be accomplished indirectly or approximated. The direct method for measuring carbon in a tree would be to cut down a tree and analyze the resulting woody material. In lieu of this sampling without replacement process, an indirect method for estimating carbon in a tree can be used to estimate carbon content given the specific gravity of a tree, the density of water, and the volume of the tree. A basic equation to estimate the dry weight of wood in a tree or stand of trees is:

Weight = (Specific gravity of wood) × (Density of water)

× (Volume of tree or stand of trees)

The carbon fraction of dry wood in the tree or stand of trees is then estimated to be about one-half of the weight of the dry wood (Smith et al., 2006).

Carbon = 0.5 × (Dry weight)

Example

Assume that the specific gravity for loblolly pine (*Pinus taeda*) is 0.47, and that the density of water is 62.4 pounds per cubic foot. If we had a stand of trees where there were 2,500 ft³ of wood per acre, how much carbon would be estimated in this stand?

Weight = (0.47) × (62.4 pounds per cubic foot)

× (2,500 cubic feet) = 73,320 pounds of wood per acre

Carbon = (0.5) × (73,320 pounds per acre)

= 36,660 pounds per acre of

carbon or 18.33 tons per acre

To estimate the amount of carbon sequestered from one period of time to the next, we would need to estimate the standing volume of the tree (or stand) at the beginning of the period, and estimate the standing volume at the end of the period. For example, if the stand in the previous example had 2,650 ft³ of wood 1 year later, the dry weight of wood at the end of the period would be:

Weight = (0.47) × (62.4 pounds per cubic foot)

× (2,650 cubic feet) = 77,719 pounds of wood per acre

and the estimate of carbon would be:

Carbon = (0.5) × (77,719 pounds)

= 38,860 pounds per acre of

carbon or 19.43 tons per acre

Thus the estimate of the amount of carbon added over the 1-year period is 1.1 tons per acre.

To be able to characterize the amount of aboveground carbon (in this case), we would need an updated inventory of the stand, and information on the average specific gravity of the trees in the stand. In cases where a landowner does not have updated inventories of their forestland, carbon tables might be utilized to assist landowners in developing rough estimates of carbon based on a stand's age (Smith et al., 2006; Georgia Forestry Commission, 2007).

N. Pine Straw

Pine straw is a valuable landscaping material in the southern United States, and an important and profitable management option for landowners with the right type of forest on an amenable landscape (Kelly et al., 2000). Pine stands with species producing longer needles, such as longleaf (*Pinus palustris*) or slash pine (*Pinus elliottii*), are the preferred pine straw source areas, although loblolly pine stands can also be used for pine straw production. Stands developed for pine straw production begin early with a control burn shortly after crown closure (before age 10). This is followed perhaps by a herbicide application to control other unwanted understory species, and perhaps by a process of clearing other material and pruning the pine trees. Straw is then raked into piles and baled. The pine straw collection process could occur annually or every 2 years, with herbicide and clearing processes added as needed over time.

As an example of the production potential, young slash pine stands can yield from 1,000 to 2,500 pounds of litterfall per acre per year (Fig. 2.9). Slash pine stands between 10 and 15 years old can yield 2,500–4,000 pounds of litterfall per acre per year. Stands over 15 years old can yield from 3,000 to 4,000 pounds of litterfall per acre per year, which slightly declines as the stand gets older (Gholz et al., 1985). Commercial baling practices vary from one operator to another due in part to differences in equipment (Kelly et al., 2000), and as a result, prices for pine straw could range from as little as $0.25 per bale to $1.00 per bale to the landowner. Stands that undergo a thinning may be unavailable for straw production until the stand has been raked (at a cost of $60–80 per acre), due to the incorporation of limbs and other unwanted logging slash with the pine straw.

O. Other Nontimber Forest Products

Nontimber forest products are those biological, medicinal, and spiritual materials from a forest, other than traditional commodities such as timber, that can be extracted from forests in one manner or another and used by humans (Svarrer and Smith Olsen, 2005). Pine straw could arguably be considered a nontimber forest product. More conventionally,

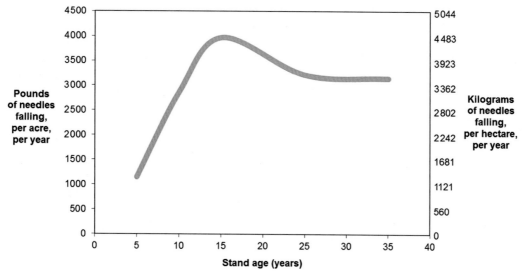

FIGURE 2.9 Pine straw production over time for a slash pine stand in the southern United States. *From Gholz, H.L., Perry, C.S., Cropper Jr., W.P., Hendry, L.C., 1985. Litterfall, decomposition, and nitrogen and phosphorus dynamics in a chronosequence of slash pine (Pinus elliottii) plantations. Forest Science 31 (2), 463–478.*

we think of resources such as mushrooms as nontimber forest products. Yet medicinal compounds made from tree bark, roots, and leaves might be considered nontimber forest products, as well as mosses and cork. More broadly, viewsheds and biodiversity might be considered nontimber forest products. How one would measure the current and future states of these resources is an important issue, particularly if they are to be considered in a forest plan.

Example

There are numerous edible mushroom species that grow in a variety of forest conditions. The impact of forest management on these resources will vary. For example, Pilz et al. (2004) describe how prescribed burning in Oregon can affect mushroom production, and suggest that prescribed fire (or lack thereof) can be used to promote different species. Since a number of mushroom species utilize downed woody material for optimal growing conditions, some activities that reduce down woody debris may discourage mushroom growth. The potential of certain mushroom species to utilize loblolly pine woody debris has also been described (Croan, 2004). Down woody debris may be able to support marketable mushroom species, such as shiitake (*Lentinula edodes*) or oyster (*Pleurotus* spp.) mushrooms, for up to 6 years (Hill, 1999). Improving mushroom production on a forest manager's property can enhance their cash flow or the rate of return on their asset. Forest managers can utilize downed material of many different species as growing media for mushroom spawn. Assuming that the downed logs are appropriately cured and of an adequate size, they can be inoculated with mushroom spawn. The first fruiting of shiitake mushrooms may take from 6 to 18 months. Ideal environments for mushroom growth are heavily shaded and moist areas. Mushrooms can be produced in both the

spring and fall seasons and must be harvested quickly. A log, 4–6″ in diameter and 37″ long, may produce up to 0.77 pounds of mushrooms every year, although this would be the production rate after first fruiting (Anderson and Marcouiller, 1990). Anderson and Marcouiller (1990) illustrate that starting with 4,000 logs (some logs are lost to production from decay over time) to grow mushrooms would start to yield 1,944 pounds of shiitakes in the second year, 5,028 pounds in the third year for a cumulated total of 10,630 pounds after the fourth year.

P. Site Quality

The term *site* in natural resource management generally refers to the various conditions present at a particular geographic location. As a result a number of factors, such as water availability and soil conditions, need to be taken into account when describing the quality of a site. There are several different perspectives on the manner in which *site quality* should be described. For example, if you were solely interested in timber production, site quality could be described by the amount of volume that can be produced over a given amount of time. An ecosystem-oriented approach to describing site quality would include describing the total annual productivity arising from all plants, animals, bacteria, and so on, and used as an expression of the potential of a site to produce biomass. A generalist approach to describing site quality suggests that you would describe the capacity of an area to produce forests or other vegetation, as it is influenced by soil type, topography, and other physical or biological factors.

Site quality can be expressed either qualitatively or quantitatively. Qualitative assessments of sites use words, rather than numeric values, to describe the appropriateness of an area for timber production, wildlife habitat, or other use. In the *Soil Survey of Saratoga County, New York*, as in other soil surveys developed by the Natural Resources Conservation Service, terms such as *unsuited, poorly suited, moderate*, and *well suited* are used to describe how a site, based on soil qualities, may accommodate various forest management activities (US Department of Agriculture, Natural Resources Conservation Service, 2004). Other qualitative systems are used to describe sites, such as those that were developed in conjunction with gypsy moth management guidelines in Wisconsin (Brooks and Hall, 1997):

> *Poor sites for forest trees include the dry to moderately dry, nutrient-poor areas, medium quality sites typically include moderately dry, nutrient-medium to nutrient-rich areas, and high quality sites include wet to moderately dry, nutrient-rich areas.*

In the early part of the 20th century, there was, as one forester suggested, "an urgent need for a simple method by which sites may be quickly and easily classified" (Frothingham, 1921). The height growth of the dominant trees in a stand was proposed as the standard of measurement, although some disagreement over the need to develop scales for different species was evident. The *site index* eventually was proposed as a quantitative measure of site quality, and it generally is reflective of the potential timber productivity of a stand of trees. Site index, as it is used in forestry and natural resources, is simply a measure of the height of the dominant and codominant trees in a stand, *at some base age*. Dominant and codominant trees are used to describe site index because they should be assumed to have been "free to grow" throughout their life; thus the growth of these trees should have been somewhat independent of other vegetation. A base age is used as a reference so that stands of different site quality can be compared. Without the base age, we would simply be communicating the average height of the dominant and codominant trees of different stands, which is an indication of their size, but not their productive potential. For example, assume we were to say that Stand A is 53 years old, and has an average dominant and codominant tree height of 83 ft, and Stand B was 21 years old and has an average dominant and codominant tree height of 54 ft. Which stand is more productive? It would be hard to determine from this limited information. However, if we were to project backward Stand A's average height to age 25, then project forward Stand B's average height to age 25, we could compare the two on using common measure (how tall the trees are, were, or will become at age 25). The taller the trees at the base age, the better the site index, and the higher the volume per unit area. Early in the twentieth century the base age proposed was 100 years (Frothingham, 1921). Today, although several base ages are used, depending on

the tree species under consideration, common base ages in the southern United States are 25 years (newer models and plantations) and 50 years (older models and natural stands). In the western United States common base ages are 50 and 100 years. Base ages generally are placed after "SI" for communication purposes. For example, $SI_{25} = 65$ suggests that the site index, base age 25, is 65 for a stand of trees. In other words, we should expect that the dominant and codominant trees in this stand will be (or were) 65 ft tall when the stand is (or was) 25 years old.

Example

$SI_{25} = 70$ indicates that at age 25, we expect the dominant and codominant trees on a site to be 70 ft tall. When a stand is greater than 25 years old, we would expect the height of the dominant and codominant trees to be greater than 70 ft. When a stand is less than 25 years old, we would expect the height of the dominant and codominant trees to be less than 70 ft.

Example

$SI_{50} = 125$ indicates that at age 50, we expect the dominant and codominant trees on a site to be 125 ft tall. When a stand is greater than 50 years old, we would expect the height of the dominant and codominant trees to be greater than 125 ft. When a stand is less than 50 years old, we would expect the height of the dominant and codominant trees to be less than 125 ft.

To determine the site index for a stand of trees, a sample of the ages and heights of dominant and codominant trees is necessary. Site index equations can be developed by sampling these characteristics of trees over a broad range of ages and site conditions. Equations are then developed to allow us to estimate the average height of the stand at any age. Site index equations are developed specifically for different tree species, and different management practices applied to stands (e.g., different site preparation methods). Some site index equations are very complex, such as the equation developed for western larch (*Larix occidentalis*) in Oregon (Cochran, 1985),

$$
\begin{aligned}
SI_{50} = {} & 78.07 + [(\text{Height} - 4.5) \times (3.51412 - 0.125483\,\text{Age} \\
& + 0.0023559\,\text{Age}^2 - 0.00002028\,\text{Age}^3 \\
& + 0.000000064782\,\text{Age}^4)] - [(3.51412 - 0.125483\,\text{Age} \\
& + 0.0023559\,\text{Age}^2 - 0.00002028\,\text{Age}^3 \\
& + 0.000000064782\,\text{Age}^4) \times (1.46897\,\text{Age} \\
& + 0.0092466\,\text{Age}^2 \\
& - 0.00023957\,\text{Age}^3 \\
& + 0.0000011122\,\text{Age}^4)]
\end{aligned}
$$

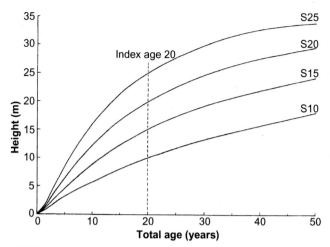

FIGURE 2.10 Site index curves for red alder. *From Harrington, C.A., Curtis, R.O., 1985. Height growth and site index curves for red alder. US Department of Agriculture, Forest Service, Pacific Northwest Research Station, Portland, OR. Research Paper PNW-358. 14 p.*

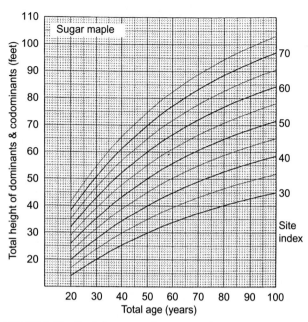

FIGURE 2.11 Site index curves for sugar maple (*Acer saccharinum*). *From Carmean, W.H., Hahn, J.T., Jacobs, R.D., 1989. Site index curves for forest tree species in the eastern United States. US Department of Agriculture, Forest Service, North Central Forest Experiment Station, St. Paul, MN. General Technical Report NC-128. 142 p.*

whereas others are relatively simple, such as the equation developed for red alder (*Alnus rubra*) in the western United States (Worthington et al., 1960):

$$SI_{50} = (0.60924 + (19.538/Age)) \times (Height)$$

Site index curves can be produced from site index equations, and provide managers with a graphical view of the height growth of a stand of trees (Fig. 2.10). Typically, the height growth progression of different sites is presented using these types of nonlinear curves. The rate of height growth can then be compared with other species. For example, in Fig. 2.10 you can observe that the rate of height growth for red alder is fast initially, but slows down and flattens by 50 years, particularly on lower quality sites. In contrast, the rate of height growth of Douglas-fir would continue to rise well after 50 years, even though both tree species may be located on the same site.

If we were presented with a set of site index curves and did not know the base age, we can determine the base age rather easily by locating the intersection of a height growth line and its associated average tree height. For example, in Fig. 2.11, the intersection of the site index 50 curve and the horizontal line that represents a height of 50 ft is directly above age 50. Therefore the base age of the curves, or the age at which the height of the dominant and codominant trees is equal to the site index curves, is 50 years. This method can be applied to any of the curves in any of the site index graphs.

As a natural resource manager, you should keep in mind that an estimate of the site index is relative, and does not provide an exact correlation to timber or biomass productivity. When developing site index values, some

error may have arisen in the tree measurements, therefore natural resource managers do not consider a 2- or 3-foot difference in site index to be very meaningful. In addition, as a natural resource manager, you should be mindful that site index can vary across the landscape, as soils, topography, and water availability change. Further, site index values can be changed for a specific site; they are not static. Since site index is simply a reflection of how tall the dominant and codominant trees will be (or were) at the base age, and since management activities can influence growth rates, the site index can be modified within a single rotation (such as with the use of early fertilization), or modified from one rotation to the next (when using a different site preparation method or when using genetically improved trees).

Q. Stocking and Density

The number of trees per unit area and the basal area of a stand are basic measures of stocking and density. Additional measures of stocking that combine the number of trees per unit area and tree size have been developed, and can be used to estimate other characteristics of a stand of trees. Stocking and density are two concepts of the condition of forests that are interrelated. Stand density is a quantitative measurement of stand conditions that describes the number of stems on a per unit area basis in either absolute or relative terms (Avery and Burkhart, 1994).

Density measures can be used as inputs for predicting growth and yield as well as guides for conducting silvicultural activities or evaluating nontimber values such as wildlife habitat. For instance, Smith and Long (1987) used a modified lodgepole pine (*Pinus contorta*) density management diagram as a tool to determine the amount of cover garnered by silvicultural activities and to evaluate the status of cover for elk (*Cervus elaphus nelsonii*) and mule deer (*Odocoileus heminonus hemionus*) in the Rocky Mountains.

Stocking is a relative concept that relates the stand density conditions of a site to an ideal condition that may not be readily achievable or identifiable. The use of stocking in a forestry context is associated with the concept of a normal forest (Avery and Burkhart, 1994), which is described in Chapter 10, Models of Desired Forest Structure. A normal forest suggests that on every unit of area, the optimum tree volume is being produced. In other words, all growing space above and below ground is being utilized to maximize timber production. However, identifying and achieving these conditions is very difficult, if not impossible. Historically, forest managers have evaluated stand conditions by judging whether the ratio of the number of trees per unit area is in line with their expectation of the ideal number of trees per unit area. In addition, this concept could be applied to nontimber related outputs. For instance, we could measure the density of deer over a squared unit area and evaluate the population stocking based on our expectation of the ideal stocking level. We could say that the measured deer population is understocked, fully stocked, or overstocked based on our ideal population density. The stocking concept can be useful for developing rules of thumb in implementing silvicultural operations, but there are many potential disadvantages of this approach, most of which are related to the need to identify the ideal stocking level. Graphical stocking guides have been developed for hardwoods in the northeastern United States (Roach, 1977). These types of charts allow us to understand both qualitative (overstocked, understocked) and quantitative (percent) levels of stocking. Given two of three measures of the structural characteristics of a stand (trees per unit area, basal area, average diameter), the third measure can be estimated, in addition to the relative stocking level of the stand.

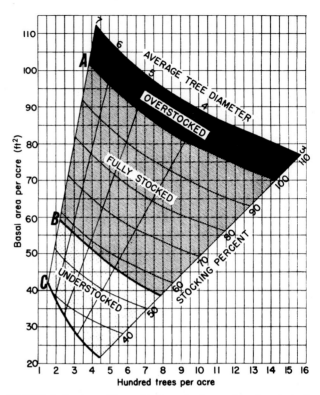

FIGURE 2.12   A stocking guide for upland central hardwoods in the United States. *From Roach, B.A., Gingrich, S.F., 1968. Even-aged silviculture for upland central hardwoods. US Department of Agriculture, Forest Service, Washington, DC. Agricultural Handbook 355. 39 p.*

represents the basal area with a vertical line representing the trees per acre, we end up in the "fully stocked" region of the chart. If we then project a line from the intersection point, diagonally and parallel to the average tree diameter lines (southwest to northeast in orientation within the chart), we find that the average diameter is about 5.7 or 5.8 inches. If we then project a curved line from the intersection point toward the southeast region of the chart, parallel to the curved "stocking percent" lines, we find that the quantitative estimate of stocking is a little less than 75%.

III. ECONOMIC EVALUATION OF NATURAL RESOURCES

We continue this chapter on valuing and characterizing forest conditions with an examination of several of the common methods for assessing forest conditions from an economic perspective. The allocation of scarce resources is a central concept of economics and inherent in forest planning and management (Pirard and Irland, 2007). As a result, typically the strongest arguments in the development of policies and plans of action involve economic analyses (Laarman, 2007). No matter what our interest

---

Example

Using the stocking guide for upland central hardwoods (Fig. 2.12), if the estimated basal area of a stand was 70 ft² per acre, and the estimated trees per acre were 400, what would be your estimate of the average stand diameter, the quantitative measure of stocking, and the qualitative measure of stocking? Intersecting a horizontal line that

in natural resource management, we should be able to understand the concepts behind economic analyses.

As an example of the need to perform an economic analysis associated with potential forest management activities, consider the following discussion of invasive species control options. Since colonial times, various persons have introduced a variety of plants and animals from around the world into the United States for many purposes. In addition, invasive species have arrived in the United States by accident. One important invasive species, cogongrass (*Imperata cylindrica*), first arrived in Alabama in the early 1900s as packing material for crates imported from Asia (Tabor, 1952; Prevost, 2007; Grebner et al., 2010). The crates and packing material were thrown out, but the cogongrass packing material and seeds were spread by wind and the displaced rhizomes were picked up and moved by mechanical equipment (Tabor, 1952; Prevost, 2007). These actions allowed it to become established in the southern United States. Later, landowners thought it could be used as animal forage, and spread it across the landscape before discovering that it caused mouth sores in animals. The plant also naturally spreads rapidly, and out-competes native vegetation. Another invasive species that has been common in the southern United States for many decades is kudzu (*Pueraria lobata*). Kudzu was first introduced into the United States in 1876 as an ornamental vine (Shurtleff and Aoyagi, 1977; Everest et al., 1991; Mitich, 2000). Later, farmers were interested in using kudzu as a potential source of forage, and various state and federal agencies promoted its use as a method to prevent soil erosion (Everest et al., 1991). The aggressive nature of its spread led to its reputation as being an invasive species.

In forest management situations, invasive species can have a negative impact on a forest manager's goals because they may remove lands from active timber production, reduce biodiversity, and degrade wildlife habitat. Ezell and Nelson (2006) developed silvicultural treatment procedures for controlling kudzu that utilize herbicides. However, forest managers are interested not only in vegetative or biological control of invasive species, but also in whether it is financially feasible to employ these treatments. Grebner et al. (2011) used this data to estimate the after-tax land expectation values (LEVs) of each silvicultural treatment for comparative purposes. In this case, they compared seven herbicide treatments that involved using Escort XP, Transline, Escort XP & Telar, Telar & Escort XP, Krenite S, and Tordon K. Their results indicated that using Escort XP had the highest after-tax LEV (an economic analysis presented in Section III, Part F).

When conducting an analysis of this type it is important to collect all the relevant cost, revenue, growth, and yield information. However, forest managers need to have information on the effectiveness of these treatments to accurately account for their expected financial outcomes. In the Grebner et al. (2011) paper, the age of kudzu patches was an important factor in deciding whether herbicide applications (and hence costs) were necessary in more than one time period. For instance, controlling older kudzu patches requires at least one follow-up treatment a year after the initial treatment. Young kudzu patches can be effectively controlled if the appropriate type and amount of herbicide is applied in the first year. In general, if forest managers do not have good information regarding a treatment's effectiveness, then future expectations of costs and revenues can be greatly distorted and lead to poor management decisions.

A. Basic Concepts: Present and Future Values

We use the term *present value* to describe the value, in today's dollars (or Euros, yen, yuan, etc.), for a future stream of benefits (revenues) and costs. The revenues and costs that are incurred in the future are discounted to today to reflect the time value of money and other risks. Two concepts underlie the concept of present value—the *time value of money* and *compound interest*. The time value of money is important in business as well as your personal life. At its basic root, some sum of money today is worth more than the same amount of money in the future. Why? Because you could take that sum of money and invest it in a savings account and earn interest. This compounding of the investment in the savings account is your payment from a bank for their use of your money. Obviously, you do not have to place your money into investments that return interest. However, you would expect that investments in activities that matter to you professionally or personally (e.g., a retirement account) would grow in value over time. Compound interest is the addition of periodic interest payments to the principal (initial investment). All subsequent interest payments are based on the initial investment, plus the interest earned up to that point in time. The frequency with which interest payments are earned should be explicitly stated. Throughout this book we use a frequency of 1 year, yet in real life these could be monthly (as in the case of basic savings accounts), quarterly, or otherwise.

A basic present value equation is:

$$\text{Present value} = \left[\frac{CF_t}{(1+i)^t} \right]$$

where:

CF_t = net cash flow (positive or negative) at time t
t = time period
i = interest rate, or discount rate

We use the term *future value* to describe the value, at some point in time in the future, of a stream of benefits (revenues) and costs. A basic future value equation for an investment made today is:

$$\text{Future value} = C_0(1+i)^t$$

where:

C_0 = an initial investment

If the interest rate were 0%, then the present value of some future flow of money would be the same as the future flow of money, since we would simply be dividing by 1 in the present value equation. However, since an organization usually uses a *hurdle rate* (cost of capital) that is something greater than 0%, the present value of some future income or future cost is generally lower than the future income or cost. Imagine, for example, that you could invest a sum of money in a certificate of deposit that earned 5% annually. If you expected to have $1,000 after 1 year of investing this money (i.e., the future value of the money), then you would need to place less than $1,000 today (about $953 in fact) into the certificate of deposit to have $1,000 a year from now. Therefore the amount you would need today (the present value) could vary considerably, depending on the interest rate assumed (Table 2.3). For example, if you expected $1,000 from an investment in a savings account or certificate of deposit 10 years from now, the amount that you would need to invest today could range anywhere from about $600 to $900, depending on the interest rate.

TABLE 2.3 Value of an Investment in a Savings Account That Yielded 1–5% Interest Over a 10-year Time Period (Year 0 Value Is the Present Value)

Year	Interest Rate				
	1%	2%	3%	4%	5%
0	905.29	820.35	744.09	675.56	613.91
1	914.34	836.76	766.42	702.59	644.61
2	923.48	853.49	789.41	730.69	676.84
3	932.72	870.56	813.09	759.92	710.68
4	942.05	887.97	837.48	790.31	746.22
5	951.47	905.73	862.61	821.93	783.53
6	960.98	923.85	888.49	854.80	822.70
7	970.59	942.32	915.14	889.00	863.84
8	980.30	961.17	942.60	924.56	907.03
9	990.10	980.39	970.87	961.54	952.38
10	1,000.00	1,000.00	1,000.00	1,000.00	1,000.00

1. Present Value of a Single Revenue or Cost

To calculate the present value of a single revenue or cost, we would use the same formula presented earlier for determining the present value. Here we are suggesting that we simply discount the value (positive or negative) from some future point in time to the present using the discount rate that is assumed.

Example

Assume that the discount rate used by a landowner in Michigan is 6%, and 7 years in the future a timber sale may yield $200,000. What is the present value of the timber sale revenue?

$$\text{Present value} = \left[\frac{\$200,000}{(1.06)^7} \right] = \$133,011.40$$

Given the time value of money, and the fact that the landowner's alternative investments might yield 6%, the value of the timber sale revenue in today's dollars is only about two-thirds of what it might be worth in 7 years. Stated another way, if the landowner placed about $133,000 today in an investment that would yield a 6% return annually, the investment would be worth about $200,000 in 7 years.

Example

Assume that 2 years from now a trail will be built in the Tillamook State Forest in Oregon, from an existing parking lot to an overlook. The cost per mile is $50,000 to build the 3-mile long trail. What is the present value of the cost of the trail if the state assumed a 5% discount rate for investments?

$$\text{Present value} = \left[\frac{\$150,000}{(1.05)^2} \right] = \$136,054.40$$

If the state could place $136,054 in an investment that would yield 5% per year, it would have about $150,000 available in 2 years to build the 3-mile long trail.

2. Present Value of a Non-terminating Annual Revenue or Cost

For analyzing the present value of annual revenues or costs, you, as an analyst, need to determine whether the costs will continue in perpetuity, or stop after some period of time. The appropriate method for calculating the present value of a non-terminating annual revenue or cost is to assume that the first payment or cost is due at the end of the first year. To calculate the present value, you simply divide the revenue or cost by the interest rate (in decimal percent terms).

$$\text{Present value} = \left[\frac{\text{Annual } CF}{i} \right]$$

TABLE 2.4 Present Value of a $4 Annual Management Cost Over a 100-year Time Frame

Year	Present Value of the Annual Cost ($)
1	3.74
2	3.49
3	3.27
4	3.05
5	2.85
....	
98	0.0053
99	0.0049
100	0.0046
Total	**57.08**

Example

Assume that the annual management costs for a forestry organization are $4 per acre per year, and that you expect that the organization will not only continue to own and manage the land in perpetuity, but also manage the land with the same amount of intensity as they do today. If their discount rate is assumed to be 7%, then what is the present value of the annual management costs?

$$\text{Present value} = \left[\frac{\$4 \text{ per acre}}{0.07} \right] = \$57.14 \text{ per acre}$$

If left unconvinced that this is the present value of all future management costs, then you can devise a system, such as that described in Table 2.4, that discounts the annual management costs incurred each year by the appropriate amount, then sum these values for a sufficiently long period of time. In the case of Table 2.4, the costs were assessed over a 100-year time frame, and the sum of the discounted costs ($57.08) is very close to our estimate using an infinite time horizon. You can also see that a $4 cost 100 years from now is worth very little today (approximately $0.005).

Example

Assume that a hunting lease in central Alabama brings in approximately $12 per acre per year for a private landowner. Assume also that the private landowner uses a discount rate of 5% for their investment analyses. If you assume that the hunting lease will not appreciate in value, what is the present value of the lease to the landowner?

$$\text{Present value} = \left[\frac{\$12 \text{ per acre}}{0.05} \right] = \$240.00 \text{ per acre}$$

In this example, as with the following examples, if you were to assume that the first revenue or cost were realized right away, rather than at the end of the first year, then the payment or cost would need to be incorporated into the calculation *without being discounted*, since time is not a factor with the first payment.

$$\text{Present value} = \text{Annual } CF + \left[\frac{\text{Annual } CF}{i} \right]$$

3. Present Value of a Terminating Annual Revenue or Cost

Some of the assumptions behind the use of a non-terminating annual revenue or cost may be unappealing to you as an analyst. For example, could you realistically assume that management costs will remain the same over a long period of time? Could you assume that hunting leases will not continue to appreciate in value as the human population increases, and people move further out into rural areas, resulting in hunting areas becoming more scarce? As a result, analyzing present values that involve annual revenues or costs that will terminate at some point may seem more reasonable. Assuming that the first payment occurs at the end of the first year, the method by which you would calculate the present value of a terminating annual revenue or cost is:

$$\text{Present value} = \left[\text{Annual } CF \left(\frac{(1+i)^t - 1}{i(1+i)^t} \right) \right]$$

Example

Analyze a modified version of the example presented earlier: Assume that the annual management costs for a forestry organization are $4 per acre per year, and you expect that the organization will not only continue to own and manage the land for 100 years, but also manage the land with the same amount of intensity as they do today. If their discount rate is assumed to be 7%, what is the present value of the terminating annual management costs?

$$\text{Present value} = \left[\$4 \text{ per acre} \left(\frac{(1.07)^{100} - 1}{0.07(1.07)^{100}} \right) \right]$$

$$= \$57.08 \text{ per acre}$$

This example results in the value we found earlier, where the sum of the individual present values of the terminating annual costs of $4 per year is $57.08 per acre.

Example

Assume that a private landowner in south Georgia leases out their property to a hunt club for $12.50 per acre per year. The length of the lease is 5 years, and the landowner uses a discount rate of 6% for all their investments. What is the present value of the hunting lease to the landowner?

$$\text{Present value} = \left[\$12.50 \text{ per acre} \left(\frac{(1.06)^5 - 1}{0.06(1.06)^5} \right) \right]$$

$$= \$52.65 \text{ per acre}$$

As with the previous example of a non-terminating annual series of revenues or costs, we have assumed that they are realized at the end of each period. However, if you were to assume that the revenues or costs are realized at the beginning of the period in the terminating series, and the first revenue or cost was made right away, at the beginning of the first year, the formula for the present value would be modified to:

$$\text{Present value} = \left[\text{Annual } CF \left(\frac{(1+i)^t - 1}{i(1+i)^t} \right) \right] (1+i)$$

Alternatively, you could assume that the first payment was made now, and did not need discounting, and that the length of the time horizon was $t-1$, resulting in the following equation:

$$\text{Present value} = \text{Annual } CF + \left[\text{Annual } CF \left(\frac{(1+i)^{t-1} - 1}{i(1+i)^{t-1}} \right) \right]$$

Example

Assume that a private landowner in south Georgia leases out his property to a hunt club for $12.50 per acre per year, yet the first payment is made right away. The length of the lease is 5 years, and the landowner uses a discount rate of 6% for all his investments. What is the present value of the hunting lease to the landowner?

$$\text{Present value} = \left[\$12.50 \text{ per acre} \left(\frac{(1.06)^5 - 1}{0.06(1.06)^5} \right) \right] (1.06)$$

$$= \$55.81 \text{ per acre}$$

or alternatively,

$$\text{Present value} = \$12.50 \text{ per acre}$$

$$+ \left[\$12.50 \text{ per acre} \left(\frac{(1.06)^4 - 1}{0.06(1.06)^4} \right) \right]$$

$$= \$55.81 \text{ per acre}$$

4. Present Value of a Non-terminating Periodic Revenue or Cost

In contrast to annual revenues or costs, some financial events reoccur over periods longer than 1 year. For example,

maintenance on a trail may occur every other year (or every third year), rather than every year. Alternatively, a landowner may devise a plan whereby they harvest timber every 5 years, and expect a given amount of revenue with each entry. These types of revenues or costs can be assumed to be non-terminating and periodic in nature. As a result, the equation that would be used to analyze the present value of these events would be:

$$\text{Present value} = \left[\frac{CF_{pl}}{(1+i)^{pl} - 1} \right]$$

Here, pl represents the period length, or the interval at which the revenues of costs reoccur, rather than the entire length of the time horizon. If the period were 1 year, representing a non-terminating annual revenue or cost, the equation becomes:

$$\text{Present value} = \left[\frac{CF_1}{(1+i)^1 - 1} \right],$$

or:

$$\text{Present value} = \left[\frac{CF_1}{(1 + i - 1)} \right]$$

$$\text{Present value} = \left[\frac{CF_1}{i} \right]$$

and reduces to the equation we presented earlier for the present value of a non-terminating annual revenue or cost.

Example

Assume that a prescribed fire regime for Douglas-fir stands in Colorado, at about 4,000–6,000 foot elevations, on moderate slopes, calls for a 10-year reentry and each entry costs $15 per acre. Assume that the prescribed fire program is on public land, where the discount rate used for investment analysis is 4%. If the first prescribed fire occurred at the end of the first decade, and was repeated every 10 years thereafter, what is the present value of the non-terminating periodic series of costs?

$$\text{Present value} = \left[\frac{\$15 \text{ per acre}}{(1.04)^{10} - 1} \right] = \$31.23 \text{ per acre}$$

As a cross-check on these results, we can construct a table whereby we indicate the cost and the timing of the costs to the fire program (Table 2.5). Simply assessing the first 100 years shows that the present value of the program ($30.61) approximates our non-terminating present value.

If you were to assume that the first periodic revenue or cost would occur today, the equation for a non-terminating periodic revenue or cost should be modified to account for the fact that the first revenue or cost need not be discounted:

$$\text{Present value} = CF_{pl} + \left[\frac{CF_{pl}}{(1+i)^{pl} - 1} \right]$$

TABLE 2.5 Present Value of a $15 per Acre per Decade Cost for Prescribed Fire in Douglas-Fir Stands in Colorado

Year	Prescribed Fire Cost ($/ac)	Present Value ($/ac)
10	15.00	10.13
20	15.00	6.85
30	15.00	4.62
40	15.00	3.12
50	15.00	2.11
60	15.00	1.43
70	15.00	0.96
80	15.00	0.65
90	15.00	0.44
100	15.00	0.30
Total		30.61

Example

Using our prescribed fire example, assume that the first treatment occurs today, and every 10 years thereafter land managers will apply a similar prescribed fire treatment. The present value of this proposed stream of activities is:

$$\text{Present value} = \$15 \text{ per acre} + \left[\frac{\$15 \text{ per acre}}{(1.04)^{10} - 1}\right] = \$46.23 \text{ per acre}$$

Alternatively, you could calculate the present value of the non-terminating periodic revenues or costs, assuming that the first revenue or cost occurs now, by using the following equation:

$$\text{Present value} = \left[\frac{CF_{pl}}{(1+i)^{pl} - 1}\right](1+i)^{pl}$$

Thus given our earlier example, assuming that the first treatment occurs now, and subsequent treatments occur every 10 years thereafter, the present value of this proposed stream of activities is:

$$\text{Present value} = \left[\frac{\$15 \text{ per acre}}{(1.04)^{10} - 1}\right](1.04)^{10} = \$46.23 \text{ per acre}$$

5. Present Value of a Terminating Periodic Revenue or Cost

The preceding discussion begs the question, how would we compute exactly the value of a terminating periodic series of revenues or costs? Given the last example, we have something with which to compare, since we solved a problem involving 10-year costs over a 100-year time frame exactly by computing the present value of each

decadal cost occurrence. Here we need to incorporate both the length of the period and the length of the time frame into the analysis.

$$\text{Present value} = \left[CF_{pl}\left(\frac{(1+i)^t - 1}{((1+i)^{pl} - 1)(1+i)^t}\right)\right]$$

Example

Use the previous example of a prescribed fire regime for Douglas-fir stands in Colorado, and assume costs of $15 per acre every decade, and a 4% discount rate. If we were to assume that the program ends after 100 years, what is the present value of the terminating, periodic series of costs?

$$\text{Present value} = \left[\$15 \text{ per acre}\left(\frac{(1.04)^{100} - 1}{(1.04^{10} - 1)(1.04)^{100}}\right)\right]$$
$$= \$30.62 \text{ per acre}$$

If you compare this to the case where the prescribed fire program was assumed to continue indefinitely, the per acre cost of the program after year 100 is only $0.61 per acre, further illustrating the need to take into account the time value of money. Keeping in mind some rounding of the values presented in the previous examples, if you were to invest $0.61 in an investment that yielded 4% interest annually, the resulting balance after 100 years would be about the same as the present value of the non-terminating periodic series of costs.

As with the other examples in this chapter, if you were to assume that the first revenue or cost occurred at the beginning of the first time period, the equation would have to be modified to reflect the fact that the first revenue or cost need not be discounted.

$$\text{Present value} = \left[CF_{pl}\left(\frac{(1+i)^t - 1}{((1+i)^{pl} - 1)(1+i)^t}\right)\right](1+i)^{pl}$$

This equation is similar to the one provided earlier for the terminating annual revenues or costs, in that we multiplied the basic equation by $(1+i)^{pl}$, and since earlier we were discussing annual revenues or costs, it reduced to $(1+i)$. Using our prescribed fire example, the present value of the terminating periodic costs is:

$$\text{Present value} = \left[\$15 \text{ per acre}\left(\frac{(1.04)^{100} - 1}{((1.04)^{10} - 1)(1.04)^{100}}\right)\right](1.04)^{10}$$
$$= \$45.32 \text{ per acre}$$

An alternative formulation is to explicitly assume that the first revenue or cost need not be discounted, and that the length of the time horizon is then $(t-pl)$ years, since the last payment will occur at the beginning of the last decade:

$$\text{Present value} = CF_{pl} + \left[CF_{pl}\left(\frac{(1+i)^{t-pl} - 1}{((1+i)^{pl} - 1)(1+i)^{t-pl}}\right)\right]$$

Carrying out the calculations for the prescribed fire example, we arrive at the same per acre present value.

Present value = $15 per acre

$$+ \left[\$15 \text{ per acre} \left(\frac{(1.04)^{90} - 1}{((1.04)^{10} - 1)(1.04)^{90}} \right) \right]$$

$$= \$45.32 \text{ per acre}$$

To assist in directly comparing the methods for calculating present values of various investments, a summary of the present value equations and forthcoming future value equations can be found in Table 2.6.

6. Future Value, One Revenue or Cost

Projecting the potential revenue or cost of an investment made today to a future monetary value is important for many professional and personal reasons. Professionally, you would be estimating the potential impact of an activity on the economic success (or failure) of an organization. Personally, you might want to understand how an investment into a savings or retirement program will be valued at various points in your future. The method by which you would calculate the future value of a one-time revenue or cost incurred today is:

$$\text{Future value} = CF_{yr}(1+i)^{t-yr}$$

where CF_{yr} is the cash flow (positive or negative) incurred during some year of the time horizon (t).

Example

Assume that the rate obtained by a long-term certificate of deposit is 4.5%, and that you invest $1,000 today and plan to keep it in the investment for 5 years. How much will

TABLE 2.6 Present Value Equations for One-Time, Annual, and Periodic Cash Flow Events

Timing of Revenue or Cost	Time Horizon	Initial Revenue or Cost Occurs...	Equation to Use
Once	–	–	$\text{Present value} = \left[\dfrac{CF_t}{(1+i)^t} \right]$
Annual	Non-terminating (infinite time horizon)	At the end of the first time period	$\text{Present value} = \left[\dfrac{\text{Annual } CF}{i} \right]$
		At the beginning of the first time period	$\text{Present value} = \text{Annual } CF + \left[\dfrac{\text{Annual } CF}{i} \right]$
	Terminating (finite time horizon)	At the end of the first time period	$\text{Present value} = \left[\text{Annual } CF \left(\dfrac{(1+i)^t - 1}{i(1+i)^t} \right) \right]$
		At the beginning of the first time period	$\text{Present value} = \left[\text{Annual } CF \left(\dfrac{(1+i)^t - 1}{i(1+i)^t} \right) \right] (1+i)$
			$\text{Present value} = \text{Annual } CF + \left[\text{Annual } CF \left(\dfrac{(1+i)^{t-1} - 1}{i(1+i)^{t-1}} \right) \right]$
Periodic	Non-terminating (infinite time horizon)	At the end of the first time period	$\text{Present value} = \left[\dfrac{CF_{pl}}{(1+i)^{pl} - 1} \right]$
		At the beginning of the first time period	$\text{Present value} = \left[\dfrac{CF_{pl}}{(1+i)^{pl} - 1} \right] (1+i)^{pl}$
			$\text{Present value} = CF_{pl} + \left[\dfrac{CF_{pl}}{(1+i)^{pl} - 1} \right]$
	Terminating (finite time horizon)	At the end of the first time period	$\text{Present value} = \left[CF_{pl} \left(\dfrac{(1+i)^t - 1}{((1+i)^{pl} - 1)(1+i)^t} \right) \right]$
		At the beginning of the first time period	$\text{Present value} = \left[CF_{pl} \left(\dfrac{(1+i)^t - 1}{((1+i)^{pl} - 1)(1+i)^t} \right) \right] (1+i)^{pl}$
			$\text{Present value} = CF_{pl} + \left[CF_{pl} \left(\dfrac{(1+i)^{t-pl} - 1}{((1+i)^{pl} - 1)(1+i)^{t-pl}} \right) \right]$

this investment be worth at the end of 5 years, including the initial amount invested and the interest that it earns?

$$\text{Future value} = \$1,000 \ (1.045)^{5-0} = \$1,246.18$$

Since the length of time (t) that the investment is made is 5 years, and since the investment is made now ($yr = 0$), compounded interest is added to the initial $1,000 at the end of each of 5 years.

Example

Assume that a stand of Douglas-fir and western hemlock (*Tsuga heterophylla*) trees was just planted, and that 5 years from now a cost will be required to perform precommercial thinning of the stand. Assume also that the rotation length of the even-aged stand of trees will be 40 years. If the cost of precommercial thinning is $125 per acre, and the interest rate that the landowner uses is 6%, what is the future value of the precommercial thinning at the end of the rotation?

$$\text{Future value} = \$125 \text{ per acre } (1.06)^{40-5} = \$960.76 \text{ per acre}$$

Another way to look at this is that the added value of the precommercial thinning investment must be at least $960.76 per acre at the time of final harvest for the landowner to make a 6% return on the precommercial thinning investment. If the landowner had invested $125 at 6% interest in some other project, they would have earned $960.76 at the end of 35 years.

7. Future Value of a Non-terminating Annual Cost or Revenue

Since the term *future value* implies that we stop at some point in the future and assess the economic success (or failure) of a project or investment, determining the future value of a non-terminating investment is not of value here.

8. Future Value of a Terminating Annual Cost or Revenue

In analyzing the future value of annual revenues or costs, we would need to use an equation that is somewhat similar to the present value of non-terminating annual revenues or costs. This equation is similar in that the denominator is the interest rate assumed. The difference is that the annual revenue or cost is compounded by $(1 + i)^t - 1$ prior to dividing through by the interest rate.

$$\text{Future value} = \left(\frac{(\text{Annual revenue or cost})((1+i)^t - 1)}{i} \right)$$

Example

Assume that the annual management costs for a forestry organization are $5 per acre per year, and you expect that the organization will not only continue to own and manage the land for the next 10 years, but also at the same intensity as they do today. If their discount rate is assumed to be 7%, what is the future value of the terminating annual management costs?

$$\text{Future value} = \left(\frac{(\$5 \text{ per acre})((1.07)^{10} - 1)}{0.07} \right)$$

$$= \$69.08 \text{ per acre}$$

To further illustrate how we arrived at this value, you can see in Table 2.7 that we invested $5 every year for 10 years, and that the interest on each $5 investment was compounded forward to the end of the time horizon. It is as if you invested $5 each year in a bank account that yielded 7% interest—how much would you have at the end of the decade? This example assumes that the first investment is made at the end of the first year, therefore the initial investment is compounded only 9 years into the future. Similarly the last investment is made at the end of the 10th year, and is not compounded. How would the method for calculating the future value of terminating annual costs change if the initial revenue or cost were assumed to be made right away?

$$\text{Future value} = \left(\frac{(\text{Annual revenue or cost})((1+i)^t - 1)}{i} \right)(1+i)^{pl}$$

TABLE 2.7 Future Value of a $5 Annual Management Cost Over a 10-Year Time Frame, Using a 7% Alternative Rate of Return

Year	Investment ($)	Years Compounded	Future Value of the Annual Cost ($)
1	5.00	9	9.19
2	5.00	8	8.59
3	5.00	7	8.03
4	5.00	6	7.50
5	5.00	5	7.01
6	5.00	4	6.55
7	5.00	3	6.13
8	5.00	2	5.72
9	5.00	1	5.35
10	5.00	0	5.00
Total			**69.08**

Since the period length (*pl*) is 1 year, the equation reduces to:

$$\text{Future value} = \left(\frac{(\text{Annual revenue or cost})((1+i)^t - 1)}{i} \right)(1+i)$$

Using the previous example, the future value becomes:

$$\text{Future value} = \left(\frac{(\$5 \text{ per acre})((1.07)^{10} - 1)}{0.07} \right)(1.07)$$

$$= \$73.92 \text{ per acre}$$

In this case, the future value is greater than the earlier example because each of the investments were compounded an extra year—the initial investment was compounded 10 years, while the final investment was compounded 1 year (i.e., it was made at the beginning of the 10th year).

Example

Assume that a private landowner in south Georgia leases her property to a hunt club for $12.50 per acre per year. The length of the lease is 5 years, and the landowner uses a discount rate of 6% for all her investments. If the hunt club pays the landowner at the beginning of each year, what is the future value of the hunting lease to the landowner?

$$\text{Future value} = \left(\frac{(\$12.50 \text{ per acre})((1.06)^5 - 1)}{0.06} \right)(1.06)$$

$$= \$74.69 \text{ per acre}$$

---

9. Future Value of a Non-terminating Periodic Cost or Revenue

As we mentioned earlier, since the term *future value* implies that we stop at some point in the future and assess the economic success (or failure) of a project or investment, determining the future value of a non-terminating investment is not relevant to this discussion.

10. Future Value of a Terminating Periodic Cost or Revenue

The future value of a project or investment with a revenue or cost stream that occurs every few years (not annually, but repeatedly) and terminates at some point in time can be calculated using the following equation:

$$\text{Future value} = \left(\frac{(CF_{pl})((1+i)^t - 1)}{((1+i)^{pl} - 1)} \right)$$

As we noted earlier, *pl* represents the period length, or interval at which the revenues of costs reoccur, rather than the entire length of the time horizon (*t*). If the period

length were 1 year, representing a terminating annual revenue or cost, the equation becomes:

$$\text{Future value} = \left(\frac{(CF_1)((1+i)^t - 1)}{((1+i)^1 - 1)} \right),$$

or:

$$\text{Future value} = \left(\frac{(CF_1)((1+i)^t - 1)}{(1+i-1)} \right),$$

or:

$$\text{Future value} = \left(\frac{(CF_1)((1+i)^t - 1)}{i} \right)$$

which reduces to the equation we used to describe the future value of a terminating annual revenue or cost.

Example

Assume that a prescribed fire regime for mixed conifer stands in California calls for a 10-year burning cycle, and each time a prescribed burn is conducted, it costs $20 per acre. Assume that the prescribed fire program is on public land, and the discount rate used for investment analysis is 4%. If the first prescribed fire occurred at the end of the first decade, then was repeated every 10 years thereafter, for 50 years, what is the future value of the terminating, periodic series of costs?

$$\text{Future value} = \left(\frac{(\$20 \text{ per acre})((1.04)^{50} - 1)}{((1.04)^{10} - 1)} \right)$$

$$= \$254.32 \text{ per acre}$$

To check these results, we can construct a table whereby we indicate the cost and the timing of the costs to the fire program (Table 2.8). Assessing the future value of each of the costs incurred each decade indicates that the sum of the individual future values equals the future value just computed.

TABLE 2.8 Future Value of a $20 per Acre per Decade Cost for Prescribed Fire in Mixed Conifer Stands in California, Assuming a 4% Alternative Rate of Return

Year	Prescribed Fire Cost ($/ac)	Future Value ($/ac)
10	20.00	96.02
20	20.00	64.87
30	20.00	43.82
40	20.00	29.60
50	20.00	20.00
Total		254.32

If we were to assume that the terminating periodic series of revenues or costs included incurring the first revenue or cost right away (at time 0), rather than at the end of the first period, then the future value equation for a terminating periodic cost or revenue becomes:

$$\text{Future value} = \left(\frac{(CF_{pl})((1+i)^t - 1)}{((1+i)^{pl} - 1)} \right)(1+i)^{pl}$$

In utilizing this adjusted formula for the future value of terminating periodic revenues or costs, the future value of the prescribed fire plan that was recently discussed becomes:

$$\text{Future value} = \left(\frac{(\$20 \text{ per acre})((1.04)^{50} - 1)}{((1.04)^{10} - 1)} \right)(1.04)^{10}$$

$$= \$376.45 \text{ per acre}$$

B. Prices and Costs

Obtaining quality prices for timber-based products as well as costs of implementing various silvicultural activities is a critical part of any financial analysis. Prices for wood-based products can be obtained from local mills. These prices are typically given after the costs of manufacture, transportation, and harvesting of the original sawtimber trees are incorporated. For example, to derive a sawtimber price

from standing stumpage, one would have to take the price of a conventional product (e.g., a $2'' \times 4''$ board in North America) and subtract out the manufacturing, transportation, and logging costs. A simpler approach would be use the delivered price of logs to a mill and subtract out the logging and transportation costs of getting that log to the mill to derive a residual price. This approach can be extremely valuable when looking a specific local market. Fortunately, in North America and elsewhere, if generalized price information is required, there are important regional and national timber price vendors that can provide this information, such as Timber Mart-South (2016), Timber Mart North (Prentice and Carlisle, 2016), Forest2Market (2016), Wood Resources International (2016), and RISI, Inc. (2016).

Consider, for example, Timber Mart-South, which has been reporting timber prices since 1976. This price reporting service provides quarterly stumpage (standing) and delivered timber prices for 11 southern states (Virginia, North Carolina, South Carolina, Georgia, Florida, Alabama, Mississippi, Louisiana, Texas, Arkansas, and Tennessee). Fig. 2.13 illustrates quarterly average stumpage prices of major timber products (pine sawtimber, pine chip-*n*-saw, pine pulpwood, hardwood sawtimber, and hardwood pulpwood) in the southern United States from the 4th quarter of 1976 to the 4th quarter of 2015. The examination of the price data indicates that stumpage prices can be quite volatile over time and that over many years even established price patterns can change. Pine

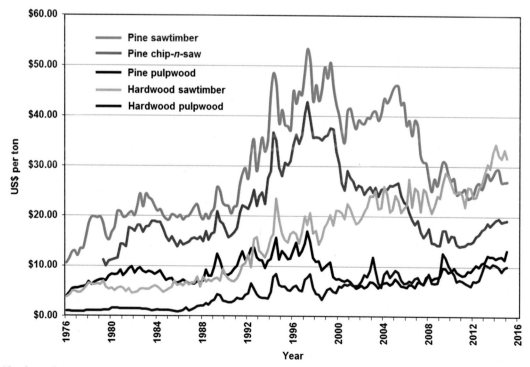

FIGURE 2.13 Quarterly average stumpage prices of major timber products in the southern United States, 1976–2015. *Courtesy of Timber Mart-South, Frank W. Norris Foundation, Athens, GA, US.*

sawtimber used to be the most valuable timber product (peaking in the 1st quarter of 1998 at $45.86 per green short ton ("per ton" from here forward)) until recently when, following multiple-year declines (to $25.60 per ton in the 4th quarter of 2015), it was overtaken by hardwood sawtimber ($30.99 per ton in the 4th quarter of 2015). Similarly, while hardwood pulpwood typically has been cheaper than pine pulpwood (approximately three times less expensive in the 4th quarter of 1976) recently this gap has not only narrowed but, at times, hardwood pulpwood prices exceeded pine pulpwood prices. In the 4th quarter of 2015 these two products traded at about or slightly above $10 per ton.

While obviously timber product prices vary over time they also vary across the southern United States as their levels depend on the strength of local timber markets. County-level, average pine sawtimber (Fig. 2.14) and pine pulpwood stumpage price (Fig. 2.15) maps also developed by Timber Mart-South describe market conditions across the southern United States in 2015. One can gather that the Coastal Plain region of the southeastern states, as well as Louisiana, Texas, and southern Mississippi were characterized by strong demand for pine sawtimber stumpage. In addition, pine pulpwood stumpage was in strong demand across the Coastal Plain region. Further, differences in stumpage prices by location could be quite striking. Depending on location, 2015 pine sawtimber prices ranged from less than $14 per ton to more than $34 per ton. Pulpwood prices ranged from less than $4 per ton to more than $22 per ton, nearly an eightfold difference. This points to the importance of understanding, selecting, and correctly using timber prices in conducting economic analyses of timber production.

Timber price reporting services are present in most countries and regions with significant forest products markets. In Finland, timber prices are reported by the Finnish Forest Industries Federation (2016). In 2015, pine sawtimber (termed as pine logs) traded on average €54.37 per m^3 (solid cubic meter (m^3) with bark), spruce at €54.22 per m^3, and birch at €41.79 per m^3. Pine pulpwood traded at €15.56 per m^3, spruce at €16.81 per m^3, and birch at €15.36 per m^3. These prices were received by forest owners while the forest company was responsible for harvesting and transportation. Further consideration of information presented in Figs. 2.16 and 2.17 illustrate that while timber prices have been recently relatively stable, they were more volatile during the period from 2010 to 2013. When considering and comparing timber prices between countries and regions it is important to recognize differences in product specifications, currencies, and units of measurement in order to draw valid analytical conclusions.

Typically, costs of silvicultural operations are commonly obtained from local contractors. In many cases, consulting foresters will have this information. An important issue

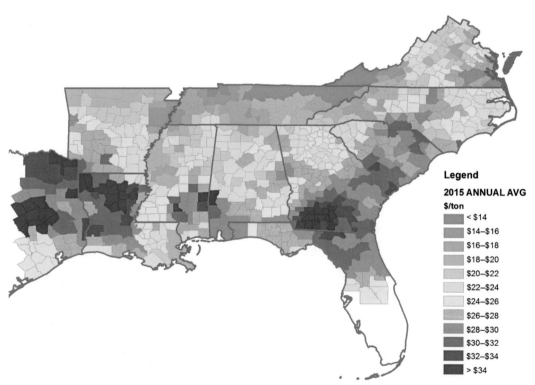

FIGURE 2.14 County-level, average pine sawtimber stumpage prices in the United States, 2015. *Courtesy of Timber Mart-South, Frank W. Norris Foundation, Athens, GA, US.*

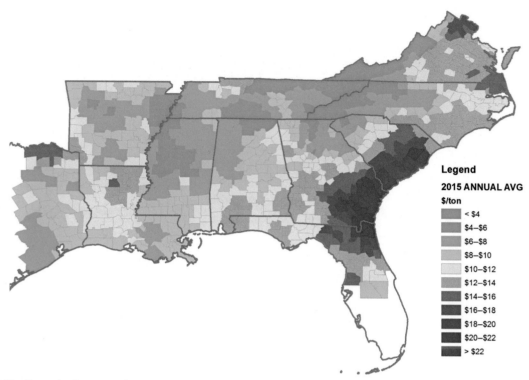

FIGURE 2.15 County-level, average pine pulpwood stumpage prices in the United States, 2015. *Courtesy of Timber Mart-South, Frank W. Norris Foundation, Athens, GA, US.*

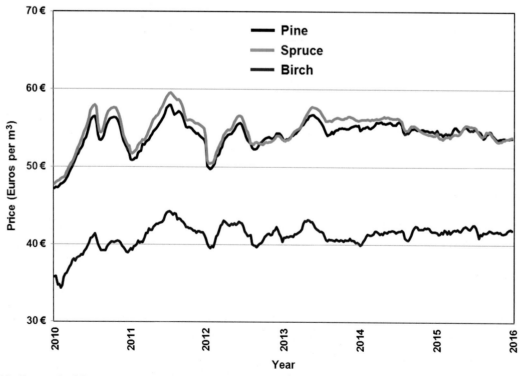

FIGURE 2.16 Four-week sliding average stumpage prices for logs in Finland, 2010–2016. *Courtesy of Finnish Forest Industries Federation, Helsinki, Finland.*

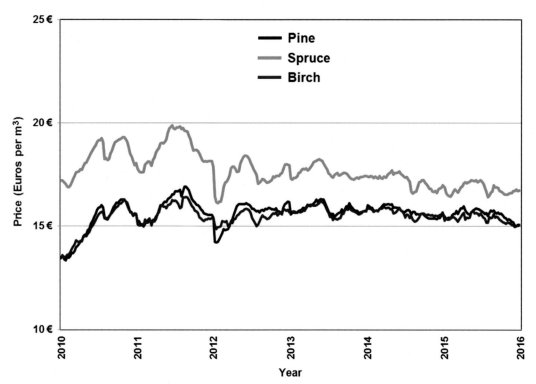

FIGURE 2.17 Four-week sliding average stumpage prices for pulpwood in Finland, 2010–2016. *Courtesy of Finnish Forest Industries Federation, Helsinki, Finland.*

when dealing with costs is to convert the available information into a per unit area basis. This may require some additional information on the type of terrain involved. One potential source of cost information for the southern United States may come from the Forest Landowners Association. They publish trends of costs every 2 years, in *Forest Landowner* magazine. Some recently reported costs include the following: mechanical site preparation, $95.78 per acre on average; hand planting of trees, $37.17 per acre on average, excluding seedlings; prescribed burning, $18.18 per acre on average; chemical applications, $29.89 per acre on average; fertilization applications, $79.49 per acre on average; point sampling for timber cruising, $4.14 per acre on average; timber marking, $29.64 per acre on average; and custodial management, $9.00 per acre on average (Barlow and Levendis, 2015). The report also indicates how these costs have changed over time, starting as far back as 1952. For example, between 2010 and 2014 the cost of fertilization increased by 26.6% while the cost of mechanical site preparation declined by 31.6%. This information can provide an insight into how these costs can very over time.

C. Net Present Value

The net present value (NPV) of an investment is the difference between the present value of revenues and the present value of costs over some period of time. NPV calculations frequently are used in budgeting analyses to sort and rank potential investments, and to determine which investments or projects should be funded. If the NPV of an investment happens to be positive, it might be considered an acceptable course of action, depending on the NPV of other alternative investments. However, if the NPV of an investment is negative, the potential investment should probably be disregarded because the future cash flows will not cover the periodic costs. Alternatively, if the NPV is equal to 0, then the investment may be viewed as neutral, since the discounted revenues will equal the discounted costs, and thus the investment earns a return equal to the assumed interest rate. One way to calculate NPV is:

$$NPV = \sum_{t=1}^{T} \left(\frac{CF_t}{(1+i)^t} \right) - C_0$$

where:

T = the time horizon, in the same units as t, the time periods

Example

Assume that we have 100 acres of bare ground in rural east central Mississippi and we are trying to decide which of three potential management options to select. First, we could prepare the site, plant 450 (10 foot × 10 foot spacing) loblolly pine seedlings per acre, and harvest the trees in 30 years. Second, we could prepare the site, plant 450 loblolly pine trees per acre, and harvest the trees in 20 years. Third, we could prepare the site, plant 450 loblolly pine trees per acre and harvest the trees in 60 years. Assume that preparing the site, purchasing the seedlings, and planting them will cost (initial investment) $200 per acre right away (in year zero). If we assume a SI_{25} is 62, and use current timber prices, we can estimate that option one will generate $4,255.65 per acre, option two will generate $1,489.73 per acre, and option three will generate $7,153.54 per acre. To calculate the NPV for each of these options we need to calculate the present value of each revenue stream using a real discount rate of 6%. For option one, our first step is to calculate the present value of $4,255.65 generated per acre in year 30.

$$\text{Present value} = \left[\frac{\$4,255.65}{(1.06)^{30}}\right] = \$741.40 \text{ per acre}$$

Our second step is to subtract from the discounted revenue of $741.40 per acre, the $200 cost per acre for preparing the site, purchasing the seedlings, and planting the trees. This subtraction yields us a NPV of $541.40 per acre. For option two, we calculate the present value of $1,489.73 per acre generated in year 20.

$$\text{Present value} = \left[\frac{\$1,489.73}{(1.06)^{20}}\right] = \$464.50 \text{ per acre}$$

Our second step is to subtract from the discounted revenue of $464.50 per acre the $200 per acre cost for preparing the site, purchasing the seedlings, and planting the trees. This yields a NPV of $264.50 per acre. For option three, we calculate the present value of $7,153.54 earned per acre in year 60.

$$\text{Present value} = \left[\frac{\$7,153.54}{(1.06)^{60}}\right] = \$216.85 \text{ per acre}$$

As in the other cases, our next step is to subtract from the discounted revenue of $216.85 per acre the $200 per acre cost of preparing the site, purchasing the seedlings, and planting the trees. This yields a NPV of $16.85 per acre. Now that we have calculated the NPV for each option we can compare them. In this example, the first option has the highest NPV compared to options two and three. Option three has a NPV relatively close to zero. Given these results, we would choose option one because it gives us the highest return on investment. However, the length of the investment periods were not consistent (20, 30, and 60 years), therefore some adjustment must be made to the analysis to recognize the different time horizons (see Section II, Part F).

D. Internal Rate of Return

The internal rate of return (IRR) is defined as the discount rate that is required to arrive at a NPV of zero. In general, the greater the IRR, the more attractive the investment, therefore IRR calculations for competing investments can allow you to rank them and develop priorities. Another way to view an IRR is to assume that it represents the interest expected from an investment. A forestry investment that results in an IRR of 10% is better than an investment in a certificate of deposit that will return around 5% (or less) over the long run. The IRR commonly is compared to an organization's cost of capital, or *hurdle rate*. For example, if an organization had a hurdle rate of 7% and a proposed investment suggested an IRR of 8%, the investment, at the very least, would be seen as favorable to the organization. Whether or not this investment would be pursued would depend on the other alternatives available to the organization for investing the same amount of money. Finally, we could view IRR as way to evaluate whether the investment will break even, given an organization's alternative investment rate or hurdle rate. For example, if an IRR of a proposed investment were 8%, and an organization's hurdle rate were 10%, then the proposed investment would not result in a positive NPV. If the hurdle rate (7%) was less than the IRR of the project (8%), the proposed investment would result in a positive NPV.

An IRR is a reflection of the efficiency of an investment, and since we can view IRR as a rate of growth of an investment, we can compare it against the rate of return for alternative investments, such as money markets, savings accounts, or certificates of deposits. However, many argue that an IRR should not be used to compare investments, but rather it should be used to decide whether an individual investment will produce the economic returns that are acceptable within an organization. One way to arrive at the IRR of an investment is to solve the following NPV equation:

$$NPV = 0 = \sum_{t=1}^{T} \left(\frac{CF_t}{(1+IRR)^t}\right) - C_0$$

Although some may prefer to evaluate decisions based on NPV, many private business decisions may be based on IRR, because it reflects the return on the investment rather than the magnitude of the value of the investment to an organization. One problem with an IRR calculation is that when cash flows fluctuate from positive to negative over time, multiple IRR values may result. For example, over a typical rotation of southern pines, we might incur a commercial thinning revenue, then a fertilization cost, then a final harvest revenue. This fluctuation from positive to negative, then back to positive, could result in multiple IRR values.

Another disadvantage of the approach is that intermediate positive revenues are not assumed to be reinvested at the IRR, therefore IRR is assumed only to be reflective of an investment's worth to an organization when there are no intermediate positive revenues. A further disadvantage of this approach, as with the NPV approach for valuing forest-related commodities, is that costs and revenues are monetized. Natural resource values such as animal unit months, stream sediment, aesthetics, and wildlife habitat quality would need to be placed in terms of dollars, euros, or other units to allow the analysis to recognize their importance in the decision-making process.

E. Benefit/Cost Ratio

The benefit/cost ratio (BCR) is the present value of the revenues associated with an investment divided by the present value of the costs. If the BCR of an investment equaled 1, then the NPV of the investment would be 0, and the IRR would equal the discount rate that was assumed. If the BCR were greater than 1, the NPV of the investment would be positive, and the IRR would be greater than the hurdle rate assumed by an organization (the discount rate assumed in the NPV calculation). Of course, if the BCR were less than 1, the present value of the benefits do not cover the present value of the costs, therefore the NPV of the investment would be negative, and the IRR would be less than the hurdle rate assumed. Obviously, when considering alternative investments, you would want to choose those with higher BCR values. The basic BCR equation is:

$$BCR = \left(\frac{\text{Present value of benefits}}{\text{Present value of costs}}\right)$$

In contrast to other economics analyses, one advantage of this method for evaluating investments is that the BCR can be applied to values that are not necessarily expressed monetarily. Therefore, a benefit/cost analysis may be applicable to natural resource management analyses involving noncommodity resources. The only assumption is that the values in both the numerator (benefits) and denominator (costs) have the same units, therefore the units will cancel each other out, and we would arrive at a unitless ratio.

F. Equal Annual Equivalent

As an alternative to making decisions based on a NPV analysis, the equal annual equivalent (EAE) can be developed for each potential investment. EAE is the net revenue (or cost) that you can obtain (or will incur) annually, over the life of an investment, given the discount rate that was applied. Conceptually, EAE is similar to the payments associated with a loan, and also is referred to as *equal annual income* and *equivalent annual cash flow*. The initial computation for EAE analysis is the NPV of

one rotation of a forest investment. The remainder of the EAE calculation determines the extent of the annual payments that would be required to equal the NPV using the discount rate that is assumed. The advantage of using EAE is that we can evaluate investments that have different time horizons. As a decision tool, we would possibly accept or invest in projects that yield a positive EAE. If several investments are being considered, then the ones with the higher EAE values would be more economically attractive than the others with lower (yet positive) EAE values.

The EAE is simply the NPV converted to an annual value paid at the end of each year (or time period) for the life of the investment. It is calculated at the appropriate discount rate using this formula:

$$EAE = NPV\left(\frac{i(1+i)^t}{(1+i)^t - 1}\right)$$

Example

Assume we have 200 acres of bare ground near Greenwood, Mississippi, and feel that there are two options to using this property productively. The first would be to afforest the site with oak trees that can be harvested in 50 years. It will cost $250 per acre to prepare the site, purchase the seedlings, and plant the oak seedlings. When the harvest occurs in year 50, we can expect to earn $7,116 per acre. The second option would be to grow soybeans, which can be planted and harvested once every year. Our expected annual total cost is $228 per acre and our expected annual revenue is $234 per acre. The net return for option two is $6 per acre per year. Option one does not have a yearly cost or revenue stream and each investment occurs at different time periods. To reconcile the cost and revenue for option one, we need to discount the harvest revenue back in time to period zero. To do this we must assume an interest rate. In this case, assume an interest rate of 5.5%.

$$\text{Present value} = \left[\frac{\$7,116}{(1.055)^{50}}\right] - 250 = \$239.34 \text{ per acre}$$

Now, to effectively compare option one and option two we need to convert the NPV for option one into an EAE. First, substitute the relevant values into the EAE formula just mentioned.

$$EAE = \$239.34\left(\frac{0.055(1+0.055)^{50}}{(1+0.055)^{50} - 1}\right) = \$14.14 \text{ per acre}$$

Second, now that we have calculated the EAE for option one, we can compare it to our annual net return for option two. As we can see, option one yields a yearly return per acre that is $8.14 greater than option two. Therefore, we would decide to afforest the 200 acres with oak seedlings.

G. Soil Expectation Value

The soil expectation value (SEV) represents the NPV of an investment in an even-aged stand from the time of planting (when the land is bare of trees), through infinite rotations of the same management regime. SEV is the same as bare land value (BLV) and LEV; however, the purchase price of the land and the revenue generated from ultimately selling the land are not included in the calculation. If the even-aged rotation lengths are the same, SEV and NPV assessments will produce the same ranking of potential investments.

All investments in even-aged rotations of trees, including the initial rotation and subsequent future rotations, involve the same length of time. To adequately compare even-aged stand management alternatives, the alternatives need to be assessed across the same time horizon. If the rotation lengths are different between competing management regimes, then using the SEV is a more appropriate way of ranking potential investments because it accounts for an infinite time line, while the NPV computation stops at the end of the rotation. As a result, simply using NPV computations to compare alternatives will ignore the opportunity to reinvest revenues (assuming one potential investment has a shorter time horizon than another) in some instances.

One equation that you could use to assess SEV involves the net revenue at the end of the first rotation (at first final harvest), and the length of the rotation:

$$SEV = \left(\frac{NR}{(1+i)^R - 1} \right)$$

where:

NR = net revenue at the end of the first rotation
R = length of the rotation

Example

Ignoring the fact that land will cost some money to purchase, assume that you decide to plant some trees on bare ground for $150 per acre today and harvest them in 25 years, earning $2,500 per acre. Assume also that your discount rate is 5%. To compute the net revenues at the end of the first rotation, we would need to determine the future value of the planting investment:

Future value of planting investment = $150(1.05)^{25} = 507.95

The net revenue at the end of the first rotation is then $2,500−$507.95, or $1,992.05. We then use this net revenue in the SEV equation to determine the SEV for the bare land:

$$SEV = \left(\frac{\$1,992.05}{(1.05)^{25} - 1} \right) = \$834.77$$

If we already manage the land being assessed, then to assess alternatives using SEV, we should interpret the SEV such that higher values represent better investments. Therefore, if one management regime (perhaps planting, herbaceous weed control, fertilization, then final harvest) yielded a higher SEV than another, then that type of management regime should be seen as a better investment for that piece of land. Alternatively, if the management regime was considered static, then we could determine the economically optimal rotation age *for that management regime*. In the case of the western forest stand that we use throughout this book (see Appendix A), assume that the management regime was simply to plant the trees, then perform a clearcut harvest at some time in the future. If the regeneration costs were $200 per acre to get the stand fully established in year 0, and that the stumpage price was $450 per MBF on average for all products, the SEV analysis suggests that the optimal rotation age is 45 years when the discount rate is 5% (Table 2.9).

TABLE 2.9 Net Present Value and Soil Expectation Value for the Western Forest Stand (From Appendix A) Using a 5% Discount Rate

Stand Age	Volume per Acre (MBF)	Value at Harvest ($/ac)[a]	NPV, One Rotation ($)[b]	SEV ($)
15	0.000	0.00	−200.00	−385.37
20	0.653	293.85	−89.25	−143.23
25	3.383	1,522.35	249.55	354.13
30	8.531	3,838.95	688.25	895.43
35	15.050	6,772.50	1,027.79	1,255.38
40	22.551	10,147.95	1,241.47	1,447.01
45	29.881	13,446.45	1,296.54	1,458.92
50	32.270	16,771.50	1,262.54	1,383.15
55	44.858	20,186.10	1,179.24	1,265.73
60	52.495	23,622.75	1,064.66	1,124.88
65	59.555	26,799.75	924.16	964.62
70	66.092	29,741.40	777.49	803.91
75	71.838	32,327.10	632.47	649.19
80	78.024	35,110.80	508.43	518.90
85	83.524	37,585.80	394.20	400.53
90	88.702	39,915.90	294.43	298.13
95	93.108	41,898.60	206.65	208.67
100	97.634	43,935.30	134.11	135.13

[a]Assuming a $450 per MBF stumpage value.
[b]Assuming a $200 per acre regeneration cost.

TABLE 2.10 Soil Expectation Value for the Western Forest Stand (From Appendix A) Using Different Discount Rates (*i*)

Stand Age	4% *i* SEV ($)	5% *i* SEV ($)	6% *i* SEV ($)	7% *i* SEV ($)
15	−449.71	−385.37	−343.21	−313.70
20	−121.21	−143.23	−157.48	−167.30
25	593.81	354.13	201.70	98.67
30	1,422.07	895.43	567.15	350.33
35	2,030.92	1,255.38	783.01	479.22
40	2,417.18	1,447.01	871.32	511.87
45	2,536.21	1,458.92	837.75	462.24
50	2,513.67	1,383.15	751.28	382.33
55	2,413.80	1,265.73	645.07	295.75
60	2,260.47	1,124.88	532.24	211.31
65	2,054.46	964.62	416.52	131.37
70	1,827.32	803.91	308.67	61.45
75	1,590.28	649.19	211.59	2.21
80	1,383.27	518.90	133.13	−43.61
85	1,182.43	400.53	65.94	−80.75
90	999.18	298.13	10.73	−109.76
95	829.31	208.67	−34.89	−132.49
100	683.45	135.13	−70.72	−149.54

In addition, you could use a different stumpage value for different log sizes, if the growth and yield projections allow you to understand the types of logs that could be produced from the stand given the diameters of the trees and their heights. Further, you could use different stumpage prices for different tree species. Last, but not least, we assumed static stumpage prices in the analysis, therefore stumpage price appreciation could be incorporated into the analysis by assuming that the prices might increase 0.5−1.0% per year.

Another way to assess SEV is to begin with the NPV of the current rotation and assume that it is a repeated value received by the landowner at an interval equal to the rotation age. As such, we would use an equation similar to the equation used for the present value of the non-terminating periodic revenues or costs, assuming that the first revenue or cost occurs right away (Section III, Part A.4).

$$SEV = \left(\frac{NPV(1+i)^R}{(1+i)^R - 1}\right)$$

The portion $NPV(1 + i)^R$ provides the future value of the net revenues at the end of the first rotation. For instance, in the previous example, the present value of earning $2,500 in 25 years is $738.26. Subtracting the planting cost from this result yields a NPV of $588.26. Substituting this value into the recent SEV equation results in the same SEV as we found earlier.

$$SEV = \left(\frac{\$588.26(1.05)^{25}}{(1.05)^{25} - 1}\right) = \$834.77$$

This also corresponds to the highest NPV for one rotation of this type of management. Therefore, for a management regime that included only regeneration and a final harvest, the economically optimal timing for the final harvest would be at 45 years using these assumptions.

As you might expect, as the discount rate increases, the value of money in the future declines. As a result, you should see the SEV for a management regime decrease for each stand age as the interest rate increases, assuming all other things being equal (Table 2.10). Here we also find that the economically optimal rotation age for our example western stand is 45 years when the discount rate is both 4% and 5%; however, it reduces to 40 years when the discount rate is 6% and 7%.

These analyses can be improved by adding more realism to the model being assessed. For example, the volumes that we have been using are the gross volume per acre at each stand age. In practice, this might be reduced by about 5−10% to account for breakage that may occur during the harvesting operation and defects that may not have been evident during the inventory.

H. Other Mixed-Method Economic Assessments

Repeatable management regimes beginning with bare land are the key to using SEV in a straightforward manner. This may not hold true in all cases of forest management, particularly in cases of uneven-aged management or in cases of even-aged management where we do not begin with bare land. In these other cases, an aggregate of NPV and SEV computations should be used to adequately determine the economic value of forest management.

In the case of uneven-aged management, some amount of time might be needed to adjust the stand structure to the point where it could sustain repeated, identical regimes. Here, an analysis of management strategies that begin with an existing stand, evolve through a transition period, and finally arrive at a single, repeatable management regime would be necessary. The management regimes used during the transition period should be assessed the same as a NPV analysis, but the repeatable regimes that carry on into infinity should be assessed as a SEV analysis (Horn et al., 1986). These portions are then added together to arrive at an economic evaluation of the management of a stand.

In the case of even-aged management beginning with an established stand of trees, a landowner may want to understand the current value of the land given that there may be revenues and costs associated with the completion of the rotation that is underway. After the completion of the current rotation, the assumption would be that subsequent rotations would utilize similar, repeated management regimes. Here, the present value of the current stand could be evaluated through the end of the current rotation. The SEV of the future management regimes could also be assessed, but the SEV would need to be discounted by the amount of time required to complete the current rotation, since the SEV reflects the management of bare land after the completion of the current rotation.

I. Selecting Discount Rates

The discount rate used in economic analyses is sometimes referred to as the *alternative rate of return*, which is the rate that can be earned by the best investment available other than the one that is being analyzed. It can also be called the *cost of capital*, which is the rate you can earn on capital in alternative uses. The discount rate is assumed to be the most appropriate one to use for calculating the NPV, SEV, BCR, and EAE. In addition, it is the appropriate rate to use as a *hurdle rate* for IRR analyses. In a real-life example of developing a discount rate, a committee of landowners and land managers was formed by the Idaho legislature to assess alternatives to Idaho's forest taxation methodology. They suggested that the guiding discount rate to value forestland for taxation purposes should be determined from recent averages of 10-year US Treasury Note constant maturity rates, minus an inflation factor, plus a risk premium. In this analysis, the resulting discount rates ranged between 4% and 5%, yet they were further negotiated down to a fixed 4% (Schlosser, 2005).

Many privately-owned and publicly-held corporations define discount rates as a weighted average cost of capital (Bullard and Straka, 1998). In other words, an organization's discount rate would be a weighted average of the amount of interest they are paying on their ratio of debt and equity. In addition, forest products corporations may specify an "exception rate" that places a strategic value on their timber-related investments (Bullard and Straka, 1998). This exception rate would be lower than a corporation's typical discount rate. Although many nonindustrial private landowners want long-term timberland investments to have higher rates of return, a high percentage of landowners place a moderate to high importance on protecting the natural environment and providing for future generations (Bullard et al., 2002). This may reflect a willingness to accept a lower rate of return to achieve these goals.

To start, let us define a discount rate. A discount rate is "a compound rate of interest that reflects the opportunity cost of capital for an individual, a firm, a government agency, etc." (Bullard and Straka, 1998). A discount rate may be either real or nominal. A *real discount rate* does not include inflation. A *nominal discount rate* does include inflation. The nominal interest rate can be calculated by incorporating both the real rate assumed by a landowner and an assumed inflation rate:

$$\text{Nominal rate} = [(1 + \text{Real rate}) \times (1 + \text{Assumed inflation rate})] - 1$$

If you knew the nominal rate used by a landowner, you could work backward to arrive at a real rate of return:

$$\text{Real rate} = [(1 + \text{Nominal rate}) / (1 + \text{Assumed inflation rate})] - 1$$

Example

Assume that you want to conduct a real analysis, but you have only a nominal interest rate of 15%, yet you know that the current inflation rate is 3.5%. How would you determine your real discount rate? We would have to substitute both pieces of information into the real rate formula just shown, then solve for the real rate.

$$\text{Real rate} = [(1 + 0.15)(1 + 0.035)] - 1 = 0.11 = 11\%$$

Alternatively, if we knew that our real discount rate was 11% and that the current inflation rate was 3.5%, what would be our nominal rate of interest? We use the nominal rate formula from earlier, and insert the appropriate pieces of information.

$$\text{Nominal rate} = [(1 + 0.11)(1 + 0.035)] - 1 = 0.15 = 15\%$$

We should notice one interesting result related to this example. We cannot just simply add the 11% real rate with the 3.5% inflation rate to derive the nominal discount rate. If added together, then the nominal rate would have been 14.5%. Just adding the two rates together would omit the interaction term between the real rate and the inflation rate. Another way to view the nominal rate equation is to algebraically rearrange it to look like the following:

$$\text{Nominal rate} = (\text{Real rate}) + (\text{Inflation rate}) + ((\text{Real rate}) \times (\text{Inflation rate}))$$

If we substitute our earlier assumptions into this equation, we find the following:

$$\text{Nominal rate} = (0.11) + (0.035) + ((0.11) \times (0.035))$$
$$\text{Nominal rate} = (0.145) + (0.004)$$
$$\text{Nominal rate} = 0.149 \approx 15\%.$$

Another component that some organizations consider when selecting an interest rate is whether or not to incorporate a risk factor. Klemperer (1996) provides several options on how to handle this situation. One conceptual approach is simply to modify the discount rate by adding a risk factor. The result would dampen future costs and revenues and shorten rotation ages. Algebraically, the relationship of a risk adjusted discount rate (RADR) is equal to the risk-free discount rate (r_f) plus a risk factor (k), and expressed as RADR $= (r_f + k)$. So if we knew that our risk-free discount rate was 15% then we could add a 4% risk factor to get an RADR of 19%. Unfortunately, no universally used risk factor (k) exists, and the choice of the risk factor is dependent on a number of considerations, such as the payoff period, the amount of risk in the revenue, and the decision-maker's degree of risk aversion (Klemperer, 1996).

An important issue when calculating any financial criteria for your analysis is being consistent with the use of inflation. If we intend to use inflation, then we must multiply all future prices by an appropriate inflation factor. These inflated values must either be compounded or discounted with a nominal interest rate. Discounting with a real interest rate will overestimate the value of your financial criteria. If we intend to conduct a real analysis, one that does not use inflation, then we need to make sure that we discount or compound all future and present values with a real discount rate. If we discount future values in "constant dollars" using a nominal discount rate, then we may underestimate the value of that particular activity. It should be noted that forest investment analyses have been traditionally conducted in real terms (excluding inflation).

Example

To illustrate some of the problems with inconsistency in conducting a financial analysis, let us borrow some of the material from our NPV example from Section III, Part C. In that example, we were interested in making a decision among three options for treating 100 acres of bare ground. The first option yielded a NPV of $541.40 per acre. The second option generated a NPV of $264.50 per acre. The third option generated $16.85 per acre. In that example, the interest rate was a real 6% discount rate. Since these NPVs were derived using a real discount rate, then what would happen if we used a nominal discount rate? If we assume a nominal discount rate of 11%, the NPV for option one would be −$14.10, the NPV for option two would be −$15.22, and the NPV for option three would be −$186.35. Given that we have mixed real values with a nominal discount rate, we can see that none of the three options would be financially feasible since their NPVs are all negative. This inconsistency would lead us not to choose any of the three management alternatives, when in fact they were actually feasible.

To summarize, each organization develops a discount rate that is used in their economic analysis of management alternatives. The discount rate may include aspects of expected price inflation and may include a perceived risk factor. In general, when a discount rate is lowered, a greater number of management alternatives will have positive NPVs. In addition, management actions that produce revenue later in a time horizon will have a greater impact on the economic results. When a discount rate is increased, fewer management alternatives will have positive NPVs, and emphasis is placed on revenues and costs that are incurred nearer to the present.

J. Forest Taxation

Up to this point in the chapter, when we talk about financial analysis, we have been conducting a before-tax analysis. Basically, a before-tax analysis may include property taxes and severance taxes as annual and single-sum costs in a discounted cash flow model, but it does not consider a landowner's marginal tax rate on ordinary income or capital gains (Bullard and Straka, 1998). Many taxes are considered as an additional cost to any activity that we intend to implement on a forest. For instance, if we want to purchase paint for marking boundary lines or buy flagging to mark plots in an inventory cruise, we may pay a sales tax on these items if our state or province imposes one. In addition, if we own forest land or manage it, then it is common to pay a land-use tax on a per unit area basis every year. For example, in Mississippi, the average land-use tax for forest land across the state is approximately $4.50 per acre. Incorporating this extra cost is pretty simple and straightforward. In effect, we just include it with the other costs in our financial time-line prior to estimating a management regime's NPV, EAE, and such. However, if we want to consider incorporating the impact of income taxes in our financial evaluation, then we need to utilize a different method that does not change despite frequent changes in our tax code.

Evaluating forest plantation or stand establishment and management can utilize either before-tax or after-tax financial criteria (Bullard and Straka, 1998; Klemperer, 1996; Gunter and Haney, 1984). Several studies such as Bettinger et al. (1989), Campbell and Colletti (1990), Bullard and Gunter (2000), and Grebner et al. (2003) suggest that considering a landowner's federal income tax bracket more fully captures the actual costs and benefits from reforestation incentives for which landowners are eligible. In addition, an after-tax analysis allows the decision-maker to incorporate into the analysis the potential tax savings, depending on whether they are an investor or a business, that may arise from deductions of expensed costs.

Bullard and Straka (1998) demonstrate that to conduct an after-tax analysis, there are four primary steps. First, we must convert all revenues to an after-tax basis. Second, we must convert all costs to an after-tax basis. Third, we must convert our discount rate to an after-tax basis. Fourth, we use these after-tax values and calculate all financial criteria of interest. The first step in the after-tax methodology is to convert all revenues to an after-tax basis. This generally is applied to revenues whether they occur from hunting leases or from a timber sale. The formula for calculating an after-tax revenue is (after-tax revenue) = (before-tax revenue) × (1−tax rate). For example, if we received $10 per acre in a year for a hunting lease and we faced a 25% marginal tax bracket, then our after-tax revenue would be $7.50 per acre.

The second step in this procedure is to calculate the after-tax costs. Calculating an after-tax cost is dependent on knowing whether the cost in question is an expensed cost or a capitalized cost. An expensed cost is a cost that can be deducted in its entirety in the year in which the expense occurs. A capitalized cost is a cost not deducted entirely in the year that it occurs. If the cost can be expensed, such as a land-use tax that occurs every year, then we can convert it using a modified version of the formula to convert an after-tax revenue. The formula for calculating an after-tax cost is (after-tax cost) = (before-tax cost) × (1−tax rate). For example, if we paid $4.50 per acre in a year for a land-use tax and we faced a 25% marginal tax bracket, then our after-tax cost would be $3.38 per acre. This lower cost reflects the $1.12 per acre tax savings we would get from deducting this cost from our federal income taxes.

If we had capital costs, then we could separate them into either a land or timber account. A land account is where we record money that we invested in the land that can be deducted from the sale of the land. A timber account is money that we invested in the timber that can be deducted through the depletion of the basis. The basis is the money we invest in reestablishing a stand or the value of the timber when we buy it. Sometimes the timber account is called the "depletion account."

As it relates to forest management, depletion is the recovery of the basis (initial investment) when the forestland (with standing trees) was purchased, and comes into play when timber products are cut and sold from the property or used within an integrated business. At its root, the depletion unit is a measure of cost basis for landowners per unit of volume in their timber accounts at some point in time. The depletion unit is computed by dividing the adjusted cost basis at any point in time by the standing timber volume available on the landowner's property. If the depletion unit were $50 per MBF, the landowner could deduct this amount from the revenue per MBF at the time of harvest, when computing the taxable income.

This taxable income also is known as the capital gain income and is used in calculating the after-tax revenue from a timber harvest. Given that this cost is a capital cost, you can then multiply the capital gain income by the capital gain tax rate to derive the after-tax revenue from a timber sale. This after-tax revenue can be calculated for the entire forest or on a per unit area basis.

The third step in the after-tax methodology is to convert before-tax discount rate to an after-tax discount rate. This generally is applied because interest paid on business-related loans is deductible (Bullard and Straka, 1998). The formula for calculating an after-tax discount rate is (after-tax discount rate) = (before-tax discount rate) × (1−tax rate). If the before-tax real discount rate is 6% and we faced a 25% marginal tax bracket, then our after-tax discount rate would be 4.5%.

The final step in this process is to use all of our after-tax revenues, costs, and the discount rate to calculate relevant financial criteria. Consistency is important here. For example, if we compound or discount future or present after-tax values using a before-tax discount rate, then we will be generating biased estimates of our financial criteria. These biased results could substantially impact our decisions made at either the stand or forest level.

IV. ENVIRONMENTAL AND SOCIAL EVALUATION OF NATURAL RESOURCES

We conclude this chapter on valuing and characterizing forest conditions with an examination of several of the common methods for evaluating forest conditions from an environmental or social perspective. More often today, the most emotional arguments in the development of policies and plans of action involve ecological or social analyses. As a natural resource manager, therefore, you should be familiar with the variety of ways in which a plan of action could affect goals other than those with an economic or commodity production aspect.

A. Habitat Suitability

Perhaps the closest tie between wildlife objectives and forest characteristics are the methods we might employ to measure wildlife habitat suitability. In the early 1980s, the US Fish and Wildlife Service developed a series of models aimed at quantifying habitat quality using vegetation and physiographic variables, many of which involve tree measurements. A number of other models have since been developed. These habitat suitability index (HSI) models are some of the most influential tools in use, even though there are a number of criticisms (lack of knowledge on habitat requirements, use of a composite score to rate the environment) about their ability to accurately portray habitat quality (Roloff and Kernohan, 1999). Habitat quality values can be determined for individual stands, or for larger areas such as a watershed or an ownership. Stand-level HSI values range from 0 to 1, where 0 represents poor habitat for the species in question and 1 represents optimal habitat. The stand-level HSI values may be multiplied by the size of the stand, and summed for a larger area, to arrive at habitat units available in a watershed or ownership. Alternatively, habitat units could be divided by the size of the watershed or ownership to arrive at an average HSI (valued between 0 and 1) for the entire area.

HSI models generally involve several variables, many of which are related to forest structure conditions. As we mentioned earlier in Section II, Part I, the HSI model for the downy woodpecker includes two variables, one that is a function of the basal area of trees in a stand, and the other that is a function of the number of snags per unit area that are greater than 6 inches DBH (Schroeder, 1983). The relationships between the forest characteristics and the habitat variable scores may be linear, as in the case of the snags:

$$\text{Snags score} = 0.2 \times (\text{Number of snags greater than}$$
$$6 \text{ inches DBH per acre})$$

Alternatively, the relationship may be nonlinear, as in the case of basal area, where low levels of basal area result in low habitat-related scores, as do high levels of basal area.

Scores for individual habitat elements also range from 0 to 1, and generally they are multiplied together in some form or fashion to arrive at the final HSI score. As a result, if one of the elements in a habitat suitability model is 0 (reflecting poor habitat conditions), the overall HSI score will be 0 (poor habitat). Some habitat suitability models have been improved since their introduction, although improvements generally imply increased complexity. Others involve spatial relationships that may suggest the use of geographic information systems. The white-tailed deer (*Odocoileus virginianus*) HSI model for the Piedmont of the southeastern United States (Crawford and Marchinton, 1989), for example, includes the following variables:

- Weight of oven dry green herbaceous plant material
- Basal area per acre of oaks 10 inches (25 cm) DBH and greater
- Number of oak species found in a stand
- Site index of loblolly pine or mixed oaks
- Percentage of the area in agricultural land
- Distance of agricultural land to forest or shrub cover

The last two variables in this list would seem to require an analysis of maps or spatial data. McComb et al. (2002)

go even further in their spotted owl (*Strix occidentalis*) model to suggest that forested conditions within three buffer zones (one as large as 1.5 miles, or 2.4 km) around each stand need to be assessed prior to arriving at a habitat value for each stand.

B. Recreation Values

Preferences for scenic beauty are indications of the values inherent in people, and thus will vary from one socioeconomic and demographic group to another (Ribe, 1989). However, the need to integrate aesthetics and scenic beauty into forest plans may be important for land managers who need to address and account for multiple uses of their forests. Scenic beauty measures often increase with increases in large trees, decreases in down woody debris, and decreases in groupings of trees (Brown and Daniel, 1986). This is in contrast to some of the wildlife suitability models that may require increases in down woody debris for increased habitat quality. And, it is in contrast with many economic models that suggest cutting trees at their economic maturity, perhaps some period of time prior to when trees become large enough to be considered the most aesthetically pleasing. Data collected during normal timber inventories or preharvest stand examinations could be used to determine scenic beauty before and after harvest, and to assess alternatives (Brown and Daniel, 1986). One model for ponderosa pine in the western United states indicates a positive contribution to scenic beauty with increases in ponderosa pine trees per acre greater than or equal to 24 inches (61 cm) DBH, and with increases in volume (pounds per acre) of herbaceous plants (Brown and Daniel, 1986):

$$\text{Scenic beauty index} = -32.47 + 4.7$$
$$\times (\text{Large pine TPA}) + 0.38$$
$$\times (\text{Herbaceous plant volume})$$

Hull and Buhyoff (1986) provide two models for estimating scenic beauty in loblolly pine stands, one based on TPA, and the other based on basal area per acre (BA):

$$\text{Scenic beauty index} = 124.96 - 0.0134$$
$$\times (\text{Std. deviation of DBH} \times \text{TPA})$$
$$- 0.02597 \times (\text{TPA}) - (9,443.4/\text{TPA})$$
$$+ 498.513 \times (\text{Average DBH}/\text{TPA})$$
$$- 86.623 \times (\text{Average DBH}/\text{stand age})$$
$$\text{Scenic beauty index} = 5.663 - 4.086 \times (\text{BA}/\text{stand age})$$
$$+ 16.148 \times (\ln(\text{BA}))$$

These scenic beauty indices range from 0 (poor) to about 100 (best). For example, using the model, assume that a loblolly pine stand was 60 years old and had a basal area per acre of 110 ft^2. Its scenic beauty index would be:

Scenic beauty index $= 5.663 - 4.086 \times (110/60)$
$$+ 16.148 \times (\ln(110))$$

Scenic beauty index $= 74.08$

If the stand were denser, or had a higher basal area per acre (say 150 ft^2 per acre), the scenic beauty index would be:

Scenic beauty index $= 5.663 - 4.086 \times (150/60)$
$$+ 16.148 \times (\ln(150))$$

Scenic beauty index $= 76.36$

Others have suggested that scenic beauty should be tied explicitly to the age of forests, since younger, managed stands usually contain a high density of trees (negatively correlated with scenic beauty). In addition, canopy closure and tree species composition have been suggested as variables that could be used to predict scenic beauty (Ribe, 1989). Evidence of management actions may also detract from scenic beauty. For instance, thinnings may reduce scenic beauty, although the effect varies by thinning intensity and context (Hull and Buhyoff, 1986).

C. Water Resources

Water resources are an important value that forest managers need to consider when evaluating different forest management activities. We are increasingly dependent on water resources for human consumption, sustainable agriculture, ecological functions, recreation opportunities, and industrial processes. Forest management activities have the potential to greatly impact water quality and water quantity, in both beneficial and adverse manners. The value of water will ultimately depend on the proximity of the water source to the users of the water, the quality of the water (e.g., turbidity), and the quantity of the water. The availability of water (quality and quantity) at various times of the year may also affect how it is valued.

In 1972, the United States Congress enacted the Federal Clean Water Act (amended in 1977 and 1987), the primary objective of which was to restore and maintain the integrity of the nation's waters. Part of this Act requires individual states to report to the US Environmental Protection Agency triennially on the nature and condition of water resources. The US Environmental Protection Agency (2007b) lists seven designated use categories: (1) fish, shellfish, and wildlife protection and propagation, (2) recreation, (3) public water supply, (4) aquatic life harvesting, (5) agriculture, (6) aesthetic values, and (7) exceptional recreational or ecological significance. Based on state assessments for the

2002 reporting period, the US Environmental Protection Agency (2007b) estimated that 45% of US rivers and 47% of US lakes, ponds, and reservoirs did not fully support intended uses and were designated as impaired.

Water quality is impacted by both point and nonpoint source pollution. The US Environmental Protection Agency (2011) describes nonpoint source pollution as the following:

Nonpoint source (NPS) pollution, unlike pollution from industrial and sewage treatment plants, comes from many diffuse sources. NPS pollution is caused by rainfall or snowmelt moving over and through the ground. As the runoff moves, it picks up and carries away natural and human-made pollutants, finally depositing them into lakes, rivers, wetlands, coastal waters, and even our underground sources of drinking water. These pollutants include: excess fertilizers, herbicides, and insecticides from agricultural lands and residential areas; oil, grease, and toxic chemicals from urban runoff and energy production; sediment from improperly managed construction sites, crop and forest lands, and eroding streambanks; salt from irrigation practices and acid drainage from abandoned mines; bacteria and nutrients from livestock, pet wastes, and faulty septic systems; and atmospheric deposition and hydromodification are also sources of nonpoint source pollution.

In addition to requiring that states designate uses to surface waters, the US Environmental Protection Agency requires that states adopt water quality standards to protect those uses and that states take measures to protect water quality. Toward that end, states have adopted total maximum daily loads, which are numerical limits for many known pollutants that do not meet the state standards. Different water quality criteria may be imposed for different water bodies within a state, and water quality monitoring is required as part of the process.

Many factors are used to measure water quality. They include such parameters as turbidity, acidity, amount of dissolved oxygen, nutrient loads, macro-invertebrates, water temperature, and bacteria (Stuart et al., 2000). Deciding on which parameter to measure depends on the intended use for that body of water, and the scale of measurement can vary from the stand-level, to the forest-level, to a regional watershed. Contaminants of concern will vary from region to region. For example, in many northern states, it is common to see signs near large reservoirs that warn commuters of slippery roadways because salt was not applied to the pavement. Although adding salt to a local road may not dramatically affect a neighboring forest stand, salt runoff across the area could have an adverse impact on the water quality of a reservoir. Further, the timing and placement of forest management activities can have an impact on water quality, and some estimates of stream sedimentation and stream temperature can be tied directly to proposed forest management activities (US Department of Agriculture, Forest Service, 1981, 1993).

The quantity of water moving through a basin is also of concern. Hydrologic benefits provided by forests include recycling of rainfall inland through transpiration, flood control, and reduction of the potential for landslides during peak-flow events (Brown, 2008). Forest management activities can greatly impact water yield and peak-flow during storm events. In the eastern United States, canopy removal has been reported to increase annual stream water yield by 40% (Hornbeck et al., 1993); thus forested landscapes can be manipulated to either increase or reduce runoff and stream flow as part of an overall basin management plan.

Management of forests to benefit water resources may be governed by state, provincial, or federal standards, regulations, or guidelines. In states with forest practices acts, such as California, Oregon, and Washington, forest managers need to adhere to regulations that mandate the appropriate procedures for different silvicultural activities and road construction. In other states, such as Alabama, Georgia, Mississippi, and Texas, forest managers voluntarily adhere to best management practices. Best management practices are guidelines for forest management operations that are used to maintain site productivity and reduce nonpoint source pollution. Some standard guidelines include marking streamside management zones on forests prior to silvicultural operations, and designing road systems that minimize sedimentation and erosion.

D. Aquatic Habitat Values

Forest managers should also be aware of the habitat values that streams provide to aquatic life when they plan various activities across a forested landscape. However, the biotic integrity of streams may not effectively be monitored for water quality using chemical and physical tests even though sustaining a balanced biotic community is one of the best indicators of a stream's ability to provide potential benefits in the future (Karr, 1981). This section will briefly describe a study (Karr, 1981), which promotes a methodology for using fish populations as an indicator of a stream's biotic integrity.

Fish have several advantages for being selected as indicator organisms. The life history of most fish species is extensive, they are easy to identify, and they are present in most perennial stream bodies. Fish communities contain a range of species covering several trophic levels such as omnivores, planktivores, piscivores, insectivores, and herbivores. Fish, especially species of salmonids, are sensitive to acute toxicity, and stress effects are easy to measure. As with any method, there are disadvantages

as well, and they include the selective nature of field sampling, high labor requirements needed for field sampling, and seasonal fish mobility.

A classification criteria system might have two major components. One deals with species composition and richness and the other deals with ecological factors (Karr, 1981). With regard to composition and richness, this component not only considers the number of species and number of individuals, but also examines the presence of intolerant and tolerant fish species. Identifying intolerant fish species is important because they are usually the first species to decline when water quality is diminished. In (Karr, 1981) other criteria examined include measuring the presence and abundance of green sunfish, and the presence of hybrids. Identifying green sunfish (*Lepomis cyanellus*) is important because they are the species most likely to be present in degraded streams. The presence of hybrids is further indication of habitat degradation because fish breeding is prevented from segregating along normal habitat gradients.

With respect to ecological factors, the first key criterion involves a measure of the proportion of individuals that are omnivores. As stream quality declines, the omnivore population increases. Another factor is the proportion of insectivorous cyprinids in the fish population. There may be a strong inverse relationship between the abundance of insectivorous cyprinids and omnivores (Karr, 1981). The last key ecological criterion is measuring the abundance of fish with tumors, deformities, parasites, or disease. In addition, age structure, growth, and recruitment rates are important in assessing stream quality.

Using an appropriate sampling scheme, it is important that the sample is a balanced representation of the fish community and the larger geographic area of interest. This methodology requires an evaluation of each of the species composition and ecological factors and the development of an index value that assesses the biotic integrity of the stream habitat. Karr (1981) uses a grading system for each evaluated criteria where a minus (−) equals 1 point, a zero (0) equals 3 points, and a plus sign (+) equals 5. Table 2.11 illustrates three examples modified from Karr's work. As you can see, the headwater #1 yields the best index value of excellent for quality habitat and headwater #3 garnered an index value of poor. The numbers in parentheses represent the number of individuals sampled.

E. Air Quality

Over the past several decades, the interaction between forests and air quality has become an increasingly important issue to society and forest managers on public and private lands. We typically think of air having good

TABLE 2.11 Example Calculation of a Biotic Integrity Index Applied to Three Hypothetical Headwater Sites

Aspect	Headwater #1	Headwater #2	Headwater #3
Number of species	+(37)	0 (6)	+(9)
Number of individuals	+(122)	+(100)	0 (51)
Number of darter fish species	+(1)	0 (1)	−(0)
Number of sunfish species	+(3)	−(0)	0 (1)
Number of sucker fish species	+(2)	−(0)	0 (2)
Number of intolerant species	+(4)	−(0)	−(0)
Proportion of omnivores	+(2)	0 (25)	−(57)
Proportion of insectivorous cyprinids (carps, etc.)	+(42)	+(66)	+(43)
Proportion of green sunfish	+(0)	+(0)	+(0)
Proportion of top carnivores	0 (3)	−(0)	−(0)
Proportion of hybrids	+(0)	+(0)	+(0)
Proportion of diseased	+(0)	+(0)	−(5)
Derived index value	58	38	34
Habitat quality class	Excellent	Fair to poor	Poor

Source: Karr, J.R., 1981. Assessment of biotic integrity using fish communities. Fisheries 6 (6), 21−27.

quality when we can't smell offensive odors and we can clearly see long distances. Air quality is threatened by air pollution. Air pollutants are materials that occur in the lowest portion of the atmosphere in quantities above normal ranges (Smith, 1990). Forests can play an important role as both a source of air pollution and as a sink for air pollution (Smith, 1990). Pollutants can occur from both natural and anthropogenic processes and be in solid, liquid, or gaseous states. These issues are particularly important when prescribed or controlled burning programs are considered for forest health, wildlife habitat development, or site preparation purposes.

Forests play in important role in these pollutants because of their interaction with carbon and sulfur. Forested areas are an important source of carbon and sulfur (Smith, 1990). Carbon resides in standing live and dead trees, shrubs, litter, the soil organic layer, and the mineral soil (Owens et al., 1999; Huang and Kronrad, 2001;

Cason et al., 2006). By dry weight, a tree is composed of about 50% carbon (Matthews, 1993). Carbon in the form of carbon dioxide (CO_2) is constantly exchanged between the atmosphere and the forest through photosynthesis, respiration, and decomposition (Kramer and Kozlowski, 1979). In an equilibrium state, the net emission of carbon in the atmosphere from forests is zero (Smith, 1990). However, natural phenomena, such as forest fires and hurricanes, and anthropogenic disturbances such as deforestation and urbanization on forests can release large quantities of carbon dioxide into the atmosphere. Other common anthropogenic sources of carbon dioxide (burning of fossil fuels) emitted into the atmosphere go beyond the scope of this book so they will not be discussed in detail. Anthropogenic disturbances in forests have led to increasing concerns over the impact of releasing forest carbon into the atmosphere on global temperatures. For instance, Brazil has been active for many years in developing portions of their Amazonia basin for economic reasons (Sirmon, 1996). This has led to policies that promoted large scale deforestation and land-use conversion for agricultural purposes. These practices have contributed to the release of carbon dioxide into the atmosphere, which is a contributing factor to global warming (Fearnside, 1996; Fearnside, 1997).

Sulfur, which can be another pollutant, is important because of its potential impact on vegetative and human health. A large portion of sulfur in the atmosphere arises from microbial activity in soil and water (Smith, 1990). Other natural mechanisms for emitting sulfur sources into the atmosphere are volcanic events and ocean spray. However, the total emission of sulfur into the atmosphere by natural sources (ignoring volcanic eruptions) is small relative to anthropogenic sources. In addition, the importance of anthropogenic sulfur released into the atmosphere is a function of the distance and location downwind from a sulfur-emitting facility. Over the last several decades, there have been numerous studies evaluating the impact of air pollution from industrial coal power generation plants in the midwestern United States. Electric power generation from coal emits a mixture of sulfur dioxide (SO_2) and nitrogen oxides (NO_x) into the atmosphere to form acid rain. *Acid rain* is a mixture of wet and dry deposition from the atmosphere containing above normal amounts of sulfuric and nitric acids (US Environmental Protection Agency, 2012). Studies have shown that acid rain adversely affects trees at high elevations, such as red spruce (*Picea rubens*), and can cause acidification of lakes and streams, which can diminish fish populations and other aquatic organisms (Smith, 1990; Pitelka and Raynal, 1989; US Environmental Protection Agency, 2007a). In addition, a study conducted at the Hubbard Brook Experimental Forest in New Hampshire observed a wood volume growth decline of 17%, coinciding with a period of increased acidity in precipitation (Smith, 1990).

Forests primarily can affect air quality through the voluntary or involuntary use of fire. Whether a forest manager has to fight a wildland fire or manage a prescribed burn, both events emit smoke and other particulate matter into the atmosphere. In North America, the public has been concerned about air pollution from forest fires since the early 1900s, and as a result, governmental agencies began developing slash burning laws in 1911 (Murphy et al., 1970). Over the past century, numerous federal and state laws have been enacted to prevent or limit the use of fire, although highly publicized fire events, such as the wildfire in Yellowstone National Park in 1988, illustrated the positive role that prescribed burning could have had in lowering fuel loads and preventing large natural catastrophes. Growing urbanization is also exposing more residential communities to potential fire hazards and its associated smoke intrusions. For forest managers, with properties near residential areas, this presents many unique problems. As we suggested here, there are several barriers to successful implementation of prescribed burning programs. Other challenges include funding limitations, availability of alternative silvicultural tools, potential liabilities, insurance availability, lack of qualified staff, excessive fuel loading, and narrow prescription windows for conducting burns (Haines et al., 2001).

Many states now have laws that require forest managers to develop smoke management plans or obtain burning permits prior to conducting prescribed burns. Plans are required even when the goal is to enhance wildlife habitat or remove unwanted slash piles before planting. For instance, in Mississippi, the 1993 Prescribed Burning Act requires that forest managers utilize a certified burn manager, obtain a notarized prescription (at least 1 day prior to burn), obtain a permit, and conduct the planned burn that is consistent with the general public interest. A burn prescription is a written plan that states the how, what, where, when, and why, and must be notarized to become a legally binding document. A permit is issued when the proper environmental conditions are met for adequate smoke dispersal. For instance, there must be a mixing height of at least 1,640 ft (500 m) and a wind speed of 7.83 miles per hour (3.5 m/s) (Londo et al., 2005). In addition, the Act defines the types of liability to which a burner is subject. Sun and Londo (2008) provide additional detail on Mississippi's legal environment for prescribed fire. Through this brief discussion the difficulties in measuring the impacts of fire on forest management, and vice versa, suggest that acknowledging these impacts in a forest plan may be challenging.

F. Income and Employment

Forest managers have a number of issues to consider when evaluating their forest plan in addition to the type of activities to implement over the planning horizon. One important consideration to both private and public forest landowners is the impact of their management activities on individuals and local economies. As they relate to local economies, forest management activities require staff to implement daily operations, such as record keeping, bookkeeping, and facility maintenance. They also require foresters or other natural resource managers to conduct inventories, monitor timber harvests, or build new logging roads. If these involve new personnel, then they could stimulate employment of other professionals in different sectors of the local economy that support their work. For instance, more people could be hired in the retail and banking industries to support the complementary services provided within a local community. These additional jobs could lead to increases in local tax revenues that may be used to improve local infrastructure and schools, which in turn could help the local community become more attractive to new businesses and future economic growth opportunities. If a forest manager's staff is small, then they may decide to hire forestry consultants and other professionals to perform many of the necessary management activities suggested by a forest plan, which could encourage many of the same growth effects. Apart from supplying employment opportunities for local communities, timber harvesting is a source of value-added revenue that yields new income for the state through severance taxes, a landowner's capital gains, and income paid to the logging contractors who then spend portions of that money within the local economy.

Another issue that is important to forest landowners is their role in providing community stability. The origins of concern for community stability can be traced back to the middle 1800s in northern Germany (Waggener, 1977). Over the years this has been an important issue for both private and public landowners big and small across many regions in the United States. However, the effectiveness of public forest management policies to promote community stability has been questioned (Waggener, 1977; Greber and Johnson, 1991). The US government has been very interested in facilitating community stability, given the long history of boom or bust cycles associated with natural resource development in the western United States (Wear et al., 1989). Even-flow harvesting of wood products may have a positive impact on community stability, but the cost of a large-scale program guided by this policy would need to be assessed. Chapter 9, Forest and Natural Resource Sustainability, addresses these types of sustainability issues in greater detail.

V. SUMMARY

There are a variety of ways in which we could characterize and value a stand of trees or a forest. We have illustrated in this chapter a number of the structural valuations that commonly are used to communicate the condition of forests. These include qualitative and quantitative measures of stand density, tree size, and tree status (live or dead). Since management activities usually involve the allocation of people's time or money, an economic analysis of alternatives is important as well. The IRR, benefit/cost analysis, NPV, and SEV assessments are used frequently to sort through the alternatives and rank them in order of economic importance. When analyzing management alternatives from an economic perspective, it is important to remember when to use present and future valuations. Further, costs or revenues do not occur on an annual basis, and whether they terminate or are assumed to continue indefinitely, it is important to use the appropriate method of projecting these values into the future, or back to the present. Forest plans typically incorporate other environmental and social assessments as well. In this chapter we have provided a few of the more common types of assessments for you to consider. As a natural resource manager, you must determine which set of measures are important to assess in order to provide land managers or landowners enough information to make a decision that approaches a rational course of action. If important measures of impact are omitted from an assessment of management actions, the subsequent decisions may lead to unintended results.

QUESTIONS

1. *Mean annual increment, periodic annual increment.* Suppose you are given the following expected biological growth information for a stand in the following table. You are interested in determining the optimal rotation age for this stand. Given the following information, what is the MAI? What is the PAI? What is the optimal biological rotation age?

Year	Total volume (ft³/ac)	Mean annual increment (ft³/ac/year)	Periodic annual increment (ft³/ac/year)
5	18		
10	355		
15	1,318		
20	2,622		
25	3,735		
30	4,395		
35	4,675		
40	4,611		
45	4,368		
50	4,060		

2. *Bare land value.* What is the value of bare land, if used with even-aged management to produce a perpetual series of identical timber rotations? You have initial stand establishment costs of $125 per acre and a harvest income of $12,500 per acre in 35 years. Assume a 5% real discount rate.

3. *Economic assessments.* Suppose that you have 250 acres of cut-over land and you want to evaluate the return from planting timber on the site. Assume that to prepare the site and plant shortleaf pine it will cost $250 an acre. Assume that incidental management costs are $3 per acre per year. Quail hunters will pay $5.35 per acre per year for hunting rights for plantations 5 years old or less. At age 17, a selection thinning can generate $335 per acre. In year 34, clear-cutting the tract can generate $2765 per acre. If your discount rate is 4%, then what is the NPV of this return? What is the BCR? In addition, if this is the optimal management regime, then what is the SEV based on this management regime?

4. *Stocking guide for upland hardwoods.* Using the stocking guide presented in Fig. 2.12, if you managed an upland hardwood stand that contained 1200 trees per acre, which represented 80 ft² per acre of basal area, what would be your estimate of the stocking level (qualitative and quantitative), and the average diameter of the trees in the stand?

5. *Stocking guide for upland hardwoods.* Again using the stocking guide presented in Fig. 2.12, if you managed an upland hardwood stand that had an average DBH of 6 inches, and a basal area of 90 ft² per acre, what would be your estimate of the stocking level (qualitative and quantitative) and the trees per acre within the stand?

6. *Rate of growth of deer populations.* As a new forest manager, you learned that the deer population on your property is currently 20 deer per square kilometer. Ten years ago, there were 10 deer per square kilometer. Twenty years ago, there were 5 deer per square kilometer. What is rate of growth for each period? What is the rate of growth for the past 20 years?

7. *After-tax net present value of a hunting lease.* Suppose you want to calculate the after-tax NPV of a hunting lease over a 10-year period. The lease yields $10 per acre per year and your yearly management costs are $3.75 per acre. If your alternative rate of return is 6.5% and your marginal tax rate is 30%, then what is your after-tax NPV? What is your tax savings?

8. *Benefit/cost ratio for road development.* Suppose your company is studying a potential contract with the federal government to construct a 3-mile road prior to harvesting timber on a public forest. You need to determine whether it is financially feasible

to build this road prior to harvesting the public timber. If calculating a BCR is your company's primary tool for evaluating any contract, then what is the ratio if the expected present value of timber revenues is $7,000,000 and the road building costs are $1,000,000 per mile? What does the ratio tell you? Is the activity feasible? What if the expected yield from future timber harvest were only $2,500,000 due to major errors in the precontract inventory? What does your BCR suggest? How does it compare to the first scenario?

9. *Assessment of corn or hybrid poplar investments.* Suppose you are an extension forester trying to give advice to a farmer in the Mississippi Delta. This farmer is trying to decide between growing corn or establishing a short-rotation hybrid poplar plantation on his 100 acres of old fields. These may be used to provide feedstock for a newly established dual feedstock bio-ethanol refinery in Vicksburg. On this land the farmer can yield 105 bushels of corn per year and garner a price of $5.00 per bushel for his corn. If the farmer plants hybrid poplar, then he can't harvest the popular for 5 years. At that time, the farmer is expected to yield 50 cords of wood per acre and garner a price of $10 per cord. Which option would you recommend the farmer to take? Why? Would it be wise to plant the entire 100 acres to only corn or trees or would it better to split up the acreage? Assume that the farmer has a hurdle rate of 5.9%.

10. *Terminating periodic net revenues.* Suppose you need to evaluate and compare the cash flows derived for a terminating periodic net revenue stream that ends in 20 years and a perpetual periodic net revenue stream. If you were evaluating these two net revenue streams on a present value basis, then what would you expect the relative values to be? In other words, would you expect them to be the same? Why or why not? Assume periodic net revenue of $500 that occurs every 5 years and a real risk-free alternative rate of return is 4.5%. After evaluating the results for the two revenue streams do you get the relationship you expected from answering the first question? What happens if you add a 3% risk factor to your alternative rate of return? Do you get the same result? What happens to the magnitude of your results?

11. *Timing of activities in a forest plan assessment.* Suppose that you have been asked by your supervisor to estimate the NPV of potential management regimes across various forest types prior to using them in a harvest scheduling analysis. If you have planning periods of 1 year, then would it make much of a difference to plan activities at the beginning or end of the period? What about the middle? Which

one would be more practical? How would it affect your return? What if your periods were longer, say 5 years or even 10 years long? What should you do?

12. *Diameter distribution.* Develop a diameter distribution of the western stand described in Appendix A, for ages 30 and 50. What can you tell about the stand from the diameter distribution? How does it change over the 20-year time period?

13. *Quadratic mean diameter.* Assume that you manage a stand of red pine in Wisconsin. The stand contains 145 ft^2 of basal area per acre and 235 trees per acre. What is the QMD of the trees in this stand?

14. *Down woody debris.* Given the following data from a 0.5 acre sample of a natural stand of 60-year old pine trees in South Carolina, what is your estimate of the down woody debris (in cubic feet per acre)? Assume that the lengths of logs are all 10 feet.

Log	Small-end diameter (inches)	Large-end diameter (inches)	Decay class
1	6	8	I
2	5	9	II
3	10	13	II
4	8	9	III
5	9	12	I
6	5	8	III
7	12	14	II

15. *Site quality.* Describe three different perspectives on site quality: the generalist, the ecosystem-oriented, and the timber-oriented. Which of the three is most closely related to the concept of a site index for a stand of trees, and why?

16. *Site index.* A 40-year old stand of ponderosa pine in eastern Oregon has a $SI_{50} = 88$. What does this indicate about the stand, in general, and about the current height of the stand?

17. *Annual hunting lease.* Assume that a landowner in north Florida leases out his or her property to a hunting club, at a rate of $12 per acre per year. If the landowner uses a 6.5% discount rate for the proposed investments, and the hunting lease covers a 5-year time period, what is the present value of the lease to the landowner? What is the future value of the investment at the end of the 5-year time period?

18. *Prescribed fire program.* A landowner in North Carolina performs prescribed burning on his land every 3 years. Assume that his alternative rate of return for investments is 5%, and the program is assumed to continue indefinitely. If the first burn

occurs now, what is the present value of the pre-scribed burning program to the landowner? If the first burn occurs 3 years from now, what is the present value of the burning program?

19. *Habitat enhancement program.* A landowner in Alabama wants to enhance the red-cockaded wood-pecker habitat in one of her older pine stands by inserting man-made cavities in some longleaf pine trees. She decides to spend $200 per acre every 2 years for the next 10 years on this project. If the first cost is incurred at the end of the first 2-year time period, what is the present value of this program to the landowner? Assume that the alternative rate of return for investments is 5%. What is the future value of this investment at the end of the decade?

20. *Future value of an investment.* Assume that you invested $1,000 today in a 3-year certificate of deposit that yields a 4.5% annual rate of return. Including the initial investment, how much money will you have at the end of the 3-year period?

21. *Selecting a discount rate.* Assume that you are a forestry consultant in southern Illinois, and you are developing a management plan for a private land-owner. In the course of developing the plan, you need to assess several alternative management prescriptions. The landowner has never really considered the matter of discount rates for investments. How would you arrive at a discount rate for use in assessing the landowner's alternatives?

22. *Ecological assessments of alternatives.* How could a typical forest inventory be used to assess the impact of proposed management activities on wild-life habitat quality, stream conditions, or recreation quality?

REFERENCES

Anderson, S., Marcouiller, D., 1990. Growing Shiitake Mushrooms. Oklahoma Cooperative Extension Service publication F-5029, 7 p.

Avery, T.E., Burkhart, H.E., 1994. Forest Measurements. McGraw-Hill, Inc, New York, NY, 408 p.

Barlow, R., Levendis, W., 2015. 2014 costs and cost trends for forestry practices in the South. For. Land. 74 (5), 22−31.

Bartelink, H.H., 1996. Allometric relationships on biomass and needle area of Douglas-fir. For. Ecol. Manage. 86 (1), 193−203.

Bettinger, P., Haney Jr., H.L., Siegel, W.C., 1989. The impact of federal and state income taxes on timber income in the South following the 1986 Tax Reform Act. South. J. Appl. For. 13 (4), 196−203.

Brooks, C., Hall, D., 1997. Gypsy Moth Silvicultural Guidelines for Wisconsin. Wisconsin Department of Natural Resources, Madison, WI, PUB-FR-123. 11 p.

Brown, L.R., 2008. Plan B 3.0: Mobilizing to Save Civilization. Earth Policy Institute, W.W. Norton & Company, New York, NY, 398 p.

Brown, T.C., Daniel, T.C., 1986. Predicting scenic beauty of timber stands. For. Sci. 32 (2), 471−487.

Bullard, S.H., Gunter, J.E., 2000. Adjusting discount rates for income taxes and inflation: a three-step process. South. J. Appl. For. 24 (4), 193−195.

Bullard, S.H., Straka, T.J., 1998. Basic Concepts in Forest Valuation and Investment Analysis (Edition 2.1). Preceda Education and Training, Auburn, AL.

Bullard, S.H., Gunter, J.E., Doolittle, M.L., Arano, K.G., 2002. Discount rates for nonindustrial private forest landowners in Mississippi: how high a hurdle? South. J. Appl. For. 26 (1), 26−31.

Campbell, G.E., Colletti, J.P., 1990. An investigation of the rule-of-thumb method of estimating after-tax rates of return. For. Sci. 36 (4), 878−893.

Carmean, W.H., Hahn, J.T., Jacobs, R.D., 1989. Site Index Curves for Forest Tree Species in the Eastern United States. US Department of Agriculture, Forest Service, North Central Forest Experiment Station, St. Paul, MN, General Technical Report NC-128. 142 p.

Cason, J.D., Grebner, D.L., Londo, A.J., Grado, S.C., 2006. Potential for carbon storage and technology transfer in the Southeastern United States. J. Extension. 44 (4), 9 p.

Cochran, P.H., 1979. Gross Yields for Even-Aged Stands of Douglas-fir and White or Grand Fir East of the Cascades in Oregon and Washington. US Department of Agriculture, Forest Service, Pacific Northwest Forest and Range Experiment Station, Portland, OR, Research Paper PNW-263. 17 p.

Cochran, P.H., 1985. Site Index, Height Growth, Normal Yields, and Stocking Levels for Larch in Oregon and Washington. US Department of Agriculture, Forest Service, Pacific Northwest Forest and Range Experiment Station, Portland, OR, Research Note PNW-424, 24 p.

Crawford, H.S., Marchinton, R.L., 1989. A habitat suitability index for white-tailed deer in the Piedmont. South. J. Appl. For. 13 (1), 12−16.

Croan, S.C., 2004. Conversion of conifer wastes into edible and medicinal mushrooms. For. Prod. J. 54 (2), 68−76.

Everest, J.W., Miller, J.H., Ball, D.M., Patterson, M.G., 1991. Kudzu in Alabama. Alabama Cooperative Extension Service, Auburn University, Auburn, AL, Circular ANR-65.

Ezell, A.W., Nelson, L., 2006. Comparison of treatments of controlling kudzu prior to planting tree seedlings. In: Connor, K.F. (Ed.), Proceedings, 13th Biennial Southern Silvicultural Research Conference. US Department of Agriculture, Forest Service, Southern Research Station, Asheville, NC, General Technical Report SRS-92, pp. 148−149.

Fearnside, P.M., 1996. Amazonian deforestation and global warming: carbon stocks in vegetation replacing Brazil's Amazon forest. For. Ecol. Manage. 80 (1−3), 21−34.

Fearnside, P.M., 1997. Greenhouse gases from deforestation in Brazilian Amazonia: net committed emissions. Clim. Change. 35 (3), 321−360.

Finnish Forest Industries Federation, 2016. Roundwood Markets in Finland. Finnish Forest Industries Federation, Helsinki. Available from: http://www.forestindustries.fi/statistics/roundwood-markets-and-forest-resources/40-Roundwood%20Markets%20in%20Finland/ (Accessed 2/22/2016).

Forest2Market, 2016. The Wood & Fiber Supply Chain Experts. Charlotte, NC. Available from: http://www.forest2market.com/ (Accessed 2/22/2016).

Frazier, J.W., Roby, K.B., Boberg, J.A., Kenfield, K., Reiner, J.B., Azuma, D.L., et al., 2005. Stream Condition Inventory Technical Guide. US Department of Agriculture, Forest Service, Pacific Southwest Region - Ecosystem Conservation Staff, Vallejo, CA, 111 p.

Frothingham, E.H., 1921. Classifying forest sites by height growth. J. For. 19 (4), 374−381.

Georgia Forestry Commission, 2007. Georgia Carbon Sequestration Registry: Registry Documents, Downloads, and Supporting Materials. Georgia Forestry Commission, Macon, GA. Available from: http://www.gacarbon.org/downloads.aspx (Accessed 2/22/2016).

Gering, L.R., May, D.M., 1995. The relationship of diameter at breast height and crown diameter for four species groups in Hardin County, Tennessee. South. J. Appl. For. 19 (4), 177–181.

Gholz, H.L., Perry, C.S., Cropper Jr., W.P., Hendry, L.C., 1985. Litterfall, decomposition, and nitrogen and phosphorus dynamics in a chronosequence of slash pine (*Pinus elliottii*) plantations. For. Sci. 31 (2), 463–478.

Gill, S.J., Biging, G.S., Murphy, E.C., 2000. Modeling conifer tree crown radius and estimating canopy cover. For. Ecol. Manage. 126 (3), 405–416.

Greber, B.J., Johnson, K.N., 1991. What's all this debate about overcutting? J. For. 89 (11), 25–30.

Grebner, D.L., Ezell, A.W., Gaddis, D., Bullard, S.H., 2003. Impacts of southern oak seedling survival on investment returns in Mississippi. J. Sustain. For. 17 (1), 1–19.

Grebner, D.L., Amacher, G.S., Prevost, J.D., Grado, S.C., Jones, J.C., 2010. Economics of cogongrass control for non-industrial private landowners in Mississippi. In: Gan, J., Grado, S., Munn, I. (Eds.), Global Change and Forestry: Economic and Policy Impacts and Responses. Nova Science, Hauppauge, NY, pp. 87–98.

Grebner, D.L., Ezell, A.W., Prevost, J.D., Gaddis, D.A., 2011. Kudzu control and impact on monetary returns to non-industrial private forest landowners in Mississippi. J. Sustainable For. 30 (3), 204–223.

Gunter, J.E., Haney, H.L., 1984. Essentials of Forestry Investment Analysis. Oregon State University Book Stores, Corvallis, OR, 337 p.

Haines, T.K., Busby, R.L., Cleaves, D.A., 2001. Prescribed burning in the South: trends, purpose, and barriers. South. J. Appl. For. 25 (4), 149–153.

Hanson, E.J., Azuma, D.L., Hisrote, B.A., 2002. Site Index Equations and Mean Annual Increment Equations for Pacific Northwest Research Station Forest Inventory and Analysis Inventories, 1985–2001. US Department of Agriculture, Forest Service, Pacific Northwest Research Station, Portland, OR, Research Note PNW-RN-533. 24 p.

Harrington, C.A., Curtis, R.O., 1985. Height Growth and Site Index Curves for Red Alder. US Department of Agriculture, Forest Service, Pacific Northwest Research Station, Portland, OR, Research Paper PNW-358. 14 p.

Hill, D.B., 1999. Farming Exotic Mushrooms in the Forest. US Department of Agriculture, National Agroforestry Center, Lincoln, NE, Agroforestry Note 13. 4 p.

Horn, J.E., Medema, E.L., Schuster, E.G., 1986. User's Guide to CHEAPO II-Economic Analysis of Stand—Prognosis Model Outputs. US Department of Agriculture, Forest Service, Intermountain Research Station, Ogden, UT, General Technical Report INT-211.

Hornbeck, J.W., Adams, M.B., Corbett, E.S., Verry, E.S., Lynch, J.A., 1993. Long-term impacts of forest treatments on water yield: a summary for northeastern USA. J. Hydrol. 150 (2–4), 323–344.

Huang, C., Kronrad, G.D., 2001. The cost of sequestering carbon on private forest lands. For. Policy Econ. 2 (2), 133–142.

Hull IV, R.B., Buhyoff, G.J., 1986. The scenic beauty temporal distribution method: an attempt to make scenic beauty assessments compatible with forest planning efforts. For. Sci. 32 (2), 271–286.

Idol, T.W., Pope, P.E., Figler, R.A., Ponder Jr., F., 1999. Characterization of coarse woody debris across a 100 year chronosequence of upland-hickory forests. Proceedings of the 12th Central Hardwood Forest Conference. US Department of Agriculture, Forest Service, Southern Research Station, Asheville, NC, General Technical Report SRS-24, pp. 60–70.

Karr, J.R., 1981. Assessment of biotic integrity using fish communities. Fisheries. 6 (6), 21–27.

Kelly, L.A., Wentworth, T.R., Brownie, C., 2000. Short-term effects of pine straw raking on plant species richness and composition of longleaf pine communities. For. Ecol. Manage. 127 (1), 233–247.

Klemperer, W.D., 1996. Forest Resource Economics and Finance. McGraw-Hill, Inc., New York, NY, 551 p.

Kramer, P.J., Kozlowski, T.T., 1979. Physiology of Woody Plants. Academic Press, Inc, San Diego, CA, 811 p.

Laarman, J.G., 2007. The World Agroforestry Centre's experience in cross-sectoral policy planning in Africa. In: Dubé, Y.C., Schmithüsen, F. (Eds.), Cross-Sectoral Policy Developments in Forestry. CABI, Oxfordshire, pp. 82–88.

Londo, A.J., Oswald, B., Dicus, C., 2005. Living on the Edge. Wildland Fire Management: A Laboratory Manual. Interactive Training Media, Tallahassee, FL, 212 p.

Matthews, G., 1993. The Carbon Content of Trees. Forestry Commission, Edinburgh, UK, Technical Paper No. 4.

McComb, W.C., McGrath, M.T., Spies, T.A., Vesely, D., 2002. Models for mapping potential habitat at landscape scales: an example using northern spotted owls. For. Sci. 48 (2), 203–216.

Mellen, K., Ager, A., 1998. Coarse Woody Debris Model, Version 1.2. US Department of Agriculture, Forest Service, Pacific Northwest Region, Portland, OR.

Mikšys, V., Varnagiryte-Kabasinskiene, I., Stupak, I., Armolaitis, K., Kukkola, M., Wójcik, J., 2007. Above-ground biomass functions for Scots pine in Lithuania. Biomass Bioenergy. 31 (10), 685–692.

Mitich, L.W., 2000. Intriguing world of weeds. Kudzu (*Pueraria lobata*) (Willd.) Ohwi. Weed Technol. 14 (1), 231–234.

Murphy, J.L., Fritschen, L.J., Cramer, O.P., 1970. Research looks at air quality and forest burning. J. For. 68 (9), 530–535.

Nova Scotia Department of Natural Resources, 2000. Ten Year Periodic Annual Increment for Nova Scotia Permanent Forest Inventory Plots 1980-85 to 1990-95. Renewable Resources Branch, Forestry Division, Forest Inventory Section, Halifax, NS, Report FOR 2000-2. 29 p.

Owens, K.E., Reed, D.D., Londo, A.J., Maclean, A.L., Mroz, G.D., 1999. A comparison of Pre-European and current soil carbon storage of a forested landscape located in Upper Michigan. For. Ecol. Manage. 113 (1–2), 179–189.

Pilz, D., Weber, N.S., Carter, M.C., Parks, C.G., Molina, R., 2004. Productivity and diversity of morel mushrooms in healthy, burned, and insect-damaged forests of northeastern Oregon. For. Ecol. Manage. 198 (1–3), 367–386.

Pinchot, G., 1905. A Primer of Forestry. Part II—Practical Forestry. US Department of Agriculture, Bureau of Forestry, Washington, DC, Bulletin No. 24. 88 p.

Pirard, R., Irland, L.C., 2007. Missing links between timber scarcity and industrial overcapacity: lessons from the Indonesian pulp and paper expansion. For. Policy Econ. 9 (8), 1056–1070.

Pitelka, L.F., Raynal, D.J., 1989. Forest decline and acidic deposition. Ecology. 70 (1), 2–10.

Prentice & Carlisle, 2016. Timber Mart North price report. Prentice & Carlisle, Bangor, ME. Available from: http://www.timbermartnorth.com (Accessed 2/22/2016).

Prevost, J.D., 2007. Financial and Biological Impacts of Cogongrass Spread in Slash Pine Forests along the Mississippi Gulf Coast. Masters of Science Thesis. Mississippi State University, Starkville, MS, 126 p.

Ribe, R.G., 1989. The aesthetics of forestry: what has empirical preference research taught us? Environ. Manage. 13 (1), 55–74.

RISI, Inc., 2016. The Leading Information Provider for the Global Forest Products Industry. RISI, Inc., Boston, MA. Available from: http://www.risiinfo.com (Accessed 2/22/2016).

Roach, B.A., 1977. A Stocking Guide for Allegheny Hardwoods and Its Use in Controlling Intermediate Cuttings. US Department of Agriculture, Forest Service, Northeastern Forest Experiment Station, Upper Darby, PA, Research Paper NE-373. 30 p.

Roach, B.A., Gingrich, S.F., 1968. Even-aged Silviculture for Upland Central Hardwoods. US Department of Agriculture, Forest Service, Washington, DC, Agricultural Handbook 355. 39 p.

Roloff, G.J., Kernohan, B.J., 1999. Evaluating reliability of habitat suitability index models. Wildlife Soc. Bull. 27 (4), 973–985.

Schlosser, W.E., 2005. Users Guide to the Timber Productivity Option's Valuation Method–2005. Northwest Management, Inc., Moscow, ID, 24 p.

Schroeder, R.L., 1983. Habitat Suitability Index Models: Downy Woodpecker. US Department of the Interior, Fish and Wildlife Service, Washington, DC, FWS/OBS-82/10.38.

Shurtleff, W., Aoyagi, A., 1977. The Book of Kudzu: A Culinary and Healing Guide. Avery Publishing Group Inc., Wayne, NJ, 104 p.

Sirmon, J.M., 1996. Facing tomorrow: Brazil's difficult forestry choices. J. For. 94 (10), 9–12.

Smith, F.W., Long, J.N., 1987. Elk hiding and thermal cover guidelines in the context of lodgepole pine stand density. Western J. Appl. For. 2 (1), 6–10.

Smith, J.E., Heath, L.S., Skog, K.E., Birdsey, R.A., 2006. Methods for Calculating Forest Ecosystem and Harvested Carbon With Standard Estimates for Forest Types of the United States. US Department of Agriculture, Forest Service, Northeastern Research Station, Newtown Square, PA, General Technical Report NE-343. 216 p.

Smith, W.H., 1990. Air Pollution and Forests: Interaction Between Air Contaminants and Forest Ecosystems. second ed. Springer-Verlag, New York, NY, 618 p.

Sousa, P.J., 1985. Habitat Suitability Index Models: Red-Spotted Newt. US Department of the Interior, Fish and Wildlife Service, Washington, DC, Biological Report 82 (10.111). 18 p.

Stuart, G.W., Edwards, P.J., McLaughlin, K.R., Philips, M.J., 2000. Monitoring the effects of riparian management on water resources. In: Verry, E.S., Hornbeck, J.W., Dolloff, C.A. (Eds.), Riparian Management in Forests of the Continental Eastern United States. Lewis Publishers, Boca Raton, FL, pp. 287–302.

Sun, C., Londo, A.J., 2008. Legal Environment for Forestry Prescribed Burning in Mississippi. Forest and Wildlife Research Center, Mississippi State University, Starkville, MS, Research Bulletin FO351. 22 p.

Svarrer, K., Smith Olsen, C., 2005. The economic value of non-timber forest products—a case study from Malaysia. J. Sustainable For. 20 (1), 17–41.

Tabor, P., 1952. Comments on cogon and torpedo grasses: a challenge to weed workers. Weeds. 1 (4), 374–375.

Thomas, J.W., 1979. Wildlife Habitat in Managed Forests in the Blue Mountains of Oregon and Washington. US Department of Agriculture, Forest Service, Washington, DC, Agricultural Handbook 553.

Timber Mart-South, 2016. Welcome to Timber Mart-South. Avaible from: http://www.timbermart-south.com/ (Accessed 2/22/2016).

US Department of Agriculture, Forest Service, 1981. Guide for Predicting Sediment Yields from Forested Watersheds. US Department of Agriculture, Forest Service, Soil and Water Management, Northern Region, Missoula, MT and Intermountain Region, Ogden, UT, 48 p.

US Department of Agriculture, Forest Service, 1993. SHADOW (Stream Temperature Management Program), Version 2.3. US Department of Agriculture, Forest Service, Pacific Northwest Region, Portland, OR, 20 p.

US Department of Agriculture, Natural Resources Conservation Service, 2004. Soil Survey of Saratoga County, New York. US Department of Agriculture, Natural Resources Conservation Service, Washington, DC, 590 p.

US Department of the Interior, Fish and Wildlife Service, 2003. Recovery Plan for the Red-Cockaded Woodpecker (*Picoides borealis*). US Department of the Interior, Fish and Wildlife Service, Washington, DC. Available from: http://www.fws.gov/rcwrecovery/finalrecoveryplan.pdf (Accessed 2/22/2016).

US Environmental Protection Agency, 2007a. Effects of Acid Rain. US Environmental Protection Agency, Washington, DC. Available from: http://www.epa.gov/acidrain/effects/index.html (Accessed 2/22/2016).

US Environmental Protection Agency, 2007b. National Water Quality Inventory: Report to Congress. US Environmental Protection Agency, Washington, DC, EPA-841-R-07-001.

US Environmental Protection Agency, 2011. What Is Nonpoint Source (NPS) Pollution? Questions and Answers. US Environmental Protection Agency, Washington, DC. Available from: http://www.epa.gov/sites/production/files/documents/whatisnonpointsourcepollution.pdf (Accessed 22/2/2016).

US Environmental Protection Agency, 2012. What Is Acid Rain? US Environmental Protection Agency, Washington, DC. Available from: http://www.epa.gov/acidrain/what/index.html (Accessed 2/22/2016).

Waggener, T.R., 1977. Community stability as a forest management objective. J. For. 75 (11), 710–714.

Wear, D.N., Hyde, W.F., Daniels, S.E., 1989. Even-flow timber harvests and community stability. J. For. 87 (9), 24–28.

Wood Resources International, 2016. Wood Prices, Market Trends, & More! Wood Resources International, Woodinville, WA. Available from: http://woodprices.com/ (Accessed 2/22/2016).

Worthington, N.P., Johnson, F.A., Staebler, G.R., Lloyd, W.J., 1960. Normal Yield Tables for Red Alder. US Department of Agriculture, Forest Service, Pacific Northwest Forest and Range Experiment Station, Portland, OR, Research Paper PNW-36. 3 p.

Chapter 3

Geographic Information and Land Classification in Support of Forest Planning

Objectives

This chapter provides a brief introduction to geographic information system (GIS) concepts and a discussion of how we might classify land to support forest management and planning. Other books (Bolstad, 2005; Wing and Bettinger, 2008; Burrough et al., 2015) provide a more focused and in-depth treatment of the concepts and applications associated with mapping and cartography. Therefore, this chapter represents an overview of several of the pertinent capabilities of GISs as a tool to support the land classification that supplies much of land data used in the forest planning process. In addition, various types of forest land classification systems are described. A land classification system may use vegetation characteristics (and perhaps potential growth rates), topography, and socioeconomic factors, such as ownership groups, to facilitate the delineation of management units for subsequent planning and analysis. The land classes may then impose limits on silvicultural regimes; for example, some landowner groups may not use large clearcuts or herbicides in certain areas. Some land classifications are strata-based, and tied to existing inventory systems and of corresponding low complexity, whereas others are stand-based and thus have higher complexity. At the conclusion of this chapter, students should understand and be able to discuss the following:

1. Spatial data sources and compilation methods.
2. Types of spatial data manipulation processes that are available.
3. The elements of the social and physical environments that could be used in a land classification system.
4. The types of land classifications that are most suitable for different phases of forest management and planning.

which this assignment can take place. These divisions or delineations can be simple: for example, a tract of land bounded on all sides by two roads and a property boundary. These divisions or delineations can also be complex, and organized to accumulate areas of common silvicultural treatments assigned at the substand level. A land classification result has a great influence on the resolution of the planning problem. In strata-based planning, where the solution to the planning problem recommends a percentage of the strata to a given treatment, the planning model is indifferent as to which stands that compose each strata are used. However, the process of grouping stands into a stratum, based on stand characteristics, and the determination of the total area of each stratum represents a key role in how land classification supports the forest planning process. As the planning becomes increasingly refined, a land classification can become more specific and describe each stand or harvest unit for which a silvicultural treatment may be assigned. The resolution of the management unit is therefore important. If our goal is to treat the individual tree, then our land classification system must be able to distinguish each tree on the landscape. Thus, the decision about how to classify land is important for the determination of the decision variables used in the formulation of the planning models (described in subsequent chapters). The land classification is thus used to organize our data for forest planning, and further, to help us understand how it may be useful in displaying our results, and to help us visualize forest plan outcomes.

I. INTRODUCTION

Much of what is accomplished in forest planning is the assignment of treatments to land over time to achieve a set of desirable outcomes. One task that is required is to be able to divide the land into management units for

II. GEOGRAPHIC INFORMATION SYSTEMS

A GIS consists of the analytical and display tools that allow one to capture, organize, manipulate, analyze, interpret, and display spatially referenced and tabular data. The broader definition of GIS involves more than just

Forest Management and Planning. DOI: http://dx.doi.org/10.1016/B978-0-12-809476-1.00003-5

hardware and software, and includes the people who interact with the system as well as the geographic databases. In addition, the parts of the organization that influence how GIS is used are considered in the broader definition; these might include departments or managers that influence the budget or impact personnel decisions. GIS skills are often in high demand and forestry organizations often compete with engineering or planning firms to attract the highest caliber people that have knowledge or experience with programming, database design, photogrammetry, image processing, and cartography.

GISs are prevalent in natural resource organizations today. Even though they were introduced about 50 years ago, they began to be used in a widespread manner in natural resource management only within the last 25 years. Although many of our maps today are created using computer mapping systems, maps may still be hand-drawn in a handful of natural resource organizations. In either case, GIS and mapping are invaluable tools for assisting with daily management of natural resources. The applications of GIS vary widely among natural resource management organizations. Management-related field maps (to support planting, herbicide applications, thinnings, and final harvests) highlight information that will be used in the field or that must be collected in the field to assist in the decision-making process. Another type of map are the thematic maps that represent the state of resources across the broader landscape (such as wildlife habitat suitability). Thematic maps are often used to illustrate the spatial arrangement of feature of interests. They can use color, shading, or texture to communicate very powerful information and can be a very effective tool in communicating issues or results surrounding a forest planning problem.

Data used in GIS can be grouped into two broad sets of structures. One set involves vector data. These are composed of lines or points that are often grouped into polygons. The other set involves raster data, most commonly represented by square pixels, yet they can be hexagonal or triangular in shape. Both are described in further detail later in this chapter.

The type of geographic information that is often needed for forest management and planning efforts includes a description of vegetation, roads, streams, ownership boundaries, home sites, water sources, past fire locations, and other physiographic (e.g., topography), socioeconomic (e.g., towns, mills), or ecological (e.g., wildlife habitat) data within and nearby the property being managed. Spatial features are used to represent these entities, and these generally are represented as points, lines, polygons, or pixels. Associated with each spatial feature may be a long list of attributes that describe qualities of each feature, or quantities contained within (or associated with) each feature. Although these attributes are contained within a relational database inside GIS, they can be exported to file structures that are compatible with spreadsheets, word processors, or harvest scheduling software programs.

Recent graduates of university-level natural resource management programs should have gained some GIS experience through their course work, and many entry-level jobs today require these skills (Brown and Lassoie, 1998; Sample et al., 1999). Merry et al. (2007) suggested that ESRI's software products are the most commonly used GIS software packages in natural resource management; however, other software products are also used (e.g., MapInfo Pro, Google Earth, DeLorme, and commercial software packages from Davey Resources). Learning how to use at least one GIS program effectively will make the process of learning and using another relatively easy. We encourage students to use the data provided with this book to practice those skills acquired in earlier courses in their curriculum. Even though we have experienced rapid advances in computer technology over the last 20 years, one of the primary challenges for natural resource management organizations relates to the design, development, and continual maintenance of GIS databases. Collecting geographic information and preparing the information for end-users is time-consuming and accounts for a large portion of the GIS budget within an organization.

A. Geographic Data Collection Processes

To utilize GIS effectively, we need high quality databases. Today, much of this data is publicly available from existing sources, although some (e.g., timber stands) is still being collected and developed on a project-by-project basis. There are a number of ways we could acquire GIS data through public agencies, many of which support GIS clearinghouses and other Internet sites devoted to the distribution of data. For example, the United States Geologic Service (USGS) makes much of the data available that is used to make their commercial maps. Also, direct contact with the appropriate people in public agencies can provide an access point to data that might not otherwise be publicly available directly over the Internet.

Attempting to acquire data from private companies or consultants may require a payment. Individual privately owned land management companies are very reluctant to share GIS databases because they view their data as proprietary, and because they have spent perhaps hundreds of thousands of dollars to collect, process, and store the data to assist them in executing their business strategy. However, many firms have maintained a library of old aerial photographs, and often these are available for purchase.

In many cases, a letter of understanding that states how, and to what extent, the data will be used may be the minimum required when acquiring data from private companies. Some consultants may develop data for sale as part of their business operation, or they may view the data they develop for the land that they manage as proprietary. Acquiring data directly from others, whether a cost was involved or not, does not assure us that we will receive enough information to effectively plan and manage land resources. In addition to acquiring data from others, GIS data can be developed, updated, or improved using a number of other computer processing methods, including (1) traditional or heads-up map digitizing, (2) field data collection using global positioning system (GPS) technology or other methods, or (3) the use of remotely sensed imagery or aerial photographs.

1. Map Digitizing

Traditional map digitizing has been a method of GIS database development for at least 50 years. In traditional digitizing, maps are affixed to a digitizing table or tablet, within which is embedded a fine mesh of copper wire. The digitizing table's puck, which is similar to a computer mouse, is used to identify the reference points on the maps by sending an electronic signal through the map and into the wire mesh within the table. After the reference points have been recognized, the landscape features that need to be created are digitized by clicking each salient or significant feature (bend in the road, change in direction of a stream, etc.) with the digitizing puck. Attribute information often then is added to the GIS data. Traditional digitizing of maps is a tedious process and, like many other tasks that require human intervention, is subject to error (Wing and Bettinger, 2008) that can be managed through continually reviewing the products. Today, traditional digitizing is still a necessary function for many natural resource management organizations, but reliance on this technique has decreased dramatically as a wider range of digital information has become available, and as heads-up digitizing (using a computer mouse and a digital aerial photograph shown on a computer screen) has become a more popular method for creating databases.

2. Field Data Collection

Field data collection methods can range from using your compass and pacing skills to collect positional data to the very common approach that uses digital instrumentation such as laser range finders and GPS technology to collect this data. The former methods are not as precise as the latter, however seasoned professionals may be able to produce high quality spatial information from field-collected positional data that can be integrated with GIS databases. At some point, data collected on field forms will require some postprocessing to transition it to a digital format. Data collected with GPS equipment should automatically be stored in a digital format, thus this data is more easily incorporated within a GIS database.

GPS receivers estimate positions by calculating the angle and distance to at least four satellites. Satellite signal quality depends on satellite geometry, and error can be introduced through atmospheric interference, the non-synchronization of satellite and GPS receiver clocks, and the bouncing of signals off nearby objects such as trees (resulting in multipath errors). One way to reduce the potential for error is to plan a data collection mission for a time period when the satellite geometry is optimal, another is to collect multiple measurements at each location, and a third is to postprocess (differentially correct) the data that is collected. GPS receivers can range in price from $100 to well over $10,000, with accuracy and precision of measurements generally increasing with price.

3. Remote Sensing

Remote sensing involves the use of a data collection instrument that does not physically touch the landscape or the feature being "sensed." Remote sensing instruments collect reflected electromagnetic energy generated by the sun, or reflected energy that was emitted by a device (such as radar). The most commonly used remote sensing information is captured by cameras that use natural light as the "sensed data." Sensors collect energy in distinct ranges of wavelengths, often called *bands*, thus there is a spectral resolution (number of bands and range of band wavelength width) associated with each sensor. In addition, there is a spatial resolution associated with each sensor that indicates the minimum ground area that is represented, often called a *grid cell*. Finally, a temporal resolution sometimes is associated with sensors that periodically, yet repeatedly, capture images of the same ground areas. LANDSAT or other satellites that orbit the earth would have associated with them a temporal resolution.

One increasingly available data product is Light Detection and Ranging (LiDAR), which is similar to RADAR. LiDAR may be collected with a ground-based device or a device mounted in a helicopter or airplane. LiDAR receivers can collect several pulses of reflected energy that they once emitted, creating point clouds, and allowing one to create three-dimensional views of the landscape and three-dimensional representations of individual trees. This technology is relatively new, and the high cost of data collection and the large amount of data that is collected can be potential disadvantages to its use in natural resource management. As the technology improves, the cost should decrease over time.

4. Aerial Photogrammetry

Aerial photogrammetry is technically a subset of remote sensing that primarily involves visible light waves in the electromagnetic spectrum. There are some excellent uses in the near-infrared applications as well. Aerial photography is perhaps the most widely used method for creating geographic databases in forestry and natural resource management. Interpretation techniques that involve geometry, trigonometry, optics, and familiarity with natural resources can allow us both to identify and estimate the size, length, or height of objects on the ground. Aerial photogrammetry requires the use of vertical aerial photographs (those where the axis of the camera was no more than 3° from vertical), and most often requires the use of stereo pairs (overlapping photos), although reasonable measurements can be made from single vertical aerial photographs if the scale of the photo can be determined. Many of the base maps used by natural resource management organizations were initially made with aerial photographs that were interpreted by natural resource managers. This information can be collected from the photographs using stereo compilers, which allow us to correct a large portion of the inherent error from sources of distortion and displacement. Alternatively, the detail associated with the interpreted photos can be transferred to maps using hand-drawing processes or traditional digitizing processes. Further, information about the camera and terrain can be combined in an analytical model that can compute the coordinates of features using softcopy (personal computer-based) photogrammetry techniques.

One product that is developed from vertical aerial photographs is the georeferenced digital orthophotograph. Digital orthophotographs (Fig. 3.1) commonly are used in GIS as a background image on top of which delineated forested stands, roads, or streams are draped or laid. To create an orthophotograph, vertical aerial photographs first are scanned using very high spatial resolution scanners. Vertical images can also be acquired from digital aerial photo cameras. Much of the topographic displacement and other distortions are removed from the vertical aerial photographs analytically. Finally, the photographs may be combined, and are then georeferenced to allow their correct placement on the landscape. Orthophotographs are hard-copy versions, commonly printed on mylar maps or glass plates. Digital orthophotographs are soft-copy versions that can be used in conjunction with computer software.

B. Geographic Data Structures

The two most basic and widely used GIS data structures are described as raster and vector data. As a natural resource manager, you should understand the differences between them and where they originate. Most natural resource management organizations rely on vector data for their basic, or corporate, databases. Raster data is useful as well for some management applications, yet

FIGURE 3.1 A digital orthophotograph of land in South Carolina.

perhaps is used more frequently in research applications. GIS not only allows us to store and manipulate raster and vector geographic data, but GIS also provides us with the ability to relate one landscape feature to another through the topology that is inherent within the system for both types of data. The following few sections provide an overview of raster and vector data structures as well as topology.

1. Raster Data

Raster GIS databases are arrangements of grid cells or pixels that are referenced by row and column positions; this type of data is sometimes referred to as a *regular data structure*. Any shape can be used that will completely fill an area; triangles, squares, or hexagons can be considered a raster data structure, although the square is the most common type of grid cell in a raster GIS database. The raster GIS databases with which you might already have become familiar include satellite images, digital elevation models (DEMs), digital orthophotographs, and digital raster graphics (DRGs). These were either developed through the use of sensors on satellites or other space vehicles, through the use of digital cameras mounted in airplanes, or through the scanning of maps.

Satellites produce images that contain reflectance values of earth features in each of the raster cells. Typically, they have a spatial resolution of 1, 2, 5, 10, or 30 m, depending on the satellite system under consideration. DEMs provide information about the topography of a landscape, typically in 3, 10 or 30 m grid cells, and many types of terrain analysis can be accommodated with this data. Digital orthophotographs, as we mentioned earlier, are digital versions of vertical aerial photographs, perhaps stitched together to represent a broader area, and registered to a coordinate system. These can be viewed in GIS in a corrected format where most of the topographic displacement and distortion have been removed, thus highly accurate aerial and linear measurements can be made from the photography. One drawback is that viewing the landscape in three dimensions is not as straightforward as with typical vertical aerial photographs, and requires the use of a digital stereo mate. DRGs are scanned versions of topographic maps, and though the resolution is often somewhat coarse, they can be used in GIS to facilitate management and planning efforts. Other types of raster GIS databases can be created in GIS using various spatial analysis and manipulation functions. Vector data, in fact, can be converted to raster data rather easily.

2. Vector Data

Vector GIS databases are compilations of points, lines, or polygons, all of which may vary considerably across a landscape and likely not cover a landscape completely. As a result, this type of data sometimes is referred to as an *irregular data structure*. Almost any landscape feature can be represented by either a point, a line, or a polygon. The vertices and nodes related to these features are represented in geographic space by X (east–west) and Y (north–south), and sometimes Z (elevation) coordinates, and each feature could have associated with it a large number of attributes. Some common vector GIS databases used in natural resource management that include point features are wells, buildings or structures, wildlife nest sites, or fire ignition points. Databases that include line features are often used to represent roads, trails, survey lines, or streams. Examples of databases containing polygon features include property boundaries, timber stands, wildlife habitat areas, logging units, fires, or watershed boundaries. Many of the GIS databases related to the hypothetical forests used in this book are vector GIS databases (Fig. 3.2). One exception is the raster digital orthophotograph associated with each forest. Vector GIS databases typically are created using traditional or

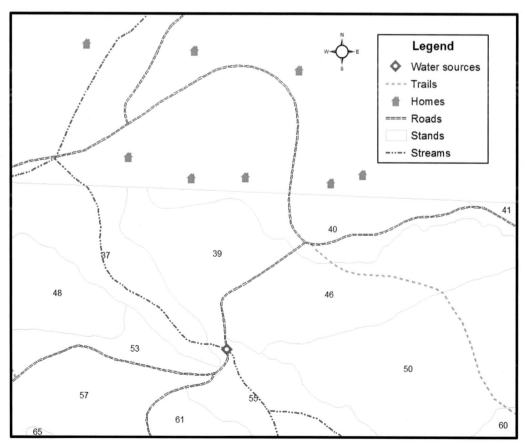

FIGURE 3.2 A map illustrating vector GIS databases for roads, streams, homesites, timber stands, and hiking trails.

heads-up digitizing processes. Some are digitized from hand-drawn maps or interpreted aerial photographs. In the case of heads-up digitizing, the interpretation of digital aerial photographs and digitizing of features may be simultaneously performed. Vector GIS databases can also be created through the multitude of GIS processes available within most GIS software, some of which are described in Section II, Part D. As we mentioned earlier, vector data can be converted to raster data. Raster data can be converted to vector data as well, although the grid cell size to which the vector data becomes related to should be considered carefully—larger grid cells can affect significantly the shape of landscape features once described by lines or polygons.

3. Topology

Topology describes the spatial relationships between (or among) spatial data, and is an important concept for some of the forest planning and management considerations we cover in this book. The topology of GIS databases allows us to understand relationships such as the distance between two features or the neighbors of each feature; the latter is important in eventually being able to control the size of harvest units or habitat patches in a forest plan. Common methods for describing topology include adjacency, connectivity, and containment of features. Adjacency allows us to understand the neighbors of each landscape feature. Connectivity allows us to understand the flow and direction of resources moving through a system of lines (usually) such as roads or streams. Containment allows us to understand which resources can be found within the boundary of other resources, the latter of which must be described by polygons.

C. Geographic Data Used in This Book

Two GIS databases are used throughout this book in various forest planning and management examples, and they are referenced in many of the end-of-chapter applications as well. One of the forests focuses on southern United States pine forest management; the other focuses on western United States conifer forest management. The southern forest is called the Putnam Tract, and contains pine plantations, natural pine stands, mixed forests, and several hardwood-dominated riparian areas. The western forest is called the Lincoln Tract, and contains a large number of conifer stands situated in rugged terrain. The GIS databases associated with each of these two forests include timber stands, roads, streams, an ownership boundary, and a digital orthophotograph. You should keep in mind that the attribute data associated with these GIS databases is hypothetical, and developed specifically for this book. Field visits to each site would undoubtedly result in more accurate and current forest and natural resource conditions. A brief

FIGURE 3.3 A color infrared image of the Putnam Tract and its 81 timber stands.

summary of the two forests, including some overview maps, is provided next.

1. Putnam Tract

The Putnam Tract is a forested area consisting of 81 timber stands covering 2,602 acres (1,053 hectares) in a contiguous block (Fig. 3.3). Pine plantations of various ages cover about 53% of the tract, while natural pine stands comprise about 25% of the forests. Some mixed pine and hardwood forests are present on the tract, but they are very limited in scope (a little less than 6% of the area). Hardwood stands occupy most of the lowlands along the streams, and account for the remaining tract area, or almost 17% of the tract area. Numerous streams are intermixed throughout the tract (Fig. 3.4), all draining into a single main stem running from the southwestern portion of the tract through the northeastern portion of the tract. About 11.8 miles (19.1 km) of intermittent and perennial streams can be found within the tract itself. As a result, there are about 2.9 miles of stream recognized per square mile of land within the tract (or about 1.8 km of stream per square km of land). Given the large stream that runs diagonally through the tract, the road system accesses the forested area from the north, east, and west, and does not connect through the middle of the tract, where a bridge or two would be required. About 10.3 miles (about 16.6 km) of native woods roads have been developed and are being maintained within the tract.

2. Lincoln Tract

The Lincoln Tract (Fig. 3.5) is also a contiguous tract of forest, composed of 87 stands covering 4,550.3 acres

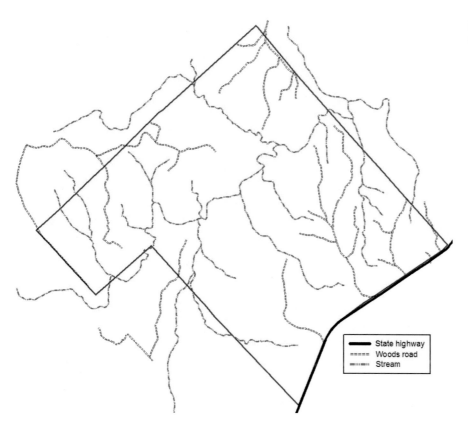

FIGURE 3.4 Roads and streams within and about the Putnam Tract.

State highway
Woods road
Stream

FIGURE 3.5   The Lincoln Tract and its 87 stands.

(1,841.5 hectares) or about 7.1 square miles. Douglas-fir (*Pseudotsuga menziesii*) stands cover most of the area (about 94%), and undoubtedly contain a minor percentage of western hemlock (*Tsuga heterophylla*) and other conifers. Some mixed conifer and hardwood stands (about 6% of the area) also are present on the southern side of the tract and along or near the stream system. A ridge crosses the tract from east to west, thus the stream system drains southward and northward from the center of the tract (Fig. 3.6). About 19.9 miles (32.1 km) of intermittent and perennial streams are contained within the tract itself. Therefore, there are about 2.1 miles of stream recognized per square mile of land within the tract (or about 1.3 km of stream per square km of land). Given the rugged terrain, and therefore the need to keep road grades below about 10% on average, the road system winds through the hills and reflects the need to provide switchbacks at various locations. About 28.1 miles (45.3 km) of rocked woods roads have been developed within the Lincoln Tract.

D. Geographic Information Processes

Depending on the GIS software that is used within a natural resource management organization, you may or may not have an extensive toolkit with which to organize,

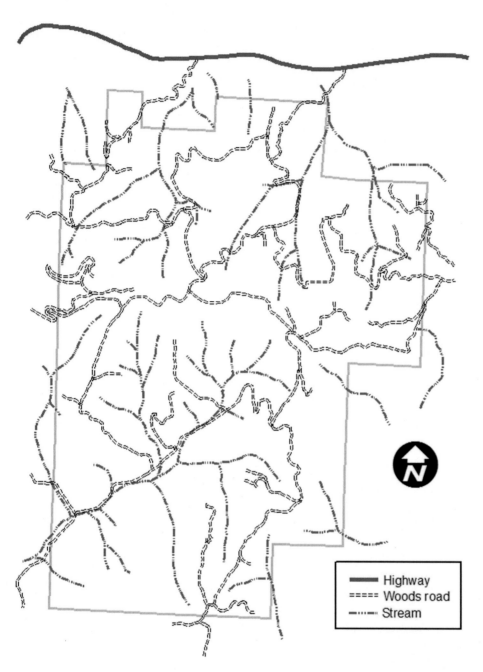

FIGURE 3.6 Roads and streams within and about the Lincoln Tract.

Highway
Woods road
Stream

manipulate, and analyze the data that is used in conjunction with the management of a landbase. The most commonly used GIS processes are editing, querying, selecting features (manually or with clipping and erasing), buffering features (performing a proximity analysis), overlaying or combining coverages, joining databases, and mapping landscape information. These processes can be applied to both vector and raster data. GIS texts provide more in-depth assessments and applications related to these processes, however we provide a brief overview of several of them in the following few sections.

1. Selecting or Querying

Selecting or querying processes help us understand specific facts about a forest, such as how much land area is composed of older forests or regenerating forests. It allows us to pose questions concerning amount or extent of features that meet certain conditions. In addition, these processes are also beneficial in helping managers and planners understand information about the extent or number of various types of features that can be found on a landscape. As a result, you could determine, for example, the types of resources that might be pervasive across an

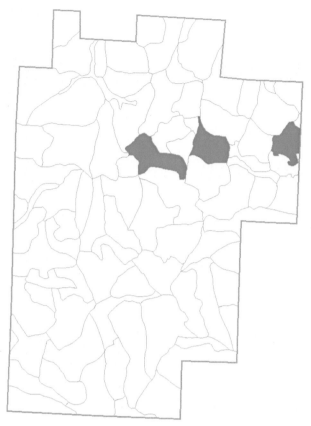

FIGURE 3.7 Potential thinning areas within the Lincoln Tract: Douglas-fir stands that are between 20 and 35 years of age, have more than 275 trees per acre, and more than 150 ft^2 of basal area per acre.

area or in short supply. For instance, you could determine how many different soil types might be found on a given property with a selection or query process, or assess which of the important types of wildlife habitat are in limited supply. Complex queries can be designed in all GIS software to determine the number or extent of resources that have several distinct characteristics. For thinning opportunities, for example, you might be concerned with understanding how many timber stands contain Douglas-fir as the dominant species, are between 20 and 35 years old, have a tree density exceeding 275 trees per acre, along with an associated basal area equal to or exceeding 150 ft^2 per acre (Fig. 3.7). Queries can be based on criteria related to the attributes within a GIS database, or based on the spatial position of landscape features.

2. Clipping and Erasing

As a natural resource manager, you may often find yourself interested in understanding the characteristics of resources contained within specific areas or outside of specific areas. One example of this interest concerns riparian areas and timber production, where a potential change in a riparian policy may extend the riparian

management zone further into the uplands. As a land manager, you may want to understand how much land area and timber volume may be affected (contained within the new riparian zone). Another example concerns homes in the wildland-urban interface. If you were actively managing a forest using prescribed fire, then you may want to schedule activities that are some distance away from homes. As a result, you may want to know how much of the forest is a certain distance away from homesites (outside of a homesite buffer).

The two GIS processes that you could use to understand these types of spatial relationships are the clipping and erasing processes. Clipping involves cutting out a portion of the landscape using a previously developed polygon GIS database. For example, if you acquired a soils database from your state or province, and you were interested only in the soil types within your property, you may want to clip the soils GIS database using a property boundary GIS database. Erasing requires a previously developed polygon GIS database as well, and this process effectively removes features from a GIS database that are found *inside* of polygons contained in the previously developed database. One must be careful not to create a variety of new databases that might confuse future processing efforts, as all data should be maintained to be subsequently reproduced as part of the documentation required in preparing a forest management plan. When using an erasing process, you are interested in understanding what lies *outside* certain landscape features, such as the land area outside riparian zones (Fig. 3.8) or outside wildlife habitat areas. As geoprocessing procedures, erasing is essentially the opposite of clipping.

3. Buffering

Buffering is one type of proximity analysis, and within GIS we use buffering processes to define enclosed areas that are within a specific distance from a point, line, polygon, or set of grid cells. There are a plethora of reasons for why you, as a natural resource manager, would want to draw boundaries around landscape features, from delineating the nesting, roosting, or foraging sites of a certain wildlife species to examining potential impacts of proposed riparian management policies. The types of buffers typically developed for natural resource management purposes utilize either a constant buffer width or a variable buffer width. Constant width buffers assume that no matter how many features are to be buffered, and no matter what characteristics they have, they will all be buffered similarly. For example, a constant stream buffer width of 100 ft assumes that all streams, no matter what their width or flow characteristics, will be used to create a 100 foot buffer. Of course, within GIS you could select a subset of streams using the query functions to buffer, and in this

FIGURE 3.8 Land area outside of riparian areas associated with the Putnam Tract.

case only the selected streams will be used to create a 100 foot buffer. Variable width buffers allow you to capitalize on the characteristics of GIS features, and use one of the numeric attributes of features as a proxy for the buffer distance. As a result, you could create buffers around different streams that vary in size based on the type of stream or stream classification (Fig. 3.9).

4. Proximity Analysis

Proximity analysis involves a set of geographic computations for understanding the nearness or closeness of one set of landscape features to another set. Nearest neighbor analysis specifically facilitates distance measurements between one feature and another, or between one feature and all other features of interest. These distances can be computed using both raster and vector GIS databases. When using a raster GIS database it typically involves measuring the distance between the centers or edges of the raster grid cells. When using vector GIS databases, it could involve measuring the distance between polygon centroids or locations of edges. For example, the distances between the centroids of polygons can be calculated to determine a proximate distance between nesting and foraging areas of spotted owls (*Strix occidentalis*). Other applications of proximity analyses might include

understanding the distance between a potential timber sale and all recently harvested timber sales, or understanding the distance between elk cover habitat and forage habitat areas. It is important that you understand what is being measured in these systems, and to make certain that the measures align with the purpose of the analysis.

5. Combining and Splitting

In a survey of graduates from a natural resource management program, Merry et al. (2007) found that basic editing processes were the most widely used GIS techniques by young professionals just a few years into their first or second job. Combining and splitting processes are basic geographical feature editing processes. In managing a GIS database, you may find that similar landscape features could be combined, resulting in fewer features to maintain. For example, if two small timber stands touch each other and have basically the same characteristics and management history, it might make sense to combine them in the GIS database. However, after combining the polygons only one record would remain, perhaps requiring editing of the resulting attributes, thus some care is needed when choosing this technique for managing spatial data. Other reasons for combining data include the need to eliminate unintended spurious features that were

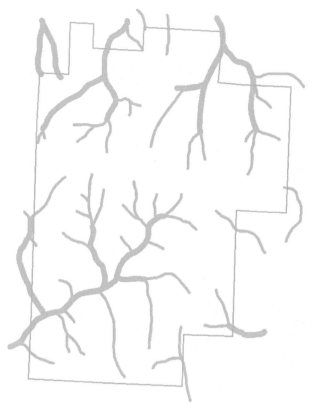

FIGURE 3.9 Variable-width riparian buffer strips on the Lincoln Tract. Perennial streams were buffered 150 ft, intermittent streams were buffered 75 ft.

created through other GIS processes, the need to reduce the number of features managed in a GIS database given some organizationally defined minimum mapping unit, or the need to combine new features with previously mapped features, perhaps as a result of a land trade or purchase. Splitting subdivides features along a line defined by the user. Changes in landscape conditions may warrant splitting features. For example, a stream survey may indicate that a stream reach should be split due to differences in habitat conditions that were not previously evident. An understanding of these editing capabilities is important for efficient management of GIS databases.

6. Joining

There are times when, as a natural resource manager, you may want to quickly, yet temporarily, associate a GIS database with either a tabular (nonspatial) database or another spatial database. This database association can be accommodated with joining processes. In performing a join process, you need a *source database*, a *target or destination database*, and a *join item*. The source database (a tabular file or a GIS database) contains the data that will eventually be associated with the features in the destination database. The join item is the attribute in

common between both the source and destination databases. Nonspatial joins, such as one-to-one join processes and one-to-many join processes assume that a tabular database will be associated with a GIS database. For example, you may have a file containing stand numbers and wildlife habitat suitability values. These may have been calculated in a spreadsheet or other computer program, but now need to be transferred to a GIS to allow you to make a map of the habitat values (Fig. 3.10). The join item in the tabular data is the stand number. Obviously, this data would be associated with a GIS database containing polygons that were also assigned stand numbers.

Spatial joins allow you to determine the characteristics of features (points, lines, or polygons) in a source GIS database that are in closest proximity to other features in a destination GIS database. As a result, you can locate the nearest road to various water sources. Spatial joins also allow you to determine, using a GIS database containing points and another containing polygons, which polygons contain each point. Here, for example, we may not only want to understand when and to what extent wildlife nest trees were last used (which would be available in the wildlife nest tree point database), but also we may want to understand the characteristics of the forest stands within which each nest tree resides. By spatially joining the two databases, the attributes of both the nest trees and the stands within which the nest trees reside will be joined together for analysis and planning. This is an example of the powerful analysis that was neither available nor reasonable for large areas until the development of GIS systems.

7. Overlaying

With overlay processes you physically are placing one map on top of another to create a third, integrated GIS database. Mapped areas as a result of an overlay process are combined, in some form or fashion, and all the attributes of both maps in the overlapping region are merged. For example, overlaying a timber stand GIS database on top of a soils GIS database creates an integrated stands/ soils GIS database, which may help you understand the site preparation or fertilization options available for an area you manage. Three of the most basic overlay processes in GIS are intersect, identity, and union processes. The differences in the three GIS overlay processes are subtle and like most GIS functions require some experimentation to be fully understood. For example, when using the intersect process, only the overlapping, or shared regions are provided in the resulting output. In contrast, when using an identity process the aim is to incorporate some information from an overlapping GIS database into a second, while maintaining the full geographic extent of the second GIS database. These outcomes are both different

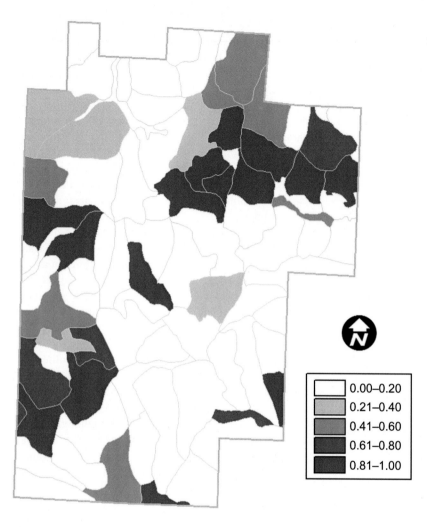

FIGURE 3.10 Habitat suitability index values for the downy woodpecker on the Lincoln Tract.

	0.00–0.20
	0.21–0.40
	0.41–0.60
	0.61–0.80
	0.81–1.00

than the result of a union process, where we would maintain the full geographic extent of both input GIS databases, while also redefining the overlapping features and merging their attributes.

8. Mapping

Cartography is the art and science of making maps. At least through the 1980s, cartography was a skill developed as a result of extensive experience in making maps with tools such as technical pens and *t*-squares, and a steady hand. With the advent of GIS, cartographic skills now mainly are developed through the manipulation of computer graphics. Maps are abstractions of the real world, yet if constructed properly, they have the ability to quickly communicate a message to a user (Fig. 3.11). As a result, a person making a map should keep in mind the following, as suggested in Wing and Bettinger (2008):

- The map message, or story that the map is telling
- The end-user of the map, and what they desire to see
- The way data are displayed on the map
- The format of the printed or digital version of the map

Maps are usually two-dimensional representations of a landscape, and the objective of making one should be to produce a graphic that communicates a message effectively. Most GIS software programs provide the capability to develop sophisticated maps with relative ease. With the advent of three-dimensional printing this may change in the near future. The components necessary for a professional map include a north arrow, the appropriate scale, and an informative legend. Annotation such as a description of the mapmaker and the date that the map was prepared are also important map elements.

III. LAND CLASSIFICATION

Land can be classified using a number of physical or socio-economic characteristics, including vegetation, soils, wildlife habitat, landform (physiography), potential forest productivity, recreation opportunities, viewsheds, wildland-urban interfaces, and forest value. While any map or table that describes areas of land can be considered a classification, formal land classifications serve as the basis for

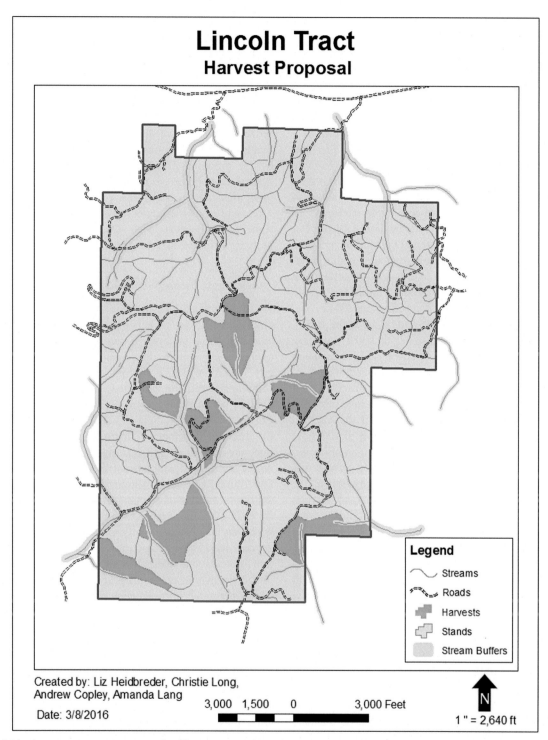

FIGURE 3.11 A management map constructed to illustrate proposed harvest areas on the Lincoln Tract for a 5-year time period.

assessing land resources and serve as a framework for scheduling and evaluating management activities (Frayer et al., 1978). These systems are the first step in allowing managers to predict outcomes of their selected treatments. To facilitate forest management and planning, a landscape is divided into management units that contain relatively homogeneous vegetation or physiographic features as the goal is for these areas is to respond similarly to each treatment applied. As you will learn later, a land classification is often related to the decision variables used in the forest planning problem. These management units, however defined, are what planners use to assign management activities over space and time. These delineations could be as simple as an aggregation of similar forested areas, or as detailed as those that include differences in age class, site class, growth pattern, management history, size of the area, and spatial location. A simple example for the western United States might be to delineate Douglas-fir from red alder (*Alnus rubra*) stands. In the southeastern United States, this might include drawing a distinction between a natural pine stand and a planted pine stand of a similar age. In the northern United States and parts of Canada, it might involve delineating mixed hardwood forests based on their dominating species composition. Land classes could also involve combining ecological aspects of the landscape, as some of the more recent US National Forest plans. For example, the 2004 Chippewa National Forest Plan provided objectives for the management of vegetation within landscape ecosystems, or classes of land based on native plant communities, ecological systems, and terrestrial inventories (US Department of Agriculture, Forest Service, 2004a).

Land classification systems should be based on professionally credible concepts (Frayer et al., 1978), and are necessary for providing policy direction and for assisting with policy implementation. In general, land managers need an organized system to understand the capability of the land to produce perhaps multiple goods and services, and thus provide a context for a plan, and this method should lead to an organized data collection plan. These classes are either developed internally within an organization for their use and guidance, or prompted by external forces, such as voluntary certification programs, the subject of Chapter 15, Forest Certification and Carbon Sequestration. For example, the Forest Stewardship Council requires maps of forest characteristics displaying general management zones, special management areas, and protected areas and should include forest types by age class for the principle associated with management plans (Forest Stewardship Council US, 2010). Similarly, the Sustainable Forestry Initiative requires an explicit land classification system as part of a long-term resource analysis related to one of the program objectives (forest management planning) and associated performance measures (Sustainable Forestry Initiative, Inc., 2015).

Classification of land is one of the first steps in developing forest management plans as it determines the decision variables and determines the resolution of the data needed for the planning problem. We must understand what each piece of land is capable of producing (timber, wildlife habitat, or otherwise) before we can develop alternatives for the land. We described earlier in Chapter 2, Valuing and Characterizing Forest Conditions, a method for estimating site quality (site index), which provides a good, positive correlation to the productive capability of sites. However, site index references the height growth of the dominant species that are present in

the stand. Site index tells us nothing about the current state of a stand (How old is it? What management has been applied thus far?). In addition, there may be many site indices present in a forest, perhaps a continuous range of values or values separated by 5 or 10 index point intervals for each reference species. As a result, although site index is a valuable characteristic of a stand of trees, it is probably one level of detail too deep for a general classification of land. Over 90 years ago, Chapman et al. (1923) proposed a simple timber-oriented classification system that involved using the dominant tree species groupings, age classes, and prior management history. This type of classification system is still used today in some form, because what Chapman et al. (1923) suggested still holds true, that the number of classes:

will vary with the individual manager, according to the aims and intensity of management and the diversity of conditions.

After a land classification has been developed, the objectives that can be accommodated and the alternatives available to a landowner can be assessed, and subsequently a plan can be developed. After classifying a property, you may decide to limit management activities in some areas, and simultaneously consider a more comprehensive set of management activities in other areas. For example, the Washington State Parks and Recreation Commission developed a land classification system that integrates physical land features with potential socioeconomic and recreational uses of the land (Washington State Parks and Recreation Commission, 2007). In addition, a discussion of allowable management activities is provided for each class. For instance, in natural forest areas, hiking trails may be developed only to the extent that they do not degrade the system of natural forest processes, and relocation of trails into natural areas may be permitted if the impact on overall resources is reduced. Scientific research also is permitted in these areas, as are some forms of nontimber forest product harvesting (mushrooms, berries, or greenery), as long as these activities do not result in the degradation of natural forest processes.

As we mentioned earlier, a map that identifies categories of land using various thematic symbols or colors is a land classification (Fig. 3.12), as is a table that presents the amount of land in each category (Table 3.1). Maps are powerful tools, if developed appropriately, for conveying information in a graphical manner and for communicating a message to a large number of people. Care must be taken to provide the relevant amount of information (spatial context, symbology, annotation) without distracting the map user from the overall message. Some of the most common mistakes involve the coloring of the classes, and the inadvertent highlighting of those that are relatively unimportant. Other mistakes include carelessly using text of various sizes and fonts. Cartography texts provide good

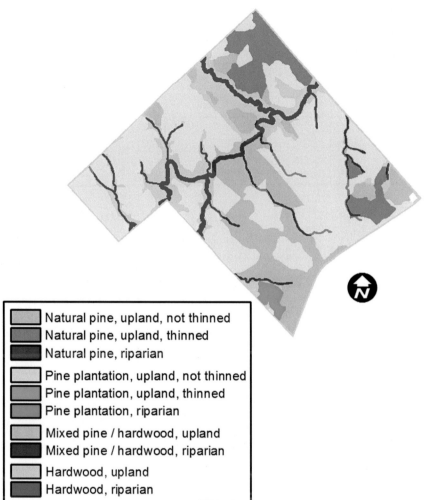

FIGURE 3.12 A land classification for the Putnam Tract.

☐ Natural pine, upland, not thinned
☐ Natural pine, upland, thinned
☐ Natural pine, riparian

☐ Pine plantation, upland, not thinned
☐ Pine plantation, upland, thinned
☐ Pine plantation, riparian

☐ Mixed pine / hardwood, upland
☐ Mixed pine / hardwood, riparian

☐ Hardwood, upland
☐ Hardwood, riparian

TABLE 3.1 A Simple Land Classification for the Putnam Tract

Land Classification	Subclass	Area (acres)
Natural pine	Upland, not thinned	401.5
	Upland, thinned	225.5
	Riparian	20.2
Planted pine	Upland, not thinned	1,321.8
	Upland, thinned	31.6
	Riparian	16.4
Mixed pine/ hardwood	Upland	121.4
	Riparian	25.6
Hardwood	Upland	308.6
	Riparian	129.9

guides for using annotation in maps, information that will add and not distract from the overall message of the map. These mistakes underscore the inattentiveness of the map maker, and detract from communicating practical management-related information. When used in conjunction with maps, tables describing land classification information can provide a wealth of information to your audience, and provide you with information that is useful in subsequent analyses.

For forest planning and management processes, there are at least three organizational methods for classifying land: (1) by strata, (2) by unit of land (stand), and (3) by unit of land and spatial position on the landscape. Each of these has it benefits and challenges (Table 3.2). For example, strata-based land classifications are simplistic, and provide rough descriptions of a property in broad classes that are often supported by existing large-scale inventory data. Assigning management activities to these classes leaves the physical implementation decisions to managers in the field, since the spatial location of land, for the most part, is ignored. Stand-based approaches require

TABLE 3.2 Level of Detail and Problem Complexity for Three Different Approaches to Land Classification

Approach	Problem Complexity	Detail Used in Recognizing the Unique Characteristics of Polygons	Detail Used in Recognizing the Unique Spatial Context of Polygons
Strata-based	Low	Low	Low
Stand-based, no spatial information	Medium	High	Low
Stand-based, spatial information	High	High	High

management of a large amount of data, but spatial location of activities is recognized in both approaches, and used in the latter to control the placement of activities. Therefore, decisions related to the location of activities are less of an issue for forest managers. Each of these three approaches are described in more detail next.

A. Strata-based Land Classifications

Strata-based land classifications group land areas with similar attributes into strata, bins, or analysis units, for management and planning purposes. For example, a 500,000 acre forest in Indiana may be described by maps and GIS databases that recognize over 10,000 forest stands (polygons). A strata-based land classification would reduce the number of recognizable analysis units to 10 or 20 by aggregating the land area of stands with very similar attributes. Land area might be aggregated by age, management history, soil type, watershed, distance to streams, slope percent, timber productivity, or another aspect of management that might be of concern to the managers of the land. These groupings are then used to help reduce the decision space, or the types of management actions considered, since the groupings themselves would inherently preclude or suggest the type of management that would be appropriate within their confines. In this type of classification, the level of spatial resolution is low because when planning occurs, we do not explicitly know where each stand is located on the landscape. Strategic forest planning (described in detail in Chapter 13: Hierarchical System for Planning and Scheduling Management Activities), however, usually is facilitated with this type of information.

Example

In 1971 the Arapaho National Forest in central Colorado developed a forest plan that was based on a land classification that focused on timber productivity (US Department of Agriculture, Forest Service, 1971). The National Forest is located in an extremely rugged area west of Denver, and most of the timber production areas are located at elevations

between 7,000 and 11,000 ft. The tree growing season is short, and the average productivity is relatively low, at about 40 ft^3 per acre per year. In addition to the timber management focus, even in 1971 there was a heavy demand for recreational opportunities on the National Forest. Current land use, land stability, and steepness of slopes were used to develop a basic land classification within which the plan of activities would operate (Table 3.3).

The Arapaho National Forest plan was developed prior to the introduction of GIS into natural resource management organizations. In the late 1960s and early 1970s, forest management organizations used hand-drawn maps and manually overlaid them on light tables to create a land classification systems. Land areas were calculated using planimeters and dot grids, two basic area computation methods that are still used today. Given the large area involved and the multitude of issues facing National Forests, it is not too difficult to understand why land classifications were relatively brief and straightforward. Today, spatial considerations such as riparian buffer zones, wildlife habitat buffers, and scenic corridors can be incorporated into land classifications much more easily given the evolution of GIS and advances in computer technology. If the criteria for land classification can be verbalized and quantified, then it most likely can be spatially represented.

Example

In 2004, the Chattahoochee-Oconee National Forest in Georgia developed a forest plan that reflected a public desire for more wilderness and recreational areas, and less timber harvesting and road building. Although the National Forest once had a timber-oriented objective, the more recently expanded recreational objectives were not unexpected, given the National Forest's proximity to the growing Atlanta metropolitan area. Achieving a balance in resource management is therefore challenging, and the land classification for the National Forest reflects the numerous demands on the management of the forest (Table 3.4).

Although it is not exactly clear how much GIS analysis was used to arrive at the estimates of land area in

TABLE 3.3 Land Classification for the Arapaho National Forest

Land Classification	Subclass	Area (acres)
Commercial timber production		336,676
Marginal timber production	Usable, stable land	17,241
	Usable, unstable, steep land	45,654
Noncommercial timber production	Unproductive	112,603
	Unsuitable and unstable	2,193
	Isolated patches of forests	9,908
	Crest Zone	21,889
	Travel-water influence	53,324
	Other conflicts	52,350
Deferred		64,885
Nonforest land		231,114
Total		**947,837**

Source: US Department of Agriculture, Forest Service, 1971. Land and Resource Management Plan, Arapaho National Forest. US Department of Agriculture, Forest Service, Rocky Mountain Region, FT; Collins, CO.

TABLE 3.4 Land Allocations Used in the Chattahoochee-Oconee National Forest Plan

Land Classification	Chattahoochee (acres)	Oconee (acres)	Total (acres)
Wilderness areas	125,530	0	125,530
Wild, scenic, and recreational river areas[a]	11,084	4,854	15,938
Water-related protection areas	27,179	0	27,179
National scenic areas	7,122	0	7,122
Scenic trail corridors	16,655	0	16,655
Other scenic areas	70,369	0	70,369
National recreation areas	25,689	0	25,689
Other recreation areas	124,993	9,368	134,361
Old-growth forests	28,657	1,617	30,274
Mid- to late-successional forests	23,693	0	23,693
Mixed successional forests	68,323	0	68,323
High elevation early successional forest	6,604	0	6,604
Management and maintenance of plant associations	172,718	35,006	207,724
Red-cockaded woodpecker habitat	0	47,108	47,108
Rare communities	505	593	1,098
Other natural areas	39,365	7,061	46,426
Experimental forests	0	9,364	9,364
Administrative areas	163	102	265
Miscellaneous	2,121	142	2,263
Total	**750,770**	**115,215**	**865,985**

[a]Designated and proposed.
Source: US Department of Agriculture, Forest Service, 2004. Land and Resource Management Plan, Chattahoochee-Oconee National Forests. US Department of Agriculture, Forest Service, Southern Region, Atlanta, GA, Management Bulletin R8-MB 113A.

each land classification associated with the recent Chattahoochee-Oconee National Forest plan, many of the classes indicate the need for processes such as buffering (e.g., scenic corridors). The forest plan acknowledged the heavy use of GIS in developing the land classes, yet suggested that some of the finer details, such as the delineation of riparian management areas, would need to be recognized at the project-level implementation of activities that would require maps produced from GIS analyses.

Fig. 3.12 and Table 3.1 illustrate a simple strata-based land classification for the Putnam Tract. The table provides no indication of where the areas are located, and if used in an analysis, the areas would be scheduled for management without the use of the spatial information or relationships suggested by the map. The classification could be further refined by incorporating, where appropriate:

- Visual quality corridors that might influence the type of management actions available near the county road or the homestead in the northeast corner of the tract
- Different silvicultural systems, such as the shelterwood (seed tree) harvest on the south side of the tract
- Soils and water information that might be influential in the selection of harvesting or site preparation systems
- Special wildlife habitat designations, perhaps for some of the older pine forests in the western half of the tract

B. Land Classification Based on Units of Land

The second type of land classification is based on recognition of each stand or management unit by its area and other physical, economic, or ecological characteristics. In this case, the level of spatial resolution is higher because when planning occurs, we know where each stand is located on the landscape. Further, we may know in which compartment or watershed each stand resides, yet not much more about the spatial relationship of each stand in relation to other landscape features (which stands are adjacent, how far a stand is from a stream, and so on). Here we could summarize the amount of land in aggregated classes, as with the previous example, but the actual analysis and planning is performed at the stand level, not the stratum level.

Table 3.5 illustrates the level of data that would be used for planning and analysis of the Putnam Tract. Here the size of each stand, its forest type, and recent management actions would be used to influence the type of activities that would be scheduled in the near future. Recent management activities may preclude scheduling other types of activities in the near term. For example, if a stand recently was thinned, it should avoid being scheduled for another thinning in the near term. Recent management activities also are important in projecting stand structure into the future. Young pine plantations that have been thinned will likely have a different growth and yield trajectory than young pine plantations of the same age that have not been thinned. This type of land classification should be contrasted with the type of data provided in Table 3.1 to better understand the difference between it and the previously described strata-based system. A cross between strategic and tactical forest planning usually is facilitated with this level of information, where strategic goals can be met while addressing some tactical stand-based planning issues.

C. Land Classification Based on Spatial Position

The third method of land classification recognizes stands or management units, as in the previous case, but also incorporates a higher level of spatial information in the process. Typical spatial information for recent forest plans

TABLE 3.5 Land Classification at the Stand Level Using Only Stand-Centric Attributes

Stand	Area (acres)	Age (years)	Forest Type	Recent Management History
1	74.0	2	Pine plantation	Seed tree harvest
2	84.5	45	Natural pine	—
3	11.9	5	Hardwood	—
4	31.6	21	Pine plantation	Thinned
5	1.2	31	Hardwood	—
....				
79	35.2	40	Natural pine	—
80	0.9	30	Hardwood	—
81	85.6	45	Natural pine	—

developed by timber companies, states, and other agencies takes the form of adjacency relationships (which stands are touching, or near, other stands). These relationships are used to control the timing and placement of clearcut harvests as well as to control the size of forest interior habitat for various species of wildlife. Although analysis and planning are performed at the stand level, as in the previous example, we also incorporate the spatial relationships between each stand and other features, and this information is used to guide the selection of activities and influence the subsequent analyses. As with the previous example of a land classification, a cross between strategic and tactical forest planning usually is facilitated with this level of information, where strategic goals can be met while addressing a wider range of tactical planning issues. However, this level of information is more commonly used in tactical or operational planning processes.

Example

Lands managed by the State of Oregon are classified prior to implementation of activities (Oregon Secretary of State, 2016a). The Oregon State forest land classification system has several categories: special use areas, high value conservation areas, focused stewardship areas, and general stewardship areas. The system is hierarchical, indicating that the special stewardship areas are the most important, and therefore should be identified first. The special stewardship areas include administrative areas, wilderness areas, rock pits, ponds, lakes, and viewsheds, among others. These areas cannot be reassigned to the other classes, and the activities allowed here include those that protect, maintain, or enhance specific resources. The high value conservation areas are of similar value and can be managed, but activities that lead to long-term adverse impacts to the specified conservation value are avoided. The focused stewardship areas

are next, and once identified, cannot be reassigned to the general stewardship areas. Focused stewardship areas include riparian areas, visual corridors around trails, and buffers related to wildlife habitat. A reduced set of management activities can be used in this land class. The general stewardship areas are basically what remain after identifying the higher levels of land classes, and the full range of management activities that meet or exceed requirements of laws and regulations can be used here.

In developing land management plans for the State of Oregon lands, the riparian management areas are explicitly delineated in GIS and activities are scheduled using this and other spatial information. Although the amount of land in each class may be aggregated and summarized in the plans themselves, the activities are scheduled at a very basic level while developing the plans, which makes this form of planning different from strata-based plans, where activities are scheduled based on the amount of land within a stratum. Adjacency relationships would then be used in developing a plan to ensure that clearcuts do not exceed 120 acres (about 49 hectares), the maximum size prescribed in state law (Oregon Secretary of State, 2016b).

For the Putnam Tract, we might recognize the land area and forest type of each stand, along with the recent management actions that have been applied, in addition to the spatial information (Table 3.6). The forest type and recent management actions will influence the types of activities that can be applied in the near future, and will influence the transition (growth and yield) from the current state to the future states. The adjacency information could be used to control the size of clearcut harvests when developing a forest plan for this area. As with the previous example of a land classification, these stands could also be subdivided to explicitly recognize the riparian areas. Two items to be mindful of while

TABLE 3.6 Land Classification at the Stand Level Using Stand-Centric Attributes and Spatial Information (Adjacent Neighbors)

Stand	Area (acres)	Age (years)	Forest Type	Recent Management History	Adjacent Neighbors
1	74.0	2	Pine plantation	Seed tree harvest	2, 22, 34, 81
2	84.5	45	Natural pine	–	1, 3, 4, 6, 7, 8, 12, 81
3	11.9	5	Hardwood	–	1, 2, 4
4	31.6	21	Pine plantation	Thinned	2, 3, 5, 6, 7, 75
5	1.2	31	Hardwood	–	4, 75
....					
79	35.2	40	Natural pine	–	15, 18, 76, 77, 78
80	0.9	30	Hardwood	–	30, 74
81	85.6	45	Natural pine	–	1, 2, 8, 11, 12, 20, 22, 34, 77, 78

explicitly recognizing the spatial location of riparian areas are (1) that the resulting number of management units (now subdivisions of stands) and their adjacency relationships may increase dramatically, and (2) that some of the resulting management units may be very small in size, many of which might be considered spurious polygons.

IV. SUMMARY

GISs are powerful tools for the storage, manipulation, and display of spatial data. Maps are very important aspects of forest management, and the ability to develop a clear and cohesive map is a reflection of your cartographic skill. Whether you develop your own maps, manage people who have responsibility for making maps, or coordinate map and database development projects with contractors, an understanding of the capabilities of GIS is important in today's management environment. Although vector-based GIS databases are very commonly used in real-world applications of GIS, raster databases are of value as well. Digital orthophotographs are perhaps the most extensively used raster GIS databases among natural resource managers, and are becoming suitable proxies for traditional vertical aerial photographs. Many land management organizations use aerial photographs, digital orthophotographs, and other GIS databases to create land classifications. A land classification can be used to develop guidelines for appropriate actions for the land areas within each class. These systems are the first step in developing suitable prescriptions and their outputs. In many instances, the classification of land is used directly in forest-level management planning processes. The display of land classifications through a well-developed thematic map will undoubtedly facilitate a discussion among the land managers and landowners regarding the appropriate management of the natural resources under their control.

QUESTIONS

1. *Strata-based land classification.* Develop a strata-based land classification for the Lincoln Tract. Use 20-year age classes of stands along with the forest vegetation type to stratify the land areas. Develop a thematic map and a table to represent the land classification. In a memorandum to the forest managers, describe the distribution of land classes within the Lincoln Tract.
2. *Stand-based land classification.* Develop a table describing the stands within the Putnam Tract. Sort the table by stand age and present the results in a memorandum to the forest manager. Take care to present the information in a professional manner. To help further understand the condition of the forest,

summarize the age class distribution of the planted and natural pine stands.
3. *Stand and spatial land classification.* Examine the Putnam Tract stands GIS database. Some stands in this database have very few adjacent neighbors, as defined by edges that touch, whereas others have an extensive list of adjacent neighbors. Describe the range of adjacent neighbors in this database. What are the characteristics of the stands that have many adjacent neighbors?
4. *Your school forest.* Arrange a meeting with your school forest manager, perhaps by inviting him or her to your class. Ask questions regarding the type of GIS databases that are used to represent the current condition of the forest's resources. List the type of data structures involved, and whether they involve points, lines, or polygons, or are raster by design. In addition, ask questions about the type of planning that he or she performs to facilitate the management of the forest. Does the forest manager use one of the land classification systems described in this Chapter? If so (or otherwise), how is a land classification used to develop the management plan? What land classifications are recognized, and how are management activities associated with each?

REFERENCES

Bolstad, P.V., 2005. GIS Fundamentals: A First Text in Geographic Information Systems. 2nd ed. Eider Press, White Bear Lake, MN, 543 p.

Brown, T.L., Lassoie, J.P., 1998. Entry-level competency and skill requirements for foresters. J. For. 96 (2), 8−14.

Burrough, P.A., McDonnell, R.A., Lloyd, C.D., 2015. Principles of Geographic Information Systems. Oxford Press, Oxford, UK, 432 p.

Chapman, H.H., Fisher, R.T., Howe, C.D., Bruce, D., Munns, E.N., Sparhawk, W.N., 1923. Classification of forest sites. J. For. 21 (2), 139−147.

Forest Stewardship Council US, 2010. FSC-US Forest Management Standard (v1.0). Forest Stewardship Council US, Minneapolis, MN.

Frayer, W.E., Davis, L.S., Risser, P.G., 1978. Uses of land classification. J. For. 76 (10), 647−649.

Merry, K.L., Bettinger, P., Clutter, M., Hepinstall, J., Nibbelink, N.P., 2007. An assessment of geographic information system (GIS) skills used by field-level natural resource managers. J. For. 105 (7), 364−370.

Oregon Secretary of State, 2016a. Department of Forestry, Division 35, Management of State Forest Lands. Oregon Administrative Rules 629-035-0055. Oregon Secretary of State, Salem, OR. Available from: http://arcweb.sos.state.or.us/pages/rules/oars_600/oar_629/629_035.html (Accessed 2/16/2016).

Oregon Secretary of State, 2016b. Department of Forestry, Division 605, Planning Forest Operations. Oregon Administrative Rules 629-605-0175. Oregon Secretary of State, Salem, OR. Available from: http://arcweb.sos.state.or.us/pages/rules/oars_600/oar_629/629_605.html (Accessed 2/16/2016).

Sample, V.A., Ringgold, P.C., Block, N.E., Giltmier, J.W., 1999. Forestry education: adapting to the changing demands. J. For. 97 (9), 4−10.

Sustainable Forestry Initiative, Inc., 2015. SFI 2015−2019 Standards and Rules. Sustainable Forestry Initiative, Inc., Washington, DC.

US Department of Agriculture, Forest Service, 1971. Land and Resource Management Plan, Arapaho National Forest. US Department of Agriculture, Forest Service, Rocky Mountain Region, FT; Collins, CO.

US Department of Agriculture, Forest Service, 2004a. Land and Resource Management Plan, Chippewa National Forest. US Department of Agriculture, Forest Service, Eastern Region, Milwaukee, WI.

US Department of Agriculture, Forest Service, 2004b. Land and Resource Management Plan, Chattahoochee-Oconee National Forests. US Department of Agriculture, Forest Service, Southern Region, Atlanta, GA, Management Bulletin R8-MB 113A.

Washington State Parks and Recreation Commission, 2007. Washington State Parks Land Classification System. Washington State Parks and Recreation Commission, Olympia, WA. Available from: http://parks. state.wa.us/DocumentCenter/Home/View/1525 (Accessed 2/16/2016).

Wing, M.G., Bettinger, P., 2008. Geographic Information Systems: Applications in Natural Resources Management. Oxford University Press, Don Mills, ON, 268 p.

Chapter 4

Estimation and Projection of Stand and Forest Conditions

Objectives

There are many phases of forest management and planning where the ability to understand the current and future structural conditions of natural resources that we manage is necessary. Key pieces of the information that are needed to determine what to do in the future include an assessment of what we currently have to manage, and an assessment of what we will likely have in the future to manage, given the management activities that are planned. From a forest management perspective, the list of planning activities that require future assessments include commercial thinning treatments, pruning options, fertilization possibilities, spacing of planted trees, and many others. The future structural conditions of forests are important for evaluating the impact of the new forestry regimes on yields and habitat values, for determining final rotation ages for existing stands, and for evaluating the susceptibility to wind damage for thinned stands, to name just a few. And as we learned in Chapter 2, Valuing and Characterizing Forest Conditions, many wildlife, ecological, and social assessments rely on current and projected forest structural conditions. Therefore, it is important for students and seasoned practitioners to understand that the assessment of these values requires understanding how forest conditions change through time. The objectives of this chapter center on our understanding of how the progression of forests over time is performed and presented by analysts, and subsequently perceived by forest managers and landowners. At the conclusion of this chapter, students should understand and be able to discuss the following:

1. The growth of forests.
2. Conceptual models of how stands grow through time.
3. Broad-scale forest transitions.
4. Volume and yield tables.
5. Types of growth and yield models.
6. Output expected from a growth and yield simulation.

I. INTRODUCTION

Left free to grow and affected solely by the forces of nature, forests change, and understanding the change that can occur is critical for forest planning efforts. The fact that forests change makes our job as forest managers dynamic and interesting, but also challenging. Management activities can also alter the character of forests. How the growth of forests evolves after a management activity is applied is central to decisions regarding the implementation of activities, as growth rates may be enhanced or harmed by human intervention. In the following sections, we describe in general how forests change, and how one can visualize and estimate potential changes and future conditions.

II. THE GROWTH OF FORESTS

The projection of the future condition of trees in a forest is generally a function of the current size and condition of the trees, and the fertility of the site. Characteristics that can improve the quality of these projections include the site index of the stand in which the trees reside (the site fertility), along with the height, crown ratio, age, and current diameter of the trees. Unfortunately, most trees are not situated in free-to-grow conditions, one where there is no competitive pressure from neighbors, and as a result there is a need to model the complex interactions between trees and their environment and how they compete for limited resources. This inter-tree competition can be acknowledged through measures of stand density (e.g., trees per unit area, basal area, and various stand density indices such as Reineke's stand density index), or through a detailed stem mapping of the location of all trees. For individual trees, the number of trees per unit area and the basal area have an effect on tree growth (Buckman et al., 2006). Ultimately, the growth rate of individual trees depends on the size and spatial location of the neighboring trees. Acknowledging neighboring trees in a growth and yield simulation may be accomplished by understanding the spatial proximity of each tree, or may involve measures that are weighted by tree size, height, and distance. These spatial dependencies may be used to influence *ingrowth*, *increment*, and *mortality* processes within a stand.

Forest Management and Planning. DOI: http://dx.doi.org/10.1016/B978-0-12-809476-1.00004-7

Ingrowth, when used to describe forest growth dynamics, usually refers to the number or volume of new, previously unreported trees that have grown into the smallest measurable diameter class. If measurements are made of the same area on two distinct visits (during the current and previous measurement periods), then these are the new trees that are measured during the current visit that were not measured during the previous visit. This can occur because the trees were either too small to measure during the previous visit or did not yet exist. Increment, or *accretion*, is the growth of trees over a measurement cycle. Accretion assumes that a measured tree was alive and measured at both ends of the cycle. *Harvest* (or cut) refers to those trees that are absent during the current measurement period, yet were present during the previous measurement period, thus were likely harvested and transported out of the forest sometime between the two measurements.

The *mortality* of trees refers to those trees that have died at some point in time between the two measurement periods. Mortality is presented as the number or volume of measurable trees that were alive during the previous measurement period, yet were dead by the time of the current measurement period. Trees reported as dead may continue to be found on the site during the current measurement period. In younger stands of trees, the amount (or volume) of mortality should represent only a small percentage of the overall growth of a stand, even though the number of trees per unit area may decline at a greater rate than when a stand is older. However, when stands are older, competition-induced mortality occurs. Here, the competitive forces disadvantage a tree until it can no longer survive, resulting in a higher volume of mortality as a percentage of the overall growth of a stand, even though the number of trees per unit area expiring may be small. As it relates to the growth of a forest, when we discuss mortality we usually are referring to competition-induced mortality or other types of natural mortality processes. Catastrophic losses from fire, ice storms, hurricanes, or insect and disease epidemics are some examples, yet usually are not considered directly by most growth and yield models.

The competition, and resulting mortality, within even-aged stands of trees can be envisioned using the −3/2 power rule of self-thinning. The −3/2 power rule defines the upper stand density in relation to average tree size, beyond which stands are incapable of growing (Buckman et al., 2006). The location of trees, their branches, their size and form, and the depth of their roots all influence stand density. Stand density is the only stand-related variable that has a general upper limit for a given tree species that is independent of other factors (Zeide, 2005). A graphical representation of the −3/2 power rule illustrates the relationship between average tree size and stand

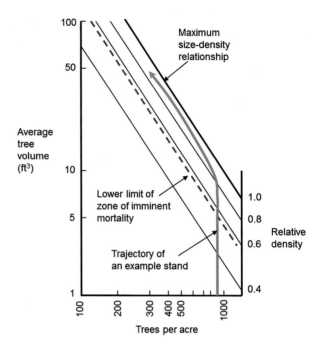

FIGURE 4.1 Stand density diagram for Douglas-fir. *Adapted from Drew, T.J., Flewelling, J.W., 1979. Stand density management: an alternative approach and its application to Douglas-fir plantations. For. Sci. 25 (3), 518−522.*

density as a straight line, using a logarithmic scale (Fig. 4.1). The line defining this upper boundary is termed the *self-thinning line* for stands encountering competition or density-related mortality, and represents the maximum average tree size that the trees can attain at a given stand density. These diagrams, depending on how they are developed, could allow you to understand the point of self-thinning, or the period of time of imminent competition mortality. The self-thinning relationship can be expressed mathematically as:

$$V = kN^{-3/2}$$

where:

V = average tree volume
N = number of live trees per unit area
k = a tree species-specific constant

After trees become established on a site, they are free to grow until subsequent growth requires more resources than can be provided by site conditions occupied by the individual tree. When this time comes, tree growth can occur only when additional resources (water, nutrients, light) are made available as a result of tree mortality. However, suboptimal tree growth can occur as long as the intake of resources is lower than the respiration requirements. This can occur for long periods in trees that can manage their respiration requirements efficiently. The negatively sloped line represents, in general, the rate of tree size growth at each level of stand density during the period of intense competition for resources. The intercept (k) of the self-thinning relationship seems to remain constant for a given tree species, and stand age does not seem to change the size-density relationship (Bégin et al., 2001).

A. Growth of Even-Aged Stands

Even-aged stands are ones where the range of tree ages within a stand do not vary by more than 20% or so. Plantation forests are the best example of even-aged stands, as often they are created using seedlings or clones from a common set of parents. The planted trees are all the same age since they are planted at the same time. Therefore even-aged stands can be created through clear-cut harvesting of trees and subsequent plantings (Fig. 4.2). Seed tree or shelterwood harvests, where a minor amount of the overstory remains to provide seed for the new stand, can be used to create an even-aged stand. Even-aged stands can also be created as a result of natural disturbances. Floods along streams and rivers that remove all the vegetation can provide fertile ground for a new crop of trees to establish. Fires create large openings, and although some remnant trees may remain, most of the new vegetation begins its life cycle at about the same time. Volcano eruptions such as Mount St. Helens in 1980 can create large expanses of open areas that will likely be revegetated with forests, either naturally or through human intervention. Hurricanes (tropical cyclones) and tornadoes are other natural disturbance events (as are disease or insect infestation outbreaks) that may create localized gaps within which new stands of trees may originate.

There is usually a small amount of variation in tree heights in even-aged stands, and since tree ages are effectively the same, the resulting forest structure is, by comparison, relatively simple. The diameter distribution will represent a bell-shaped curve (see Fig. 2.1), and with time a stand will have associated with it more variation as competition influences the growth rate of individual trees. Management actions associated with even-aged stands may include planting, commercial thinning, fertilization practices, competition control, and final harvests. In stands that are intensively managed, weed control and precommercial thinnings at early ages, and fertilization processes at early and middle ages might be used to enhance the growth rate of the desired trees and to control the unwanted vegetative competition. A typical time-line of volume accumulation within an even-aged stand is represented by a non-linear curve (Fig. 4.3). The volume growth rate could vary depending on the intermediate stand management activities that are scheduled (Fig. 4.4). Often, measures of wildlife habitat quality will vary depending on the structural characteristics of trees. These measures can reflect the suitability of habitat for nesting, roosting, or foraging (Fig. 4.5). For example, some wildlife species require open areas containing grasses and shrubs for foraging. These conditions may be available only during the early years of an even-aged stand. Other species may require a certain number of large trees for nesting and roosting, and these conditions may be available only during the later years of an even-aged stand.

Stand age and site quality have been used to predict the growth and yield of even-aged stands for over 100 years; it is only in the last 60 years that stand density has been included to further refine growth and yield

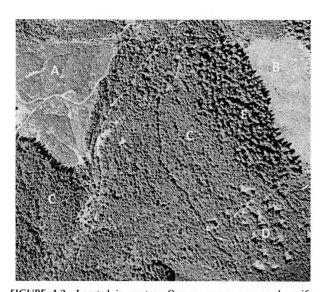

FIGURE 4.2 Located in western Oregon, young even-aged conifer stands (A and B), two older even-aged conifer stands (C), a stand where group selection harvests are used to promote an uneven-aged forest (D), and an old growth, uneven-aged stand (E).

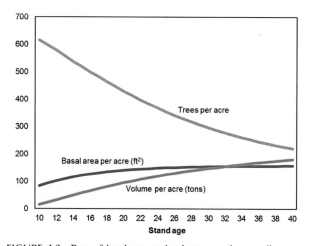

FIGURE 4.3 Rate of basal area and volume growth, as well as trees per acre decline, in an even-aged southern pine stand.

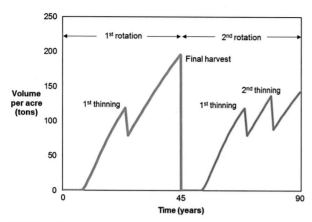

FIGURE 4.4 Development of standing volume in a privately owned southern pine even-aged stand, over two rotations.

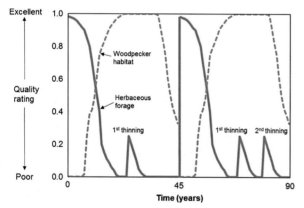

FIGURE 4.5 Development of habitat quality in a privately owned southern pine even-aged stand, over two rotations.

TABLE 4.1 A Comparison of Several Growth and Development Characteristics of Even-Aged and Uneven-Aged Stands

Growth and Development Characteristic	Even-aged Stands	Uneven-aged Stands
Trees per unit area	Decreases with age	Varies through time
Mortality rate of stems	Decreases with age	Stays relatively constant over time
Mortality of volume	Increases with age	Stays relatively constant over time
Height of canopy	Increases with age, then plateaus	Stays relatively constant over time
Canopy cover	Ranges from none to full	Ranges from full to one containing gaps
Average tree diameter	Increases with age	Fluctuates with harvest entries and mortality
Diameter distribution	Bell-shaped curve	Reverse J-shaped curve
Basal area	Increases with age, then plateaus	Fluctuates with harvest entries and mortality
Timber growth rate	Rises, peaks, then declines	Stays relatively constant over time
Timber yield	Increases with age, then plateaus	Fluctuates with harvest entries and mortality

forecasts. Site index is one of the most influential indicators of the growth and subsequent productivity of a stand of trees and economic profitability; areas with high site index value often receive more intensive and costly vegetation management, because one can demonstrate higher returns for these treatments. In young stands, growth is strongly affected by the number of trees per acre, but at some point, basal area is often a more appropriate expression of stand density for stands of trees (Buckman et al., 2006).

Several characteristics of the growth dynamics in even-aged stands set them apart from uneven-aged stands and other types of stand structures. Whether an even-aged stand was planted or whether it began as natural regeneration following a disturbance, the number of trees per unit area will generally decline over time (Table 4.1). Initially, the mortality of trees is related to competition from herbaceous vegetation, shrubs, and other volunteer trees and remnants from the previous stand. In addition, some regenerated trees may suffer from being browsed

by various species of wildlife (e.g., deer or elk) and subsequently die. As the even-aged stand canopy closes, inter-tree competition for resources (light, water, nutrients) will lead to the weakening and mortality of some trees. As a result, mortality rates are generally high in younger even-aged stands than they are in older even-aged stands. As we suggested earlier, the reduction or loss in standing volume from mortality generally increases as stands get older.

If we view the time of initiation of an even-aged stand as being the time of artificial regeneration or the onset of natural regeneration, it is obvious that the height of the planted or regenerated trees will increase over time. If, however, an even-aged stand is not harvested prior to the culmination of height growth for a particular tree species, the height of the canopy is likely to reach a plateau at some level and remain constant for a long period of time. Canopy cover in an even-aged stand can range from little to none (at the time of regeneration), to completely closed. In even-aged stands, the average tree diameter will likely increase over time as the residual live trees continually add annual growth rings. This assumes that the larger trees in a stand will continue to survive over time, whereas the smaller trees (in diameter and height), perhaps shade-intolerant trees, will have a higher mortality probability over time. The diameter distribution of an even-aged stand, as we discussed in Chapter 2, Valuing and Characterizing Forest Conditions, should resemble a bell-shaped curve. This distribution will generally flatten out over time, and the range of diameters will widen, as trees either express dominance or suffer from competitive or site-related circumstances.

Since the average diameter of trees in an even-aged stand usually tends to increase over time, the average basal area also increases, even though the number of trees per unit area decreases. Basal area is one measure of stand density, and generally increases, if left unencumbered, to a maximum plateau. Maximum basal area per acre for even-aged stands in the southern United States may be around $250 \, \text{ft}^2$ per acre ($57 \, \text{m}^2$ per hectare), whereas in the western United States it could climb above $500 \, \text{ft}^2$ per acre ($115 \, \text{m}^2$ per hectare) in forests containing redwoods (*Sequoia sempervirens*). Volume growth rates rise quickly in young even-aged stands, peak at some point, then decline.

The periodic annual increment, as we saw in Chapter 2, Valuing and Characterizing Forest Conditions, is a reflection of the rate of growth of a stand of trees, and as we discussed, when the periodic annual increment declines to the point where it equals the mean annual increment, the biological rotation age has arrived. When volume measurements that incorporate lumber utilization standards (e.g., Scribner or Doyle), the intersection of the periodic and mean annual increments generally increases with stand age in even-aged stands, in the absence of any intermediate silvicultural activities or other disturbances to the stand structure.

B. Growth of Uneven-Aged Forests

Uneven-aged stands are sometimes referred to as all-aged stands, but generally are considered those containing more than two or three distinct age classes or age cohorts. Management of these stands is often associated with the maintenance of the targeted diameter distribution. While a harvest schedule may determine the number and species to remove, it is often left to the field forester to select the best candidates for removal in the stand, based on tree vigor ratings. The target forest structure of an uneven-aged stand is one where equilibrium in volume growth has been reached. The equilibrium is a theoretical state where a sustainable timber increment (accretion) is

developed over a period of time using a diameter distribution of trees that remains roughly constant. One of the assumptions of this silvicultural system is that an adequate amount of regeneration is provided, which will maintain the system indefinitely.

Uneven-aged stands can be created from even-aged stands either using a selective harvesting program that treats individual trees, or using patch selection harvesting program where the patches are between about 0.5 and 4 acres (0.2 and 1.6 hectares) in size (i.e., group selection patches) (Fig. 4.2). Some of the challenges facing the conversion of even-aged forests to uneven-aged forests were recently described for coniferous stands in the United Kingdom (Mason, 2015). These include creating the sufficient wind-firmness among the patches, and maintaining the canopy in such a way as to allow sufficient light to be transmitted through to facilitate growth of the regenerated stands.

The rules that are used to manage uneven-aged stands are not based on age, as they might be with even-aged stands, but rather are based on measures of stand density and a desired diameter distribution. The structural trajectory of an uneven-aged stand will depend on the treatments applied to different diameter classes within the stand. An understanding of the shade-tolerance of desired tree species may help determine gap size needed during each logging entry to promote regeneration and growth. As a result, an extensive examination of the distribution of shade tolerant and shade-intolerant trees, and the advanced regeneration that may be present in the stand, would seem necessary prior to scheduling management activities in uneven-aged stands. Harvest entries, or cutting cycles, generally are scheduled at 5−30 year intervals (Fig. 4.6), and with each entry a range of trees of various sizes are extracted. As a result of either individual tree removals or patch cuttings, openings are created in the canopy and regeneration (natural or artificial) is facilitated. In addition, wildlife habitat quality may vary depending on the structural

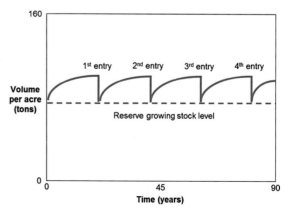

FIGURE 4.6 Development of standing timber volume in an uneven-aged southern pine stand, with 20-year harvest entry cycles.

characteristics of the residual live trees and the coarse woody debris that remains in the stand. These fluctuating structural conditions influence habitat suitability and the use of the stand by wildlife for nesting, roosting, or foraging purposes (Fig. 4.7).

One increasingly socially positive characteristic of uneven-aged stands is that the continuous forest cover is maintained, an immediate benefit to most in that the aesthetic quality of the forest is maintained. In addition, stands with several age classes may be less prone to insect and disease outbreaks. On the negative side, harvesting systems using these types of silvicultural practices (individual tree selection, group selection patches) may be less economically efficient than when employed in larger even-aged stand clearcuts. In addition, residual live trees may be damaged by logging machinery, since logging systems must maneuver around them during harvesting operations. And, the species distribution may not be consistent with a landowner's goals (Fig. 4.8). As low vigor trees are often the ones selected for harvest,

natural mortality can be low in these stands, and the result may be a shortage of snags and down wood. On the other hand, the periodic income associated with repeated entries may be more desirable for certain landowners. Thus, this type of management may involve a full trade-off analysis that makes forestry an exciting and challenging endeavor.

Whether an uneven-aged stand began as a planted stand or whether it evolved through natural processes, the number of trees per unit area will fluctuate through time as gaps occur and regeneration takes hold (Table 4.1). However, in an uneven-aged stand, the number of trees per unit area should never be reduced to very low levels, which suggests that the transition to an even-aged stand structure may have occurred. The mortality of trees in an uneven-aged stand is related to competition from other living trees and disturbances such as windfall. Some regeneration in canopy gaps may suffer due to regeneration-related competition issues, and from deer and elk browsing activities. As the uneven-aged stand canopy progresses through time, inter-tree competition for resources (light, water, nutrients) will lead to the weakening and mortality of some trees, which will eventually die and create other canopy gaps. When in equilibrium, the mortality rates in uneven-aged stands generally are assumed to be constant over time.

The height of the largest trees within the canopy of an uneven-aged stand should stay relatively constant over time. However, canopy cover ranges from fully closed stands to stands with partial canopy gaps. The gaps in uneven-aged stands are created where individual trees have died, or where silvicultural practices designed for uneven-aged stands (single-tree selection harvests or group selection harvests) have been implemented. The average tree diameter in an uneven-aged stand will likely fluctuate over time with harvest removals, ingrowth, and mortality. The diameter distribution of an uneven-aged stand, as we discussed in Chapter 2, Valuing and Characterizing Forest Conditions, should resemble a reverse J-shaped curve. Although we present the reverse J-shaped distribution as a conceptual model to represent trees per unit area by diameter class, uneven-aged stands can be represented by a variety of diameter distributions that are not as easy to classify as they are in even-aged stand systems. In many cases, there are gaps and bumps in the reverse J-shaped distribution that represent a lack of certain sized trees or a clump of similar cohorts. Since diameters of trees fluctuate with harvest, ingrowth, and mortality, the average basal area also fluctuates. Volume growth rates can remain relatively constant in an uneven-aged stand if the stand continues to contain healthy trees. However, yields may fluctuate given the distribution and quality of trees selected during each entry.

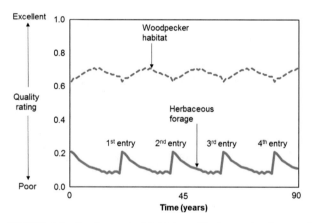

FIGURE 4.7 Development of habitat quality in a privately owned southern pine uneven-aged stand, assuming a 20-year harvest entry cycle.

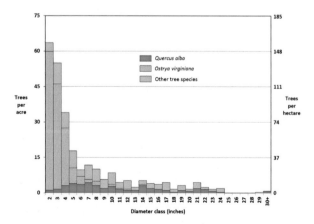

FIGURE 4.8 Diameter distribution for an uneven-aged deciduous forest in the State of Georgia, US.

Ideally, as Meyer (1952) suggested, an uneven-aged stand can be characterized as:

A balanced uneven-aged forest is one in which the current growth can be removed annually or periodically while maintaining at the same time the structure and initial volume of the forest.

Unfortunately, many uneven-aged forests either do not contain a balanced structure, or are in such a state that the current volume contained in the forest is not sustainably maintainable because it is either depleted or overstocked. The normal growing stock volume in an uneven-aged stand varies, and may fluctuate widely to suit the needs of managing these stands in changing economic and social conditions (Meyer and Stevenson, 1943). In the ideal case, a balanced uneven-aged stand can be sustainable, where current growth is offset by an equal amount of mortality; however, this is not a universal rule. As an example, in hardwood forests of Pennsylvania, Meyer and Stevenson (1943) observed growth rates ranging between about 44 and 67 ft^3 per acre (3.1 and 4.7 m^3 per hectare) per year, with an average of about 51 ft^3 per acre (3.6 m^3 per hectare) per year. These gross growth rates were offset by some amount of mortality, however the annual increment exceeded mortality. As stands transition from a low volume per unit area condition to a high volume per unit area condition, the gross increment should exceed mortality. In the latter stages of development, however, mortality may equal or exceed gross increment.

Meyer and Stevenson (1943) developed a method to estimate the number of trees that may be found in each diameter class of a theoretical uneven-aged stand, and suggest that each diameter distribution can be characterized by an exponential function:

$$TPA_x = ke^{-aDBH}\,dDBH$$

where:

TPA_x = trees per acre in diameter class x
k, a = constants that characterize a certain structure
e = natural logarithm
$dDBH$ = width of the diameter class (inches)

Using the data associated with tracts 1, 27, and 38 from Meyer (1952), the estimated diameter distributions of the stands are illustrated in Fig. 4.9 to show the variety of nonlinear curves that represent a decreasing number of trees per unit area by diameter class in these uneven-aged stands.

Using this theoretical model, we know that the number of trees per unit area in a class changes from one class to the next in a constant fashion, by a factor of q, which is referred to as the *diminution quotient*. This relationship often is referred to as the *Law of de Liocourt*, after the French forester who originally proposed the concept

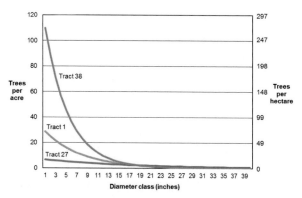

FIGURE 4.9 Projected uneven-aged stand diameter distributions for three tracts described by Meyer (1952). Tract 1: $a = 0.143$, $k = 33$; Tract 27: $a = 0.060$, $k = 7$; Tract 38: $a = 0.221$, $k = 137$.

(De Liocourt, 1898). The diminution quotient can be determined either by calculating it:

$$q = e^{a\,dDBH}$$

or through inspection of the data (Table 4.2). For Meyer's Tract 38, for example, $q = e^{0.221(1)}$, or 1.247 for 1-inch diameter classes, and $q = e^{0.221(2)}$, or 1.556 for 2-inch diameter classes. When we inspect the stand table, we find that the number of trees per acre begins at about 110 in the 1-inch diameter class, and declines to about 88 by the 2-inch class. The ratio for this 1-inch difference is 109.835/88.057, or 1.247. When performing this computation on any two adjacent diameter classes, the ratio remains the same. Stated another way, the number of trees in each subsequently greater diameter class is 1.247 times less than the number of trees in the previous diameter class.

The diminution quotient for 2-inch diameter classes can be found in a similar manner by inspecting Table 4.2. Here we need to use the trees per unit area in one class, and the trees per unit area in a class two inches away. For example, the number of trees per acre is about 110 in the 1-inch diameter class, and about 71 in the 3-inch class. The ratio for this 2-inch difference is 109.835/70.597, or 1.556. In a manner similar to what we noted a moment ago, the number of trees in each subsequently greater 2-inch diameter class is 1.556 times less than the number of trees in the previous diameter class.

Given the work of Meyer (1952), the following thoughts about the nonlinear diameter distributions of uneven-aged stands can be concluded:

- Values of the variable q can fluctuate widely, but generally range between 1.0 and 2.0.
- Values of the variable q over 2.0 are suggestive of balanced diameter distributions of under-stocked stands as they will have fewer large trees.
- High values of the variable a suggest rapid decreases in trees per unit area in subsequent diameter classes.

TABLE 4.2 Trees per Acre for Tract 38, a Spruce-Fir-Beech Selection Forest in Emmenthal, Switzerland

Diameter Class (inches)	Trees per Acre
1	109.835
2	88.057
3	70.597
4	56.598
5	45.376
6	36.379
7	29.165
8	23.382
9	18.746
10	15.029
11	12.049
12	9.660
....	
25	0.546
26	0.438
27	0.351
28	0.281
29	0.226
30	0.181

Source: Meyer, H.A., 1952. Structure, growth, and drain in balanced even-aged forests. J. For. 50(2), 85−95.

This relationship is evident in Fig. 4.9, where Tract 38 had the highest value of the variable a (0.221) of the three example stands, and Tract 27 had the lowest (0.060).

- Values of the variable k can vary considerably, indicating different degrees of stand density or stocking. This relationship is also evident in Fig. 4.9, where Tract 38 had the highest value of the variable k (137) of the three example stands, and Tract 27 had the lowest (7).

C. Growth of Two-Aged Forests

A two-aged silvicultural system maintains two distinct age classes throughout a planning horizon for a stand. This may occur in conjunction with management activities that result in a small number of large trees in the upper canopy, and a large number of smaller trees surrounding these larger ones. The trees in the upper canopy may account for perhaps 30 ft^2 per acre (6.9 m^2 per hectare) of basal area, and are arranged in a manner that they have a negligible effect on the growth of the smaller trees. In this system, the trees in the upper canopy are not intended to facilitate the regeneration of the second age class. This objective distinguishes this type of silvicultural system from even-aged shelterwood harvests, where the residual overstory trees are needed to help regenerate the area, and subsequently are removed once regeneration has been established. In a two-aged stand, the reserved trees in the upper canopy may reside on the site for multiple rotations. Imagine the younger trees that have been established growing to commercial timber size. During a logging entry, some of these younger trees will remain, whereas others will be harvested along with some of the older trees left behind from the previous harvest entry. The trees that remain (of both age classes) may not be removed until the next entry into the stand. Advantages to this system relate to the aesthetic quality of the site (a continuous cover of trees) and to the maintenance of higher levels of wildlife habitat for some species. As larger trees senescence, their tops may break, scars from fire may form, and ultimately these trees may become important structural features for wildlife habitat, such as snags.

D. Growth Transition Through Time

At the heart of discussions about the sustainability of forest values is the issue of forest growth and the ability maintain it through time (Beers, 1962). An understanding of the changes in the structural condition of stands is important for land managers and decision-makers. How forests transition through time is important in understanding whether goals and objectives can be met, sustained, or lost through the management actions suggested in a forest plan. Several concepts are central in understanding forest transitions. First, there is *survivor growth*, or the growth of trees that were alive both at the beginning of a measurement period and at the end. If individual trees in a stand were not cut, nor did they die, then they likely grew in size (e.g., diameter and height, which relates to volume). From the perspective of a stand's diameter distribution, there will likely be a shift in the diameter distribution to the right as a result of survivor growth (Fig. 4.10) from the beginning to the end of the measurement period. To be more realistic about the transition of a stand of trees, however, three other factors must be incorporated into the assessment: ingrowth of seedlings into the smallest diameter class, death of trees (mortality), and harvest (cutting) of trees from the stand. (Fig. 4.11). When these transitions are considered at the stand-level, ingrowth is the entry of trees into the smallest diameter class that is recognized. As we inferred earlier, ingrowth may include seedlings that were below a minimum merchantable size prior to the first measurement,

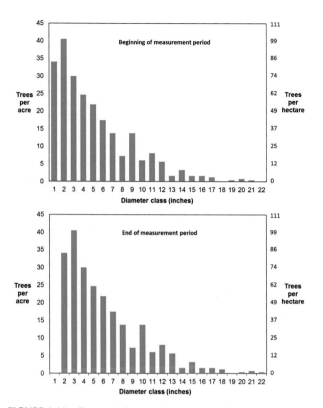

FIGURE 4.10 Change in diameter distribution of an uneven-aged stand of oaks in Missouri, assuming that the trees grew, on average, 1 inch between beginning and ending measurement periods.

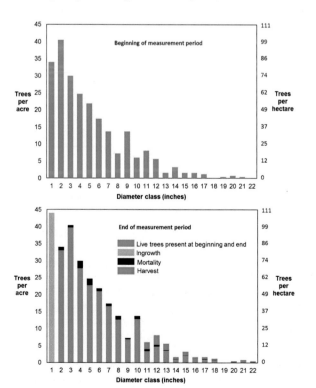

FIGURE 4.11 Change in diameter distribution of an uneven-aged stand of oaks in Missouri, assuming ingrowth, mortality, and harvest.

or may include seedlings that had not begun their life at the time of the first measurement.

1. Stand-Level Volume Estimates

When considering the growth dynamics of individual stands of trees, we could use the concepts of increment, ingrowth, mortality, and harvest to describe the transition of the stand volume through time. Rather than simply reporting the current condition of a stand of trees, if previous information about the stand were available, then we could examine the change that had occurred between the two measurement periods. This allows us to understand the transition of forest conditions in a stand along with the current status of the stand. The following discussion of growth dynamics follows the instructional analysis provided over 50 years ago by Beers (1962). If we assume that subscript 1 refers to the initial set of measurements captured at the beginning of a measurement period, and that subscript 2 refers to the second set of measurements captured at the end of the measurement period, the following terms can be used to describe the transition of a stand's condition from the beginning to the end of a measurement period:

V_1 = volume of a stand at the beginning of a measurement period

V_2 = volume of a stand at the end of a measurement period

$M_{1 \rightarrow 2}$ = volume of trees that died between measurement periods 1 and 2

$C_{1 \rightarrow 2}$ = volume of trees that were harvested (cut) between measurement periods 1 and 2

I_2 = ingrowth of volume that is recognized at the end of a measurement period, and that has grown into the smallest diameter class that is recognized

V_2 inherently includes ingrowth (I_2), since V_2 represents the condition of the live trees contained in a stand at the end of the measurement period. The net change (increase or decrease) of volume over the measurement period can easily be estimated by subtracting the initial volume estimate from the ending volume estimate:

$$\text{Net change (increase or decrease)} = V_2 - V_1$$

The net change estimate inherently includes the ingrowth, since it is included in V_2, yet it recognizes the loss of volume as a result of mortality and harvest because those volumes are excluded from V_2. To estimate the gross growth of the volume of a stand over a measurement period, which would include ingrowth, it can be represented by the equation:

$$\text{Gross growth, including ingrowth} = (V_2 + M_{1 \rightarrow 2} + C_{1 \rightarrow 2}) - V_1$$

Here we assume that ingrowth is again included in V_2, yet we add to V_2 the volume that was extracted during the interval between measurements ($C_{1 \to 2}$) and the mortality that occurred during the measurement period ($M_{1 \to 2}$). If we were interested only in the gross growth of the trees that were measured at the beginning of the measurement period, we would need to remove the ingrowth from V_2:

$$\text{Gross growth, excluding ingrowth} = (V_2 + M_{1 \to 2} + C_{1 \to 2} - I_2) - V_1$$

Measures of gross growth include the contribution of trees that had died during the measurement period. These trees may become snags or down logs, and therefore may become important structural features for wildlife habitat purposes, yet potentially can become fuel that contributes to fire risk. However, from a timber production perspective, we might be interested only in the volume that can be utilized for manufacturing or livelihood purposes (e.g., fuel wood), thus we may be interested only in the volume contained in the live trees. This net growth can be estimated by removing the mortality values from the gross growth equations. For example, to estimate the net growth of a stand of trees including ingrowth, we would use:

$$\text{Net growth, including ingrowth} = (V_2 + C_{1 \to 2}) - V_1$$

Further, if we were interested only in the net growth of the trees that were measured at the beginning of the measurement period (i.e., excluding ingrowth), we would also remove the contribution provided by the ingrowth into the smallest diameter class:

$$\text{Net growth, excluding ingrowth} = (V_2 + C_{1 \to 2} - I_2) - V_1$$

Each of these estimates are conservative because we may not have considered the growth of the trees before they died, or before they were cut, between the time of the first measurement to the time that they no longer were considered live trees. Mortality and harvest can assume to be represented by the condition of the trees at the beginning of the measurement period, since they likely were not measured at the end of the measurement period. However, in today's multiresource inventory processes, some trees that may have died between subsequent measurements may actually have been measured at the end of the period. In addition, volumes of trees in the period between measurements may be estimated from growth and yield relationships.

Example

Using the uneven-aged stand data presented in Table 4.3, let us now view growth dynamics using tree-level inventory data. Assume that the stand is comprised of tree

records, where each record represents a number of trees per unit area, and that the length of the measurement period is 10 years. Assume as well that the smallest merchantable volume class is five inches, even though our inventory data indicates that there are trees in smaller size classes. The net change in volume between the two measurement periods is:

$$\text{Net change (increase or decrease)} = 3,948 \text{ ft}^3 - 3,844 \text{ ft}^3$$
$$= 104 \text{ ft}^3 \text{ per acre}$$

The gross growth of the stand, including the ingrowth, the volume extracted during the harvest that occurred, and the mortality of the stand, is:

$$\text{Gross growth, including ingrowth} = (3,948 \text{ ft}^3 + 82 \text{ ft}^3 + 371 \text{ ft}^3) - 3,844 \text{ ft}^3$$
$$= 557 \text{ ft}^3 \text{ per acre}$$

The gross growth of the stand using the trees that were initially five inches and greater at the beginning of the measurement period is:

$$\text{Gross growth, excluding ingrowth} = (3,948 \text{ ft}^3 + 82 \text{ ft}^3 + 371 \text{ ft}^3 - 30 \text{ ft}^3) - 3,844 \text{ ft}^3$$
$$= 527 \text{ ft}^3 \text{ per acre}$$

The net growth of the stand, including ingrowth, becomes:

$$\text{Net growth, including ingrowth} = (3,948 \text{ ft}^3 + 371 \text{ ft}^3) - 3,844 \text{ ft}^3$$
$$= 475 \text{ ft}^3 \text{ per acre}$$

Finally, the net growth of the stand using only the trees that were five inches and greater at the beginning of the measurement period is:

$$\text{Net growth, excluding ingrowth} = (3,948 \text{ ft}^3 + 371 \text{ ft}^3 - 30 \text{ ft}^3) - 3,844 \text{ ft}^3$$
$$= 445 \text{ ft}^3 \text{ per acre}$$

Example

Using the western forest stand in Appendix A as an example, let us now view growth dynamics using stand-level summary data. First, assume that the stand is 30 years old at the beginning of a measurement period, and that the end of the period is 5 years later. Assume as well that the smallest merchantable volume class is eight inches in diameter, even though our inventory data indicates there are trees in smaller size classes. Since the growth rate of trees upward into higher diameter classes is undetermined from the data we have, let's assume that ingrowth into the 8-inch class over the 5-year period is 500 board feet (300 ft^3). Finally, even though the data in Appendix A does not indicate a harvest, let's assume that over the 5-year period a thinning occurred, and some of the volume was extracted (Table 4.4). As a result of these assumptions and

TABLE 4.3 Dynamics of Growth as Viewed at the Tree-Record Level for an Uneven-Aged Stand of Mixed Conifers in Eastern Oregon

Tree Record	Initial		Ending		Tree Species	V_1 (ft^3/ac)	V_2 (ft^3/ac)	$M_{1 \rightarrow 2}$ (ft^3/ac)	$C_{1 \rightarrow 2}$ (ft^3/ac)	I_2 (ft^3/ac)
	Trees per Acre	DBH (inches)	Trees per Acre	DBH (inches)						
1	5.3	16.9	4.2	17.7	WL	340	317	8	68	–
2	5.3	16.4	4.2	17.4	WL	283	267	6	57	–
3	5.3	26.4	5.2	27.5	WL	854	905	17	–	–
4	20.1	3.0	19.7	4.7	ES	–	–	–	–	–
5	20.1	3.5	19.7	4.6	ES	–	–	–	–	–
6	20.1	12.8	15.8	13.9	ES	590	557	11	118	–
7	20.1	4.3	19.7	5.2	DF	–	30	1	–	30
8	5.3	21.9	5.2	22.6	WL	553	610	12	–	–
9	5.3	18.4	4.2	19.5	WL	404	351	8	81	–
10	5.3	22.2	5.2	24.0	ES	585	668	13	–	–
11	5.3	14.1	4.2	15.6	WL	235	243	6	47	–
12	–	–	422.0	0.1	GF	–	–	–	–	–
13	–	–	142.0	0.1	DF	–	–	–	–	–
14	–	–	150.7	0.1	LP	–	–	–	–	–
Total						3,844	3,948	82	371	30

WL, Western larch (*Larix occidentalis*); ES, Engelmann spruce (*Picea engelmannii*); DF, Douglas-fir (*Pseudotsuga menziesii*); GF, Grand fir (*Abies grandis*); LP, Lodgepole pine (*Pinus contorta*).

TABLE 4.4 Initial Stock Table for a 30-Year-Old Douglas-Fir Stand, Stock Table at Age 35, and Mortality and Harvest (Cut) Over the 5-Year Measurement Period, All on a Per-Acre Basis

DBH Class (inches)	Age 30		Age 35		Mortality		Cut	
	MBF[a]	Cubic Feet	MBF[a]	Cubic Feet	MBF[a]	Cubic Feet	MBF[a]	Cubic Feet
4	–	9	–	–	–	0.2	–	–
5	–	14	–	–	–	0.3	–	–
6	0.313	192	0.162	78	0.003	1.4	–	–
7	0.469	287	0.242	117	0.005	2.4	–	–
8	1.922	644	1.261	394	0.011	3.4	0.541	169
9	1.774	594	1.164	364	0.010	3.2	0.499	156
10	1.690	537	2.552	763	0.004	1.2	1.094	327
11	1.382	439	2.088	624	0.003	0.9	0.895	267
12	0.589	174	1.556	444	0.001	0.3	0.667	190
13	0.392	116	1.038	296	–	–	0.445	127
14	–	–	0.307	83	–	–	0.131	35
15	–	–	0.286	70	–	–	0.122	30
Total[b]	8.531	3,006	10.656	3,233	0.037	13.4	4.394	1,301
Total[c]	7.749	2,504	10.252	3,038	0.029	9.0	4.394	1,301

[a]MBF, thousand board feet.
[b]Using all diameter classes.
[c]Only the 8-inch class and greater.

management actions, our estimate of the net change in volume for this stand over the 5-year period is:

$$\text{Net change (increase or decrease)} = 10.252 \text{ MBF} - 7.749 \text{ MBF}$$
$$= 2.503 \text{ MBF per acre}$$

The gross growth of the stand, including the ingrowth, the volume extracted during the harvest that occurred, and the mortality of the stand, is:

$$\text{Gross growth, including ingrowth}$$
$$= (10.252 + 0.029 + 4.394) - 7.749$$
$$= 6.926 \text{ MBF per acre}$$

The gross growth of the stand using the trees that were initially eight inches and greater at the beginning of the measurement period (age 30) is:

$$\text{Gross growth, excluding ingrowth}$$
$$= (10.252 + 0.029 + 4.394 - 0.500) - 7.749$$
$$= 6.426 \text{ MBF per acre}$$

The net growth of the stand, including ingrowth, becomes:

$$\text{Net growth, including ingrowth} = (10.252 + 4.394) - 7.749$$
$$= 6.897 \text{ MBF per acre}$$

And the net growth of the stand using only the trees that were eight inches and greater at the beginning of the measurement period is:

$$\text{Net growth, excluding ingrowth}$$
$$= (10.252 + 4.394 - 0.500) - 7.749$$
$$= 6.397 \text{ MBF per acre}$$

2. Broader-Scale Volume Estimates

Broad-scale volume estimates, such as those found in state, provincial, regional, or national reports, involve an assessment of forest stand conditions that are aggregated across large regions. As a result, the ingrowth, mortality, and harvest include contributions from even-aged, uneven-aged, and other types of managed stand structures. Many broader-scale analyses of forest conditions use terms slightly different than those used at the stand-level, to reflect socioeconomic dynamics of the larger system. Accretion, for example, does not include the growth of trees that were harvested or that died between the measurement periods. Accretion can be defined as:

$A_{1 \to 2} =$ Volume growth of trees that were alive at the beginning and the end of measurement periods 1 and 2

Ingrowth, for broader-scale purposes, may include trees that have grown to a size that represents the smallest size reported (5 inches (12 cm) DBH for growing stock trees in

a recent Georgia analysis), and the volume of trees on land that has been reclassified from noncommercial forest or nonforest to timberland during the measurement cycle (Brandeis, 2015). Mortality estimates, in a manner similar to the stand-level analysis, represent the net volume of trees that were alive during the first measurement period, yet died sometime before the second measurement period. Removals of volume may be broader in scope, and include volume that was harvested, or volume destroyed during land clearing operations. In addition, removals may include the net volume of trees that were neither harvested nor destroyed, but simply reside on land that has been reclassified to something other than timberland during the measurement cycle.

$R_{1 \to 2} =$ Volume of trees that were removed due to logging, land conversion, or land reallocation during the measurement cycle

Some broad-scale analyses result in estimates of gross growth, which is the sum of accretion and ingrowth. In other words, the gross growth measurement includes the growth of live trees that were present at the beginning of the measurement cycle and the live trees that were not measured at the beginning of the measurement period, but have grown into the smallest classes measurable during the measurement period.

$$\text{Gross growth} = A_{1 \to 2} + I_2$$

Net growth is represented as the change in growing stock volume across the broader area due to natural causes, and typically is defined as gross growth less the mortality that occurred during the measurement cycle.

$$\text{Net growth} = A_{1 \to 2} + I_2 - M_{1 \to 2}$$

Many recent broad-scale reports (e.g., state-level forest resource reports) indicate that net growth incorporates accretion, ingrowth, and mortality. When only net growth and mortality estimates are provided, one can determine gross growth by adding mortality estimates to net growth estimates. However, it is difficult to determine the two individual pieces of growth estimates (accretion and ingrowth) when neither is reported separately.

The net change that often is reported in broader scale analyses is the difference in volume between the two successive measurement periods, and this takes into account removals:

$$\text{Net change} = A_{1 \to 2} + I_2 - M_{1 \to 2} - R_{1 \to 2}$$

Example

In a recent assessment of the forest conditions in Georgia (Brandeis, 2015), the change in forest conditions were presented by tree species groups for the entire state, as well as

TABLE 4.5 Average Annual Change in Volume for Certain Tree Species Groups in Georgia

Tree Species	Gross Growth		
	Ingrowth + Accretion (million ft³)	Mortality (million ft³)	Removals (million ft³)
Longleaf-slash pine	354.7	33.9	350.2
Loblolly-shortleaf pine	1,053.3	108.1	687.6
Oak-pine	238.3	47.6	94.0
Oak-hickory	394.8	82.2	159.3
Oak-gum-cypress	240.3	103.7	105.7

Longleaf pine (*Pinus palustris*); slash pine (*Pinus elliottii*); loblolly pine (*Pinus taeda*); shortleaf pine (*Pinus echinata*); oak (*Quercus* spp.); hickory (*Carya* spp.); gum (*Nyssa* spp.); cypress (*Taxodium distichum*).
Source: Brandeis, T.J., 2015. Georgia's forests, 2009. US Department of Agriculture, Forest Service, Southern Research Station, Asheville, NC. Resource Bulletin SRS-207. 59 p.

for subdivisions of the state (Table 4.5). Estimates of ingrowth, accretion, mortality, and removal were provided to allow us to understand gross growth, net growth, and net change of the forest resources. To simplify the discussion, we will ignore here the changes in cull (damage or defect) volume associated with the report. As a result, with this data we can estimate gross growth of the Loblolly-shortleaf pine (*Pinus taeda* and *Pinus echinata*) resource over the measurement cycle:

$$\text{Gross growth} = 1,053.3 \text{ million ft}^3 \text{ of loblolly and shortleaf pine}$$

Net growth takes into account the mortality that occurred over the measurement cycle:

$$\text{Net growth} = 1,053.3 \text{ million ft}^3 - 108.1 \text{ million ft}^3$$
$$= 945.2 \text{ million ft}^3 \text{ of loblolly and shortleaf pine}$$

Finally, net change factors removals into the analysis:

$$\text{Net change} = 1,053.3 \text{ million ft}^3 - 108.1 \text{ million ft}^3$$
$$- 687.6 \text{ million ft}^3$$
$$= 257.6 \text{ million ft}^3 \text{ of loblolly and shortleaf pine}$$

Example

Using hundreds of permanent plots distributed across the Province of Nova Scotia in Canada, Townsend (2007) was able to estimate changes in forest structure for the province as a whole in a manner similar to the changes we

discussed related to a single stand. In addition, given the wide distribution of regularly measured permanent plots, the analysis can be disaggregated down to forest type, landowner, diameter class, and age class. Some of the results provided by Townsend (2007) are illustrated in Table 4.6. In this example, the gross growth, net growth, and change of softwood volume *per unit area* on public land in the province are:

$$\text{Gross growth} = (\text{Accretion} + \text{Ingrowth})$$
$$= (1.666 + 0.201)$$
$$= 1.867 \text{ m}^3/\text{ha}/\text{year}$$
$$\text{Net growth} = (\text{Gross growth} - \text{Mortality})$$
$$= (1.666 + 0.201 - 0.753)$$
$$= 1.114 \text{ m}^3/\text{ha}/\text{year}$$
$$\text{Net change} = (\text{Net growth} - \text{Harvest})$$
$$= (1.114 - 0.450) = 0.664 \text{ m}^3/\text{ha}/\text{year}$$

3. Broad-Scale Habitat Estimates

In addition to the projected timber volumes and the areas to be applied various silvicultural treatments, there is an increasing desire by decision-makers to understand through forest plans the estimates of other projected landscape conditions, such as those related to wildlife habitat and recreational opportunities. Many decision-makers tend to focus mainly on the net change in conditions from one period of time to the next. In fact, habitat levels for certain wildlife species in many forest plans are reported using the net area amount or the change in quality over time. For example, the 2012 amendment to the Huron-Manistee National Forest plan (US Department of Agriculture, Forest Service, 2012) suggests that 1,600 acres (648 hectares) of essential breeding habitat for the Kirtland's warbler (*Setophaga kirtlandii*) would be created each year, and 15,960 acres (6,459 hectares) would be available at any one time into the foreseeable future in Management Area 4.2 (roaded natural sandy plains and hills). This basic data helps land managers understand the impact of the plan on projected habitat conditions, and sheds light on some of the transitional aspects of the habitat over time. To understand these transitions, terms similar to those used for the transition of volume within a stand or forest could be applied to the transition of habitat as well (Bettinger, 2006). Measures of the net change, the gross increment, and the net increment can be developed using terms such as ingrowth, mortality, and harvest (cut) of habitat.

$I_{t \rightarrow 1}$ = The amount of land transitioning into suitable habitat from an unsuitable state between periods t and $t + 1$

TABLE 4.6 Periodic Annual Increment for Forests of Nova Scotia, Over a 10-Year Period Ending in 2005

Owner	Accretion	Ingrowth	Mortality	Harvest	Gross Growth	Net Growth	Net Change
Softwood (m³/ha/year)							
Public (Crown)	1.666	0.201	0.753	0.450	1.867	1.114	0.664
Private, small	1.848	0.163	0.979	1.696	2.011	1.031	−0.665
Private, large	2.144	0.218	0.765	1.523	2.361	1.596	0.073
Hardwood (m³/ha/year)							
Public (Crown)	0.680	0.055	0.232	0.085	0.736	0.504	0.419
Private, small	0.927	0.067	0.255	0.353	0.993	0.738	0.384
Private, large	0.580	0.046	0.269	0.622	0.626	0.357	−0.265

Source: Townsend, P., 2007. Ten Year Periodic Annual Increment for Nova Scotia Permanent Forest Inventory Plots 1990−95 to 2000−05. Nova Scotia Department of Natural Resources, Renewable Resources Branch, Forestry Division, Forest Inventory Section, Truro, NS. 38 p.

$M_{t \to t+1}$ = The amount of land naturally transitioning out of a suitable habitat class between periods t and $t+1$

$C_{t \to t+1}$ = The amount of land transitioning out of a suitable habitat class between periods t and $t+1$ primarily due to a management activity

The transition in suitable habitat from one time period to the next can be expressed as:

$$\Delta SH_{t \to t+1} = I_{t \to t+1} - M_{t \to t+1} - C_{t \to t+1}$$

This differs from the dynamics associated with timber characteristics of a stand in that there is no increment (growth) of suitable habitat in a single stand; a stand either is suitable habitat or it is not. The gross and net increments of the suitable habitat can then be expressed as:

Gross increment of SH_t including ingrowth
$$= SH_{t+1} + M_{t \to t+1} + C_{t \to t+1} - SH_t$$

Net increment of SH_t including ingrowth
$$= SH_{t+1} + C_{t \to t+1} - SH_t$$

The net change ignores the fact that some habitat may no longer be suitable (the amount of land that transitioned out of a suitable class). Because we implicitly included ingrowth in these assessments in SH_{t+1}, in fact we were examining how the landscape-level suitable habitat changed through time. However, if we were interested in assessing only how the initial areas classified as suitable habitat changed through time, we would have

to ignore the implicit ingrowth from the gross and net increment equations:

Gross increment of SH_t ignoring ingrowth
$$= SH_{t+1} + M_{t \to t+1} + C_{t \to t+1} - I_{t \to t+1} - SH_t$$

Net increment of SH_t ignoring ingrowth
$$= SH_{t+1} + C_{t \to t+1} - I_{t \to t+1} - SH_t$$

Example

The current level of downy woodpecker (*Picoides pubescens*) habitat for the Lincoln Tract was estimated, then projected into the future 20 years, in 5-year increments, with the assumption that no management activities would occur over this period of time. Suitable habitat was defined (albeit subjectively by the authors) as any areas with an HSI score of 0.400 or higher. Initial suitable habitat levels, ingrowth, mortality, and cut are presented in Table 4.7. The net increment is simply the change (column 2) from one time period to the next. What is the gross increment, including ingrowth, of the suitable habitat at the end of periods 1 through 4?

Gross increment for period 1 = 507 + 258 + 0 − 744
$$= 21 \text{ acres}$$

Gross increment for period 2 = 691 + 322 + 0 − 507
$$= 506 \text{ acres}$$

Gross increment for period 3 = 1,103 + 163 + 0 − 691
$$= 575 \text{ acres}$$

Gross increment for period 4 = 1,690 + 130 + 0 − 1,103
$$= 717 \text{ acres}$$

When examining broad-scale habitat changes, besides the net change, the gross increment of habitat may be the most informative measure of landscape change because it

TABLE 4.7 Land Classified as Suitable Habitat (HIS ≥ 0.400), Ingrowth Transitioning Into the Suitable Habitat Class, and Mortality and Harvest Transitioning Out of the Suitable Habitat Class for Downy Woodpecker Habitat on the Lincoln Tract

Decade (t)	SH_t (acres)	$\Delta SH_{t \to t+1}$ (acres)	$I_{t \to t+1}$ (acres)	$M_{t \to t+1}$ (acres)	$C_{t \to t+1}$ (acres)
0 (initial)	744				
		−237	21	258	0
1	507				
		184	506	322	0
2	691				
		412	575	163	0
3	1,103				
		587	717	130	0
4	1,690				

includes the amount of land naturally transitioning out of a suitable habitat class between two periods of time. What is the gross increment, ignoring ingrowth, of the suitable habitat at the end of period 1?

Gross increment for period 1, ignoring ingrowth
$$= 507 + 258 + 0 - 21 - 744 = 0 \text{ acres}$$

In fact, if you perform the calculations for each of the other time periods, the gross increment, when ingrowth is ignored, is always zero. Further, the net increment measures the change to the suitable habitat, so what would be the net increment, ignoring ingrowth, of the suitable habitat at the end of period 1?

Net increment of SH_t ignoring ingrowth
$$= 507 + 0 - 21 - 744 = -258 \text{ acres}$$

As we can see, the net increment, when we ignore land that transitions into suitable habitat, is simply the amount of the mortality. In other words, this is the amount of land that transitions out of suitable wildlife habitat.

III. PROJECTING STAND CONDITIONS

One of the main considerations in forest management and planning is whether we can locate a growth and yield system that will allow us to appropriately project forest structural characteristics into the future. In today's management environment, these systems need to be flexible enough to allow a number of management alternatives to be modeled. Growth and yield tables initially were developed for natural resource managers to allow them to understand how forests will transition through time. As computer technology evolved, the underlying functional relationships of accretion, ingrowth, and mortality have been incorporated into computer programs. Many of the contemporary growth and yield models allow a user to simulate the effect of various management activities and report the impact on the structure of forests.

A. Growth and Yield Tables

Natural resource managers in North America have used published tables to estimate tree volumes and yields for over 90 years. Some of the classic sources of broad-scale forest yield information include McArdle and Meyer (1930) for Douglas-fir (*Pseudotsuga menziesii*) often referred to just as Bulletin 201, Barnes (1962) for western hemlock (*Tsuga heterophylla*), Johnson (1955) for a broader range of western tree species, and the US Department of Agriculture (1929) and Forbes and Bruce (1930) for southern pines. Later in the development of North American forestry, others began to produce tables to represent more specific management approaches and narrower geographical areas, such as information for site-prepared loblolly pine (*Pinus taeda*) in the lower Coastal Plain of the southeastern United States (Clutter et al., 1984).

When using yield tables for volume estimation, the grade of wood might be taken into account; however, in some tables representing younger second-growth forests it is assumed that the trees will contain a large number of knots (as compared to older trees), and less heartwood. As a result, in these cases a single grade of wood may be assumed. Should a stand being assessed contain a large amount of rot or defect, an additional allowance should be made to reduce the volumes you might use from the published tables. This is the type of professional judgment you will gain with experience. Further, it is important to use the appropriate table for the conditions of the stand being assessed. For example, older stands have less stem taper, and therefore may contain more volume per tree than younger stands of the same average diameter. The two most common types of growth and yield tables are the volume table (for individual trees) and the yield table (for stands of trees).

1. Volume Table

A volume table is one in which the volume of wood contained in trees of various diameters and heights is provided (Forbes and Bruce 1930). Volumes can be represented in terms of cords, board feet, cubic feet, cubic meters, or tons, among other measures, and are provided as *per-tree estimates*. An upper merchantable diameter usually is assumed with each table, and it is further assumed that trees are free from rot and defect, and have an average amount of crook. In addition, since volume

TABLE 4.8 A Volume Table for Longleaf Pine, Providing Board Feet (International 1/8-inch Rule), to a 5-inch Merchantable Top Diameter

Diameter at Breast Height (inches)	Number of 16-foot logs						Basis (trees)
	1	2	3	4	5	6	
7	19	29	44	–	–	–	77
8	21	37	59	82	–	–	89
9	23	45	75	106	137	–	45
10	26	53	91	129	168	210	35
11	29	61	107	154	201	248	27
12	32	69	124	180	235	291	11
13	–	78	140	207	270	336	9
14	–	86	158	234	306	382	5
15	–	94	177	263	345	432	15
16	–	103	197	293	387	484	9
17	–	112	218	324	431	539	11
18	–	121	239	356	475	595	1
Basis (trees)	46	134	86	60	8	–	334

Source: Forbes, R.D., Bruce, D., 1930. Rate of Growth of Second-Growth Southern Pines in Full Stands. US Department of Agriculture, Forest Service, Washington, DC. Circular No. 124. 77 p.

tables are based on measurements of trees in experimental plots, volume tables will frequently indicate, using ruled lines that create a block, the average volumes associated with sizes of trees that were actually measured (Table 4.8). Volumes presented outside the block therefore are assumed to be estimates not based on field samples. When a volume table provides a *basis* column or row, it allows you to understand the number of trees that were sampled by diameter or height class.

Example

Assume that a longleaf pine (*Pinus palustris*) tree was 13 inches in diameter, and three merchantable 16-foot logs could be obtained from its bole when harvested. The volume of this tree could be estimated to be about 140 board feet using the volume table presented in Table 4.8.

2. Yield Table

A yield table is a tabular record illustrating the expected volume of wood using a combination of measurable stand characteristics such as age, site quality, and stand density (Palahí et al., 2003). When presented, a yield table illustrates the amount of wood that would be available *per unit area* from a stand of trees at a given age. This differs from a volume table in that all the trees in a stand are represented in the volume estimates, whereas the volume table allows you to estimate the volume of individual trees. Yield tables enable you to understand the capacity of a site to produce wood volume, and may be used to help estimate future volumes at subsequent stand ages. Yield table values generally include only living trees, do not account for growth responses to intermediate treatments, such as thinnings, and generally do not provide an allowance for logging damage or tree defect. In addition, yield tables generally provide values for a fully stocked, "average" stand for a given class of stands, whereas in practice stands of trees are rarely average. In normal applications of yield tables, proper allowance must be made for these unaccounted variables (McArdle and Meyer, 1930). In addition, yield tables may present volume by age class, the average stocking (trees per acre), diameter at breast height, and basal area along with estimated volumes.

Example

Assume that a western hemlock stand in southeastern Alaska had a site index (base age 100) of 120, and was 60 years old on average. An estimate of the cubic foot volume per acre could be made using the information provided in Table 4.9 by Barnes (1962), and suggests that stands on these types of sites, at this age, should contain around 9,500 ft^3 per acre of merchantable wood.

Yield tables generally are based on samples from well-stocked stands throughout, perhaps, a broad geographic range. Locally derived yield tables may provide more accurate estimates of yields, particularly if the yield table is based on data concentrated in a particular area. However, the construction of a local yield table may require a considerable amount of time and effort. A correction value applied to broad area yield tables may be as equally satisfactory as developing and using a local yield table (US Department of Agriculture, 1929). To develop a correction value, we would need a sample of well-stocked stands in a local area and a subsequent analysis of the difference in volumes between the samples and the previously developed yield tables.

Because yield tables generally are based on samples from well-stocked stands, there is an assumption that these represent fully stocked (i.e., 100% stocked) conditions. As a result, these are considered *normal yield tables*. Normal yield tables take into account variations in yield due to site quality and tree age. *Empirical yield tables* are based on average stocked stands. The volume in stands that are not fully stocked can be estimated with these tables by comparing the basal area of the stand being represented in a yield table with the basal area of a stand measured in the field.

TABLE 4.9 Cubic Foot Volume per Acre Yield Table for Even-Aged Western Hemlock Stands in Alaska

Site Index (Base Age 100)

Stand Age	60	80	100	120	140	160
20	0	500	850	1,300	1,700	2,100
30	800	1,550	2,300	3,050	3,800	4,600
40	1,850	2,900	4,200	5,500	6,900	8,000
50	2,700	4,250	6,000	7,800	9,500	10,900
60	3,400	5,300	7,300	9,500	11,650	13,300
70	4,000	6,100	8,400	10,900	13,200	15,200
80	4,500	6,850	9,350	12,000	14,500	16,500
90	5,000	7,500	10,250	13,000	15,700	17,600
100	5,350	8,000	11,000	13,800	16,550	18,500

Source: Barnes, G.H., 1962. Yield of Even-Aged Stands of Western Hemlock. US Department of Agriculture, Forest Service, Washington, DC. Technical Bulletin No. 544. 52 p.

The ratio of the two basal area estimates could be used to adjust the estimate of volume for the stand measured in the field.

Example

Expanding on the previous example, assume that your 60-year old stand had a basal area of 220 ft^2 per acre. If a well-stocked, normal stand at this age and on these types of sites should have 289 ft^2 per acre of basal area, then what would your adjusted volume per acre estimate be?

$$\text{Volume per acre} = \left(\frac{220 \text{ ft}^2 \text{ per acre}}{289 \text{ ft}^2 \text{ per acre}} \right) 9,500 \text{ ft}^3 \text{ per acre}$$

$$= 7,232 \text{ ft}^3 \text{ per acre}$$

B. Growth and Yield Simulators

A growth and yield simulator allows you to project into the future the structural characteristics of a stand of trees, and forecast the likely characteristics of the stand under varying management regimes. Growth and yield simulators are computer programs that allow the user to create and evaluate how management activities may change the character of a forest under different circumstances. They are similar to yield tables in that an estimate of the potential characteristics of a stand can be obtained. However, yield tables are limited by the number and resolution of categories from which you can directly arrive at stand volumes or densities, whereas growth and yield simulators are generally not. Modeling stand growth development

requires a number of assumptions about individual stands, and may give the illusion that stands grow in a nice, predictable manner (see the western forest stand presented in Appendix A). In actuality, a number of factors that may affect stand growth are not inherent in some growth and yield models, such as climate and precipitation variations. In fact, stands may grow somewhat irregularly when compared to the smooth growth curves provided by some modeling systems. However, growth and yield simulators are useful for developing estimates of projected future conditions of forests, and given the generally broader suite of factors involved in predicting growth dynamics, provide an advance over the use of volume or yield tables.

Our treatment of growth and yield simulators is not all-inclusive. Ritchie (1999) has gone much further, and described the capabilities of 31 growth and yield models for the west coast of North America. An interesting description of current and past approaches for modeling forests in Spain (Bravo et al., 2011) delved deeper into the variety of approaches one might use to estimate forest conditions. All growth and yield models have their limitations, some of which include making projections beyond the range of data that were used to create the growth and yield relationships. As a result, there will ultimately be some combination of site class, tree species, and management action for which the outcomes of a projection may have a limited (yet perhaps realistic) basis. The legitimacy of the output from a growth and yield model must ultimately be determined by the user and it is their responsibility to ensure that the projections are reasonable (Ritchie, 1999). Managers and analysts who need to project stand conditions into the future should consider the geographic location, management history, and composition of the data prior to deciding which model is more appropriate for the objectives of the effort.

1. Individual Tree, Distance-Independent Models

The basic modeling unit for these types of growth and yield simulators is the individual tree, or the tree record. Individual tree measurements (DBH, height, etc.) differ from tree records in that they represent only one tree per unit area. A tree record contains the same type of data (DBH, height, etc.), yet represents more than one tree per unit area. The number of trees that a tree record represents is sometimes referred to as the *expansion factor*. The entire list of trees or tree records is called a *tree list*. The periods of time that are projected by these models are usually in 5- or 10-year intervals; however, some models allow annual projections of tree growth, especially those that were developed for intensive plantation forestry purposes. Mortality is simulated by applying a probability of death for a given projection period. The expansion factor associated with each tree record therefore is adjusted accordingly each time period. For planners and managers interested in the production of snags, the projected mortality by diameter class can be determined by assessing the change in tree record expansion factors from one time period to the next.

The growth and mortality of individual trees is a function of the size and location of trees in a stand with respect to other vegetation with which it will compete for light, water, and nutrients. However, the actual distance from one tree to the next is not used as a variable in these models. Distance-independent growth and yield models use measures of stand density, such as basal area, as a proxy for competition among trees. Competition can also be implied given the diameter, height, and crown characteristics of a tree in relation to other trees being modeled in the stand. To determine stand-level characteristics using individual tree models, each tree record first is grown and perhaps subjected to a mortality probability function, then the volume of all the trees of a certain status (e.g., still alive) is determined and the appropriate expansion factor is applied (trees per unit area). The sum of the contribution of each individual tree or tree record for each stand-level characteristic is then used to produce stand-level estimates.

2. Individual Tree, Distance-Dependent Models

Distance-dependent growth and yield models use detailed measurements of the spatial position of each tree in relation to their neighbors to model competition among trees. These types of models attempt to use this spatial information to account for the competition for light, water, and nutrients among trees. Some of these types of growth and yield models emulate three-dimensional structures of tree attributes (tree location, height, diameter, and crown characteristics) to derive a three-dimensional view of the stand structures. This three-dimensional view then serves as the basis for measuring competition for every tree, and the allocation of resources. Tree diameter growth, height growth, and changes to crowns are all controlled in this manner. When using individual tree models, the potential growth of each tree is projected into the future, particularly in mixed-species stands, and can be very specific. Models such as these can be used in pure and mixed stands of all age combinations, thus are of value in projecting the growth and yield of uneven-aged stands (Hanewinkel and Pretzsch, 2000). As with distance-independent models, to determine stand-level characteristics using individual tree models, each tree or tree record is first grown and perhaps applied a mortality probability, then the volume of all the trees of a certain status (e.g., still alive) are determined and applied

the appropriate expansion factor (trees per unit area). The sum of the contribution of each individual tree record to each stand-level characteristic is then used to produce the stand-level estimates.

3. Whole-Stand Models

Yield tables are one form of whole-stand models. For instance, some whole-stand models are essentially normal yield tables that are derived from measurements of natural stands. However, normal yield tables do not take into account differences in stand density, as they were developed to represent fully stocked stands of trees. Site index and stand age generally are used to determine an estimate of stand volume when using these tables. Some of the difficulties in using normal yield tables include accounting for differences in stand structure between the stand for which an estimate is desired and the stands that were used to create the tables, and differences in growth rates among stands with different densities. However, normal yield tables allow one to estimate maximum yields for stands of trees at various ages and site qualities, and to estimate how these yields may change given some estimate of a stand's relationship to potential full stocking.

Another type of whole-stand model uses empirical yield tables that are derived from measurements of stands that have been managed, and these models then reflect the average conditions expected throughout the life of a stand. Empirical yield tables thus take into account an average stand density. One of the problems with using empirical yield tables is the fact that some data are from measurements of young stands, and other data are from measurements of older stands, and each may not have been applied the same history of management, creating a distortion in growth projections (Davis et al., 2001).

Whole-stand simulators use stand-level data as input, yet ignore much of the detail associated with individual tree simulators (Ritchie, 1999). Common stand-level data that are needed to utilize a whole-stand simulator include stand age, site index, stand density, and quadratic mean diameter. In some cases, transition probabilities, matrices of growth probabilities for various stocking and density classes, can be used to project whole-stand conditions through time. Whole-stand simulators provide stand-level output of value to land managers, such as the basal area per unit area and other measures of stand density, as well as volume. However, tree-level data generally is not provided, but tree-level information in the form of a diameter distribution (number of trees per unit area, by diameter class) can be imputed from the stand-level information by some of these models, based on probability density functions such as the Weibull curve. These types of diameter distribution models grow a theoretical distribution, rather than classes of diameters. The growth function relates to the characteristics of the stand, and diameter class information is generated by disaggregating the distribution down to diameter classes. Although we suggested that the diameter distribution of even-aged stands tends to be described by a bell-shaped curve (a normal distribution), the Weibull function commonly is used to represent a distribution of trees in an even-aged stand (Bailey and Dell, 1973). The Weibull function allows the representation of extreme values (limiting values), such as the largest tree in a stand, by assuming that the distribution does not continue off into infinity, but rather becomes truncated at some maximum tree diameter. Some disaggregative simulators also allocate growth from whole-stand models to a tree list (Ritchie, 1999). For the most part, whole-stand models are relatively easy to use in comparison to individual tree models, but they may not provide information as reliable as individual tree models for stands with mixed species (Sironen et al., 2001).

4. Diameter Class Models

Diameter class models use more detail than whole-stand models in projecting forest conditions through time, and rather than project the entire stand condition at once, they project the development of each diameter class within a stand separately. These models sometimes are referred to as stand table projection systems, and they represent a compromise between the whole-stand models and individual tree models. The projection of stand tables is a technique used to determine the future structural condition of both even-aged and uneven-aged forests. When the diameter classes become very large, the models tend to behave in a manner similar to whole-stand models. When the diameter classes become very small and finite, they behave in a manner more similar to individual tree models. These methods estimate the structural condition of a future stand table from the current condition of a stand table by adjusting each diameter class accordingly using an increment and a mortality probability.

As we suggested, diameter class models simulate the growth of trees in each diameter class; therefore, the size of the classes needs to be defined. The number of trees in each class also are defined and independently modeled as a class of trees. A key methodology inherent in diameter class models involves the process of projecting the growth of the individual diameter classes. A diameter class may contain 50 trees. These 50 trees can either be projected using (1) the average growth rate for trees of the mid-point of the size class, (2) the actual growth rates from individual trees of different sizes to grow the classes, or (3) a growth rate appropriate for different tree sizes within a single class. The latter approach assumes that the trees within a diameter class are more finely

distributed within the class at the time of projection using an assumed distribution, such as a uniform distribution. Estimated diameter increments for each class can be determined through the use of regression models that were developed from field studies, estimates of mean increments observed from field studies, or even educated guesses. Some methods increment diameter classes in such a way that the same trees (less mortality) move forward one class, other methods assume that not all trees in a diameter class are alike, and the number of trees that transition upward to the next diameter class is probabilistic. As with whole-stand models, transition probabilities can be used here to project diameter classes through time. More recent approaches in diameter class models involve using growth and mortality rates to estimate more closely the transition of trees from one diameter class to the next. To determine volumes using diameter class models, the diameter classes are grown, then expanded by the trees per unit area that are represented by the class (where necessary) and subjected to the appropriate volume computation methods.

5. Gap Simulators

Gap simulators are very similar to individual tree models in that trees serve as the basis for simulation. Each tree is represented spatially in the model by the gap that it might occupy in the canopy over a given space such as an acre or hectare. Forest dynamics are then simulated based on the light made available from gaps in the canopy caused by mortality. Mortality is modeled using a probability of death; the greater the competitive forces on the tree the higher the probability of death. Subsequently, ingrowth is modeled in the gap given assumptions of the different species reactions to changes in sunlight and nutrient availability. Given that these models were initially developed for ecological modeling purposes rather than timber production purposes, some gap simulators remove the trees from the simulation process when they are assumed to have died or been harvested, and timber volume may not be reported. Other gap simulators allow the reporting of tree records that have been managed in this manner. Gap models include stochastic elements, and therefore may need to be run multiple times to develop a pattern of forest growth behavior. The average pattern of behavior then is reported. Some concern has been noted previously over possible invalid assumptions of growth rates in these models; other concern has been noted regarding assumptions of the regeneration processes within the gaps (Yaussy, 2000).

6. Snag and Coarse Woody Debris Models

As we mentioned in Chapter 2, Valuing and Characterizing Forest Conditions, snags are dead, standing trees, and coarse woody debris includes those parts of former snags that now reside on the ground, including the stumps. Projecting the amount of each through time may be important for various wildlife habitat suitability models. Snags generally are reported in the same manner as live trees (number of snags per unit area); however, estimates can be further refined by diameter class and by decay class. Coarse woody debris generally is reported in the same manner as standing tree volume (amount per unit area), and estimates again may be further refined by size and decay class. The decay of snags and logs is highly variable, and dependent on local site conditions, regional weather patterns, and tree species. On the same site, some tree species, such as eastern red cedar (*Juniperus virginiana*), may take longer to decay than others, such as shortleaf pine (*Pinus echinata*). For example, a recently deceased 12-inch diameter shortleaf pine snag might fall and become coarse woody debris in less than 5 years, whereas a recently deceased eastern red cedar of the same size may stay standing several decades. It goes without saying that snags and coarse woody debris in arid or cold climates will likely take longer to decay than snags and coarse woody debris in warm, humid climates.

Snags can be reported from some growth and yield models either directly (through mortality) or as the difference in live trees from one measurement period to the next. How snags transition from standing dead trees to coarse woody debris to, eventually, soft forest floor duff, is another matter entirely. Some snag models (Mellen and Ager, 1998) include relationships for breakage (height loss) and subsequent falling of pieces of wood to the forest floor. Decay rates can be applied to both the standing snags and the coarse woody debris on the forest floor to transition them to the softer decay classes. As you may have gathered, projecting the condition of snags and coarse woody debris decay is essentially the opposite of that related to projecting the condition of live trees. A number of minor differences can be found between the two, however. First, there may be ingrowth of snags into any diameter class in any projection period, not just into the smallest diameter class (as in the case of live trees). Also, accretion is modeled as the input or decay of a snag or down log, rather than as growth of a tree. Even though each are technically physiologically dead, mortality of snags and coarse woody debris can be assumed because functionally; for example, they may be of value for many wildlife species only for a limited amount of time (until they have become too soft, or have decayed too much).

C. Brief Summary of Some Growth and Yield Simulators

There are too many growth and yield simulators to provide a thorough treatment of each in this text. The following summary is of a number of commonly used models throughout North America. We suggest that, as a land manager, you work with your colleagues and nearby educational and public land management institutions to determine the appropriate simulator for the forest conditions in your area, and for the purpose of your analyses.

1. Forest Vegetation Simulator

The Prognosis model was an individual-tree, distance-independent model developed over 20 years ago for application in the interior northwest. It has since been renamed the Forest Vegetation Simulator (FVS), and has been modified to allow the projection of forest growth and yield for a number of geographical regions in the United States, and has been adapted for use in British Columbia as well. The different processes that provide users this flexibility are called *variants* in the FVS system. Each variant involves different methods for projecting stand development, and extensions can be applied to the FVS model to simulate the impact of insect and disease outbreaks as well as fire events. Although the output from FVS is consistent from one variant to the next, each utilizes different volume tables and may quantify site productivity differently (Ritchie, 1999). A large number of tree species can be modeled with FVS variants, which distinguishes this system from many of the others where modeled geographic areas overlap. Users of FVS need to provide it with a species code, DBH, height, and expansion factor for each tree record. A variety of management actions can be simulated in FVS, from a number of different thinning regimes within a stand to the timing of a final harvest. The FVS modeling system is available at no cost from the Forest Management Service Center of the US Forest Service (2013b).

2. California Conifer Timber Output Simulator

For modeling mixed-conifer stands in California, the California Conifer Timber Output Simulator (CACTOS) may be of value. CACTOS is an individual-tree, distance-independent model that provides projections not only for conifers, but also for several hardwood species commonly found in California, such as chinkapin (*Castanopsis chrysophylla*) and tanoak (*Lithocarpus densiflorus*). As with the FVS model, users provide the model with a tree list that indicates a species code, DBH, and expansion factor for each tree record. The addition of a measured tree height, age, and crown ratio would improve the projections. If these are not available, then height and crown ratio are imputed. And as with FVS, a variety of management actions can be simulated in CACTOS, from a number of different manners to thin a stand, to a final harvest. Details of the CACTOS model can be obtained from the California Department of Forestry and Fire Protection (2012).

3. ORGANON

The ORGANON model is an individual-tree, distance-independent growth and yield projection system that has been used by many land management organizations for simulating the growth and yield of mixed-conifer stands in Oregon and Washington. There are at least three versions of the model that can be applied to different parts of the Pacific Northwest region. A limited number of hardwood species, such as Oregon white oak (*Quercus garryana*) can also be modeled in conjunction with the conifer species. As with the other models described thus far, a species code, DBH, height, and expansion factor are required for each tree record represented in a stand. Crown ratio would help improve the quality of projections, yet if it is not available, it is imputed for each tree record. An estimate of site index (base age 50, either Douglas-fir or western hemlock) is required for each stand. A number of management actions can be modeled with ORGANON, from thinnings to fertilization treatments. ORGANON is available at no cost from the Organon Growth and Yield Project at Oregon State University (Organon Growth and Yield Project, 2013).

4. Zelig

Zelig is a growth model that simulates each forest gap as it may progress through different phases of stand development (Urban, 1990). These types of gap models emphasize the influence of environmental factors on the growth of stands of trees. Three fundamental forest processes are simulated: accretion, mortality, and regeneration. An estimate of the maximum potential behavior of trees is made first, which is then tempered by constraints related to a variety of resources that may be limiting, such as sunlight, water, and characteristics of the soil resources. Leaf area index is the driver of tree growth. One strength of Zelig is its ability to modify tree growth based on the interaction of climate and site conditions. Annual precipitation and growing degree days interact with soil conditions to influence tree growth. Tree growth is further influenced by the ability of each tree to obtain sunlight, nutrients, and water (Yaussy, 2000). Mortality is based on both density-dependent relationships and density-independent relationships (stand density).

General information about each stand that the gap model requires are tree age, species, and DBH. Maximum age, maximum height, and reproductive success of each tree species need to be provided by the user, as do a number of soil, climate, and weather parameters. Tree height is estimated using each tree's DBH. The output from Zelig includes stand-level data reporting measures of trees

per unit area, basal area, average DBH, biomass, and leaf area index. Given some adjustments to the model, local volume equations can be incorporated to allow the development of timber volume estimates through time. Zelig has been used in a number of regions of North America, and simulations can be performed for very long time periods (hundreds of years).

5. DFSIM

For simulating the structure of even-aged, managed stands of Douglas-fir in the Pacific Northwest, the Douglas-fir simulator (DFSIM) model is available. DFSIM is a whole-stand simulator that allows the modeling of a number of thinning options along with fertilization and precommercial thinning activities. One limitation of the model is that it fails to model ingrowth of trees in gaps created by thinnings or mortality. For each stand, a site index and stand age are required, along with some indicators of stand density (basal area, trees per acre). A wide variety of output is produced from DFSIM, including average stand structural characteristics (height, diameter), stand density measures, stand increment (mean annual and periodic), as well as timber volumes (Ritchie, 1999). The model can be obtained from the Pacific Northwest Research Station of the US Forest Service in Olympia, Washington (US Forest Service, 2013a).

6. ASPEN

The ASPEN model is a whole-stand growth and yield simulator designed for aspen (*Populus tremuloides*) stands in the northern regions of North America. Stand age, dominant height, quadratic mean diameter, and product specifications are used as inputs to the model, and projections are made in annual increments. Outcomes include total number of live trees, basal area, and merchantable volume and biomass per unit area. A number of management alternatives can be explored for aspen stands. The model can acquired from the Natural Resources Research Institute of the University of Minnesota (Host and Perala, 1996).

7. PTAEDA 4.0

To simulate and project the structure of loblolly pine plantations in the southern United States, PTAEDA 4.0 is available. This model is an individual-tree, distance-dependent growth and yield model that accounts for the effects of biological and physical variables on the photosynthesis and respiration processes of trees (Burkhart et al., 2008). Growth projections are made for individual trees, then they are summed to produce stand-level estimates. Stands can be modeled from inception (time of planting) or from some intermediate point in a rotation based on recently acquired inventory data. Growth of trees is projected using theoretical growth potential relationships, and then adjusted to reflect the competitive status of each tree. Mortality is incorporated by assessing the probability of survival each year. This simulator is available from the Loblolly Pine Growth and Yield Research Cooperative within the Department of Forestry at Virginia Tech (Burkhart et al., 2008).

8. Tree and Stand Simulator

The Tree and Stand Simulator (TASS) is an individual-tree, distance-dependent growth and yield model developed by the British Columbia Forest Service. This model projects the characteristics of eight types of even-aged, pure conifer stands. Spatially-explicit tree maps are required to allow the model to simulate height growth, foliage and branch responses, suppression, and ultimately mortality. Some insect and disease impacts can be accommodated within the projection system. Although widely used in western Canada, some of the limitations of the model include the inability to apply the model to mixed species stands or stands that are not even-aged, and the inability to accommodate hardwood tree species. The model can be obtained from the British Columbia Ministry of Forests and Range (2007).

9. Simulator for Intensively Managed Stands

In the southern United States, growth and yield for planted pine, natural pine, and hardwood stands can be accomplished using the Simulator for Intensively Managed Stands model (SiMS). This model is a diameter distribution projection model that allows the simulation of a number of intensive management actions. Required inputs include site index, tree species, DBH, height, and trees per unit area. The trees per unit area can be in the form of a diameter distribution, or if a stand projection begins at the time of planting, in the form of site index and number of trees planted. In the latter case a diameter distribution is imputed at the time of the first activity, using a Weibull distribution. Afterward, the residual diameter distribution is projected through time. The amount of herbaceous and woody competition can be factored into the projections, and release and fertilization treatments along with a variety of site preparation options can be modeled. These options can be modeled along with the ability to report yields by product type (pulpwood, chip-n-saw, sawtimber, poles, etc.), which distinguishes this model from many of the others. Information regarding the SiMS model can be obtained from ForesTech International, LLC.

10. Landscape Management System

The landscape management system (LMS) began as a method for efficiently performing many of the repetitive tasks that are required in forest planning and management for projecting a stand through time under a variety of management alternatives. In the course of forest planning, developing the databases requires more than 50% of the effort of planners and analysts. In some cases the large number of alternatives combined with the large number of stands makes projecting alternatives into the future a cumbersome data management process. As a result, the amount of data that might be generated in the development of a management plan could become prohibitively large. LMS was created as a software program that could associate more closely stand-level data, geographic information, and growth and yield models to make some parts of the planning effort more efficient. LMS incorporates the capabilities of ORGANON and FVS individual-tree, density-independent growth and yield models to enable the projection of stands into the future. In addition, LMS contains models to visualize, structurally, stands in three dimensions, as well as a stand's place on a landscape. LMS can be obtained from the cooperative formed by the College of Forest Resources at the University of Washington and the School of Forestry and Environmental Studies at Yale University (McCarter, 2013).

IV. OUTPUT FROM GROWTH AND YIELD MODELS

The data that can be generated from whole-stand models is generally more limited than output from individual tree or gap models. In whole-stand models, total stand estimates may be provided for various measures of stand density (trees per unit area, basal area), average stand conditions (DBH, height), and measures of volume. Data that can be made available from individual tree models include estimates of individual tree (or tree record) characteristics through time, including the amount of mortality. In addition, this data can be aggregated to produce stand tables (trees per unit area by diameter class) and stock tables (volume per unit area by diameter class), and tables that represent stand-level volumes and structural characteristics. Other types of output that may be generated by various models include the culmination of mean annual increment, forest products and other economic or commodity production estimates, and coarse woody debris estimates. The amount and type of output will vary by model, and in some cases additional programs may need to be developed to produce the data desired in forest management and planning.

V. MODEL EVALUATION

Users of growth and yield models must consider how they will apply the model and how they will assess its performance, and ultimately decide whether the model meets their needs (Brand and Holdaway, 1983). Evaluation of a model is not a simple process, and is partially subjective and partially a function of the objectives of a planner or analyst. Each growth and yield model has its strengths and weaknesses; as a result, Buchman and Shifley (1983) suggested that when a decision regarding the adoption of a model arises, some broad areas of concern should be taken into consideration: those related to the application environment, the performance of the model, the biological realism, and the design of the modeling process. Some of the main questions to ask are as follows.

Concerning the application environment:

- Is the model well-documented?
- Is assistance available from the developer?
- Is there a user's group that can provide assistance?
- If the model is a computer program, is it user-friendly and intuitive?
- Who will support the model over time?
- What data are required?
- Are the data requirements compatible with the data currently being managed?
- If data are missing, can the model impute (estimate) the data?
- Are illegal data values caught and is the user alerted?
- Is model calibration necessary for local conditions?
- What type of computer is needed to run the model?
- How fast does the model run on a new computer?

Concerning the performance of the modeling system, the biological realism, and the design:

- How accurate and precise are the projections?
- Do the projections contain bias under changing management circumstances?
- Does accuracy deteriorate with longer projections?
- Is the system easily modified for changing management circumstances?
- Can some parts of the modeling system be avoided if not needed?
- Are there conditions that would cause the model to project unreasonable yields or forest structures?

Henderson et al. (2013) compared the growth predictions from five southern United States loblolly pine (*P. taeda*) growth and yield simulators, using a graphical approach to compare and evaluate various relationships: comparing tree height at each tree age, comparing tree height at each tree diameter, and comparing tree volume at each tree age. In addition, the Sukachev effect (comparing trees per unit area at each age), Reineke's rule (comparing mean diameter at each age), a tree spacing relationship (comparing trees per unit area and tree height), and a yield-density effect (comparing trees per unit area and tree volume) were evaluated amongst the

five growth and yield models. The results indicated that differences can exist between models designed to emulate the growth of the same tree species, with some outcomes perhaps being inconsistent with the biological relationships ingrained in forest stand dynamics. The selection and potential use of a model can therefore have important managerial ramifications.

More difficult to assess are the costs associated with changing from one system to another or simply adopting a system for the first time, and with the training and personnel needed to make it work effectively. A hybrid benefit/cost analysis that allows both qualitative and quantitative aspects to be assessed as part of the decision process might be of value (Bettinger et al., 2010). As a result of these issues, natural resource professionals who must evaluate and select from among the various growth and yield models may face a challenging task (Buchman and Shifley, 1983).

VI. SUMMARY

Projections of future forest conditions result in a necessary set of information for the assessment of alternative management plans. A large portion of the forest management and planning effort should be placed on careful and reasonable projections of current and future conditions. The growth and development of forest structures can be modeled using a number of methods, from volume and yield tables to growth and yield simulation models. The outcomes from these modeling processes are critical in assessing the economic, environmental, and social aspects of management alternatives. Although it is obvious that future commodity production plans are closely tied to the projections of forest conditions, a number of wildlife habitat relationships are also contingent on forest stand density, tree species configurations, and snag and coarse woody debris loads that are derived from natural mortality and decay projections. The methods used to project forest conditions into the future need to be considered carefully, as some regional models may not be applicable over broad areas. In addition, the input required, process emulated, and output desired may influence the selection of a modeling process.

QUESTIONS

1. *Forest growth dynamics.* Given the following data regarding recent annual changes in forest land volume in Maine, what are the gross growth, net growth, and net change for the balsam fir (*Abies balsamea*), red spruce (*Picea rubens*), and red maple (*Acer rubrum*) resources? The average annual change in volume on forest land

in Maine, for three select species was derived from McWilliams et al. (2005).

Tree species	Ingrowth (1000 ft³)	Accretion (1000 ft³)	Mortality (1000 ft³)	Removals (1000 ft³)
Balsam fir	43,696	82,158	83,752	69,108
Red spruce	14,967	108,875	41,585	84,330
Red maple	16,477	87,335	14,659	76,450

2. *Nonlinear diameter distribution relationships of uneven-aged stands.* For Tract 38 from Meyer (1952), what would you expect the trees per acre to be in the 13-inch diameter class?

3. *Nonlinear diameter distribution relationships of uneven-aged stands.* For Tract 38 from Meyer (1952), what would you expect the trees per acre to be in the 14-inch diameter class? How is this estimate of trees per acre different from the answer provided for question 2, and why?

4. *Nonlinear diameter distribution relationships of uneven-aged stands.* Develop a stand table for Tract 41 from Meyer (1952), where $a = 0.163$, $k = 66$, and $q = 1.385$ for a 2-inch diameter class table, and 1.177 for a 1-inch diameter class table. Use diameters ranging from about 1 to 40 inches.

5. *Forest growth dynamics.* What is the net annual change in basal area of a fully stocked, even-aged, 33-year old upland oak site in Kentucky described using the following table? The growth transition of a fully stocked, even-aged, 33-year old upland oak site in Kentucky, over 7 years, was derived from Dale (1972).

	Basal area at beginning (ft²/ac)	Basal area at end (ft²/ac)
White oaks	73.3	75.3
Red oaks	18.0	16.5
Walnut, yellow-poplar, ash, and others	1.1	1.0
Hickory, gum, maple, and others	5.5	5.5
Dogwood, sourwood, sassafras, and others	2.2	1.7

6. *Even-aged versus uneven-aged management.* Assume that you are a forestry consultant in Missouri, and are advising a landowner who owns 350 acres of 50−60-year old mixed hardwood stands. The landowner is a bit confused about the even-aged and uneven-aged

approaches to the management of the forest. Prepare for them a short memorandum that describes the main similarities and differences between the two management approaches to their forest.

7. *Yield and stock tables.* Assume that you are working for the Bureau of Indian Affairs in northern Arizona, and are involved in the planning of a forested area. When describing how to model the growth and yield of forests, members of the planning team have thrown around the terms "volume table" and "yield table," and as a result you determine that they may be unfamiliar with the characteristics of each. Prepare for the planning team a short memorandum that describes the similarities and differences between the two approaches for estimating tree and stand volumes.

8. *Growth and yield models.* Assume that you work for a forestry consulting firm in north Florida and are given the task of projecting the growth of forests 20 years into the future. These analytical efforts will support the development of forest plans for private landowners that have come to your firm for assistance. Your managers are unsure which approach is more appropriate for estimating forest conditions. Describe in a short report the similarities and differences between distance-independent models, distance-dependent models, whole-stand models, and gap simulators.

REFERENCES

Bailey, R.L., Dell, T.R., 1973. Quantifying diameter distributions with the Weibull function. For. Sci. 19 (2), 97−104.

Barnes, G.H., 1962. Yield of Even-Aged Stands of Western Hemlock. US Department of Agriculture, Forest Service, Washington, DC., Technical Bulletin No. 544. 52 p.

Beers, T.W., 1962. Components of forest growth. J. For. 60 (4), 245−248.

Bégin, E., Bégin, J., Bélanger, L., Rivest, L.-P., Tremblay, S., 2001. Balsam fir self-thinning relationship and its constancy among different ecological regions. Can. J. For. Res. 31 (6), 950−959.

Bettinger, P., 2006. Measures of change in projected landscape conditions. J. For. Plan. 12 (2), 39−47.

Bettinger, P., Lowe, T., Siry, J., Merry, K., Nibbelink, N., 2010. Analytical decisionmaking considerations for upgrading or changing geographic information system software. J. For. 108 (5), 238−244.

Brand, G.J., Holdaway, M.R., 1983. Users need performance information to evaluate models. J. For. 81 (4), 235−237, 254.

Brandeis, T.J., 2015. Georgia's Forests, 2009. US Department of Agriculture, Forest Service, Southern Research Station, Asheville, NC, Resource Bulletin SRS-207. 59 p.

Bravo, F., Alvarez-Gonzalez, J.G., del Rio, M., Barrio, M., Bonet, J.A., Bravo-Oviedo, A., et al., 2011. Growth and yield models in Spain: historical overview, contemporary examples and perspectives. For. Sys. 20 (2), 315−328.

British Columbia Ministry of Forests and Range, 2007. Research Branch Extension Note 80. British Columbia Ministry of Forests and Range, Victoria, BC. Available from: https://www.for.gov.bc.ca/hfd/pubs/Docs/En/En80.htm (Accessed 2/10/2016).

Buchman, R.G., Shifley, S.R., 1983. Guide to evaluating forest growth projection systems. J. For. 81 (4), 232–234, 254.

Buckman, R.E., Bishaw, B., Hanson, T.J., Benford, F.A., 2006. Growth and Yield of Red Pine in the Lake States. US Department of Agriculture, Forest Service, North Central Research Station, St. Paul, MN, 114 p.

Burkhart, H., Amateis, R.L., Westfall, J.A., Daniels, R.F., 2008. PTAEDA 4.0: Simulation of Individual Tree Growth, Stand Development and Economic Evaluation in Loblolly Pine Plantations. Department of Forestry, Virginia Tech, Blacksburg, VA.

California Department of Forestry and Fire Protection, 2012. Growth. California Department of Forestry and Fire Protection, Sacramento, CA. Available from: http://www.fire.ca.gov/resource_mgt/resource_mgt_stateforests_forestry_growth (Accessed 2/10/2016).

Clutter, J.L., Harms, W.R., Brister, G.H., Rheney, J.W., 1984. Stand Structure and Yields of Site-Prepared Loblolly Pine Plantations in the Lower Coastal Plain of the Carolinas, Georgia, and North Florida. US Department of Agriculture, Forest Service, Southeastern Forest Experiment Station, Asheville, NC, General Technical Report SE-27. 173 p.

Dale, M.E., 1972. Growth and Yield Predictions for Upland Oak Stands. 10 Years After Initial Thinning. US Department of Agriculture Forest Service, Northeastern Forest Experiment Station, Upper Darby, PA, Research Paper NE-241.

Davis, L.S., Johnson, K.N., Bettinger, P., Howard, T.E., 2001. Forest Management. McGraw-Hill, Inc, New York, NY, 804 p.

De Liocourt, F. 1898. De l'aménagement des Sapinières. Bul. de la Siciétié Forestière de Franch-Conté et Belfort, Besancon.

Drew, T.J., Flewelling, J.W., 1979. Stand density management: an alternative approach and its application to Douglas-fir plantations. For. Sci. 25 (3), 518–522.

Forbes, R.D., Bruce, D., 1930. Rate of Growth of Second-Growth Southern Pines in Full Stands. US Department of Agriculture, Forest Service, Washington, DC, Circular No. 124. 77 p.

Hanewinkel, M., Pretzsch, H., 2000. Modelling the conversion from even-aged to uneven-aged stands of Norway spruce (*Picea abies* L. Karst.) with a distance-dependent growth simulator. For. Ecol. Manage. 134 (1–3), 55–70.

Henderson, J., Roberts, S.D., Grebner, D.L., Munn, I.A., 2013. A graphical comparison of loblolly pine growth and yield models. Southern J. Appl. For. 37 (3), 169–176.

Host, G.E., Perala, D.A., 1996. ASPEN—A Circumboreal Growth and Yield Model for *Populus Tremuloides* and *P. tremula*: User's Guide. University of Minnesota, Natural Resources Research Institute, Duluth, MN. Available from: http://www.d.umn.edu/~ghost/FMSASPEN.HTM (Accessed 2/10/2016).

Johnson, F.S., 1955. Volume Tables for Pacific Northwest Trees. US Department of Agriculture, Forest Service, Washington, DC., Agriculture Handbook No. 92.

Mason, W.L., 2015. Implementing continuous cover forestry in planted forests: experience with Sitka spruce (*Picea sitchensis*) in the British Isles. Forests. 6 (4), 879–902.

McArdle, R.E., Meyer, W.H., 1930. The Yield of Douglas Fir in the Pacific Northwest. US Department of Agriculture, Forest Service, Washington, DC, Technical Bulletin No. 201. 64 p.

McCarter, J., 2013. Landscape Management System Home Page. 2013. College of Forest Resources, University of Washington, Seattle, WA, and School of Forestry and Environmental Studies, Yale University, New Haven, CT. Available from: http://landscapemanagementsystem.org/index.php (Accessed 2/10/2016).

McWilliams, W.H., Butler, B.J., Caldwell, L.E., Griffith, D.M., Hoppus, M.L., Laustsen, K.M., et al., 2005. The Forests of Maine: 2003. US Department of Agriculture, Forest Service, Northeastern Research Station, Newtown Square, PA, Resource Bulletin NE-164. 188 p.

Mellen, K., Ager, A., 1998. Coarse Woody Debris Model, Version 1.2. US Department of Agriculture, Forest Service, Pacific Northwest Region, Portland, OR.

Meyer, H.A., 1952. Structure, growth, and drain in balanced even-aged forests. J. For. 50 (2), 85–95.

Meyer, H.A., Stevenson, D.D., 1943. The structure and growth of virgin beech-birch-maple-hemlock forests in northern Pennsylvania. J. Agr. Res. 67 (12), 465–484.

Organon Growth and Yield Project, 2013. Model Description. Organon Growth and Yield Project, College of Forestry, Oregon State University, Corvallis, OR. Available from: http://www.cof.orst.edu/cof/fr/research/organon/orginf.htm (Accessed 2/10/2016).

Palahí, M., Pukkala, T., Miina, J., Montero, G., 2003. Individual-tree growth and mortality models for Scots pine (*Pinus sylvestris* L.) in north-east Spain. Ann. For. Sci. 60 (1), 1–10.

Ritchie, M.W., 1999. A Compendium of Forest Growth and Yield Simulators for the Pacific Coast States. US Department of Agriculture, Forest Service, Pacific Southwest Research Station, Albany, CA, General Technical Report PSW-GTR-174. 59 p.

Sironen, S., Kangas, A., Maltamo, M., Kangas, J., 2001. Estimating individual tree growth with the k-nearest neighbour and k-most similar neighbor methods. Silva Fennica. 35 (4), 453–467.

Townsend, P., 2007. Ten Year Periodic Annual Increment for Nova Scotia Permanent Forest Inventory Plots 1990–95 to 2000–05. Nova Scotia Department of Natural Resources, Renewable Resources Branch, Forestry Division, Forest Inventory Section, Truro, NS, 38 p.

US Department of Agriculture., 1929. Volume, Yield, and Stand Tables for Second-Growth Southern Pines. US Department of Agriculture, Forest Service, Office of Forest Experiment Stations, Washington, DC, Miscellaneous Publication No. 50. 202 p.

US Department of Agriculture, Forest Service, 2012. Huron-Manistee National Forests Land and Resource Management Plan (as Amended January 2012). US Department of Agriculture, Forest Service, Eastern Region, Milwaukee, WI.

US Forest Service, 2013a. DFSIM With Economics. US Forest Service, Pacific Northwest Research Station, Olympia Forestry Sciences Laboratory, Olympia, WA. Available from: http://www.fs.fed.us/pnw/software/DFSIM14/DFSIM.htm (Accessed 2/10/2016).

US Forest Service, 2013b. Forest Vegetation Simulator (FVS). US Forest Service, Forest Management Service Center, Fort Collins, CO. Available from: http://www.fs.fed.us/fmsc/fvs/ (Accessed 2/10/2016).

Urban, D.L., 1990. A versatile model to simulate forest pattern: a user guide to ZELIG, version 1.0. Environmental Sciences Department, University of Virginia, Charlottesville, VA.

Yaussy, D.A., 2000. Comparison of an empirical forest growth and yield simulator and a forest gap simulator using actual 30-year growth from two even-aged forests in Kentucky. For. Ecol. Manage. 126 (3), 385–398.

Zeide, B., 2005. How to measure stand density. Trees. 19 (1), 1–14.

Chapter 5

Optimization of Tree- and Stand-Level Objectives

Objectives

In this chapter, we explore methods for optimizing the management of a tree or stand of trees. The goal here is to derive the optimal management of an individual unit that in turn may contribute to the overall management of the forest. A number of resource managers and analysts dislike the term *optimize*, perhaps because it implies one or more goals will override other goals, even if the implied importance of goals are determined by the landowner. Instead, *simulated* management actions may be more palatable to many resource managers and analysts. At its basic level of understanding, simulation of management actions refers to the projection of a stand or forest through time given a predefined set of management actions. If stochastic (random) processes are involved in the representation of the system being projected, then multiple simulations of the same set of management actions may lead to different results. The distribution of these results can then facilitate the development of a confidence interval around which likely (or unlikely) outcomes might be realized. Whether or not stochastic processes are included in the representation of a system, we could infer that simply projecting a stand into the future, given a set of management actions, represents a form of simulation. As a result, simulation may be a valuable tool for the forest management and planning process, since what we hope to observe with repeated simulations are the consequences of a set of management actions applied to a stand or forest. In effect, we are bringing the stand or forest into our office to study a number of alternatives (Buongiorno and Gilless, 1987). On the downside, there is no guarantee that the simulated alternatives are optimal given the resources available.

Exploring different management options for trees or stands implies that by some ad hoc, haphazard, or systematic method, you look into or investigate different management regimes for a stand of trees. As with simulation, exploring options for trees or stands in this manner does not imply that the best schedule of management activities will be found, it simply suggests that you examined some alternatives. Since resources—time, energy, and money—usually are limited, many landowners place a high level of importance on reducing or eliminating their wasteful use. As a result, there is the need to optimize, or make as efficient as possible, the schedule of activities that should be applied to trees or stands of trees. Upon completion of this chapter, you should be able to:

1. Understand how and why optimization would be considered for determining plans of action for stands or individual trees.
2. Determine the optimal timing to cut an individual tree based on economic criteria.
3. Discuss the various methods by which you could develop an estimate of the optimal timber rotation length for an even-aged stand of trees.
4. Discuss the factors that should be considered when assessing the optimal thinning or partial cutting entries into an even-aged stand or an uneven-aged stand.
5. Discuss the management issues related to developing an optimal stand-level management plan that emphasizes stand density or stocking levels.
6. Understand the basic structure of dynamic programming, its recursive methods, and other issues related to its use in stand-level optimization.

I. INTRODUCTION

Each of us, in both our personal and professional lives, encounters numerous situations on a daily basis that require selecting actions and making decisions with some forethought of how future conditions will be affected. Some decisions can be difficult to make, and can require considerable time for pondering or analysis. If we were to act rationally, then we would need to think through the effects of all the potential choices and understand how they may affect other people or resources. Bellman (1957) once painted a colorful picture of personal and professional decision-making:

> in modern life, in economic, industrial, scientific and even political spheres, we are continually surrounded by multi-stage decision processes. Some of these we treat on the basis of experience, some we resolve by rule-of-thumb, and some are too complex for anything but an educated guess and a prayer.

Though ad-hoc methods for making forest management decisions are widespread even today, our natural resource management decision environment is too complex, economically, ecologically, and socially, and there are too many important goals to consider simultaneously (e.g., returns on investments, habitat quality, jobs) to make decisions in a haphazard fashion. The process of making decisions is not trivial, and is the source of great consternation among natural resource managers. Some decisions we make are associated with considerable economic and ecological uncertainty, whereas others will have a collateral effect on economic, ecological, or social conditions or processes. If a management situation can formally be described and sufficiently quantified, using variables and decisions associated with moments in time, then optimization methods can be helpful in sorting through the alternatives and suggesting courses of action that will provide the most benefit to the landowner. Optimization of objectives is meaningless only when the criteria for judging the value of the potential outcomes are unclear or lacking (Whittle, 1982).

Unfortunately, when the number of variables in a management problem is large, the determination of an optimum combination of management actions becomes difficult. As a consequence, many problems suffer from the *curse of dimensionality*. To fully enumerate a management problem, or evaluate every option, the number of options increases exponentially with each additional variable. For example, assume that you manage four stands of trees, and that over the next decade you have the opportunity to thin each of them once. To determine the number of possible combinations of actions for simple problems such as these is:

$$(\text{Number of choices})^{\text{Number of management units}}$$

In our case, there are two choices for each stand (thin or do not thin), so $(2)^4 = 16$ different plans of action. Imagine a more realistic case, such as the Putnam Tract described in Chapter 3, Geographic Information and Land Classification in Support of Forest Planning, where you have 81 stands of trees. Assume that each stand, on average, has potentially three management activities (including doing nothing) that could be applied over the next 15 years. The number of different management plans that would need to be assessed for full enumeration of the management situation would be $(3)^{81}$, or 4.4×10^{38}, an excruciatingly large number of alternatives. Similarly, if at the stand-level we were considering three thinning options for a stand during five different time periods, the number of distinct stand-level alternatives we would need to evaluate are:

$$(\text{Number of choices})^{\text{Number of time periods}}$$

or $(3)^5$, or 243. As a result, optimization techniques can assist us with sorting through the options in an efficient manner. The theme of this chapter is to explore options for scheduling a set of management activities for individual trees or stands. The theme of future chapters is to explore options for forests composed of multiple stands and various options for each stand.

II. TREE-LEVEL OPTIMIZATION

Optimal management decisions at the tree-level are perhaps the finest scale at which decisions are made in natural resource management. The main question facing a land manager concerns the purpose of each tree: How does it contribute to the landowner's objectives? Depending on the purpose of the tree, we may or may not be able to quantify its value. For example, landowners may indicate that purposes such as aesthetics, recreation, and wildlife habitat are important, yet these are difficult to value at the tree-level. How much value would you put on a single tree that can produce mast for deer, or on a tree that acts as your favorite deer stand? These assessments may require a subjective valuation on the part of the analyst or the landowner. However, economic analyses related to timber values can be much more objectively valued at the tree-level, and these types of assessments will be illustrated in this section.

To assess the economic value of a tree, a number of characteristics of the tree need to be ascertained, including its current growth rate, the presence of defect, and any other quality that might increase or reduce it market value. The economic value of trees can increase rapidly if a tree has a high growth rate, or has the potential to grow into higher lumber or product grades. Market prices for pulpwood are significantly less than sawtimber, which are lower than the prices for plywood or peeler logs. Understanding the dimensions and conditions that allow movement upward in grade or product are therefore important to a landowner. However, there are a number of potential characteristics that can be seen on a tree that can lower its value, such as cracks, splits, or other internal or external defects that will act to reduce the value of logs derived from a tree. As timber prices change, the value of a tree will change; therefore an assessment of future markets may be equally important.

The optimum economic solution to a tree-level management problem is one where the marginal cost of holding the tree is equal to the marginal revenue from cutting the tree. When expressed in terms of capital value, the marginal cost is represented by the alternative rate of return, and the marginal revenue is represented by the value growth rate of the tree (Chappelle and Nelson, 1964). Therefore, to optimize the economic value of a tree, we need to compare the rate of growth in value of the tree to the alternative rate or return (discount rate) specified by the landowner. If a tree is growing in value at a rate higher than the alternative rate of return, then the decision should be to allow the tree to grow and add value. If a tree is growing in value, due to a combination of quality and volume growth, at a rate less than the alternative rate of return, then the decision should be to sell the tree, and reinvest the money in opportunities that earn at least the alternative rate of return. To calculate the value growth percent of an individual tree, we need to know only two items: the current value of the tree, and the projected future value of the tree. The rate of growth can be determined using the following equation:

$$\text{Rate of growth} = \sqrt[\text{Years}]{\left(\frac{\text{Future value of tree}}{\text{Present value of tree}}\right)} - 1$$

Example

Assume you own a small stand of hardwoods in central Tennessee, and you are considering the options for a specific tree, a 100-foot tall yellow-poplar (*Liriodendron tulipifera*) that is currently 22 inches in diameter. Let's assume that the current stumpage prices for average grade yellow poplar logs might be about $200 per thousand board feet, after harvest cost, haul cost, and buyer profit are removed from the delivered log price (Tennessee Department of Agriculture—Division of Forestry, 2015). You measured the growth of the tree by extracting an increment core, and note that its rate of growth is about 0.5 inches in diameter per year. You estimate that it currently contains 614 board feet of wood, and that in 2 years it should contain 671 board feet of wood. If you have selected an alternative rate of return of 5%, then should you harvest the tree now or wait?

The present value of the tree is (0.614 thousand board feet) × ($200 per thousand board feet), or $122.80. In 2 years the future value of the tree will be (0.671 thousand board feet) × ($200 per thousand board feet), or $134.20, assuming that the tree remains in good health and that the markets have not changed. The rate of growth of the value of the tree is then:

$$\text{Rate of growth} = \sqrt[2]{\left(\frac{\$134.20}{\$122.80}\right)} - 1 = 0.045, \text{ or } 4.5\%$$

What we find is that the value growth of the tree over a 2-year period is 4.5%. Given your alternative rate of return of 5%, your decision would be to harvest the tree now.

Example

Assume that in this same stand of hardwoods, you also are considering the options for a 90-foot tall yellow-poplar tree that is currently 14 inches in diameter. Again, current stumpage prices for average grade yellow poplar logs in your area are about $200 per thousand board feet. You measured the growth of the tree by extracting an increment core, and note that its rate of growth is about the same as the larger tree (0.5 inches per year). You estimate that it currently contains 174 board feet of wood, and that in 2 years it should contain 208 board feet of wood. If your alternative rate of return is 5%, then should you harvest the tree now or wait?

The present value of the tree is (0.174 thousand board feet) × ($200 per thousand board feet), or $34.80. In 2 years the future value of the tree will be (0.208 thousand board feet) × ($200 per thousand board feet), or $41.60, again assuming that the tree remains in good health and that the markets have not changed. The rate of growth of the value of the tree is then:

$$\text{Rate of growth} = \sqrt[2]{\left(\frac{\$41.60}{\$34.80}\right)} - 1 = 0.093, \text{ or } 9.3\%$$

What we find with this tree is that the value growth of the tree over a 2-year period is about 9.3%. Given your alternative rate of return of 5%, your decision would be to let the tree grow a few more years.

Example

Using a modified example from Jacobson (2008), suppose that you work in Pennsylvania and are interested in determining the value of black cherry growth over time. Let's assume that the average prices are $835 per thousand board feet (International ¼). For instance, imagine that you have to decide whether to cut a single black cherry tree or leave it to grow for another few years. It is currently 12 inches in diameter with a volume of 62 board feet. You know that it will grow two inches in diameter over a 7-year period. The volume of a 14 inch tree is 117 board feet. If your alternative rate of return is 6%, then should you harvest the tree now or wait?

The present value of the tree is (0.062 board feet) × ($835 per thousand board feet), or $51.77. In 7 years the future value of the tree will be (0.117 thousand board feet) × ($835 per thousand board feet), or $97.70, again assuming that the tree remains in good health and that the markets have not changed. The rate of growth of the value of the tree is then:

$$\text{Rate of growth} = \sqrt[7]{\left(\frac{\$97.70}{\$51.77}\right)} - 1 = 0.094, \text{ or } 9.4\%$$

What we find with this tree is that the value growth of the tree over a 7-year period is about 9.4%. Given your alternative rate of return of 6%, your decision would be to let the tree grow a few more years. If the rate of growth was less than 6%, then the decision would be to harvest the tree and put the money into the alternative use.

Once the determination to cut or leave each tree has been made, the number of trees to remove from a stand

and the resulting total harvest volume for a tract arises from a summation of the decisions applied to each individual tree. Unfortunately, the efficiency of tree-level management decisions is based on other operations that will likely occur within the stand. Harvesting of individual trees may not be economically viable for both the landowner and the logger unless the trees are very highly valued. As a result, although a decision to harvest an individual tree can be determined, it may not be economically efficient (or possible) to extract the tree from the forest when the costs for logging the tree are considered.

As we suggested, the value of a tree for aesthetics, recreation, or wildlife may be subjectively assigned by the analyst or the landowner. Another way to approach the valuation problem is to assess the potential economic value of these trees, assuming that the opportunity cost of reserving a tree for other uses is equal to the potential economic value. In a rational decision-making environment, this type of analysis may make sense; however you should recognize some people do not agree with this type of comparison (economic vs ecological or social value), and may subsequently assign a value to trees at a level higher (or lower) than the current market price.

III. STAND-LEVEL OPTIMIZATION

The next level up from tree-level optimization is stand-level optimization, which involves developing the very best management plan for a stand of trees. The planning process may involve analyzing a number of intermediate treatments as well as final harvest decisions, all scheduled to achieve one or more objectives of the landowner. Constraints may also need to be considered, such as those related to the timing of activities, or related to the structure of the stand that should remain after an activity has been implemented. Objectives and constraints for stands within forests will vary according to the desires and needs of a landowner, the condition of each stand, and the larger socioeconomic and geographic context within which the stand resides and the landowner operates. Objectives and constraints can be associated with economic values (cash flow, net present value), commodities (timber volume), ecological values (habitat), or social concerns (aesthetic, jobs). It is difficult, but not impossible, to accommodate multiple goals or services within individual stands of trees. The challenges facing the production of multiple services from a single stand or property were recognized nearly 80 years ago, as von Ciriacy-Wantrup (1938) noted:

On a single unit of land some uses are complementary or supplementary under certain economic conditions, but more often they are competitive.

As a result, many of the potential uses of a stand of trees may compete with the main use(s), as suggested by the landowner. Therefore, optimizing the use(s) of the land can lead to (1) simply managing the land with respect to the main use, (2) managing the land with respect to the main use, yet accommodating other uses as long as they do not interfere with the attainment of the main use, or (3) managing the land with respect to several uses of high importance.

Two problems have captured the attention of forest managers over the last 50 years: the problem of the optimal timber rotation for even-aged stands, and the problem of optimal timing of intermediate treatments for both even-aged and uneven-aged stands. The former frequently is called the *rotation problem*, and the latter the *thinning problem*, and each of these was dealt initially as an economic or commodity production issue. More recently, problems concerning biodiversity and forest health have driven the need for similar types of stand-level optimization problem structures in an effort to control optimal stand density or stocking levels.

A. Optimum Timber Rotation

A rotation of trees is the number of years between the establishment of an even-aged stand and the final harvest. In classical systems of management and desired forest structure, which we cover in Chapter 10, Models of Desired Forest Structure, the rotation decision is important for moving the structure of a land ownership to one of a "normal" forest, which contains the same amount of area in each age class. The rotation age is less critical for other systems of management, yet still important when we seek to optimize economic or commodity production goals. Williams (1988) describes seven types of rotation ages for even-aged stands of trees:

1. The *physical rotation age*, or the life span of a species of tree. This rule may be problematic in localized conditions, as natural disturbances may shorten the physical length of the life of a stand of trees (Williams, 1988). The main point, it seems, is that the rotation age is defined by the length of the average life span of trees. For a given site, this obviously will vary by species, since some tree species are more long-lived than others. Coastal redwoods (*Sequoia sempervirens*), for example, could live more than 1,000 years (Olson et al., 1990), whereas red alder (*Alnus rubra*) growing on the same site would have a much shorter natural life, perhaps with a maximum age of 100 years (Harrington, 1990).

2. The *technical rotation age*, or the length of time required to grow a stand of trees to certain dimensions to best meet the needs of various commercial markets. For example, for oak trees to be of value for wine barrel uses, they must be a certain age and have minimum annual ring count. Landowners may want to provide a reasonable amount of these types of trees to

meet market needs while earning a profit on the management of their forest. In relatively stable markets, this could be seen as a reasonable approach for developing a rotation age. However, product demand, inventories at various mills, mill availability, competition from other suppliers, and other factors create a large amount of uncertainty regarding future product needs. Many of these factors are not discernable until you near a final harvest decision, and as a result, making long-range plans based on a technical rotation may be difficult. One challenge for landowners is not to hold a stand of trees past its ability to provide a reasonable amount of suitable wood for commercial markets, because at some point losses due to decay, diseases, insects, and other factors will occur.

3. The *silvicultural rotation age*, or the age at which the maximum seed production is obtained to facilitate natural regeneration. Although seed production is an important issue for some types of silvicultural systems (seed tree harvest, shelterwood, and group selection systems), the production of seed may vary by physiographic region, climatic factors, and stand condition, and seed production can be stimulated by intermediate silvicultural treatments. In addition, seed production may be lower and more erratic in some portions of the natural range of a tree species (Baker and Langdon, 1990). As a general example of this method for determining the optimal rotation age, in the case of loblolly pine (*Pinus taeda*), Baker and Langdon (1990) suggest the following: seed production, of individual trees,

increases with tree age, size, and freedom from crown competition. By age 25, enough seeds may be produced in widely spaced trees to regenerate a stand; however, trees at 40 years generally produce 3−5 times more. Rotations shorter than 30 years usually do not lend themselves to natural regeneration.

4. The rotation age that provides the maximum volume production. This concept, commonly termed the *biological rotation age*, suggests that a stand of trees should be harvested at the point where average production (mean annual increment) is at its highest. As we mentioned in Chapter 2, Valuing and Characterizing Forest Conditions, the mean annual increment is simply the volume (or weight) per unit area of products divided by a stand's age. In theory it represents the average growth rate of products, but the relationship to stand age is not linear (Fig. 5.1). Graphically, where the mean annual increment and the periodic annual increment cross is the suggested rotation age. The gross or net volume of products can be used to determine the mean annual increment (Fig. 5.2). Net volumes are those where defect, breakage, and cull volume are removed from the gross volume. However, biological rotation ages may differ based on these assumptions, as net volumes generally suggest slightly shorter rotations than gross volumes. Further, if specific products are considered, the mean annual increment may differ considerably (Fig. 5.3).

5. The *income generation rotation age*, or the one that produces the highest average income, as determined

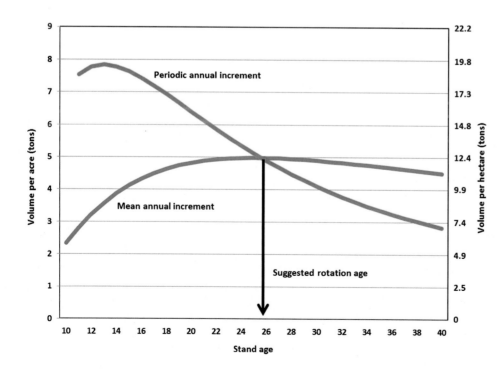

FIGURE 5.1 Mean and periodic annual increment for a southern pine stand.

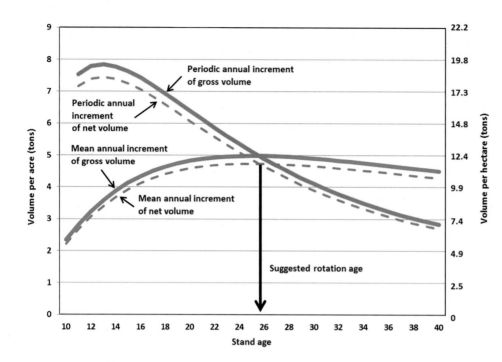

FIGURE 5.2 Mean and periodic annual increment for a southern pine stand using gross and net volumes.

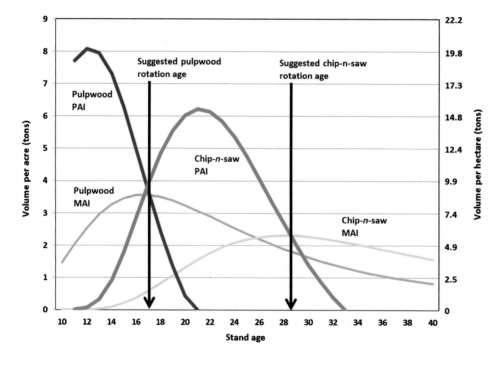

FIGURE 5.3 Mean and periodic annual increment for pulpwood and chip-*n*-saw products grown in a southern pine stand.

by dividing the potential harvest revenue in each time period by the age of the stand. This often is referred to as *forest rent*, where the average annual revenue minus cost is at its greatest level. This method, however, does not recognize the cost of locking up the capital in the forest investment. In addition, this method for determining the optimal rotation age is dependent in part on the assumptions made about future stumpage prices. Uncertainty is inherent in future prices levels, and poses a problem for any projection of future economic conditions and outcomes, thus this method suffers from this assumption as much as other economic methods. However, the other interesting aspect of this method (and others) is the different results that might be obtained if a single composite price is assumed rather than an individual price for each product that is projected to be available over time (Fig. 5.4).

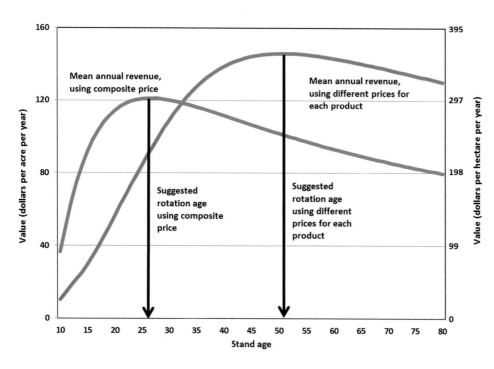

FIGURE 5.4 Mean annual revenue using a composite price and a price specific to the products grown in a southern pine stand.

6. The *economic rotation age* that produces the maximum discounted net revenue. This method seems to involve the determination of the maximum net present value for a stand of trees over time. Inherent in this policy is that only the initial rotation age is considered, rather than future uses of the land over an infinite time horizon. However, if you were to include in the analysis successive rotations of the same type of management, and produce a soil expectation value, you would likely find that the economic rotation age is slightly shorter using this rule than when simply valuing the initial rotation of trees (Fig. 5.5).

7. The *value growth percent rotation age* of a stand. In a manner similar to the valuing of current and future states of individual trees, and subsequently making a decision of whether to cut or leave them standing, you can assess the value growth rate of a stand of trees. This can be accomplished by estimating the value increase from 1 year to the next, then dividing that by the previous value of the stand.

$$\text{Value growth rate} = \left(\frac{\text{Value increase}_{t \to t+1}}{\text{Value of stand}_t} \right)(100)$$

For example, if a stand were valued at $906.71 per acre at age 19, and valued at $1,031.39 per acre at age 20, the increase in value is:

$$\text{Value growth rate} = \left(\frac{1,031.39 - 906.71}{906.71} \right)(100) = 13.75\%$$

The results in this example represent the percent value growth over a 1-year period, but they can be assessed over longer time intervals. The decision rule would be to keep the stand growing as long as the value growth percent is above the alternative rate of return (discount rate) that is assumed by the landowner (Fig. 5.6). If the value growth rate is greater than the landowner's alternative rate of return, then the decision would be to let the stand grow. If the value growth rate is less than the landowner's alternative rate of return, then the decision would be to harvest the stand.

As we suggested, one consequence of determining the length of the rotation period is that decisions far off in the future are subject to a considerable amount of uncertainty. For example, the chance of losing a stand of trees due to fire, insect, and disease outbreaks, and to other risks, increases as the length of a rotation increases and this assumes that the occurrence of the agent remains constant in time. In addition, long-term projections of the structure of stands, as well as trends in prices and costs are all highly uncertain the longer we look into the future. Generally when we examine the rotation problem, we assume that some economic or commodity production assumptions are made that suggest prices, costs, and management intensity are held constant. We do this because it simplifies the problem. However, methods can be devised to add stochastic (random) elements to the process, or to increase (or decrease) prices and costs based on a market analysis.

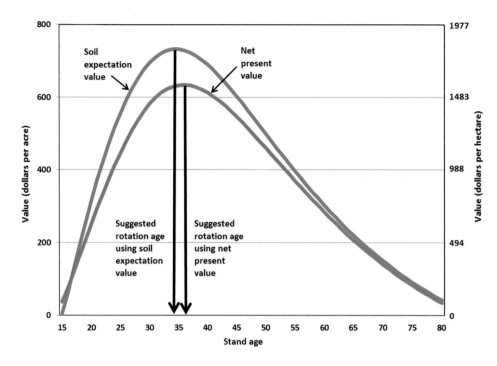

FIGURE 5.5 Optimum rotation age when using net present value and soil expectation value for a southern pine stand.

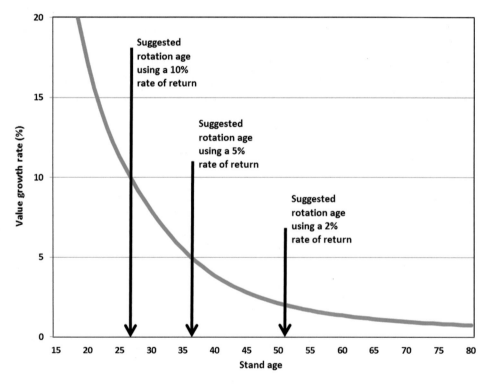

FIGURE 5.6 Optimum rotation ages using the value growth percent of a southern pine stand.

B. Optimum Thinning Timing

The optimum schedule of thinnings or partial cuts within a stand of trees is a difficult problem to solve due to a number of factors, including:

- The potential mixture of tree species within a stand.
- The varying growth rate of trees within a stand.

- The varying growth rates with different stand ages.
- The potential to thin from above (removing the larger, taller trees), from below (removing the smaller, shorter trees), or to remove trees in proportion to the diameter distribution.
- The need to assess a large number of residual stocking or density levels. For example, we could remove 40 ft^2

of basal area per acre in a thinning, or 50 ft^2 per acre, or 60 ft^2 per acre, and so on during a thinning entry. This sometimes is referred to as the *intensity of thinning*.

- The potential to perform a thinning at any time period in the analysis. This often is referred to as the *timing of thinning*, the *cutting cycle*, or the *entry period*.

In addition, a number of interactions occur among these factors. The intensity of thinning at any one point in time can affect the intensity of subsequent thinnings as well as the potential final harvest volumes (Amidon and Akin, 1968). The type of thinning will affect the diameter distribution and thus the value of the residual stand, and the growth response of the residual stand depends on the stocking and stand density that remains. For even-aged stands, the thinning choices are confined to those that can be implemented within a rotation; however, thinnings may extend the optimum rotation age. For uneven-aged stands, the entries are assumed to continue forever, therefore the number of entries represented in an uneven-aged management problem is usually greater than those found in an even-aged management problem. Optimum decisions for uneven-aged stands involve determining the following:

- The optimal sustainable diameter distribution
- The optimal species mix
- The optimal entry cycle
- When converting even-aged stands to uneven-aged stands, the optimal conversion strategy and length of time required to adequately convert an even-aged stand to an uneven-aged stand

When assessing the thinning options for even-aged stands, the discounted net present value or the soil expectation values generally are used. The present value of a non-terminating series of harvest revenues has been suggested as the best measure for assessing the optimum stand-level growing stock for uneven-aged stands (Rideout, 1985).

C. Optimum Stand Density or Stocking

In many forests, maintaining the stand density within a preferred range may be more highly desired than maximizing an economic or commodity production value. Expressing the density of a stand of trees is a way to communicate the amount of vegetation occupying a unit of land (Smith et al., 1996), and as we have shown in Chapter 2, Valuing and Characterizing Forest Conditions, it can be expressed as an index. Riitters and Brodie (1984) suggested over 30 years ago that that optimal thinning strategies for stands should include some knowledge of stand density, and the density limits applicable for

different tree species. Maximum stand density indices have been suggested for a number of tree species in the interior northwest of the United States (Reineke, 1933; Long, 1985). Given the mixed species nature of uneven-aged stands, a stand density index can be weighted proportionally by the basal area of each species to produce an overall stand density. After accounting for the mixture of tree species, bounds can then be placed on the achievement of stand densities over time, effectively keeping stand densities within an ecologically appropriate or managerially preferred range.

The management problem for optimal stand density becomes one of determining which trees to harvest in each time period to best maintain stand density within a predefined range. The activities considered could be influenced by operational constraints that act at the tree-level, such as limiting the treatments to trees or tree records that are of a certain quality (e.g., trees smaller than a 25 inches DBH, or greater than a 5 inches DBH). The activities considered also could be influenced by operational constraints at the stand-level, such as policies that limit harvest entries until a minimum harvest level can be obtained. A number of variations on operational constraints can reasonably be pursued, particularly as the mixture of tree species and tree sizes change over time in an uneven-aged stand.

Much of the early work in determining optimal management regimes for uneven-aged stands involved an assessment of the optimal stocking to carry in a stand at each point in time. The decisions associated with optimal stocking rates involve periodic harvesting or partial cuts, and the need to leave a certain amount of residual stocking (basal area, trees per acre, or volume) in a stand after the treatments were applied.

In the management of even-aged stands, two questions face the forest landowner: what level of stand density should be maintained during the management of the stand, and in response, how long should the rotation last. To optimize the management of stands, stand densities should be kept near an optimal economic or biological level throughout the rotation. If a rotation age is fixed, then it exerts an influence on the stocking and density levels that need to be maintained (Chappelle and Nelson, 1964). Methods for solving these types of problems are similar to the thinning problem described in the previous section (III.B.). The optimal growing stock of a stand of trees, from an economic perspective, is indicated when the marginal unit revenue is equal to the marginal unit cost (Duerr and Bond, 1952). As a result, the additional value accrued through time is a necessary piece of information for making this decision. If multiple products are considered, or when a significant change in grade of expected products occurs, then the optimal stocking levels may change.

Optimum stocking of an uneven-aged forest is defined as the quantity of growing stock that permits the maximum net return to the landowner over time (Duerr and Bond, 1952). In doing so, we need to determine the optimum point at which to stop the accumulation or build-up of the growing stock. There are three general alternatives to managing an uneven-aged forest: (1) allow the growing stock to increase, (2) keep the growing stock constant with periodic harvest entries or management activities, or (3) reduce the growing stock. From an economic perspective, growing stock should be increased as long as the marginal value growth rate is greater than the alternative rate of return.

In either case of stand management, optimum growing stock volume will increase with increasing rates of growth. Better sites, lower mortality, fertilization activities, and a more productive stand composition all may contribute to a higher rate of growth. Any management-related costs that increase as growing stock levels increase (such as property taxes) effectively act to reduce the optimum growing stock level. Finally, as the alternative rate of return is lowered, the optimum growing stock level increases, since the need to produce greater amounts of revenue to cover the marginal costs is reduced (Duerr and Bond, 1952).

D. Recent Developments in the Scientific Literature

Addressing the three main questions in stand-level optimization can be accommodated with a variety of methods. Recently, a review (Kaya et al., 2016) of existing literature was conducted, concerning optimization tools used to examine questions such as the optimal timber rotation, optimal time to thin, and optimal stand density. Some approaches that have been in use for a long time, along with a few newer ones, include the Faustmann optimization, Markov decision models, dynamic programming, the Hooke and Jeeves method, heuristics, and the Escalator Boxcar Train method. The Faustmann approach has been in use in one form or another since its inception by the German Forester Martin Faustmann in 1849. It is commonly used to optimize the value and management of forest stands and identify the optimal rotation age of timber stands. Recent research expanded this concept to help create diversity in uneven-aged stands (Duduman, 2011), to assess the financial feasibility of sequestering forest carbon (Nepal et al., 2012), and to identify optimal control measures for invasive plant species (Grebner et al., 2011). Markov decision models use matrices to evaluate the probabilities of forest stands to transition across different states. This type of model has been applied to consider harvesting

decisions for Douglas-fir (*Pseudotsuga menziesii*) and western hemlock (*Tsuga heterophylla*) forests in the Pacific Northwest using variable or stochastic interest rates (Zhou and Buongiorno, 2011). Another approach discussed in greater detail later in this chapter is dynamic programming. Dynamic programming focuses on mapping the time periods and potential conditions of forests between time periods. This type of problem-solving approach can be used to assess very interesting stand-level issues, such as short-rotation coppice management options for *Eucalyptus globulus* and potential wildfire losses (Ferreira et al., 2012). Pattern search processes such as the Hooke and Jeeves method or the Nelder and Mead method attempt to reduce the feasible region in search of an optimal solution. As with other models, some interesting stand-level management issues can be assessed, such as the optimal combination of timber and honey production (de-Miguel et al., 2014). Heuristic methods, also discussed in greater detail later in this book, use rules and logic for finding a near optimal solutions to problems. The literature on this approach is extensive, and examples of recent work (Niinimäki et al., 2012; Ahtisoski et al., 2013) utilized heuristic methods to evaluate optimal economic concerns of forests. Lastly, the Escalator Boxcar Train approach maximizes a system constrained by a set of ordinary differential equations. This type of problem-solving method has been used, for example, to assess carbon sequestration options for stands of trees (Goetz et al., 2010). In sum, the types of problem-solving methods are extensive. When a problem can be described mathematically, these methods can be applied to produce optimal or near-optimal solutions that can guide forest managers in their efforts to sustainably manage resources.

IV. DECISION TREE ANALYSIS

Another consideration when evaluating alternative stand-level decisions is to assess the implications of risk that a manager may face. It is not uncommon for managers to evaluate alternative choices assuming that all future events occur with 100% certainty, but in reality managers know that is not the case. One important tool for analyzing alternative outcomes is the use of decision trees. Decision trees allow managers to lay out alternative courses of action and to inject some level of risk which will have an impact on the potential outcome of the alternative courses of action. Before constructing a decision tree, one needs to first understand the idea behind calculating an expected value. An expected value is basically the sum of the products of each potential outcome multiplied by their probability of occurrence, where all of the probabilities sum up to

1.0. For instance, an expected value equation would have a structural form like this:

Expected value = (Outcome 1) × (Probability of occurrence)

+ (Outcome 2)

× (Probability of occurrence) + (Outcome 3)

× (Probability of occurrence)

If outcome 1 had a 30% chance of yielding $5,000, outcome 2 had a 20% chance of yielding $7,000, and outcome 3 had a 50% chance of garnering $3,500, then the expected value would be $4,650.

$$\text{Expected value} = (\$5,000) \times (0.30) + (\$7,000) \times (0.20)$$
$$+ (\$3,500) \times (0.50) = \$4,650$$

The expected value therefore represents a most likely outcome amount that considers the risk of different chance events.

A decision tree allows managers to map out alternative courses of actions and identify chance events that each potential action may face. Typically, decision trees are drawn out by moving (conceptually) from left to right, and their solutions are calculated by moving right to left. Fig. 5.7 illustrates a square where the manager needs to make a decision between two alternative courses of action, depicted by straight lines branching out from the decision node. Each alternative course of action faces a set of risky outcomes whose probability of occurrence sum to 1. Each possible outcome (good and bad) is first estimated as if it will occur with complete certainty. The manager will calculate the expected value for both alternatives and will typically select the one that has the highest value. In other words, they are maximizing expected values.

For example, suppose a manager had to decide on whether to plant 726 pine seedlings or 450 pine seedlings, and do nothing else before a final harvest. Either alternative faces a chance of being affected by drought during the life of the trees, and lower wood production and value may occur as a result. Alternative 1 is where the manager plants 726 seedlings and faces a 30% chance that the net present value of that stand will yield $800 per acre, a 20% chance that the net present value will be $200 per acre, and a 50% chance that the net present value is $600 per acre. Alternative 2 is where the manager plants 450

seedlings per acre and faces a 30% chance that the net present value of that stand will yield $700 per acre, a 20% chance that the net present value will be $400 per acre, and a 50% chance that the net present value is $500 per acre (Fig. 5.8).

To solve this problem, the manager needs to work their way from the right hand side of the decision tree to the original decision box on the left. For each alternative, the manager needs to calculate the expected value. For the alternative to plant 726 seedlings, the expected value is:

$$\text{Expected value}_{726} = (\$800) \times (0.30) + (\$600) \times (0.50)$$
$$+ (\$200) \times (0.20) = \$580$$

For the alternative to plant 450 seedlings, the expected value is:

$$\text{Expected value}_{450} = (\$700) \times (0.30) + (\$500) \times (0.50)$$
$$+ (\$400) \times (0.20) = \$540$$

If the manager is interested in maximizing their expected values, then they would select the alternative where they plant 726 seedlings per acre.

Other decision rules, besides locating the highest expected value, can be used. These include the most likely outcome, a risk and return comparison, a maxi-min analysis, and break-even probabilities (Kay and Edwards, 1999). The most likely outcome rule is simple because it allows one to choose the outcome with the highest likelihood of occurring without considering other possibilities. The risk and return comparison rule is one where managers weigh the expected values of alternative courses of action, along with their variability, when deciding on what to do. Depending on their risk tolerance, a manager may select an option with a lower expected value because there is less variability in the possible outcomes. The maxi-min rule is one where managers are more concerned with losses, so they choose the best option amongst the worst possible outcomes. Lastly, the break-even rule examines the possibility of financial loss from different alternatives, but focuses on the probability of not covering all production costs. However, this method has to include expected returns.

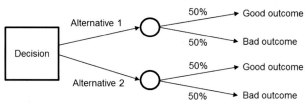

FIGURE 5.7 A conceptual decision tree representing two opportunities, each with two potential outcomes.

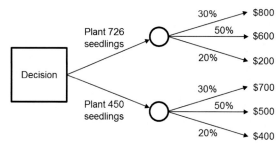

FIGURE 5.8 A decision tree representing a potential planting opportunity.

Decision trees can be an effective method for communicating risk to stakeholders and other interested parties. The challenge is obtaining good probability information that can accurately account for the various risks that managers face when managing their forests.

V. MATHEMATICAL MODELS FOR OPTIMIZING STAND-LEVEL MANAGEMENT REGIMES

One of the fundamental components of the management direction provided to (or by) field foresters is the appropriate stocking level or stand density to maintain over a course of time (Adams and Ek, 1974). Other attributes of a stand, such as the number of snags or amount of down wood, could also serve as measures from which objectives will be assessed. In developing a recommendation for the management of a stand of trees, we typically consider a lengthy time horizon. How one would schedule activities to meet stand density or stocking goals over time can be a complex planning problem, given the larger (perhaps infinite) number of options generally available. Fortunately, a number of mathematical models have been developed in the last 60 years to assist in our ability to incorporate the many options. Mathematical models for stand-level optimization can be grouped into three broad categories:

- The Hooke and Jeeves (1961) method, and other non-linear programming approaches (Kao and Brodie, 1980; Bare and Opalach, 1987)
- Heuristics or meta models (Buongiorno and Michie, 1980; Valsta, 1990; Wikstrom, 2001)
- Dynamic programming approaches (Hool, 1966; Brodie and Kao, 1979)

The Hooke and Jeeves method is a direct search process, where a sequential analysis of alternatives to a stand-level problem are assessed in a rational manner. The method also consists of strategies for assessing the next potential alternatives in a stream of actions, based on previously examined results. This type of problem-solving structure is similar to that used by modern heuristic techniques. Nonlinear programming is similar to linear programming, which we cover in Chapter 7, Linear Programming. These techniques are based on locating the optimal solution to an equation set. In linear programming, the equations contain only linear relationships (i.e., $X + Y$, not XY or X^2Y). In nonlinear programming, the objective function and constraints can be represented by nonlinear functions, which is a useful characteristic since many of the growth and yield relationships are nonlinear. Heuristics are based on logic and rules-of-thumb, and are designed to explore larger areas of the solution space, either randomly or deterministically.

Dynamic programming is perhaps the most widely used stand-level optimization process. From the 1970s through the 1990s, significant advances occurred in the application of dynamic programming to forest management problems. Early research mostly emphasized economic or commodity production goals. However, significant work continues today exploring the use of the approach to recognize and accommodate environmental and social objectives. Dynamic programming facilitates the examination of a large number of alternatives for the management of a stand of trees by reducing the range of options explored. Enhancements to the dynamic programming process by Yoshimoto et al. (1990, 1988) provide efficiencies in the computational burden associated with solving a complex management problem.

VI. DYNAMIC PROGRAMMING

Dynamic programming often has been used to maximize biological potential (mean annual increment) and economic returns (soil expectation value) related to a stand of trees, and represents a technique for the systematic determination of optimal combinations of decisions. Dynamic programming is also a method for numerically solving a dynamic system of equations. The range of types of problems that can be solved is extremely wide, and encompasses not only natural resource management issues, but many business and industrial applications as well (Kennedy, 1986). Since the management of forests and the resulting growth responses of forests are both sequences of actions that may follow similar pathways at various points in time, sorting through a group of alternatives and selecting the optimum course of action may require multiple passes through the same data. What makes the dynamic programming interesting is that you can work iteratively through the sequence of decisions in a normal forward-flowing fashion, or work backward through the sequence from the ending condition to the beginning condition, without having to make the same calculations twice.

Stages within dynamic programming are the positions in the problem where a number of different conditions of the problem can exist. In stand-level optimization, stages could be, for example, the age of a stand. Alternatively, if dealing with an uneven-aged stand where age is irrelevant for decision-making purposes, stages could be defined as the number of years from the present point in time. In the network illustrated in Fig. 5.9, the stages contain vertical columns of nodes. At each stage a policy decision must be made. In the case of stand-level management, this involves selecting a management action (which includes not doing anything). The number of stages should reflect or encompass those that are needed to adequately

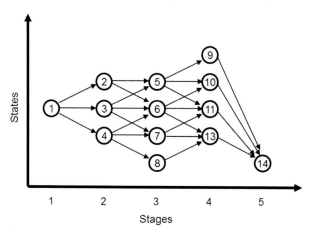

FIGURE 5.9 Example states and stages for a network of choices related to a management problem.

represent the reasonable options for the management of a stand of trees.

States are the conditions of the problem that are recognized at each stage (the nodes at each stage). For stand-level optimization problems, these might refer to different levels of residual growing stock volume or different residual stand densities at each stand age. In other words, states are the various possible stand structural conditions that might exist if a certain course of action is chosen. Each stage has a specific number of states. For example, in Fig. 5.9 stage one has only one state, whereas stage two contains three states. States need to be defined rationally, because using more states suggests recognizing finer-scale ranges of values, and requires additional computation time. However fewer, coarser-resolution states may misrepresent important differences in stand structure. States should describe the characteristics of the stand that have an effect on the objective function value. The number of states needs to be controlled, therefore states that are described using continuous numbers need to be approximated by a range of values. For example, basal area can reasonably range from 0 to 200 ft^2 per acre in the southern United States. States that contain 5- or 10-ft^2 per acre ranges would seem sufficient to represent the problem rather than explicitly representing each square foot of basal area as a state in a dynamic programming problem.

Should the number of stages and states that are used to represent a problem expand to the point of very fine delineations in ages and stand structure, the problem may suffer from the *curse of dimensionality*. Thus, there is a conflict in dynamic programming between the ability to solve a problem and the desire to formulate it with using the highest resolution of data. As a result, there may be a need to limit the number of stages and states within stages when defining the problem. However, this may lead to situations where the decisions made at one stage may not lead nicely to a state defined at the next (or preceding)

stage. The nonlinear growth of trees is one of the primary reasons, and the fact that some growth models acknowledge accelerated growth as a result of intermediate stand treatments is another. Methods for overcoming the rounding errors associated with fitting a variety of stand conditions into states with broad descriptors (e.g., basal areas incrementing by 10 ft^2 per acre) have been proposed (Brodie and Kao, 1979).

A decision must be made at each stage of the analysis that is guided by the notion that the best action for the stand will be chosen. Decisions, in effect, involve transforming a state associated with one stage to a state associated with the next (or preceding, if working backward) stage. *Nodes* reflect the entire set of decisions across the time horizon, and can be represented by a value that has been accumulated from the beginning stage of the problem to the end, or vice versa. *Branches* are the transitions that are possible from nodes at one stage to nodes at the next stage. A value is assigned to each branch indicating the benefit or cost associated with the transition from one state to another (e.g., fully-stocked to clearcut, fully-stocked to 50% stocking, etc.). Since the nodes and branches form a network of sorts, the goal of the dynamic programming process is to find the shortest or longest path through the network. Within a network, shortest paths relate to minimizing an objective, longest paths relate to maximizing an objective.

A. Recursive Relationships

Dynamic programming uses either forward recursion or backward recursion to solve a management problem. *Forward recursion* involves moving in a direction from the first stage to the last stage. *Backward recursion* is the opposite, where the problem is solved from the last stage backward to the first stage. Forward recursion is advantageous for problems that involve uncertain time horizons (Kennedy, 1986). Backward recursion is advantageous for solving problems that contain options with the same time horizon. In forest management, backward recursion of an even-aged management problem would start with the oldest possible stand age recognized and determine the optimal management plan backward to the youngest stand age recognized. Forward recursion starts with the youngest age recognized to determine for each possible rotation length the optimal treatment path (Hann and Brodie, 1980). For uneven-aged management problems, *age* is replaced with *time*. One advantage of forward recursion is that the integration with stand growth models is facilitated because stand growth functions are a forward recursion process.

The recursion process itself requires very little memory. For example, if we were moving forward through the network described in Fig. 5.9, we might know how we reached a particular node, but the details are unimportant.

In recursive problems, the process works by finding the optimal policy for states at stages earlier than those that have already been assessed. For example, in Fig. 5.9, if we understood the optimal policies related to states 9 through 13 at stage 4, we could determine which of the states at stage 3 (5−8) would be used to get to states 9 through 13 at stage 4. For example, we might have found that the best course of action was to use state 5 to go to states 9 through 11, and state 7 to go to state 13.

B. Caveats of Dynamic Programming

Two aspects of dynamic programming are important to understand and are helpful in interpreting the results that are generated:

1. The optimal path through the network of states and stages does not necessarily include the highest cumulative reward at each age of the stand for an even-aged management problem, or each year of the plan for an uneven-aged management problem. The most efficient decision to make regarding the management of a stand, for example, may include treatments that reduce the standing volume or delay the revenues.

2. Assuming we are moving forward through a network of options, at any given state within a stage, the optimal path from this point forward is independent of how we arrived at the state. An optimal solution has the property that whatever the previous decisions were, the remaining decisions will always constitute an optimal management regime, regardless of the state or stage between the beginning and end of the time horizon of the problem. As a result, the decisions that remain will lead to the optimal solution and are independent of any previous decisions (Bellman, 1957). This concept often is referred to as the *Principle of Optimality* (Hillier and Lieberman, 1980). For example, if an optimal solution to the management of an even-aged stand were:

> Age 0, Site prepare and plant
> Age 1, Herbaceous weed control
> Age 15, Commercial thin
> Age 16, Fertilize
> Age 23, Commercial thin
> Age 30, Clearcut

then the optimal pathway of a similar stand on a similar site that is age 16 must be the same (fertilize, then commercial thin at age 23, and clearcut at age 30), regardless of previous management actions.

C. Disadvantages of Dynamic Programming

Although we have stressed the advantageous aspects of dynamic programming for addressing the need to optimize stand-level decisions, as with any process for making decisions there are some disadvantages. The disadvantages associated with using dynamic programming for stand-level optimization include the following:

- The lack of shadow prices or other measures of sensitivity. Shadow prices usually are provided by other traditional mathematical optimization techniques such as linear programming, the focus of Chapter 7, Linear Programming. Shadow prices help you understand how much the objective function would change with the addition or subtraction of one more unit of a constraint. For example, if a constraint were designed to limit the residual basal area in a stand to $100\,\text{ft}^2$ per acre, then a shadow price might allow you to understand how the objective function might change if one more unit (square foot) of the constraint were added or subtracted.

- Stages and states need to be reasonably defined. If stage intervals are long (many years) and state ranges are large (e.g., $30-50\,\text{ft}^2$ residual basal area ranges) the true optimal solution might not be located (Arthaud and Klemperer, 1988).

- Uncertainty of future conditions and events, as with any modeling process, could cause a problem. One of the major drawbacks of many optimization processes is that a number of important assumptions are considered static. Prices, costs, interest rates, and taxes are all examples of information that are frequently held static in an analysis. We say this not to discourage you from analyzing alternatives, for most important decisions would seem to require a rigorous analysis of the options or risks. However, some uncertainty always is associated with projections into the future. In some of these cases, a stochastic version of dynamic programming could be used, but it would require more information about the problem, and may also be a more extensive computational process.

D. Dynamic Programming Example— An Evening Out

To illustrate how we might use dynamic programming, let's take a diversion from growing trees and explore an example that may be similar to your daily lives. Assume that on one particular evening you plan to leave from home in your car, pick up a coffee or soft drink, and then study for a few hours at either the library, student union, or forestry school. Afterward, you plan to visit with some friends for a little while and watch your favorite television show. Later in the evening, you will need to drive home. The potential options available to you can be described by the network illustrated in Fig. 5.10, where node 0 is

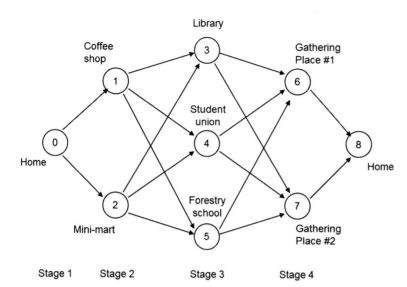

Library
Coffee shop
Gathering Place #1
Student union
Home
Home
Forestry school
Mini-mart
Gathering Place #2

Stage 1   Stage 2   Stage 3   Stage 4

FIGURE 5.10 A network consisting of the possible choices related to an evening's worth of activities.

your home, nodes 1 and 2 are the potential places to get coffee or a soft drink, nodes 3 through 7 are the potential places you will visit in your travel, and node 8 is once again your home. Let's also assume that you want to minimize the cost of your evening's worth of activities. As a result, we need to place a cost on the actions associated with each arc leading from node 0 to node 8 (Table 5.1). For this exercise, let's assume that the costs associated with the gasoline, insurance, and wear and tear on your car are estimated to be about $0.40 per mile.

One way to solve the problem is to select the cheapest cost at each stage of the trip. For example, the lowest cost alternative from home to the coffee shop or mini-mart is $2.61. From the mini-mart, the cheapest cost alternative is to study at the student union ($1.08). From the student union, the cheapest cost alternative is to meet at gathering place #1 ($1.52), from where you will need to go home eventually ($1.60). The total cost is $6.81, but is this the lowest cost alternative for your evening's activities? The route taken using this approach was $0 \rightarrow 2 \rightarrow 4 \rightarrow 6 \rightarrow 8$.

To help understand how the dynamic programming process works, some terminology needs to be defined:

From-node = the node from where a branch originates. It precedes the to-node in a forward recursion process. It follows the to-node in a backward recursion process.

To-node = the node where a branch ends. It precedes the from-node in a backward recursion process. It follows the from-node in a forward recursion process.

Cost = the accumulated cost associated with the route.

Route = the path through the network represented by the cost.

TABLE 5.1 Costs Associated With an Evening's Worth of Activities

From-node	To-node	$r_{a \rightarrow b}$	Comment
0	1	$2.85	Coffee ($1.65) + 3 miles @ $0.40 per mile
0	2	$2.61	Soft drink ($1.29) + 3.3 miles @ $0.40 per mile
1	3	$1.20	3.0 miles @ $0.40 per mile
1	4	$1.00	2.5 miles @ $0.40 per mile
1	5	$1.60	4.0 miles @ $0.40 per mile
2	3	$1.40	3.5 miles @ $0.40 per mile
2	4	$1.08	2.7 miles @ $0.40 per mile
2	5	$1.44	3.6 miles @ $0.40 per mile
3	6	$1.60	4.0 miles @ $0.40 per mile
3	7	$1.88	4.7 miles @ $0.40 per mile
4	6	$1.52	3.8 miles @ $0.40 per mile
4	7	$1.56	3.9 miles @ $0.40 per mile
5	6	$2.08	5.2 miles @ $0.40 per mile
5	7	$2.24	5.6 miles @ $0.40 per mile
6	8	$1.60	4.0 miles @ $0.40 per mile
7	8	$1.12	2.8 miles @ $0.40 per mile

$r_{a \rightarrow b}$ = The "reward" for going from node a (from-node) to node b (to-node) on a branch in the dynamic programming network. In this case, it represents the cost associated with each action.

$r_{a \to b}$ = the "reward" for going from node a (from-node) to node b (to-node) on a branch in the dynamic programming network.

R_b = the maximum reward possible for moving along a particular path to node b. It equals the maximum value of $R_a + r_{a \to b}$ for all nodes a that lead to node b.

P_b = the previous node that provided the path for the maximum reward possible.

Using the reverse method of dynamic programming, let's evaluate the options. First, from node 8 (at the final stage) backward to nodes 6 and 7 (at stage 4) we find the following:

Stage	From-node	Cost	To-node	Route
4	6	1.60	8	6→8*
	7	1.12	8	7→8*

As a result, $R_6 = \$1.60$, since $r_{6 \to 8} = \$1.60$, and $R_7 = \$1.12$, since $r_{7 \to 8} = \$1.12$. R_8 initially had no value, so it was not added to the two reward values. In addition, P_6 = node 8, and P_7 also = node 8. These results are not too interesting. They simply say that the best route from node 6 or 7 is to node 8 (denoted with an asterisk (*)). When we begin to consider the other stages, we find more intriguing results.

Stage	From-node	Cost	To-node	Route
3	3	3.20	6	3→6→8
	3	3.00	7	3→7→8*
	4	3.12	6	4→6→8
	4	2.68	7	4→7→8*
	5	3.68	6	5→6→8
	5	3.36	7	5→7→8*

As a result of this second iteration, we find $R_3 = \$3.00$, since $r_{3 \to 7} = \$1.88$, and R_7 was \$1.12.

At this juncture, we needed to minimize the value of the two options related to from-node 3:

Minimize
$R_6 + r_{3 \to 6} = (\$1.60 + \$1.60) = \$3.20$
$R_7 + r_{3 \to 7} = (\$1.12 + \$1.88) = \$3.00$ (which becomes R_3)

as well as minimize the value of the options related to from-nodes 4 and 5:

Minimize
$R_6 + r_{4 \to 6} = (\$1.60 + \$1.52) = \$3.12$
$R_7 + r_{4 \to 7} = (\$1.12 + \$1.56) = \$2.68$ (which becomes R_4)

Minimize
$R_6 + r_{5 \to 6} = (\$1.60 + \$2.08) = \$3.68$
$R_7 + r_{5 \to 7} = (\$1.12 + \$2.24) = \$3.36$ (which becomes R_5)

Acting as a tracking process, P_3 = node 7, P_4 = node 7, and P_5 = node 7. Here we find that the best routes from any states at stage 3 (3, 4, or 5) take us through nodes 7 and 8, which is not the route we chose earlier by simply summing the lowest cost alternatives available at each stop along the way. At this point, we need to remember that the lowest cost from node 3 forward is \$3.00, that the lowest cost from node 4 forward is \$2.68, and that the lowest cost from node 5 forward is \$3.36.

Stage	From-node	Cost	To-node	Route
2	1	4.20	3	1→3→7→8
	1	3.68	4	1→4→7→8*
	1	4.96	5	1→5→7→8
	2	4.40	3	2→3→7→8
	2	3.76	4	2→4→7→8*
	2	4.80	5	2→5→7→8

As a result of this third iteration, we find $R_1 = \$3.68$, since $r_{1 \to 4} = \$1.00$, and R_4 was \$2.68.

In this analysis, we needed to minimize the value of the three options related to from-node 1:

Minimize
$R_3 + r_{1 \to 3} = (\$3.00 + \$1.20) = \$4.20$
$R_4 + r_{1 \to 4} = (\$2.68 + \$1.00) = \$3.68$ (which becomes R_1)
$R_5 + r_{1 \to 5} = (\$3.36 + \$1.60) = \$4.96$

Can you determine how we arrived at the value of R_2? Again, acting as a tracking process, P_1 = node 4 and P_2 = node 4. Using the information gathered from this analysis, the best route from the coffee shop forward is through nodes 4, 7, and 8 (student union, gathering place #2, then home). The best route from the mini-mart forward is also through nodes 4, 7, and 8. An assessment of the final stage (#1) will provide us with the least-cost alternative, and determine whether to go to the coffee shop or the mini-mart first. We need to remember that the least-cost route forward from node 1 is \$3.68, and the least-cost route forward from node 2 is \$3.76.

Stage	From-node	Cost	To-node	Route
1	0	6.53	1	0→1→4→7→8
	0	6.37	2	0→2→4→7→8*

As a result of this final iteration, we find $R_0 = \$6.37$, since $r_{0 \to 2} = \$2.61$, and R_2 was \$3.76.

In this analysis, we needed to minimize the value of the final two options, which were related to from-node 0:

Minimize

$R_1 + r_{0 \to 1} = (\$3.68 + \$2.85) = \$6.53$

$R_2 + r_{0 \to 2} = (\$3.76 + \$2.61) = \$6.37$ (which becomes R_0)

In addition, $P_0 = $ node 2. Although each of these last two alternatives costs less than our initial attempt at solving the problem, the alternative that takes you to the mini-mart, student union, gathering place #2, then back home is the optimal, costing $6.37 for the evening's activities. This could have been determined by building the path using the appropriate P_b values:

$P_0 = $ node 2 (implying moving from node 0 to node 2)
$P_2 = $ node 4 (implying moving from node 2 to node 4)
$P_4 = $ node 7 (implying moving from node 4 to node 7)
$P_7 = $ node 8 (implying moving from node 7 to node 8)

As you might have noticed, dynamic programming works by accumulating the optimal objective function values as it moves from one stage to the next. Suboptimal decisions also are ignored from one stage to the next. For example, once we determined in stage 3 that travel through node 6 (gathering place #1) was suboptimal, this opportunity was no longer considered during stages 1 or 2. This allows dynamic programming to continually reduce the solution space, and avoid complete enumeration of the problem. Two important pieces of information are produced at the conclusion of the last stage of the analysis:

1. The objective function value has been identified. In this case, the objective function was to minimize the cost of an evening's activity, which was found to be $6.37.
2. The optimal solution to the problem has been located. Here, it involved driving to the mini-mart and picking up a soft drink, driving to the student union and studying for a few hours, driving to gathering place #2 and watching your favorite television show with your friends, then returning home. In our network (Fig. 5.10) this was represented by route $0 \to 2 \to 4 \to 7 \to 8$.

E. Dynamic Programming Example— Western Stand Thinning, Fixed Rotation Length

Using the western forest stand introduced earlier in this book, several alternatives for the management of the stand can be explored under the assumption that the landowner had determined not to clearcut the stand until it was

55 years old and that the landowner desired to maximize the value of the investment. The alternatives for this example include the following:

- Doing nothing prior to clearcutting
- Thin at age 35, from below, to a residual basal area of 90 ft² per acre
- Thin at age 35, from below, to a residual basal area of 100 ft² per acre
- Thin at age 35, from below, to a residual basal area of 110 ft² per acre
- Thin at age 35, from below, to a residual basal area of 90 ft² per acre, then thin again at age 45, from below, to a residual basal area of 100 ft² per acre
- Thin at age 35, from below, to a residual basal area of 90 ft² per acre, then thin again at age 45, from below, to a residual basal area of 110 ft² per acre
- Thin at age 35, from below, to a residual basal area of 100 ft² per acre, then thin again at age 45, from below, to a residual basal area of 100 ft² per acre
- Thin at age 35, from below, to a residual basal area of 100 ft² per acre, then thin again at age 45, from below, to a residual basal area of 110 ft² per acre
- Thin at age 35, from below, to a residual basal area of 110 ft² per acre, then thin again at age 45, from below, to a residual basal area of 100 ft² per acre
- Thin at age 35, from below, to a residual basal area of 110 ft² per acre, then thin again at age 45, from below, to a residual basal area of 110 ft² per acre

These 10 alternatives could be assessed individually to determine which one would provide the highest return. In practice, perhaps hundreds of options might be assessed for individual stands. However, we limited our analysis to these 10 to provide a balance between illustrating the technical detail of dynamic programming and providing a realistic management example.

The alternatives can be assessed using the backward recursion (or reverse) method of dynamic programming. A network of the states and stages related to the problem can be designed to better envision the transitions related to the structure of the stand through time (Fig. 5.11). Here, we decided to define the stages as some period of time within which a decision is made. For example, at stage 3 we implement a clearcut harvest. At stages 1 and 2 we need to determine whether or not to thin the stand. At stage 1 we will also account for the site preparation and planting costs related to the stand. Let's begin the analysis at stage 3 using node 11. The states consist of residual basal area levels, or the approximate basal area that would remain after the treatments were applied. Using the process we described earlier, and using the data provided in

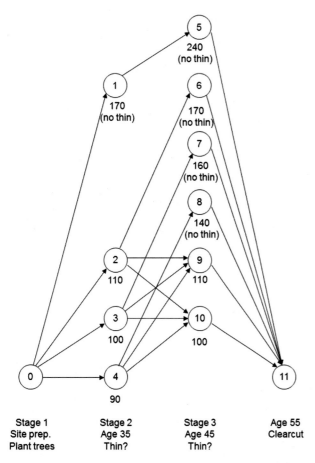

FIGURE 5.11 Network of options for a managed Douglas-fir stand. Nodes 2–4 are first thinning (from below) options, with residual basal areas noted beside each node in ft² per acre. Nodes 9–10 represent second thinning options.

Stage 1	Stage 2	Stage 3	Age 55
Site prep.	Age 35	Age 45	Clearcut
Plant trees	Thin?	Thin?	

TABLE 5.2 Discounted Revenues and Costs Associated With Some Management Alternatives for a Managed Douglas-Fir Stand

From-node	To-node	Volume Harvested (MBF)	$r_{a \to b}$
0	1	–	(250.00)
0	2	3.378	(5.04)
0	3	4.360	66.17
0	4	5.352	138.11
1	5	–	0.00
2	6	–	0.00
2	9	7.931	353.08
2	10	9.343	415.94
3	7	–	0.00
3	9	6.188	275.48
3	10	7.636	339.94
4	8	–	0.00
4	9	4.347	193.52
4	10	5.762	256.52
5	11	44.858	1,225.99
6	11	36.620	1,000.85
7	11	33.804	923.88
8	11	31.031	848.09
9	11	24.500	669.60
10	11	23.000	628.60

$r_{a \to b}$ = The "reward" for going from node a (from-node) to node b (to-node) on a branch in the dynamic programming network. In this case, it represents the discounted net revenue. Values in parentheses represent negative cash flows here and elsewhere.

Table 5.2, where the discount rate is 5% and the stumpage price is $400 per MBF. At stage 4 we find the following:

Stage	From-node	Net Revenue	To-node	Route
3	5	1,225.99	11	5→11*
	6	1,000.85	11	6→11*
	7	923.88	11	7→11*
	8	848.09	11	8→11*
	9	669.60	11	9→11*
	10	628.60	11	10→11*

As a result of this analysis:

$R_5 = \$1,225.99$	$P_5 =$ node 11
$R_6 = \$1,000.85$	$P_6 =$ node 11
$R_7 = \$923.88$	$P_7 =$ node 11
$R_8 = \$848.09$	$P_8 =$ node 11
$R_9 = \$669.60$	$P_9 =$ node 11
$R_{10} = \$628.60$	$P_{10} =$ node 11

As we suggested in the example describing your activities during an evening of study and relaxation, these results are not that interesting. They simply indicate that the best route from nodes 5 to 10 is to node 11, the clearcut activity (best route from each state is denoted with an asterisk (*)). Although we used costs to sort out the alternatives earlier, here net revenue is used. What this provides is the cumulative net revenue of the route. Moving backward one stage, we find the following results:

Stage	From-node	Net Revenue	To-node	Route
2	1	1,225.99	5	1→5→11*
	2	1,000.85	6	2→6→11
	2	1,022.58	9	2→9→11
	2	1,044.54	10	2→10→11*
	3	923.88	7	3→7→11
	3	945.08	9	3→9→11
	3	968.54	10	3→10→11*
	4	848.09	8	4→8→11
	4	863.12	9	4→9→11
	4	885.12	10	4→10→11*

At this point in the analysis:

$R_1 = \$1,225.99$	$(R_5 + r_{1 \to 5})$	$P_1 = $ node 5
$R_2 = \$1,044.54$	$(R_{10} + r_{2 \to 10})$	$P_2 = $ node 10
$R_3 = \$968.54$	$(R_{10} + r_{3 \to 10})$	$P_3 = $ node 10
$R_4 = \$885.12$	$(R_{10} + r_{4 \to 10})$	$P_4 = $ node 10

The net revenue, as we mentioned earlier, is cumulative, thus the calculations involve the net revenue associated with the decision at this stage and the net revenue associated with the best path from the to-node to the final destination. For example, the net revenue for moving from node 3 to node 10 involves the net revenue associated with this thinning option ($339.94 per acre) plus the net revenue associated with the best path from node 10 to the final destination ($628.60 per acre). Here we find that the best routes from the states at stage 2 are the following:

- If we are at node 1, proceed to node 5, then to node 11 (no thin, no thin, clearcut)
- If we are at node 2, proceed to node 10, then node 11 (thin to 110 ft^2 per acre at age 35, thin to 100 ft^2 per acre at age 45, then clearcut)

- If we are at node 3, proceed to node 10, then node 11 (thin to 100 ft^2 per acre at age 35, thin to 100 ft^2 per acre at age 45, then clearcut)
- If we are at node 4, proceed to node 10, then node 11 (thin to 90 ft^2 per acre at age 35, thin to 100 ft^2 per acre at age 45, then clearcut)

Each of the thinning options suggests that the second thinning involves a residual basal area of 100 ft^2 per acre.

Stage	From-node	Net Revenue	To-node	Route
1	0	975.99	1	0→1→5→11
	0	1,039.50	2	0→2→10→11*
	0	1,034.71	3	0→3→10→11
	0	1,023.23	4	0→4→10→11

Nearing the final determination of the optimal thinning regime(s), we find that:

$R_0 = \$1,039.50$	$(R_2 + r_{0 \to 2})$	$P_0 = $ node 2

Each of these last four alternatives involves the cost of the site preparation and planting ($250 per acre) along with the discounted cost of the thinning options (where appropriate). As a result, the best alternative for this stand is:

Site prepare and plant the stand
Thin the stand to 110 ft^2 per acre at age 35
Thin the stand to 100 ft^2 per acre at age 45
Clearcut the stand at age 55

This management regime could have been determined by building the path using the appropriate P_b values:

$P_0 = $ node 2 (implying moving from site preparation and planting to thinning the stand to 110 ft^2 per acre at age 35)
$P_2 = $ node 10 (implying moving from thinning to 110 ft^2 per acre at age 35, to thinning to 100 ft^2 per acre at age 45)
$P_{10} = $ node 11 (implying moving from thinning to 100 ft^2 per acre at age 45 to a final harvest at age 55)

As you might have noticed, this example of dynamic programming accumulated the discounted net revenue from one stage to the next. The site preparation and planting costs were not discounted because they were assumed to have been incurred immediately at the start of the investment. In addition, since all the investment alternatives began and ended with the same time horizon, no adjustment for varying investment lengths was required.

F. Dynamic Programming Example— Southern Stand Thinning, Varying Rotation Lengths

Determining the optimal rotation length using dynamic programming is somewhat different than determining the optimal thinning policies for even-aged stands, since the latter usually assumes a fixed rotation length. Here, we leave the rotation length open-ended at first, and use $r_{a \to b}$ values that reflect their contribution to soil expectation value. The contributions associated with each branch in a dynamic programming network are compounded to the end of the rotation, then discounted to the present to reflect incurring the cost (or revenue) perpetually in future rotations of the same set of treatments. Each $r_{a \to b}$ value that is not associated with a final harvest is adjusted using the equation:

$$r'_{a \to b} = r_{a \to b} \left[\frac{(1+i)^{R-t}}{(1+i)^R - 1} \right]$$

where:

R = the rotation age assumed in the analysis
t = the time period in which the cost or revenue is incurred

Final harvest values are adjusted using the equation:

$$r'_{a \to b} = r_{a \to b} \left[\frac{1}{(1+i)^R - 1} \right]$$

Take, for example, an even-aged stand located in the southern United States. Several basic management regimes (Fig. 5.12) were considered that included thinning at age 18 and possibly age 25, with potential clearcut harvest ages of 25, 30, and 35. Which of these management alternatives would you suggest to a landowner who had the intention of maximizing his or her investment? Given a set of yields for the thinnings and final harvests, and assumptions regarding the stumpage prices for the mixture of products (which varies by age given assumptions of the mixture of pulpwood, chip-n-saw, and sawtimber volume), we could develop the "rewards" associated with each branch in the dynamic programming network (Table 5.3). Site preparation, planting, and herbaceous weed control costs are assumed to be $205 per acre in this example.

There is one caveat to this problem, however, that makes it slightly complicated: the thinning revenues could be associated with final harvest ages ranging from 25 to 35. As a result, we would need to determine the revenue associated with each thinning occurring within a different final harvest regime. We redefine the reward to reflect this fact.

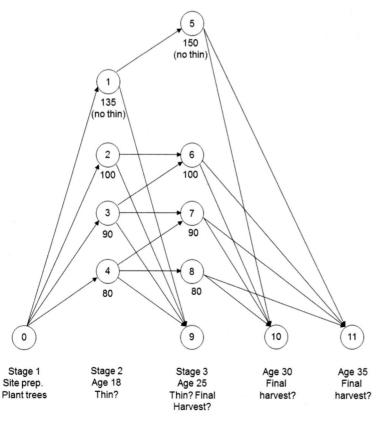

FIGURE 5.12 Network of options for a managed southern pine stand. Nodes 2–4 are first thinning (from below) options, with residual basal areas noted beside each node in ft² per acre. Nodes 6–8 represent second thinning options.

TABLE 5.3 Discounted Revenues and Costs Associated With Some Management Alternatives for a Managed Southern Pine Stand

From-node	To-node	Volume Harvested (tons)	Price ($/ton)	Final Harvest Age (xx)	$r_{xx,\,a \to b}$
0	1	–	–	25	(290.91)
0	1	–	–	30	(266.71)
0	1	–	–	35	(250.39)
0	2	19	9.00	25	(190.08)
0	2	19	9.00	30	(174.27)
0	2	19	9.00	35	(163.60)
0	3	24	9.00	25	(163.55)
0	3	24	9.00	30	(149.94)
0	3	24	9.00	35	(140.76)
0	4	30	9.00	25	(131.71)
0	4	30	9.00	30	(120.75)
0	4	30	9.00	35	(113.36)
1	5	–	0.00	–	0.00
1	9	124	12.00	25	623.55
2	6	22	12.00	30	101.43
2	6	22	12.00	35	95.22
2	9	109	12.00	25	548.12
3	6	16	12.00	30	73.77
3	6	16	12.00	35	69.25
3	7	24	12.00	30	110.65
3	7	24	12.00	35	103.88
3	9	103	12.00	25	517.94
4	7	17	12.00	30	78.38
4	7	17	12.00	35	73.58
4	8	26	12.00	30	119.87
4	8	26	12.00	35	112.54
4	9	96	12.00	25	482.74
5	10	147	15.00	30	663.77
5	11	166	18.00	35	661.65
6	10	119	15.00	30	537.34
6	11	150	18.00	35	597.87
7	10	110	15.00	30	496.70
7	11	140	18.00	35	558.01
8	10	100	15.00	30	451.54
8	11	129	18.00	35	514.17

$r_{xx,\,a \to b}$ = The "reward" for going from node a (from-node) to node b (to-node) on a branch in the dynamic programming network, with an associated final harvest age of xx. In this case, it represents the discounted net revenue.

$r_{xx,\ a \to b} =$ the "reward" for going from node a (from-node) to node b (to-node) on a branch in the dynamic programming network, with an associated final harvest age of xx.

For example, the stand represented at node 6, which represents a thinning at age 25 to a residual basal area of 100 ft^2 per acre, could be clearcut at age 30 or age 35. Therefore, the contribution to the present value of this perpetual series of subsequent final harvest revenues needs to be calculated twice:

1. If the clearcut age is 30, the perpetual series of revenues begins at age 30, and repeats every 30 years (60, 90, 120, etc.). The associated reward for this option is:

$$r_{30,6 \to 10} = r_{6 \to 10}\left[\frac{1}{(1.05)^{30} - 1}\right] = \$1,785(0.301029)$$
$$= \$537.34$$

2. If the clearcut age is 35, the perpetual series of revenues begins at age 35, and repeats every 35 years (70, 105, 140, etc.). The associated reward for this option is:

$$r_{35,6 \to 11} = r_{6 \to 11}\left[\frac{1}{(1.05)^{35} - 1}\right] = \$2,700(0.221434)$$
$$= \$597.87$$

The thinning options require more extensive analysis. Let's focus on node 2, which was thinned at age 18 to a residual basal area of 100 ft^2 per acre. From node 2 forward there are two choices, thin it again or clearcut the stand. The final harvest option terminates at age 25, thus the reward for this choice is:

$$r_{25,2 \to 9} = r_{2 \to 9}\left[\frac{1}{(1.05)^{25} - 1}\right] = \$1,308(0.419049)$$
$$= \$548.12$$

However, if the stand is thinned, it could have either a final harvest age of 30 or 35. Therefore, the contribution to the present value of this perpetual series of subsequent final harvest revenues also must be calculated twice:

$$r_{30,2 \to 6} = r_{2 \to 6}\left[\frac{(1.05)^{30-25}}{(1.05)^{30} - 1}\right] = \$264(0.384197) = \$101.43$$

$$r_{35,2 \to 6} = r_{2 \to 6}\left[\frac{(1.05)^{35-25}}{(1.05)^{35} - 1}\right] = \$264(0.360693) = \$95.22$$

The options that involve moving from node 0 to nodes 1 through 4 each can involve regeneration costs. The regeneration cost itself is \$205, but the regeneration cost associated with repeated rotations needs to be factored into the analysis. As a result, moving from node 0 to node 2, for example, and assuming a 25-year rotation age, involves calculating the regeneration cost associated with a 25-year rotation age:

$$\$205\left[\frac{(1.05)^{25-0}}{(1.05)^{25} - 1}\right] = \$205(1.419049) = \$290.91$$

Moving from node 0 to node 2 is also associated with a potential thinning value:

$$\$171\left[\frac{(1.05)^{25-18}}{(1.05)^{25} - 1}\right] = \$171(0.589644) = \$100.83$$

Combining the revenue and the cost produces the reward for the route:

$$r_{25,0 \to 2} = \$100.83 - \$290.91 = -\$190.08$$

The alternatives for the southern thinning problem can be assessed using the backward recursion (or reverse) method or forward recursion methods of dynamic programming. We begin the analysis at stage 3 by assessing all the opportunities that result in the states described at stage 3. The states consist of residual basal area levels, and leading back to these states are final harvests (nodes 10 and 11) that occur at ages 30 and 35:

Stage	From-node	Net Revenue	To-node	Route
3	5	663.77	10	5→10*
	5	661.65	11	5→11
	6	537.34	10	6→10
	6	597.87	11	6→11*
	7	496.70	10	7→10
	7	558.01	11	7→11*
	8	451.54	10	8→10
	8	514.17	11	8→11*

As a result of this analysis:

$R_5 = \$663.77$	$P_5 = $ node 10
$R_6 = \$597.87$	$P_6 = $ node 11
$R_7 = \$558.01$	$P_7 = $ node 11
$R_8 = \$514.17$	$P_8 = $ node 11

With this initial analysis, we find that the best route from the unthinned state (5) is to perform a final harvest at age 30, and the best routes from the thinned states (6–8) is to wait an extra 5 years and perform the final harvest at age 35. Moving backward one stage, we find the following results:

Stage	From-node	Net Revenue	To-node	Route
2	1	663.77	5	$1 \to 5 \to 10^*$
	1	623.55	9	$1 \to 9$
	2	693.09	6	$2 \to 6 \to 11^*$
	2	548.12	9	$2 \to 9$
	3	667.12	6	$3 \to 6 \to 11^*$
	3	661.89	7	$3 \to 7 \to 11$
	3	517.94	9	$3 \to 9$
	4	631.59	7	$4 \to 7 \to 11^*$
	4	626.71	8	$4 \to 8 \to 11$
	4	482.74	9	$4 \to 9$

At this point in the analysis:

$R_1 = \$663.77$	$(R_5 + r_{1 \to 5})$	$P_1 = $ node 5
$R_2 = \$693.09$	$(R_6 + r_{35,\ 2 \to 6})$	$P_2 = $ node 6
$R_3 = \$667.12$	$(R_6 + r_{35,\ 3 \to 6})$	$P_3 = $ node 6
$R_4 = \$631.59$	$(R_7 + r_{35,\ 4 \to 7})$	$P_4 = $ node 7

The final stage of the analysis will reveal the appropriate management regime for this stand.

Stage	From-node	Net Revenue	To-node	Route
1	0	397.06	1	$0 \to 1 \to 5 \to 10$
	0	529.49	2	$0 \to 2 \to 6 \to 11^*$
	0	526.36	3	$0 \to 3 \to 6 \to 11$
	0	518.23	4	$0 \to 4 \to 7 \to 11$

Nearing the final determination of the optimal thinning regime(s) and rotation age, we find that:

$R_0 = \$529.49$	$(R_2 + r_{35,\ 0 \to 2})$	$P_0 = $ node 2

Each of these last four alternatives included the cost of the site preparation and planting along with the discounted cost of the thinning options (where appropriate). As a result, the best alternative for this stand seems to be to:

- Prepare the site and plant the stand
- Thin the stand to 100 ft^2 per acre at age 18
- Thin the stand to 100 ft^2 per acre at age 25
- Clearcut the stand at age 35

This management regime could have been determined by building the path using the appropriate P_b values:

$P_0 = $ node 2 (implying moving from site preparation and planting to thinning the stand to 100 ft^2 per acre at age 18)
$P_2 = $ node 6 (implying moving from thinning to 100 ft^2 per acre at age 18, to thinning to 100 ft^2 per acre at age 25)
$P_{10} = $ node 11 (implying moving from thinning to 100 ft^2 per acre at age 25 to a final harvest at age 35)

It has been suggested that the backward recursion method of dynamic programming can be used only to solve management problems with a single rotation length time, and that a number of backward recursion runs are needed to fully evaluate a problem (Hann and Brodie, 1980). However, we have shown here that by assessing the best choice at each stage (in terms of discounted perpetual future values), the backward recursion method is just as effective at locating the optimum solution as the forward method. As a check on this assertion, the best forward recursion paths for each potential rotation length is:

Clearcut age 25:

Route: $0 \to 2 \to 9$
Value: $(r_{25,\ 0 \to 2} + r_{25,\ 2 \to 9}) = (-190.08 + 548.12) = $ \$358.04 per acre

Clearcut age 30:

Route: $0 \to 2 \to 6 \to 10$
Value: $(r_{30,\ 0 \to 2} + r_{30,\ 2 \to 6} + r_{30,\ 6 \to 10}) = (-174.27 + 101.43 + 537.34) = $ \$464.50 per acre

Clearcut age 35:

Route: $0 \to 2 \to 6 \to 11$
Value: $(r_{35,\ 0 \to 2} + r_{35,\ 2 \to 6} + r_{30,\ 6 \to 11}) = (-163.60 + 95.22 + 597.87) = $ \$529.49 per acre

As you can see, the optimum solution is to clearcut the stand at age 35, after two intermediate thinnings. The resulting soil expectation value (\$529.49 per acre) is exactly the same as what we found with the backward recursion method.

VII. SUMMARY

Managers have to make decisions and be able to justify their selection of management alternatives for individual trees or stands of trees. The tools that are used to provide information are only as good as the data upon which they were created, therefore in some cases care should be taken to ensure that high quality data are used. Blame for poor decisions sometimes usually is placed on the analytical methodology, yet ultimately, it is the interpretation of the results by the manager and the judgment of the manager that matters (Williams, 1988). Being able to assess a number of alternatives for the management of trees and stands is a fundamental step in the management of forests, and demonstrates your competence as a manager of resources to your supervisor or client. Stand-level optimization is important for sorting through the options available, and in cases where landowners desire to use resources efficiently, assists in the decision-making process. There is one distinct drawback in optimization at this fine of a scale, however. When you consider the management of a tree or stand, generally these decisions are not impacted by the decisions assigned to other trees or stands, nor by the condition of the surrounding landscape. However, decisions made at this scale may be useful for the development of alternatives for larger forest- or landscape-level problems. For example, higher-level goals that involve resources from more than one stand can influence decisions made at the tree or stand-level. These include organizational wood-flow objectives as well as wildlife habitat objectives, each of which may require a large spatial context to assess. However, stand-level optimization is used very often in the management of forests, and attempts to integrate stand-level optimization with forest-level goals are pervasive in forest planning. For example, we may want to include the optimal rotation for each stand as one silvicultural option in a forest-level plan of action that is designed to produce maximum volume for forest landowner. Another approach is to have available not only the optimal stand-level management plan, but other stand-level regimes that may include facilitating the development of a particular type of habitat, or that provide some flexibility with regard to volume flows. We will cover a variety of issues related to forest-level planning in Chapter 7, Linear Programming; Chapter 8, Advanced Planning Techniques; Chapter 9, Forest and Natural Resource Sustainability; Chapter 10, Models of Desired Forest Structure; Chapter 11, Control Techniques for Commodity Production and Wildlife Objectives; and Chapter 12, Spatial Restrictions and Considerations in Forest Planning.

QUESTIONS

1. *Curse of dimensionality.* Consider an uneven-aged hardwood stand in Ohio. You are tasked with developing a management recommendation for the stand that will encompass actions to be undertaken over the next 30 years. Breaking the timeline down into three decades, you determine that two types of thinning alternatives can be prescribed for the stand in each decade. Knowing that prescribing none of the thinnings is an option, and that prescribing all of the thinnings is an option, how many different alternatives would need to be evaluated? How would this change if the number of thinning options changed to six?

2. *Tree-level optimization.* Assume that you manage a small parcel of land in upstate New York, and that you have a 20-inch black cherry (*Prunus serotina*) on your property that you estimate contains 364 board feet of sound wood. Current stumpage prices for your area are about $450 per thousand board feet for black cherry, and you expect that they will remain at about this level for the next couple of years. If in 3 years the tree might contain 411 board feet, then should you wait to cut it then, or cut it now? Your alternative rate of return is 5%.

3. *Tree-level optimization.* Assume that you manage a small parcel of land in Indiana, and that you have a 29-inch northern red oak (*Quercus rubra*) on your property that you estimate contains about 1,024 board feet of no. 1 common lumber. Current stumpage prices for your area are about $600 per thousand board feet for no. 1 common red oak, and you expect that they will remain at about this level for the next couple of years. If in 5 years the tree might contain 1,223 board feet, then should you wait to cut it then, or cut it now? Your alternative rate of return is 5%.

4. *Optimum timber rotation.* One of your landowner clients is interested in setting a rotation age for their 30-year-old stand of eastern hardwoods. Prepare for the landowner a short, one-page summary of the various approaches you would use to define a rotation age.

5. *Optimum thinning schedule.* A landowner in Arkansas owns a sizable stand of older pines with some hardwoods scattered throughout. They want to manage the stand as an uneven-aged forest, maintaining a continuous overstory, providing habitat for wildlife, and maintaining a visually pleasing landscape. They want to understand how they can keep this forest structure for the next 30−40 years. Prepare for the landowner a one-page summary of the factors that should be considered when contemplating the development of an optimum thinning schedule.

6. *Dynamic programming.* You recently have been hired by a consultant in Maine, and they are interested in understanding more about stand-level optimization using dynamic programming. During your interview you mentioned that you had experience with quantitative forest management techniques, so they naturally consider you an expert in the subject area. Prepare for your supervisor a short one- to two-page discussion on the basic methodology behind dynamic programming as a method for developing stand-level optimal management regimes, and point out its strengths and weaknesses.

7. *Dynamic programming.* Assume that you manage an even-aged stand of red pine (*Pinus resinosa*) in Minnesota. A number of thinning options can be applied to the stand over the intended rotation age (40 years). If the objective were to maximize the production of basal area over the intended rotation age, then what course of action would you take? Use the following data and figure, and solve the problem using dynamic programming.

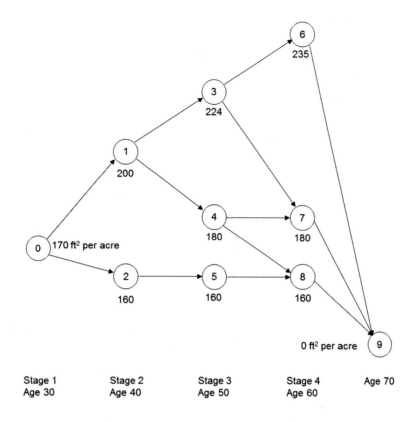

| Stage 1 | Stage 2 | Stage 3 | Stage 4 | Age 70 |
| Age 30 | Age 40 | Age 50 | Age 60 | |

From-node	To-node	$r_{a \to b}$
0	1	0
0	2	40
1	3	0
1	4	44
2	5	28
3	6	0
3	7	55
4	7	27
4	8	47
5	8	29
6	9	246
7	9	206
8	9	185

REFERENCES

Adams, D.M., Ek, A.R., 1974. Optimizing the management of uneven-aged forest stands. Can. J. For. Res. 4 (3), 274–287.

Ahtisoski, A., Salminen, H., Ojansuu, R., Hynynen, J., Kärkkäinenen, K., Haapanen, M., 2013. Optimizing stand management involving the effect of genetic gain: preliminary results for Scots pine in Finland. Can. J. For. Res. 43 (3), 299–305.

Amidon, E.L., Akin, G.S., 1968. Dynamic programming to determine optimum levels of growing stock. For. Sci. 14 (3), 287–291.

Arthaud, G.J., Klemperer, W.D., 1988. Optimizing high and low thinnings in loblolly pine with dynamic programming. Can. J. For. Res. 18 (9), 1118–1122.

Baker, J.B., Langdon, O.G., 1990. *Pinus taeda* L. Loblolly pine. In: Burns, R.M., Honkala, B.H. (Eds.), Silvics of North America, Volume I: Conifers. US Department of Agriculture, Forest Service, Washington, DC, Agriculture Handbook 654.

Bare, B.B., Opalach, D., 1987. Optimizing species composition in uneven-aged forest stands. For. Sci. 33 (4), 958–970.

Bellman, R., 1957. Dynamic Programming. Princeton University Press, Princeton, NJ, 337 p.

Brodie, J.D., Kao, C., 1979. Optimizing thinning in Douglas-fir with three-descriptor dynamic programming to account for accelerated diameter growth. For. Sci. 25 (4), 665–672.

Buongiorno, J., Gilless, J.K., 1987. Forest Management and Economics. Macmillan Publishing Company, New York, NY, 285 p.

Buongiorno, J., Michie, B.R., 1980. A matrix model of uneven-aged forest management. For. Sci. 26 (4), 609–625.

Chappelle, D.E., Nelson, T.C., 1964. Estimation of optimal stocking levels and rotation ages of loblolly pine. For. Sci. 10 (4), 471–502.

de-Miguel, S., Pukkala, T., Yeşil, A., 2014. Integrating pine honeydew honey production into forest management optimization. Eur. J. For. Res. 133 (3), 423–432.

Duduman, G., 2011. A forest management planning tool to create highly diverse uneven-aged stands. Forestry. 84 (3), 301–313.

Duerr, W.A., Bond, W.E., 1952. Optimum stocking of a selection forest. J. For. 50 (1), 12–16.

Ferreira, L., Constantino, M.F., Borges, J.G., Garcia-Gonzalo, J., 2012. A stochastic dynamic programming approach to optimize short-rotation coppice systems management scheduling: an application to eucalypt plantations under wildfire risk in Portugal. For. Sci. 58 (4), 353–365.

Goetz, R.-U., Hritonenko, N., Mur, R.J., Xabadia, A., Yatsenko, Y., 2010. Forest management and carbon sequestration in size-structured forests: the case of *Pinus sylvestris* in Spain. For. Sci. 56 (3), 242–256.

Grebner, D.L., Ezell, A.W., Prevost, J.D., Gaddis, D.A., 2011. Kudzu control and impact on monetary returns to non-industrial private forest landowners in Mississippi. J. Sustain. For. 30 (3), 204–223.

Hann, D.W., Brodie, J.D., 1980. Even-aged Management: Basic Managerial Questions and Available or Potential Techniques for Answering Them. US Department of Agriculture, Intermountain Forest and Range Experiment Station, Ogden, UT, General Technical Report INT-83. 29 p.

Harrington, C.A., 1990. Alnus rubra Bong. Red Alder. In: Burns, R.M., Honkala, B.H. (Eds.), Silvics of North America, Volume II: Hardwoods. US Department of Agriculture, Forest Service, Washington, DC, Agriculture Handbook 654.

Hillier, F.S., Lieberman, G.J., 1980. Introduction to Operations Research. Holden-Day, Inc., San Francisco, CA, 829 p.

Hooke, R., Jeeves, T.A., 1961. "Direct Search" solution of numerical and statistical problems. J. ACM. 8 (2), 212–229.

Hool, J.N., 1966. A dynamic programming–Markov chain approach to forest production control. For. Sci. Monogr. 12, 26 p.

Jacobson, M., 2008. Forest Finance 8: To Cut or Not Cut—Tree Value and Deciding When to Harvest Timber. Pennsylvania State University, College of Agricultural Sciences, Cooperative Extension, University Park, PA, 8 p.

Kao, C., Brodie, J.D., 1980. Simultaneous optimization of thinnings and rotation with continuous stocking and entry intervals. For. Sci. 26 (3), 338–346.

Kay, R.D., Edwards, W.M., 1999. Farm Management. McGraw-Hill, New York, NY, 465 p.

Kaya, A., Bettinger, P., Boston, K., Akbulut, R., Ucar, Z., Siry, J., Merry, K., Cieszewski, C., 2016. Optimisation in forest management. Cur. For. Rep. 2 (1), 1–17.

Kennedy, J.O.S., 1986. Dynamic Programming. Applications to Agriculture and Natural Resources. Elsevier Applied Science Publishers, London, 341 p.

Long, J.N., 1985. A practical approach to density management. For. Chron. 61 (1), 23–27.

Nepal, P., Grala, R.K., Grebner, D.L., 2012. Financial feasibility of increasing carbon sequestration in harvested wood products in Mississippi. For. Policy Econ. 14 (1), 99–106.

Niinimäki, S., Tahvonen, O., Mäkelä, A., 2012. Applying a process-based model in Norway spruce management. For. Ecol. Manage. 265, 102–115.

Olson Jr., D.F., Roy, D.F., Walters, G.A., 1990. Sequoia sempervirens (D. Don) Endl. Redwood. In: Burns, R.M., Honkala, B.H. (Eds.), Silvics of North America, Volume I: Conifers. US Department of Agriculture, Forest Service, Washington, DC, Agriculture Handbook 654.

Reineke, L.H., 1933. Perfecting a stand-density index for even-aged forests. J. Agric. Res. 46 (7), 627–638.

Rideout, D., 1985. Managerial finance for silvicultural systems. Can. J. For. Res. 15 (1), 163–166.

Riitters, K., Brodie, J.D., 1984. Implementing optimal thinning strategies. For. Sci. 30 (1), 82–85.

Smith, D.M., Larsen, B.C., Kelty, M.J., Ashton, P.M.S., 1996. The Practice of Silviculture: Applied Forest Ecology. 9th ed. John Wiley & Sons, New York, NY, 537 p.

Tennessee Department of Agriculture—Division of Forestry, 2015. Tennessee Forest Products Bulletin, July–September 2015 (Volume 41, Number 3). Tennessee Department of Agriculture—Division of Forestry, Nashville, TN.

Valsta, L.T., 1990. A comparison of numerical methods for optimizing even aged stand management. Can. J. For. Res. 20 (7), 961–969.

von Ciriacy-Wantrup, S., 1938. Multiple and optimum use of wild land under different economic conditions. J. For. 36 (7), 665–674.

Whittle, P., 1982. Optimization Over Time. Dynamic Programming and Stochastic Control, Vol. 1. John Wiley & Sons, Ltd., Chichester, UK, 317 p.

Wikstrom, P., 2001. Effect of decision variable definition and data aggregation on a search process applied to a single-tree simulator. Can. J. For. Res. 31 (6), 1057–1066.

Williams, M.R.W., 1988. Decision-Making in Forest Management. 2nd ed. Research Studies Press, Ltd., Letchworth, UK, 133 p.

Yoshimoto, A., Haight, R., Brodie, J.D., 1990. A comparison of the pattern search algorithm and the modified PATH algorithm for optimizing an individual tree model. For. Sci. 36 (2), 394–412.

Yoshimoto, A., Paredes, V., Brodie, J.D., 1988. Efficient Optimization of an Individual Tree Growth Model. US Department of Agriculture, Forest Service, Rocky Mountain Research Station, Ft. Collins, CO, pp. 154–162., General Technical Report RM-161.

Zhou, M., Buongiorno, J., 2011. Effects of stochastic interest rates in decision making under risk: a Markov decision process model for forest management. For. Policy Econ. 13 (5), 402–410.

Chapter 6

Graphical Solution Techniques for Two-Variable Linear Problems

Objectives

Linear models are widely used for developing contemporary strategic forest management plans. Although we begin coverage of addressing planning models with linear programming in the next chapter, to understand how forest planning models work, it may be necessary to view linear problems in graphical form first. Viewing linear problems in two dimensions can help one to comprehend multidimensional space when moving beyond two or three dimensions. Therefore, we focus on simple two-variable problems in this chapter and provide graphical and algebraic representations of the constraints so that the solution space, where feasible solutions to a problem reside, can be examined. Upon completion of this chapter, you should be able to:

1. Understand how a problem described verbally or in writing could be translated into a set of linear equations.
2. Graph the constraints associated with two-variable linear problems.
3. Identify the feasible region and the optimal solution for two-variable linear management problems.
4. Algebraically identify the optimal solution to linear problems.
5. Understand the terms feasibility, infeasibility, efficiency, inefficiency, and optimality.

I. INTRODUCTION

For many people, the most difficult aspect of solving problems is taking a written or spoken representation of them (from a colleague or supervisor) and translating it into a set of quantitative relationships. Therefore, this chapter is provided as a review of interpreting and solving word problems. Since it may have been some time since students have used these skills, we provide an introduction (or review) of their application in natural resource management. With this in mind, we walk through three management problems very explicitly (in Sections II.A, II.B, and II.C), then one other problem less explicitly (Section II.D), with the hope that students will think about

how the problems were developed and solved. In addition, this chapter serves as an introduction to the forthcoming chapters on forest-level planning.

Following are a set of steps we might logically use to effectively solve a problem that is either communicated verbally or in writing.

1. *Understand the management problem.* If the problem is verbally given to you, then after listening carefully, you should write down the problem and review it with the land manager or landowner. Thoroughly reading a written problem, and carefully listening to a spoken problem, are both important so that you can avoid considering solutions that may not be appropriate, and thus prevent wasting time or forgetting to consider one the important constraints of the landowner. For example, if a landowner, suggests the periodic attainment of revenue is not important, then plans that explicitly provide this may be irrelevant and lead to suboptimal solutions for the landowner. In addition, when developing an understanding of a problem, some information that is not given initially, but ultimately is necessary to solving the problem, might be identified through clarification of the problem with the landowner. For instance, when clarifying a management problem, you may determine that one of a set of desires expressed by the landowner is actually more important than others.

2. *Translate the management problem into mathematical equations.* Although many people don't consider themselves to be "quantitative," the need to quantify relationships is important in all fields of natural resource management, particularly when determining how to allocate budgets to projects or how to allocate people to management activities. Thus this second step is needed to force you to work in an organized fashion. After understanding the problem, you will need to identify the potential decisions that are being considered, develop variables and coefficients, and assign them to the potential decisions. For example, if

Forest Management and Planning. DOI: http://dx.doi.org/10.1016/B978-0-12-809476-1.00006-0

a decision involves determining how many acres to assign to treatment #1 within stand #1, then the variable developed to represent this decision might be designated S1T1. And, if each acre of stand #1 can produce 12,000 board feet of timber volume (12 MBF) when managed under treatment #1, the coefficient 12 may be associated with the variable. Further, you must clearly understand what types of values can be assigned to each variable (i.e., acres of land, money, miles of road, number of snags, etc.). In the cases presented in this chapter, we will identify the objective of each problem (maximize or minimize some value) and the constraints on the activities that are being scheduled. Each of these relationships will be described by linear equations that are made up of the variables, their coefficients, and other terms that place limits on those decisions. Finally, keep in mind that in these examples we need to represent all relationships in a linear manner, thus there are no nonlinear relationships (e.g., X^2, XY, DBH^2H). However, these may be necessary in more complex planning problems.

3. *Solve the problem using mathematical or graphical methods.* The set of linear equations that are developed for a management problem can be solved in such a way that allows you to locate the optimal solution. In the examples provided in this chapter, we will demonstrate how to use graphical methods to draw the linear relationships, identify the feasible region of the solution space, and locate the optimal solution to the problem. In addition, each example will be solved algebraically to provide greater solution precision. Of course, to visualize a management problem graphically, we reduce the problem to a number of variables (2) and use a small set of linear equations that can easily be drawn on graph paper. In practice, you will deal with numerous competing decisions for each natural resource management issue. However, as we have noted, visualizing three or more dimensions (for three or more decisions) is difficult.

4. *Check the solution to the management problem.* Once a management problem has been solved, and the optimal solution has been located, you should perform some analysis to check the solution values. It is not unreasonable to find, in doing so, that the problem was not specified correctly. One method is to take the resulting optimal values of the decision variables and insert them into the constraints to see whether the constraints still hold (i.e., are not violated). Another method is to view a graphical representation of the solution to see whether the solution and the constraints were developed correctly. Either way, verification of the results is important. This step could occur concurrently with the previous step.

The examples that we present in this chapter were designed to develop your skills in creating mathematical relationships from a written description of a problem. The linear relationships should then be graphed in two-dimensional space and solved. In addition, we provide some guidance to help you understand how a problem can be checked for feasibility. Ultimately, the solution to the management problem should be translated back into a written description of the activities necessary to optimize the resources available.

II. EXAMPLE PROBLEMS IN NATURAL RESOURCE MANAGEMENT

The following four problems (a road construction plan, a snag development plan, a fish structure plan, and a hurricane clean-up plan) were developed to help students visualize how a word problem can be graphed to define the *solution space* of a management problem. The solution space is a general area within which managers can make decisions about how to allocate resources. These examples utilize only two variables, so a simple *X-Y* graph can be developed. In addition, the relationships within each problem are linear, meaning that straight lines can be drawn to identify the *feasible region* of the solution space, or the area of the solution space within which all combinations of *X* and *Y* lead to solutions that can be implemented without violating a constraint. We make this assumption again about two variables being involved in linear relationships to simplify the discussion. In addition, many functional relationships are nonlinear, thus much more difficult to graph by hand than linear equations. Additionally, we solve three of the problems algebraically to illustrate how the precision of the solution differs with that provided by the graphical method.

A. The Road Construction Plan

Assume that you work for a medium-sized forestry company in the interior western United States, and are developing a plan of action related to road construction opportunities over the next year. Your annual budget is $300,000, and you want to build as much road as you can within the limitations of your system. Based on recent road construction contracts, building a standard woods road, one without an aggregate (rock) surface, will cost about $30,000 per mile. Building a road with a layer of aggregate (rock) will cost about $50,000 per mile. After doing some preliminary reconnaissance, you decide that you need at least 1.5 miles of rocked road. And, you decide that at most, you need 4 miles of rocked road. At this point, your organization has already signed a contract with a local road construction company to develop 2.5 miles of woods road next year. You have decided that at most, up to 6 miles of woods road need to be built next year. The question is, how many miles of each type of road should be built given the budget constraint, your assessment of the road system needs, and the other obligations?

1. Understand the Management Problem

As mentioned in the written description of the problem, you are interested in building as much road as possible next year. This suggests that you need to maximize the number of miles of road that will be built over the next year. As a result, two decision variables are needed:

WR = miles of woods road that will be constructed over the next year
RR = miles of rocked road that will be constructed over the next year

2. Translate the Management Problem Into Mathematical Equations

Since the problem is to maximize the sum of the miles of road that are built, the objective function is relatively straightforward:

$$\text{Maximize } WR + RR$$

The objective function represents the sum of the miles of woods roads and rocked roads to be constructed next year. One obvious constraint relates to the road construction budget:

$$30,000 \, WR + 50,000 \, RR \leq 300,000$$

Here, we will multiply the miles of each type of road to be built by the cost coefficient associated with each type of road. As you can see, the total cost of building both types of road must be less than or equal to your budget. Based on your preliminary analysis, you determined that at least 1.5 miles of rocked roads are needed, therefore another relatively simple constraint to the problem is:

$$RR \geq 1.5$$

In addition, you determined that at most, 4 miles of rocked road are needed, thus a subsequent constraint to the problem is:

$$RR \leq 4$$

Given the contractual obligation to a road construction company, you must build 2.5 miles of woods road. As a result, a fourth constraint on the management problem is:

$$WR \geq 2.5$$

Finally, you determined that, at most, 6 miles of woods road are needed. Therefore the final constraint to the problem is:

$$WR \leq 6$$

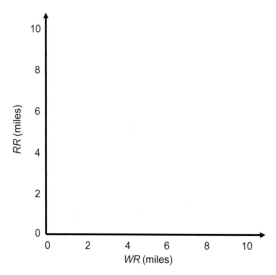

FIGURE 6.1 Initial development of a two-variable graph for the road construction problem, with the miles of rocked roads to be built on the *Y*-axis, and the amount of woods roads to be built on the *X*-axis.

3. Solve the Problem and Check the Solution

To visualize this two-variable problem, we can develop a graph with two axes that represent the amount of woods and rocked roads to be built, and label them WR and RR (Fig. 6.1). The extent of the scale on each axis can be limited to 10 units (miles of road) or so, which is slightly more than the maximum number of miles of road that we already know can be scheduled for construction. On this graph we will draw lines representing the constraints. To begin the graphing of this two-variable linear model, we will first consider the budget constraint. Our goal is to draw a straight (linear) line that represents this constraint, and to do so we can determine the point where the budget constraint line will pass through each axis. For example, if the extent of rocked roads constructed was assumed to be 0 miles, the maximum length of woods roads that could be constructed would be 10 miles ($300,000 budget/$30,000 per mile of woods road). This indicates that one end of the budget constraint line is where $WR = 10$ and $RR = 0$.

Conversely, if the number of woods road miles constructed was assumed to be 0 miles, the maximum length of rocked roads that could be constructed would be 6 miles. This tells us that the other end of the budget constraint line can be plotted where $WR = 0$ and $RR = 6$. We then draw a straight line (Fig. 6.2) to connect the two *X-Y* (*WR-RR*) pairs (10,0 and 0,6). One interpretation of the resulting graph is that every point on or below the budget constraint line is considered *feasible* (i.e., does not violate the budget constraint). Another interpretation is that every point above the line is considered *infeasible*. As an easy check on this assertion, a feasible solution is one where the length of rocked roads to be built is 0 miles

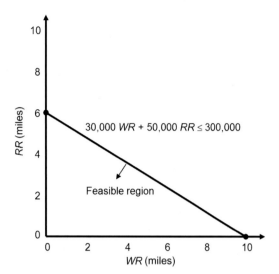

FIGURE 6.2 The budget constraint for the road construction problem.

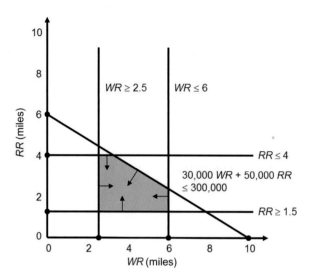

FIGURE 6.3 A graph of the entire set of constraints to the road construction problem, and the areas related to the constraints where solutions are feasible.

($RR = 0$) and the length of woods roads to be built is 0 miles ($WR = 0$). The resulting point that represents this combination of choices is below the budget constraint line, therefore all points below the budget constraint line are feasible.

The other four constraints to the road construction problem suggest the use of lines that are drawn either horizontally or vertically. Since RR is represented by the Y-axis of the graph, the two remaining RR-related constraints ($RR \geq 1.5$ and $RR \leq 4$) are horizontal lines. Feasible solutions to the road construction problem are combinations of WR and RR that can be found in the area between these two constraints (Fig. 6.3). Since WR is represented by the X-axis, we draw one line that rises vertically where $WR = 6$ miles to represent the maximum amount of woods roads that can be built. Feasibility is maintained here by all points to the left of the line, since we can build at most 6 miles of WR. The final constraint is represented by a second vertical line drawn where $WR = 2.5$ miles. Feasibility is maintained here by all points to the right of the line, since we needed at least 2.5 miles of WR to fulfill the contractual obligation. Once all the constraints have been plotted on the graph, and the direction of feasibility is noted for each of the constraints, the entire *feasible region* of the solution space can be visualized (Fig. 6.3; purple area).

Various combinations of woods roads and rocked roads can be assessed for feasibility either by viewing the X-Y (WR-RR) location on the graph, or by placing values of WR and RR back into each constraint and assessing whether the logic related to each constraint is still true. For example, given the assumptions noted earlier, we might ask whether building 4 miles of each type of road result in a feasible solution. If you were to plot on the graph the location where $WR = 4$ and $RR = 4$, you would

find that it is slightly outside of the feasible region. Inserting $WR = 4$ and $RR = 4$ into each of the constraints (where appropriate), you would also find that the budget constraint would be violated, since it would require $320,000 to build 4 miles of each type of road, and this exceeds our budget ($300,000).

To graphically locate the optimal solution to the road construction problem, you could create a series of lines (parallel to each other) that represent the slope of the objective function, by adding a test value to the right-hand side (beyond an equality sign) of the objective function (Fig. 6.4). These test values represent the total number of miles constructed of both types of road, since the objective was to maximize the sum of the roads constructed ($WR + RR$). Another way to graph the objective function is to determine its slope by algebraically rearranging the objective function to resemble $Y = MX + B$ where M equals the slope, Y represents the decision variable on the Y-axis, X represents the decision variable on the X-axis, and B equals the test value mentioned earlier. After calculating the value for M, you can use the rise/run ratio to construct your first objective function line. Since an objective function line equaling one test value is parallel to the same function equaling another test value, you can easily draw a whole family of lines with a ruler or simple straight edge.

The goal here is to find the line that has a test value that rests on the last corner of the feasible region of the solution space pointing outward toward the maximum values of RR and WR, since this is a maximization problem. In other words, we are looking for a solution that increases toward the northeast corner of a two-dimensional graph but is also feasible. In this case, we

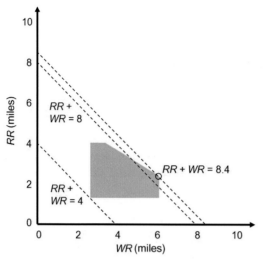

FIGURE 6.4 Identification of the optimal solution to the road construction problem using a family of objective functions.

find that the graphical solution for equation $RR + WR$ is approximately 8.4 because it touches the last corner of the solution space (Fig. 6.4), where WR and RR are approximately 6.0 and 2.4, respectively. However, the graphical solution gives only an approximate answer due to the scale and resolution of the graph. Regardless of this limitation, this is your estimate of the optimal solution to the problem (build 6.0 miles of woods roads, and 2.4 miles of rocked roads).

The graphical method for solving the road construction problem is very useful in conceptually understanding how the constraints form the feasible region that contains the possible solutions. In addition, the graphical approach allows the construction of objective function lines that are used to identify the solution that is not only feasible, but also optimal. In other words, the optimal solution is one with the highest objective function value. Although the graphical method is very useful from a conceptual standpoint, its ability to find the optimal combination of decision variables and objective function value is not very precise. Interpreting the solution from the graph is dependent on the scale of the X and Y axis as well as the grid size of the graph paper. Each of these can lead to a lack of precision in your solution. A simple mathematical approach to solving the road construction problem is to algebraically solve the problem for the combination of decisions that best meet the objective.

To solve the problem algebraically, we first summarize the road construction problem described earlier. The objective is to maximize miles of road built subject to the constraints. This is illustrated by Eq. (6.1).

$$\text{Maximize } WR + RR \qquad (6.1)$$

The objective faces six constraints, which limit how many miles of road can be built. These constraints are illustrated by Eqs. (6.2) through (6.6).

$$30{,}000\,WR + 50{,}000\,RR \le 300{,}000 \qquad (6.2)$$

$$RR \ge 1.5 \qquad (6.3)$$

$$RR \le 4 \qquad (6.4)$$

$$WR \ge 2.5 \qquad (6.5)$$

$$WR \le 6 \qquad (6.6)$$

To achieve the objective of building as many miles of road as possible, we need to find the combinations of WR and RR that are feasible, or satisfy the constraints of Eqs. (6.2) through (6.6). To solve for two unknowns, you need to use two equations. To start, you can use Eqs. (6.2) and (6.3) and assume that all of the inequality signs are equalities.

$$30{,}000\,WR + 50{,}000\,RR = 300{,}000$$
$$RR = 1.5$$

Then, you can take Eq. (6.2) and isolate WR by moving the RR to the right-hand side of the equation:

$$30{,}000\,WR = 300{,}000 - 50{,}000\,RR$$

and divide both sides of the equation by 30,000. This results in the following:

$$WR = 10 - 1.67\,RR$$

Now, since we know from above that $RR = 1.5$ you can substitute Eqs. (6.3) into (6.2) to get:

$$WR = 10 - 1.67(1.5)$$

This yields a potentially feasible combination of WR equaling 7.5 and RR equaling 1.5. To evaluate whether this combination of WR and RR is feasible, you need to substitute these values into Eqs. (6.4), (6.5), and (6.6). For Eq. (6.4) you can clearly see that $RR = 1.5$ is less than 4, so you conclude that this constraint is satisfied. For Eq. (6.5), $WR = 7.5$ satisfies the constraint that WR has be equal to or greater than 2.5. For Eq. (6.6), we see that a WR of 7.5 does not satisfy the constraint of being less than 6. Given that the combination of $WR = 7.5$ and $RR = 1.5$ does not satisfy all the constraints it is deemed an infeasible solution.

This is obviously only one possible solution, however. Since there are five constraint equations, you could envision that there are 10 unique possible combinations of WR and RR that can be derived (Table 6.1). In this problem, there are actually only eight unique possible combinations of WR and RR, since two constraint combinations involve only one decision variable (constraint combinations 3 and 4, and 5 and 6). Remember, not all combinations will be feasible. Once you determine that a

TABLE 6.1 Possible Combinations of Woods Roads and Rocked Roads by Combining Algebraically the Constraints Related to the Road Construction Problem

Constraint Combination	Woods Roads (miles)	Rocked Roads (miles)	Total (miles)	Budget ($)	Feasible?
(2) and (3)	7.50	1.50	9.00	300,000	No
(2) and (4)	3.32	4.00	7.32	299,600	Yes
(2) and (5)	2.50	4.50	7.00	300,000	No
(2) and (6)	6.00	2.40	8.40	300,000	Yes
(3) and (4)	–	–	–	–	–[a]
(3) and (5)	2.50	1.50	4.00	150,000	Yes
(3) and (6)	6.00	1.50	7.50	255,000	Yes
(4) and (5)	2.50	4.00	6.50	275,000	Yes
(4) and (6)	6.00	4.00	10.00	380,000	No
(5) and (6)	–	–	–	–	–[a]

[a]One variable undetermined through the constraint combination.

combination is not feasible then you do not consider it as a potential solution. For the road construction problem, the constraints suggest that there are eight unique combinations of WR and RR that could solve the problem, however only five combinations yield unique feasible solutions (Table 6.1). If you examine Fig. 6.4 closely, you will see that these five solutions actually represent the corners of the feasible region of the solution space.

The first unique solution using a combination of constraints in Eqs. (6.2) and (6.4) yields an objective function value of 7.32 miles of road constructed. The second combination, Eqs. (6.2) and (6.6), yields an objective function value of 8.4 miles of road constructed. The third combination, Eqs. (6.3) and (6.5), yields 4.0 miles of road built. The fourth unique combination, Eqs. (6.3) and (6.6), yields 7.5 miles of road built. The last unique combination, Eqs. (6.4) and (6.5) yields 6.5 miles of road built. Since the goal is to build as many miles of road as possible, your optimal solution is the second combination, which generates 8.4 miles of road to be built, by constructing 6 miles of woods road and 2.4 miles of rocked road.

B. The Plan for Developing Snags to Enhance Wildlife Habitat

In this example, assume that you work as a wildlife biologist for the US Forest Service in Sweet Home, in the Oregon Cascade Mountains. In your position, you need to develop an annual plan of action to develop snags (dead trees) in clearcuts to enhance habitat conditions for pileated woodpeckers (*Dryocopus pileatus*) and purple martins (*Progne subis*). Your goal is to develop as many snags as possible over the next year. You decide, based on the available local expertise, that you will either have someone blast (with dynamite) the tops out of trees, or have someone top (cut) the trees with a chainsaw. All of this, of course, happens about 100 ft off the ground, thus the cost for blasting the trees is estimated to be $100 per tree, whereas the cost for chainsaw cutting is estimated to be $50 per tree. The budget for this work is, however, only $80,000. Prior to doing this analysis, you already have entered into a contract with a local logger to top 250 trees with a chainsaw, and you also have entered into a contract with a local consultant to blast 100 trees with dynamite. However, you also have determined that at most, 600 of the trees can be blasted due to the proximity of the trees to nearby homes.

1. Understand the Management Problem

As you may have gathered in the problem just described, you are interested in creating as many snags as possible, which implies that you seek to maximize the number of snags created. Two decision variables therefore are needed, such as the following:

CS = number of snags created using a chainsaw
DS = number of snags created using dynamite

2. Translate the Management Problem Into Mathematical Equations

The objective function to the problem is to maximize the total number of the snags created over the next year.

Maximize $CS + DS$

As with the previous road construction example, one constraint can be developed that relates to the annual budget:

$$100\, DS + 50\, CS \leq 80{,}000$$

Here, each snag created using dynamite will cost $100, and each snag created using a conventional chainsaw will cost $50. The total cost for both snag-creation processes must be less than the budget. Based on your previous commitments, there are minimum numbers of each type of snag that need to be included in your final plan. These include at least 250 chainsaw-created snags and at least 100 dynamite-created snags:

$$CS \geq 250$$
$$DS \geq 100$$

Finally, only a limited number of snags can be created using the blasting method. Therefore, the problem must

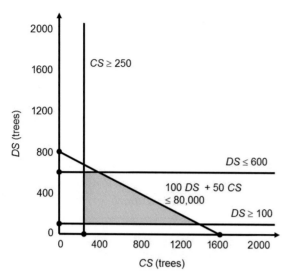

FIGURE 6.5 The graphed constraints to the snag development problem, and the identification of the feasible region (orange area).

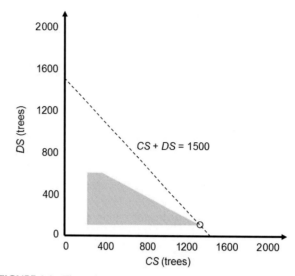

FIGURE 6.6 The optimal solution to the snag development problem.

include a constraint limiting the number of snags created in this manner:

$$DS \leq 600$$

3. Solve the Problem and Check the Solution

To visualize this two-variable problem, we develop a graph with two axes, and label them DS and CS (Fig. 6.5). The budget constraint and constraints related to existing agreements are drawn using techniques described in the previous section, along with the resource-related constraint that concerns the maximum number of snags created from blasting trees with dynamite. Once all the constraints have been plotted on the graph, and the direction of feasibility is noted as it relates to each constraint, we can visualize the feasible region of the solution space. The optimal solution to this problem is one where $DS = 100$, and $CS = 1,400$, or where we would create 100 snags using dynamite, and 1,400 snags using a chainsaw (Fig. 6.6).

To solve the problem algebraically, you can begin by summarizing the snag development. The objective is to maximize the sum of snags created. This is illustrated by Eq. (6.7).

$$\text{Maximize } CS + DS \qquad (6.7)$$

The objective faces four constraints that limit how many snags can be created. These constraints are illustrated by Eqs. (6.8) through (6.11).

$$100\,DS + 50\,CS \leq 80,000 \qquad (6.8)$$

$$CS \geq 250 \qquad (6.9)$$

$$DS \geq 100 \qquad (6.10)$$

$$DS \leq 600 \qquad (6.11)$$

To achieve the objective of creating as many snags as possible, we need to find the combinations of DS and CS that are feasible or satisfy the constraints in Eqs. (6.8) through (6.11). To start, you can use Eqs. (6.8) and (6.9). First, you can take Eq. (6.8) and isolate DS by moving the CS to the right-hand side of the equation, and divide both sides of the equation by 100. This results in the following:

$$DS = 800 - 0.5\,CS$$

Now, you can substitute Eqs. (6.9) into (6.8) to get:

$$DS = 800 - 0.5(250)$$

This substitution yields a potentially feasible combination of DS equaling 675 and CS equaling 250. To evaluate whether this combination of DS and CS is feasible, you need to substitute these values into Eqs. (6.10) and (6.11). For Eq. (6.10) you can clearly see that $DS = 675$ is greater than 100, so we conclude that this constraint is satisfied. For Eq. (6.11), you see how $DS = 675$ is greater than 600, which does not satisfy the constraint. Given this information, you should not consider this combination of DS and CS for any further evaluation. The next combination to consider is Eqs. (6.8) and (6.10). Following the same process as before but now isolating CS, you now get CS equal to 1,400 and DS equals 100. Taking this combination, you evaluate Eqs. (6.9) and (6.11) to see that they are satisfied. This combination of DS and CS creates 1,500 snags. Another combination of DS and CS to evaluate can be solved by using Eqs. (6.8) and (6.11). Solving for DS and CS using these equations yields a CS of 400 and a DS of 600, which creates 1,000 snags. Eqs. (6.9) and (6.10) yield a CS and DS that creates only 350 snags. Solving Eqs. (6.9) and (6.11) for CS and

DS creates a feasible solution that generates 850 snags. Since the goal is to create as many snags as possible, our optimal solution is the second combination that generates 1,500 snags; 1,400 by use of chain saws and 100 by use of dynamite.

C. The Plan for Fish Habitat Development

For the third example of graphical and algebraic techniques applied to a two-variable linear problem, assume that you are a fisheries and wildlife consultant, and have entered into a contract with the US Forest Service to develop fish structures in streams within a specific watershed in Idaho. These fish structures are designed to facilitate the development of pools in the stream system. The structures themselves can be developed using either logs or boulders. Assume that the Forest Service would like you to develop these structures across as many miles of stream as possible within the limit of the budget. After reviewing maps and aerial photographs, you decide that at least 5 miles of the stream system should be treated by placing logs in various places, and that at least 2.5 miles should be treated using boulders. Given the limited availability of large boulders in this area, you have determined that only up to 7.5 miles of streams can be treated with boulders. The budget for doing this work is $250,000. It will cost $10,000 per stream mile to create fish structures using logs, and $21,000 per stream mile to create structures using boulders.

1. Understand the Management Problem

After reading through the problem statement, you may have gathered that the objective of the problem is to treat as many miles of stream as possible with either logs or boulders. This implies that you need to maximize the miles of stream treated. To simplify the problem, we will also assume that the same stream reach will not be treated with both logs and boulders. Two decision variables therefore are needed for this analysis:

Logs = number of miles of streams where logs will be installed to create pools
Boulders = number of miles of streams where boulders will be installed to create pools

2. Translate the Management Problem Into Mathematical Equations

The objective function to the problem is to maximize the sum of the miles of stream that are treated with either logs or boulders.

$$\text{Maximize } Logs + Boulders \qquad (6.12)$$

The constraints associated with this problem include the following:

1. (Budget) $10,000 \, Logs + 21,000 \, Boulders \le 250,000$
$$(6.13)$$

2. (*Logs* minimum) $\qquad Logs \ge 5 \qquad (6.14)$

3. (*Boulders* minimum) $\quad Boulders \ge 2.5 \qquad (6.15)$

4. (*Boulders* maximum) $\quad Boulders \le 7.5 \qquad (6.16)$

After working through the previous two problems, you should be able to understand how these four constraints arose from the written description of the management problem.

3. Solve the Problem and Check the Solution

For this two-variable problem, we develop a graph (Fig. 6.7) and label the two axes *Logs* and *Boulders*. All the constraints related to the budget and the resources (both the availability of resources and the limitations related to resources) are drawn, and once the direction of feasibility is noted (as it relates to each constraint), you should be able to visualize the entire feasible region of the solution space. The optimal solution to this problem is where *Logs* is approximately 20, and *Boulders* = 2.5. This suggests that to best utilize the budget, about 20 miles of the stream system should be enhanced with logs to create pools for fish habitat, and about 2.5 miles of the stream system should be enhanced with boulders to create pools for fish habitat.

To achieve the objective of treating as many miles of stream as possible with logs and boulders, you need to find the combinations of *Logs* and *Boulders* that are feasible, or that satisfy the constraints in Eqs. (6.13) through (6.16). To start, you can use Eqs. (6.13) and (6.14). First, you can take Eq. (6.13) and isolate *Boulders* by moving

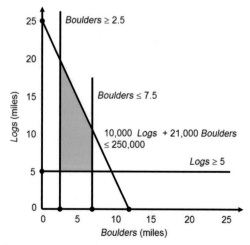

FIGURE 6.7 The constraints and feasible region (green area) associated with the fish habitat problem.

the *Logs* to the right-hand side of the equation, and divide both sides of the equation by 21,000. This results in the following:

$$Boulders = 11.9 - 0.48 \, Logs$$

Now, you can substitute Eq. (6.14) into Eq. (6.13) to get:

$$Boulders = 11.9 - 0.48(5)$$

This yields a potentially feasible combination of *Boulders* equaling 9.52 and *Logs* equaling 5. To evaluate whether this combination of *Boulders* and *Logs* is feasible, you need to substitute these values into Eqs. (6.15) and (6.16). For Eq. (6.15) you can clearly see that *Boulders* = 9.52 is greater than 2.5 so you conclude that this constraint is satisfied. For Eq. (6.16), you can see how *Boulders* = 9.52 is greater than the upper limit of 7.5 so it does not satisfy the constraint. Given this information, you should not consider this combination of *Boulders* and *Logs* for any further evaluation.

The next combination to consider is Eqs. (6.13) and (6.15). Following the same process as before but isolating *Logs*, we now get *Logs* equal to 19.75 and *Boulders* equal to 2.5. Taking this combination, we evaluate Eqs. (6.14) and (6.16) see that they are satisfied. This combination of *Boulders* and *Logs* treats 22.25 miles of stream. Another combination of *Boulders* and *Logs* to evaluate can be solved by using Eqs. (6.13) and (6.16). Solving for *Boulders* and *Logs* using these equations yields *Logs* of 9.25 and *Boulders* of 7.5, which treats 16.75 miles of stream. Eqs. (6.14) and (6.15) yield *Logs* and *Boulders* that treat only 7.5 miles of stream. Solving Eqs. (6.14) and (6.16) for *Logs* and *Boulders* creates a feasible solution that treats 12.5 miles of stream. Since the goal is to treat as many miles of stream as possible, our optimal solution is the combination of activities that treats 22.25 miles of stream by installing logs in 19.75 miles of streams and laying boulders in 2.5 miles of streams, all to create pools to enhance fish habitat quality.

D. The Hurricane Clean-up Plan

In this example, assume that you work as a government forester in Hancock County, Mississippi. You have been assigned the task for developing a hurricane clean-up plan for 750 acres of forested property adversely affected by Hurricane Katrina (Fig. 6.8). The agency you work for wants to minimize the cost of the clean-up process. You decide that you will either have someone cut and haul the debris off the site to a nearby landfill, or have someone cut, pile, and burn the debris on the site. The cost for cutting and hauling the debris is estimated to be $2,000 per acre, and the cost for cutting,

FIGURE 6.8 Hurricane damage to a pine stand after Hurricane Katrina in 2005. *Photo courtesy of Andrew J. Londo.*

piling, and burning is estimated to be $1,000 per acre. The budget for this work is $1,000,000 for the 750 acres. Prior to doing this analysis, you already have entered into a contract with a local logging contractor to cut and haul 200 acres of debris costing $400,000, and you also have entered into an agreement with the Mississippi Forestry Commission to burn at least 300 acres, and along with the cutting and piling activities, this will cost $300,000. However, you also have determined that the emission of smoke particulates into the atmosphere could pose a significant health risk to local communities, so no more than 500 acres, which will cost $500,000, can be cut, piled, and burned.

1. Understand the Management Problem

After having worked through the previous problems, you quickly surmise that you are interested here in reducing the cost of clean-up as much as possible, which implies that you are facing a cost minimization problem. Two decision variables therefore are needed, such as the following:

CH = cost of cutting and hauling woody debris to nearby landfills
CPB = cost of cutting, piling, and burning woody debris on site

2. Translate the Management Problem Into Mathematical Equations

The objective function to the problem is to minimize the cost of cleaning up the 750 acres in Hancock County, MS.

$$\text{Minimize } CH + CPB \tag{6.17}$$

The constraints associated with this problem include the following:

$$CH + CPB \leq 1,000,000 \tag{6.18}$$

$$CH \geq 400,000 \tag{6.19}$$

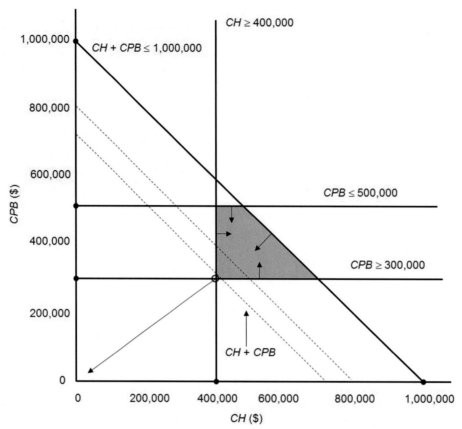

FIGURE 6.9 Identification of the feasible region and optimal solution to the hurricane clean-up problem.

$$CPB \geq 300,000 \qquad (6.20)$$
$$CPB \leq 500,000 \qquad (6.21)$$

3. Solve the Problem and Check the Solution

To visualize this two-variable problem, we develop a graph with two axes, and label them CH and CPB (Fig. 6.9). The budget constraint and constraints related to existing agreements are drawn using techniques described in the previous section. Once all the constraints have been plotted on the graph, and the direction of feasibility is noted as it relates to each constraint, we can visualize the feasible region of the solution space. The optimal solution to this problem is one where $CH = 400,000$ and $CPB = 300,000$, or where you would treat only 500 acres of the property (Fig. 6.9).

If we were to insist that all 750 acres of land are treated with one or both of the two methods, then we could rearrange the problem so that the decision variables are acres treated, and the decisions are related to assigning acres to each treatment.

$$\text{Minimize } 2,000\, CH + 1,000\, CPB \qquad (6.22)$$

where CH and CPB represent acres for each treatment. The constraints associated with this modified problem include the following:

$$2,000\, CH + 1,000\, CPB \leq 1,000,000 \qquad (6.23)$$
$$2,000\, CH \geq 400,000 \qquad (6.24)$$
$$1,000\, CPB \geq 300,000 \qquad (6.25)$$
$$1,000\, CPB \leq 500,000 \qquad (6.26)$$

And the following additional constraint,

$$CH + CPB = 750 \qquad (6.27)$$

Since the last constraint is an equality, if we were to solve the problem we would find a single solution, rather than a feasible region of solutions, where $CH = 250$ acres and $CPB = 500$ acres.

III. OPTIMALITY, FEASIBILITY, AND EFFICIENCY

When we assign a value to a set of decision variables associated with a mathematical problem, we say we have developed a *solution* to the problem. Whether the solution is feasible, infeasible, optimal, or suboptimal is another matter, however. As you may have gathered through the discussion of the examples presented in this chapter, a *feasible solution* is one where the values for each decision variable are such that all the constraints are satisfied (i.e., not violated). For example, in Fig. 6.7, a solution that requires 5 miles of stream to be treated with boulders and 10 miles of stream to be treated with logs in an effort to create better fish habitat is a feasible solution. A solution that requires treating 10 miles of streams with boulders and 10 different miles of streams with logs violates two constraints, the budget constraint and the constraint related to the maximum length of stream that could be treated with boulders. This latter case is considered to be an *infeasible solution*, since one or more constraints are violated.

The *optimal solution* to a management problem is one where a set of values assigned to the decision variables produces the highest (in the case of a maximization problem) or lowest (in the case of a minimization problem) possible objective function value without violating any of the constraints. Given that there are usually numerous feasible solutions to a problem, this is the one that produces the most favorable and efficient plan for a landowner. Any feasible solution, or combination of management choices, that leads to an objective function value that is not the optimal value is known as a *suboptimal solution*. Suboptimal solutions are not necessarily bad, and often are implemented in practice (perhaps without understanding why they are suboptimal). In many cases, you may find that there numerous suboptimal solutions that are not very different from the optimal solution. A common example involves switching the harvest timing of a management unit with that of another, because of some issue (perhaps related to equipment logistics) that was not recognized in the original mathematical problem. Also keep in mind that suboptimal, feasible solutions may be far from the optimal solution, as quantified by the objective function value. At the extreme, sometimes doing nothing at all is a suboptimal, feasible solution. Graphically, when the origin (0,0) is located on the edge of the feasible region of the solution space, doing nothing at all may represent a suboptimal feasible solution.

There are times, however, when no optimal solutions can be located for a management problem. This situation could arise if there are no feasible solutions to a problem. For example, in the snag development problem of Section II.B, if the constraints were simply:

$$100\ DS + 50\ CS \leq 80,000$$

$$CS \geq 250$$

$$DS \geq 800$$

then we would find that no feasible solution to the problem can be developed, because the budget required to create 250 snags using a chainsaw and 800 snags using dynamite would be $92,500, which would violate the budget constraint. In other cases, numerous feasible, optimal solutions may be available for a management problem. For example, if we added one last constraint to the fish structure management problem from Section II.C, that indicated that no more than 15 miles of stream could be treated,

$$Boulders + Logs \leq 15$$

then we would find, when we graphed the problem, that a number of solutions are both feasible and optimal (those along the line from points A to B in Fig. 6.10) since this last constraint has the same slope as various versions of the objective function (e.g., *Boulders + Logs* = 10, *Boulders + Logs* = 20).

Solutions to problems are said to be *efficient* when there are no other solutions that can produce more of one of the outputs, while producing the same amount of the other outputs. Two examples will help illustrate this point. First, from the problem described in Section II.A, one feasible solution could be where we construct 2.4 miles of rocked roads and 4 miles of woods roads (*RR* = 2.4 and *WR* = 4). We know, of course, that this is not the optimal solution to the management problem. It is also not the most efficient solution to the problem. We know this because we could have constructed the same amount of rocked roads (2.4 miles), yet also could have constructed two more miles of woods roads, creating the solution *RR* = 2.4 and *WR* = 6. Alternatively, we could create a solution with the same amount of woods roads (4 miles), yet with 1.2 more miles of rocked roads (*RR* = 3.6 and *WR* = 4). Each of these solutions would be considered

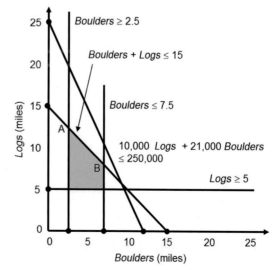

FIGURE 6.10 A modified fish habitat problem, with multiple optimal solutions.

more efficient, within the framework of the problem, than the solution $RR = 2.4$ and $WR = 4$.

As a second example of efficiency, try to imagine a broader problem of scheduling forest management activities that will simultaneously produce timber volume and affect the quantity of wildlife habitat. The feasible solutions to the problem might be graphed (Fig. 6.11) in such a way to show the trade-offs among the goals. What is provided, of course, is a generic solution space, but it illustrates nicely the concept of efficiency. Point A on the graph illustrates an inefficient solution. That is, the combination of activities that produced the solution described by point A could be improved so that the same amount of wildlife habitat is available, yet with a larger timber harvest volume (point B), or that the same amount of timber volume could be produced, yet with a larger amount of wildlife habitat (point C). In either of the latter two cases, a rearrangement of the temporal or spatial distribution of activities could result in a more efficient forest plan. The plan that is described by point D is infeasible, since one or more constraints must be violated for the objective function value to be located outside of the feasible region. To get to point D, we would either need more resources or need to relax one or more constraints.

The curved line that describes the outer edge of the solution space in Fig. 6.11 is sometimes called a *production possibility frontier*. This curve represents the efficient solutions to the problem under consideration. If resources are allocated efficiently to the management of the forest, then the possible solutions and the trade-offs among competing efficient solutions can be visualized very clearly. The production possibility frontier curve can be convex with respect to the origin of the graph (curving or bulging outward), but not necessarily so. The relationships are

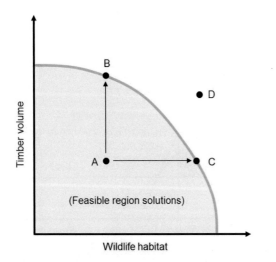

FIGURE 6.11 An example of efficient, feasible, inefficient, and infeasible solutions to a broad timber harvest and wildlife habitat management problem.

interesting in that we can determine the maximum amount of one output (wildlife habitat) that is possible given any other amount of the other output, and the given amounts of inputs (land, labor, budget, etc.). The relationship of the two levels of output generally indicates that increases in the production of one output will usually result in decreases in the production of the other output.

The production possibility frontier also plays an important role in identifying what economists call "pareto optimal" outcomes. Points along the frontier are continuous and are technically efficient because increases in any one solution cannot occur without a reduction in the other solution while maintaining the same feasible output level. This allows for the output level found on the production possibility frontier to be the same all along the curve despite changes in the inputs or solutions. We can also think of the different combinations of solutions as having the same marginal rate of technical substitution. Simplifying Boadway and Bruce's (1984) definition, a pareto optimal solution is a state where no one aspect of a plan can be made better without hurting another aspect of a plan. Production possibility frontiers (curves that reflect the optimal combinations of outcomes) provide insight into these relationships, however they are just a first step in determining whether pareto optimal solutions exist when using multiobjective optimization techniques. The techniques used by Diaz-Balteiro et al. (2014) provided a production possibility frontier, where each point on a curve represented possible pareto optimal outcomes for plans that had rotation age, fire risk, timber production, and carbon sequestration goals. A more detailed description is beyond the scope of this book, but some related topics on multiobjective optimization are discussed in later chapters.

Although the specific location and timing of activities is not presented, the alternatives that are considered when viewing a production possibility frontier can provide information that is relevant to a planning process. Broadly speaking, planners and managers might be able to better understand the fact that efficiently managing resources that are not complementary with respect to their achievement requires that increasing the value of one output may result in decreased values of another output. When resources are not being managed efficiently, one output could be increased without a subsequent decrease in other outputs.

IV. SUMMARY

Understanding the goals, objectives, and constraints of a landowner or a management situation is a necessary aspect of forest planning. This process is instrumental in determining how to address a management problem, and in determining the aspects of the problem that need a thorough assessment. The problems we presented in this chapter involved only two variables, which we acknowledge simplifies many modern forest management and planning challenges. However, two-variable linear problems are easy to visualize and provide an entry point into understanding how other quantitative methods approach developing feasible forest plans. With three or more decision variables, a problem becomes difficult or impossible to visualize, yet the concepts of the solution space, the feasible region, efficiency, and infeasibility remain the same. In addition, nonlinear relationships will likely be introduced into conventional forest management and planning problems, and graphical solutions involving these may become difficult to visualize. Nonlinear relationships include the growth behavior of trees, the development and decay of snags and down wood, the adjacency of harvests, stream-related impacts (e.g., temperature concerns), wildlife habitat relationships, and others. As we move forward, we will introduce a number of forest- or landscape-level management planning techniques that can be used to address these more realistic problems.

QUESTIONS

1. *Pruning Contract.* Assume that you work for Continental Pacific Timberlands in western Washington, and you are in charge of the pruning program, which has a budget of $130,000. After reviewing the forest inventory, you determine that there are a sufficient number of acres of well-stocked Douglas-fir (*Pseudotsuga menziesii*) in your district in need of pruning this year. Pruning eliminates the lower branches on trees to promote the development of knot-free wood that can be of high value. There are two available contractors for this work, and you want them to prune as many acres as possible; however, you only want one contractor on the district at any one time. You want all the work to be completed in 70 days. You determine through conversations with the contractors that Crew #1 requires 0.16 days per acre, and Crew #2 requires 0.13 days per acre. On average, you will have the crews prune 100 trees per acre. Crew #1 indicates that their work will cost $1.95 per tree. Crew #2 will cost $2.05 per tree. Finally, to compare their work, you want to give at least 50 acres to each of the two crews. To visualize the problem and the associated solution space:
   a. Write out the problem formulation (objective function and constraints).
   b. On a graph, draw and label the lines to describe each constraint.
   c. Identify the feasible region on the graph.
   Would it be feasible to allocate 100 acres to crew #1 and 300 acres to crew #2? Would allocating 250 acres to each crew lead to a feasible solution?
2. *Developing a trail system.* Assume that you work for a National Park in Colorado. As part of your job you

need to develop an estimate for the development of a new trail system. You would like to develop as many trails (defined by their length) as possible, however your budget is only $150,000. Two types of trails can be built: (1) easily traveled trails that are initially cut with a small bulldozer, then hand-raked and graded, and (2) rougher, more natural trails developed using picks, axes, and shovels. In each case, the average grade of the trails will be maintained at, or below, 8%, and aligned to suit the topography of the area. All debris within 25 ft of the trails will be removed as well. You estimate that the first type of trail (1) can be developed at a rate of 5 miles per month, at a cost of about $7,500 per mile. Since hand tools are primarily used the second type of trail (2) can be developed at a slower rate, 4 miles per month, yet cost about $4,000 per mile. At a minimum you decide that you want to develop at least 2 miles of each type of trail. Ideally, you would like all the work to be completed in 6 months, and you would like the work to be completed using only one contractor and one crew. The rates of trail development can be converted to (1) 0.2 months per mile, and (2) 0.25 months per mile to make the problem more readily solvable. To visualize the trail development problem and the associated solution space:
   a. Write out the problem formulation (objective function and constraints).
   b. On a graph, draw and label the lines to describe each constraint.
   c. Identify the feasible region on the graph.
   Optimally, about how many miles of each type of trail should be built over the next 6 months? Would it be feasible to build 10 miles of trail type 1 and 20 miles of trail type 2?
3. *Cruising a Potential Land Purchase Area.* Assume that you work for a timber investment management organization (TIMO) in Louisiana. You need to develop an estimate of the timber resources on 5,000 acres that your organization is considering buying. You need to have cruised as much of this land as possible, however your budget is only $36,000. Two local consulting foresters can do the timber cruising for you; consultant #1 prefers to use fixed plot sampling, consultant #2 prefers to use point (prism) sampling. You want to give as much work as you can to each of them. Consultant #1 is somewhat busy, but can cruise up to 3,000 acres within your time frame, however they need a guarantee of 1,000 acres before they will agree to do the work. You have agreed (by contract) to provide this. Consultant #2 also requires a guarantee of 1,000 acres, but can cruise as many acres as are possible within the time frame. Consultant #1 can do the work for $8.50 per acre. Consultant #2 can do the

work for $6.50 per acre. To visualize the problem and the associated solution space:

   a. Write out the problem formulation (objective function and constraints).
   b. On a graph, draw and label the lines to describe each constraint.
   c. Identify the feasible region on the graph.

   Would it be feasible to ask consultant #1 to cruise 3,000 acres, and consultant #2 to cruise 2,000 acres? Would giving 2,500 acres to each consultant lead to a feasible solution?

4. *Snag development.* Table 6.1 is instructive in understanding the combinations of choices that can be used to evaluate solutions from an algebraic manipulation of the constraints. Use the constraints suggested for problem II.B to develop a table similar to Table 6.1 that illustrates the options for *CS* and *DS*. Highlight the combinations that are infeasible, and make a note of the optimal solution.

5. *Hurricane clean-up plan.* Given the information provided in Section II.D, solve algebraically the hurricane clean-up plan problem.

6. *Cruising the Putnam Tract.* You need to develop an estimate of the timber resources on the 2,602 acres that your organization manages within the Putnam Tract. You need to have cruised as much of this land as possible, however your budget is only $20,000. As with the Louisiana problem, two local consulting foresters are available who can do the timber cruising for you; consultant #1 prefers to use point (prism) sampling, and consultant #2 prefers to use fixed plot sampling. You want to give as much work as you can to each of them. Consultant #1 is somewhat busy, but can cruise up to 1,500 acres within your time frame; however, they need a guarantee of 500 acres before they will agree to do the work. You have agreed (by contract) to provide this. Consultant #2 also requires a guarantee of 500 acres, but can cruise as many acres as are possible within the time frame. Consultant #1 can do the work for $7.20 per acre. Consultant #2 can do the work for $8.50 per acre. To visualize the problem and the associated solution space:

   a. Write out the problem formulation (objective function and constraints).
   b. On a graph, draw and label the lines to describe each constraint.
   c. Identify the feasible region on the graph.
   d. What is the optimal solution to the problem?
   e. How would you allocate the areas on a map of the Putnam Tract?

7. *Stream enhancement project on the Lincoln Tract.* Assume that as a land manager for the Lincoln Tract, you want to improve fish habitat on about 1 mile of stream in the southwest portion of the property (see the following figure).

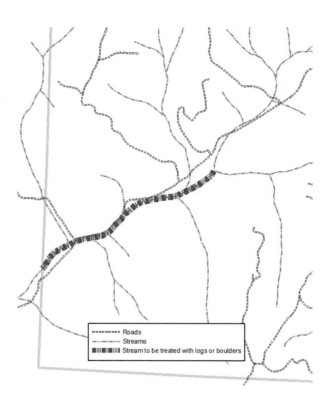

The fish habitat structures to facilitate the development of pools can be developed using either logs or boulders. Assume that you would like to develop these structures within the entire mile of stream, if possible, within the limit of your budget ($15,000). After reviewing the site, you decide that at least 0.25 miles of the stream system should be treated by placing logs in various places, and that at least 0.30 miles should be treated using boulders. It will cost about $11,000 per stream mile to create fish structures using logs, and $19,000 per stream mile to create structures using boulders.

   a. Write out the problem formulation (objective function and constraints).
   b. On a graph, draw and label the lines to describe each constraint.
   c. Identify the feasible region on the graph.
   d. What is the optimal solution to the problem?
   e. How would you allocate the areas to treat on a map of the Lincoln Tract?

REFERENCES

Boadway, R.W., Bruce, N., 1984. Welfare Economics. Basil Blackwell, New York, NY, 344 p.

Diaz-Balteiro, L., Martell, D.L., Romero, C., Weintraub, A., 2014. The optimal rotation of a flammable forest stand when both carbon sequestration and timber are valued: a multi-criteria approach. Natural Hazards 72 (2), 375–387.

Chapter 7

Linear Programming

Objectives

In developing a natural resource management plan, we often either maximize or minimize some set of values; otherwise we are simply simulating management activities with little regard to the overall efficiency (wise use of resources) of the plan. Although simulation is often used in science to test hypotheses or scenarios, we concentrate here on optimization processes, which allow us to separate, rank, and evaluate different plans of actions. Ultimately, we would prefer to develop the best plan possible to use as a guide in managing our natural resources. We say a *guide* here because we usually fail to or are unable to incorporate into a planning process all the characteristics of a management problem. Weather conditions, for example, could force a land manager to deviate from a plan, as could natural disasters, price fluctuations beyond their forecasted values, and other unforeseen or unrecognized circumstances at the time the plan was developed. The next few sections of this chapter describe the aspects of a mathematical programming problem-solving method, linear programming, and illustrate how a linear programming model can be developed to represent a forest planning problem. After solving a linear programming problem, we discuss the concepts of feasibility and efficiency once again. Upon completing this chapter, you should be able to:

1. Understand the components of linear programming that are necessary to allow you to develop a solution to a natural resource management problem.
2. Construct an objective function, resource and policy constraints, and accounting rows associated with a natural resource management problem.
3. Interpret the solution values generated, once a linear programming problem has been solved.
4. Apply the concepts of feasibility, infeasibility, efficiency, and inefficiency to a linear programming problem and its associated solution.

I. INTRODUCTION

Mathematical programming is a generic term for a set of methods that can be used to optimize an objective in light of a set of constraints imposed on management activities or constraints imposed on the allocation of land to various uses (Bell, 1977). Linear programming is one of these mathematical programming techniques. Although initially described over 70 years ago, one of the first papers illustrating its potential to assist with the development of natural resource management plans came in 1962 (Curtis, 1962). Many natural resource management plans today, and some developed from over a century ago, call for the sustainability of resources, and involve imposing constraints on a problem to ensure that the use of resources does not surpass a desired rate. Mathematical programming methods such as linear programming can accommodate these problems. As a result of the need to efficiently use resources, and because of improvements in computer technology, linear programming has been used successfully in natural resource planning (Weintraub and Romero, 2006; Kaya et al., 2016). Linear programming, much to the consternation of students in natural resource management disciplines, continues to be used widely throughout North America today.

A linear programming model for a management problem is composed of an objective function, one or more constraints, and perhaps some accounting rows. These equations form the basis for a quantitative method of solving problems. Two key elements of these models are (1) that all the relationships needed to develop a plan must be quantifiable, and (2) that all the relationships in the model are linear. The output, or solution, to a linear programming problem (the plan) provides a quantitative assessment of the proposed management activities.

II. FOUR ASSUMPTIONS INHERENT IN STANDARD LINEAR PROGRAMMING MODELS

Each linear programming model, in its most basic form, is guided by the four main assumptions of proportionality, additivity, divisibility, and certainty. Some variations on these assumptions have been explored over the past four decades. In addition, we typically assume that all solution variables have either a zero or positive value. This prevents variables from providing a negative contribution to the optimal solution. However, our work will adhere to

Forest Management and Planning. DOI: http://dx.doi.org/10.1016/B978-0-12-809476-1.00007-2

the four main assumptions above as we develop and solve natural resource management problems in Chapter 8, Advanced Planning Techniques and Chapter 9, Forest and Natural Resource Sustainability.

A. The Assumption of Proportionality

Each variable in a linear programming problem is associated with a coefficient (e.g., revenue per unit produced, cost per mile of road built). The contribution of each product produced (or road built, etc.) to the objective function is directly proportional to the number of units of each product produced (or number of miles of roads of each type developed). For example, if 10 units of product A are produced, and if each is worth $50, then the contribution to the objective function is $500. If 20 units of product A are produced, then the contribution to the objective function is $1,000. The contribution of product A to the objective function is thus directly proportional to the value (potential revenue) of product A ($50), and its per-unit contribution does not change (remains constant) if more or less of the product is produced.

B. The Assumption of Additivity

Each variable in an objective function contributes to the objective function value in a way that is independent of the other variables. For example, the value of product A ($50 of revenue per unit produced) does not increase or decrease with the number of units of products B and C that are produced.

C. The Assumption of Divisibility

The value assigned to each variable in a linear programming model is assumed to be a continuous real number, thus it can be assigned fractional values. For example, a solution to a maximization problem could be $A = 3.24$, $B = 112.94$, and $C = 12$. Each value must be zero or greater. In some real-life problems, this assumption may not be valid, and may require another type of solution technique to be employed. For example, if we were attempting to maximize the number of fish structures (let's assume either logs or boulders) inserted into a stream system, then a solution of 12.56 logs and 9.23 boulders would seem unreasonable. In this case, we may need to ensure that the value assigned to each variable in the model be an integer, which would require a solution technique such as mixed integer programming (i.e., some variables can be assigned integer values, some assigned continuous numbers), or integer programming (i.e., all variables are assigned integer values). These are subjects of Chapter 8, Advanced Planning Techniques.

D. The Assumption of Certainty

With this linear programming assumption, the coefficients associated with each variable are assumed to be known with certainty. This implies that there are no stochastic (random) variations in the coefficients within a linear programming model. For example, the revenue for each unit of product A is $50, not $50 ± some random interaction (e.g., $49.35 in some instances, $52.94 in others, and so on). In addition, when we discuss the constraints to linear models in a later section, all of the right-hand side (RHS) values of the equations (the goals) are known with certainty. Papps (2000) provides an example of stochastic linear programming in forest management planning, where this assumption is relaxed.

III. OBJECTIVE FUNCTIONS FOR LINEAR PROGRAMMING PROBLEMS

An *objective function* for a linear programming model is a linear function that helps you evaluate the quality of a solution to a problem. It is used by the model to evaluate all potential combinations of management actions, essentially it is used to rate each solution. Associated with objective functions is the notion that something is either being *maximized* or *minimized*. For example, imagine a landowner is interested in creating high quality wildlife habitat. The landowner may require a plan that maximized habitat quality. Alternatively, they may require a plan that minimized the timeframe necessary to create a sufficient amount of habitat. If a landowner were interested in producing timber volume at the lowest cost possible, then he or she may require a plan that minimized management-related costs. As an example of an objective function, assume that the intent of a management problem is to generate as much revenue as possible, and that there are three decisions (produce product A, B, or C). If each unit of product A generates $50 of revenue, each unit of product B produces $60 of revenue, and each unit of product C produces $100 of revenue, then the objective function would become:

$$\text{Maximize } 50\,A + 60\,B + 100\,C$$

Obviously, some constraints would be required to ensure that the number of units that are suggested to be produced are consistent with the resources, labor, and productive capacity of the system. Otherwise the solution to this problem would be $A = \infty$, $B = \infty$, and $C = \infty$, which is an unbounded solution. Alternatively, assume that you are managing a forest road construction budget, and that the management problem is not only to build roads, but to minimize the costs incurred. Three activities are possible: road type A costs $20,000 per mile, road type B costs $15,000 per mile, and road type C costs

$12,500 per mile. The objective function then would become:

$$\text{Minimize } 20,000\,A + 15,000\,B + 12,500\,C$$

In this case, of course, some constraints would be necessary to ensure that some roads are built; otherwise the solution would become $A = 0$, $B = 0$, and $C = 0$.

In real-life natural resource planning models, the decision variables necessary to specify the model can number into the hundreds, thousands, or millions as in the case of a southern United States timber company (McTague and Oppenheimer, 2015). In this section of the book, to provide a practical example that can be easily followed by students, we present a small forest planning problem that consists of 20 stands (management units) of various sizes. The planning horizon is assumed to be 15 years, and there are three planning periods that are each 5 years long. The landowner's objective is to maximize the net present value (NPV) of the scheduled management activities.

Volumes contained in the stands of trees are similar to the average yields (cords) for planted loblolly pine (*Pinus taeda*) stands in south Georgia, on medium sites ($SI_{25} = 65$) found in the Service Forester's Handbook (US Department of Agriculture and Forest Service, 1986). Some variation was incorporated, and the values were then converted to tons per acre using 5,600 pounds per cord as a standard (Gunter, 1987). The interest rate assumed is 5%, as is a stumpage price of $25 per ton for pine products. Table 7.1 illustrates the basic characteristics of each stand that will allow us to develop the objective function and subsequent constraints.

Since there are 20 stands, and three time periods in which the land area (acres) of each stand could be harvested, there are 60 potential decisions that could be made, assuming only one prescription is allowed per stand per time period. Therefore, 60 decision variables are needed, each of which will contain the land area assigned to each decision. To create a code name for decision variables, we use a combination of letters and numbers to describe the stand and the potential harvest period. The variables will take the form $S1P1$, where S indicates that the stand number follows (in this case, stand 1), and P indicates that the time period of harvest follows (in this case, time period 1). This is just one of many ways we could define decision variables to represent the choices available.

To begin the development of the objective function, first we must remember what we are attempting to do—maximize NPV. Each of the 60 decision variables will appear in the objective function, and each needs a coefficient that describes the potential NPV per acre for each decision. It is important to remember that both aspects of the linear objective function (the coefficients and

TABLE 7.1 Characteristics of 20 Loblolly Pine Stands in South Georgia

Stand	Size (acres)	Age[a]	Period 1 Volume (tons/ac)	Period 2 Volume (tons/ac)	Period 3 Volume (tons/ac)
1	12.3	15	40	62	84
2	35.6	20	64	87	104
3	34.6	30	104	117	127
4	85.4	30	102	115	126
5	65.4	25	87	104	117
6	69.1	15	42	64	87
7	78.3	20	65	89	106
8	71.0	10	20	42	64
9	19.6	25	85	101	114
10	34.8	30	101	114	125
11	81.6	25	89	106	119
12	90.2	20	62	85	101
13	45.6	30	106	120	131
14	67.2	15	43	65	88
15	38.6	10	22	44	67
16	49.1	15	44	66	88
17	58.3	20	63	87	103
18	26.8	30	103	117	126
19	53.0	20	65	89	107
20	46.0	10	18	40	61

[a]*Stand age in the middle of planning period 1.*

the associated decision variables) must be compatible. For example, the coefficients in this case are dollars per unit area (acres), and the decisions involve the land area (acres) to assign to each choice. When multiplied together, the result is dollars (NPV), which needs to be maximized. The coefficient for decision variable $S1P1$ (harvest stand 1 in time period 1), is:

$$((40 \text{ tons per acre}) \times (\$25 \text{ per ton}))/(1.05^{2.5})$$
$$= \$885.17 \text{ per acre}$$

Here, the volume per acre is multiplied by the stumpage price, then divided by the interest rate raised to a period of time that represents the mid-point of the first time period (2.5 years from now). The coefficient for decision variable $S1P2$ (harvest stand 1 in time period 2) is:

$$((62 \text{ tons per acre}) \times (\$25 \text{ per ton}))/(1.05^{7.5})$$
$$= \$1,075.01 \text{ per acre}$$

and the coefficient for decision variable $S1P3$ (harvest stand 1 in time period 3) is:

$$((84 \text{ tons per acre}) \times (\$25 \text{ per ton}))/(1.05^{12.5})$$

$$= \$1,141.18 \text{ per acre}$$

Following this example through the entire set of 20 stands, we should be able to develop an objective function that resembles the one shown here:

```
Maximize
      885.17   S1P1   +   1075.01   S1P2   +   1141.18   S1P3
+   1416.27   S2P1   +   1508.48   S2P2   +   1412.89   S2P3
+   2301.44   S3P1   +   2028.65   S3P2   +   1725.35   S3P3
+   2257.18   S4P1   +   1993.97   S4P2   +   1711.77   S4P3
+   1925.25   S5P1   +   1803.24   S5P2   +   1589.50   S5P3
+    929.43   S6P1   +   1109.69   S6P2   +   1181.93   S6P3
+   1438.40   S7P1   +   1543.16   S7P2   +   1440.06   S7P3
+    442.59   S8P1   +    728.23   S8P2   +    869.47   S8P3
+   1880.99   S9P1   +   1751.22   S9P2   +   1548.74   S9P3
+   2235.05   S10P1  +   1976.63   S10P2  +   1698.18   S10P3
+   1969.50   S11P1  +   1837.92   S11P2  +   1616.67   S11P3
+   1372.01   S12P1  +   1473.80   S12P2  +   1372.13   S12P3
+   2345.70   S13P1  +   2080.66   S13P2  +   1779.69   S13P3
+    951.56   S14P1  +   1127.03   S14P2  +   1195.52   S14P3
+    486.84   S15P1  +    762.91   S15P2  +    910.22   S15P3
+    973.69   S16P1  +   1144.36   S16P2  +   1195.52   S16P3
+   1394.14   S17P1  +   1508.48   S17P2  +   1399.30   S17P3
+   2279.31   S18P1  +   2028.65   S18P2  +   1711.77   S18P3
+   1438.40   S19P1  +   1543.16   S19P2  +   1453.64   S19P3
+    398.33   S20P1  +    693.55   S20P2  +    828.71   S20P3
```

IV. ACCOUNTING ROWS FOR LINEAR PROGRAMMING PROBLEMS

Accounting rows are used to aggregate values that can be useful for reporting purposes, although the values accumulated through an accounting row also could be used to constrain the solution to a problem. As an example, a natural resource management organization may be developing a plan that maximizes the NPV of activities over a 20-year time horizon, subject to a variety of constraints. The management organization may also desire to understand how much habitat, or how much timber volume is produced each time period. To do so, they would need to somehow add up the habitat or volume potentially produced in each time period. Accounting rows are used for this purpose. To develop an accounting row, you must understand what it is that you are trying to accumulate. Three brief examples are helpful here, a land area accounting row, a commodity production accounting row, and a habitat-related accounting row.

A. Accounting Rows Related to Land Areas Scheduled for Treatment

Using the model we began building in this chapter, assume that you are interested in determining how much land area is being scheduled for treatment (harvest) each

time period. Variables such as $S1P1$, as you recall, indicate the stand and period of harvest, and will be assigned a real (continuous) number that represents the land area (acres) assigned to the activity. If $S1P1 = 12.56$, then the linear programming solution indicates that 12.56 acres of stand 1 should be harvested in time period 1. Similarly, if $S2P1 = 4.75$, then the linear programming solution indicates that 4.75 acres of stand 2 should be harvested in time period 1. To understand how much total land area was scheduled for treatment in time period 1, you would naturally add together the values assigned to the decision variables that relate to time period 1:

$S1P1 + S2P1 =$ Area scheduled for harvest in time period 1

To understand how much land area was scheduled for treatment in time period 2, you would add together the values assigned to the decision variables that relate to time period 2:

$S1P2 + S2P2 =$ Area scheduled for harvest in time period 2

With other similar equations, you could understand how much area is scheduled for treatment in each time period of the forest plan. To make the relationships useable in a linear programming model, we must first change the RHS of each equation to a variable (rather than a set of words). Therefore, assume that the area scheduled for treatment in time period 1 will be, when summed, contained in the variable $AC1$, that the area scheduled for treatment in time period 2 will be contained in the variable $AC2$, and so on. The accounting rows then become:

$$S1P1 + S2P1 = AC1$$
$$S1P2 + S2P2 = AC2$$

Finally, linear programming models require that the RHS of the equation be a value (a number), not a variable, so we need to adjust the equations using a little algebra to ensure that all variables are found on the left-hand side of each equation and constants are on the RHS:

$$S1P1 + S2P1 - AC1 = 0$$
$$S1P2 + S2P2 - AC2 = 0$$

When the linear programming model has been solved, the total area scheduled for harvest will be easily accessed by simply consulting the solution for the values assigned to variables $AC1$, $AC2$, and so on.

The full land area accounting rows for the 20-unit problem we began developing in Section III are now presented. First is the accounting row related to the total area scheduled for harvest in time period 1:

```
S1P1 + S2P1 + S3P1 + S4P1 + S5P1 + S6P1 + S7P1 + S8P1
+ S9P1 + S10P1 + S11P1 + S12P1 + S13P1 + S14P1 +
S15P1+ S16P1+ S17P1+ S18P1+ S19P1+ S20P1 - AC1 = 0
```

As we noted earlier, we are using a new variable ($AC1$) to represent the total area scheduled for harvest in time period 1. The accounting row related to the total land area scheduled for harvest in time period 2 is similar to the one created for time period 1:

```
S1P2 + S2P2 + S3P2 + S4P2 + S5P2 + S6P2 + S7P2 + S8P2
+ S9P2 + S10P2 + S11P2 + S12P2 + S13P2 + S14P2 +
S15P2+ S16P2+ S17P2+ S18P2+ S19P2+ S20P2 - AC2 = 0
```

And finally, the accounting row related to the total land area scheduled for harvest in time period 3 is:

```
S1P3 + S2P3 + S3P3 + S4P3 + S5P3 + S6P3 + S7P3 + S8P3
+ S9P3 + S10P3 + S11P3 + S12P3 + S13P3 + S14P3 +
S15P3+ S16P3+ S17P3+ S18P3+ S19P3+ S20P3 - AC3 = 0
```

We could use the knowledge gained with this information to constrain a plan such that minimum or maximum land areas are harvested during each time period, or to ensure that an equal land area is harvested during each time period. At a minimum, the variables $AC1$, $AC2$, and $AC3$ can be used simply to report the area scheduled for harvest in each time period.

Accounting rows can also be developed for other uses, such as for tracking the amount of older forest harvested in each time period. Assume that the older forest includes those stands that are at least 30 years old. If you wanted to know how much area of older forest was harvested in each time period, then you could design an accounting row to add up the areas assigned to the decision variables that are used to represent the older forests. For example, the accounting rows related to the harvest of older forest during each of the three time periods could be:

Time period 1
```
S3P1 + S4P1 + S10P1 + S13P1 + S18P1 - OF1 = 0
```

Time period 2
```
S3P2 + S4P2 + S5P2 + S9P2 + S10P2 + S11P2 + S13P2 +
S18P2 - OF2 = 0
```

Time period 3
```
S2P3 + S3P3 + S4P3 + S5P3 + S7P3 + S9P3 + S10P3 +
S11P3+ S12P3+ S13P3+ S17P3+ S18P3+ S19P3 - OF3 = 0
```

These equations are more extensive for time periods 2 and 3 since some younger stands become "older forest" later in the management of the property. You could use the knowledge gained with this information to constrain a plan such that a limited area of older forest is scheduled for harvest during each time period. Otherwise, the plan may, based on the contribution of each stand to the NPV of the forest, suggest harvesting all the older forest first (not an uncommon practice). Alternatively, you can use the information developed here to ensure that an equal amount of older forest is present in each of the time periods of the management plan. At a minimum, the variables $OF1$, $OF2$, and $OF3$ can be used simply to report the area of older forest scheduled for harvest in each time period.

B. Wood Flow-Related Accounting Rows

To further build upon the model introduced in this chapter, assume that you are now interested in understanding how much timber volume is being scheduled for harvest during each time period of a forest plan. Wood flow accounting rows increase the usability of the linear programming solution, as the aggregation of volume scheduled for harvest can be used to constrain a problem. In addition, the summation of potential wood flow is usually an important outcome for decision-makers to consider. If wood flows are not aggregated within the linear problem, a report writing program may be necessary to summarize these results. Again, the decision variables (such as $S1P1$) will be assigned a real (continuous) number that represents the land area (acres) scheduled for harvest. To understand how much timber volume is scheduled for harvest you would need to know the volume per unit area (per acre) associated with each stand in each time period of the management plan. With this information, an equation can be constructed, such as the one shown here, that accumulates the total volume scheduled for harvest:

$$V1P1 \times S1P1 + V2P1 \times S2P1$$
$$= \text{Volume scheduled for harvest in time period 1}$$

Here, $V1P1$ is a technical coefficient that represents the timber harvest volume per unit area for stand 1 during time period 1. Similarly, $V2P1$ represents the timber harvest volume per unit area for stand 2 during time period 1. These coefficients must be developed prior to attempting to solve the linear programming model.

As with the area-related accounting rows, to make the relationships useable in a linear programming model, we must change the RHS to a variable (rather than a set of words). Therefore assume here that the total volume scheduled for harvest in time period 1 will be contained in the variable $HV1$. The accounting row then becomes:

$$V1P1 \times S1P1 + V2P1 \times S2P1 = HV1$$

As we noted earlier, linear programming models require that the RHS be a value (not a variable), so we need to adjust the equation to show all variables on the left-hand side of the equation:

$$V1P1 \times S1P1 + V2P1 \times S2P1 - HV1 = 0$$

The wood flow accounting rows for the 20-unit problem we began developing in Section III are now presented. First is the accounting row related to the total timber volume scheduled for harvest during time period 1:

```
40 S1P1 + 64 S2P1 + 104 S3P1 + 102 S4P1 + 87 S5P1 +
42 S6P1 + 65 S7P1 + 20 S8P1 + 85 S9P1 + 101 S10P1 +
89 S11P1 + 62 S12P1 + 106 S13P1 + 43 S14P1 + 22 S15P1 +
44 S16P1 + 63 S17P1 + 103 S18P1 + 65 S19P1 + 18 S20P1 −
HV1 = 0
```

As we noted earlier, we are using a new variable ($HV1$) to represent the total harvest volume scheduled in time period 1. The accounting row related to the total volume scheduled for harvest during time period 2 is similar to the one created for time period 1, yet uses different coefficients and variables to represent the volume per unit area related to the potential decisions for time period 2:

```
62 S1P2 + 87 S2P2 + 117 S3P2 + 115 S4P2 + 104 S5P2 +
64 S6P2 + 89 S7P2 + 42 S8P2 + 101 S9P2 + 114 S10P2 +
106 S11P2 + 85 S12P2 + 120 S13P2 + 65 S14P2 + 44 S15P2
+ 66 S16P2 + 87 S17P2 + 117 S18P2 + 89 S19P2 +
40 S20P2 − HV2 = 0
```

Finally, the accounting row related to the total volume scheduled for harvest in time period 3 is:

```
84 S1P3 + 104 S2P3 + 127 S3P3 + 126 S4P3 + 117 S5P3 +
87 S6P3 + 106 S7P3 + 64 S8P3 + 114 S9P3 + 125 S10P3 +
119 S11P3 + 101 S12P3 + 131 S13P3 + 88 S14P3 +
67 S15P3 + 88 S16P3 + 103 S17P3 + 126 S18P3 +
107 S19P3 + 61 S20P3 − HV3 = 0
```

The knowledge gained with this information can be used to constrain a forest plan such that either (1) a minimum amount of volume is harvested in each time period, (2) a maximum amount of volume is harvested in each

time period, or (3) an equal amount of volume is harvested in each time period. In addition, constraints can be designed to ensure that the harvest levels do not decline from one time period to the next. At a minimum, the variables $HV1$, $HV2$, and $HV3$ can be used simply to report the volume scheduled for harvest in each time period.

C. Habitat-Related Accounting Rows

As we mentioned in Chapter 3, Geographic Information and Land Classification in Support of Forest Planning, there are a number of quantitative methods for evaluating the quality of habitat for fish and wildlife species. As an example of how we might account for the habitat of a wildlife species, assume that the species in question prefers older forest habitat. In conjunction with more detailed data regarding the current and future forest structure of each stand, you arrived at the *potential* habitat suitability index values for each stand, *given that the stand is not harvested* (Table 7.2).

To be able to determine the quality of habitat related to forest plans using this example, we need to calculate the nonharvested portion of each stand in each time period, as harvested areas of each stand are assumed (in this example) to have a habitat suitability index value of 0. This is not to imply that all harvested areas (clearcuts) lead to habitat scores of 0; in fact some wildlife species require and thrive in early successional forests or open spaces. However, our example suggests that the habitat scores shown in Table 7.2 relate to uncut areas. To determine how much of each stand is not harvested in each time period, 60 new variables are introduced. These variables will take the form $N1P1$, where N indicates that the stand number follows (in this case, stand 1), and P indicates that the time period of harvest follows (in this case, time period 1). However, the values assigned to these variables (acres) will indicate how much of each stand's land area *has not been harvested* each time period.

For example, if $S1P1$ is assigned the amount of land harvested in stand 1 during time period 1, then it makes sense that $N1P1$ should be assigned a value that represents the balance of the area of stand 1, up to the size of the stand:

$$S1P1 + N1P1 = 12.3$$

Therefore, if 5.0 acres of stand 1 are scheduled for harvest during time period 1 ($S1P1 = 5.0$), the amount of stand 1 that is not harvested at the end of time period 1 would have to be 7.3 acres ($N1P1 = 7.3$). Twenty of these equations are needed to determine the amount of land of each stand not scheduled for harvest during time period 1.

To determine how much land area of each stand remains uncut in time period 2, we need to understand how much of each stand is scheduled for harvest during

TABLE 7.2 Habitat Suitability Index (HSI) Values for 20 Loblolly Pine Stands in South Georgia

Stand Stand	Size (acres)	Age[a]	Period 1 HSI (per ac)	Period 2 HSI (per ac)	Period 3 HSI (per ac)
1	12.3	15	0.304	0.450	0.629
2	35.6	20	0.482	0.663	0.768
3	34.6	30	0.805	0.903	0.904
4	85.4	30	0.765	0.882	0.972
5	65.4	25	0.683	0.797	0.892
6	69.1	15	0.377	0.526	0.692
7	78.3	20	0.514	0.690	0.789
8	71.0	10	0.181	0.352	0.501
9	19.6	25	0.669	0.738	0.812
10	34.8	30	0.788	0.869	0.903
11	81.6	25	0.629	0.810	0.910
12	90.2	20	0.447	0.609	0.779
13	45.6	30	0.814	0.905	0.995
14	67.2	15	0.397	0.505	0.654
15	38.6	10	0.218	0.331	0.498
16	49.1	15	0.361	0.474	0.649
17	58.3	20	0.478	0.707	0.766
18	26.8	30	0.799	0.872	0.962
19	53.0	20	0.500	0.688	0.821
20	46.0	10	0.141	0.363	0.505

[a]*Stand age in the middle of planning period 1.*

both time periods 1 and 2. The resulting equation uses the decision variables for time periods 1 and 2:

$$S1P1 + S1P2 + N1P2 = 12.3$$

Thus if 5.0 acres of stand 1 are scheduled for harvest during time period 1 ($S1P1 = 5.0$), and 3.3 acres are scheduled for harvest during time period 2 ($S1P2 = 3.3$), then the balance (4.0 acres) is assigned to variable $N1P2$ to represent the amount of land that remains uncut in stand 1 at the end of time period 2.

Finally, to determine how much land area of each stand remains uncut at the end of time period 3, we need to understand how much of each stand is scheduled for harvest during time periods 1, 2, and 3. The equation utilizes decision variables for time periods 1−3:

$$S1P1 + S1P2 + S1P3 + N1P3 = 12.3$$

With the 60 equations (20 stands, 3 time periods) that it takes to determine how many acres of each stand are

left uncut in each time period, we can now build an accounting row to add up the habitat units associated with each time period of the management plan. Using the HSI values provided in Table 7.2 as coefficients, the accounting row for time period 1 becomes:

```
0.304 N1P1 + 0.482 N2P1 + 0.805 N3P1 + 0.765 N4P1 +
0.683 N5P1 + 0.377 N6P1 + 0.514 N7P1 + 0.181 N8P1 +
0.669 N9P1 + 0.788 N10P1 + 0.629 N11P1 + 0.447 N12P1 +
0.814 N13P1 + 0.397 N14P1 + 0.218 N15P1 + 0.361
N16P1 + 0.478 N17P1 + 0.799 N18P1 + 0.500 N19P1 +
0.141 N20P1 − HU1 = 0
```

The new variable introduced ($HU1$) is considered the sum of "habitat units" because the HSI coefficient is a per-unit area value, and the values contained in the decision variables describe the nonharvested areas (acres). The result is a number not between 0 and 1, as most habitat suitability indices are represented, but between 0 and the size of the forest, since theoretically, if the HSI values were all 1.0, and no stands were scheduled for harvest, the sum would match the total area of the forest. Using the HSI values from Table 7.2, the accounting row related to habitat units present after scheduled harvests during time period 2 is:

```
0.450 N1P2 + 0.663 N2P2 + 0.903 N3P2 + 0.882 N4P2 +
0.797 N5P2 + 0.526 N6P2 + 0.690 N7P2 + 0.352 N8P2 +
0.738 N9P2 + 0.869 N10P2 + 0.810 N11P2 + 0.609 N12P2 +
0.905 N13P2 + 0.505 N14P2 + 0.331 N15P2 + 0.474
N16P2 + 0.707 N17P2 + 0.872 N18P2 + 0.688 N19P2 +
0.363 N20P2 − HU2 = 0
```

Finally, using the HSI values from Table 7.2, the accounting row related to habitat units present after scheduled harvests during time period 3 is:

```
0.629 N1P3 + 0.768 N2P3 + 0.904 N3P3 + 0.972 N4P3 +
0.892 N5P3 + 0.692 N6P3 + 0.789 N7P3 + 0.501 N8P3 +
0.812 N9P3 + 0.903 N10P3 + 0.910 N11P3 + 0.779 N12P3 +
0.995 N13P3 + 0.654 N14P3 + 0.498 N15P3 + 0.649
N16P3 + 0.766 N17P3 + 0.962 N18P3 + 0.821 N19P3 +
0.505 N20P3 − HU3 = 0
```

We could use the knowledge obtained with this information to constrain a plan such that the total number of habitat units does not fall below a minimum level determined by the wildlife biologist in your organization. The values $HU1$, $HU2$, and $HU3$ could also be converted to an average HSI value for the entire property (between 0 and 1) by multiplying them by the inverse of the size of the property. The size of our example forest is 1,062.5 acres, thus the inverse is 0.00094118. To convert the habitat units to HSI values, new variables could be introduced ($HSI1$, $HSI2$, and $HSI3$), and the relationships would become:

```
0.00094118 HU1 = HSI1
0.00094118 HU2 = HSI2
0.00094118 HU3 = HSI3
```

Since all variables must be located on the left-hand side of the equation, the linear programming equations become:

```
0.00094118 HU1 − HSI1 = 0
0.00094118 HU2 − HSI2 = 0
0.00094118 HU3 − HSI3 = 0
```

Here, the variables *HSI*1, *HSI*2, and *HSI*3 can be used simply to report the average habitat quality during each time period, or they can be used to constrain the forest plans generated by mandating minimum or maximum levels. However, a linear programming solver may suggest that when using these types of equations, the problem is poorly scaled. That is, the coefficients used are small relative to the values assigned to the variables. Fortunately, this usually does not prevent the problem from being solved, but it should cause the analyst to ponder the choice of variables to determine if a better set should be used.

V. CONSTRAINTS FOR LINEAR PROGRAMMING PROBLEMS

Two types of constraints generally are used in linear programming models: resource and policy (managerial) constraints. Resource constraints ensure that no more of some resource at our disposal can be used in a plan of action. Resources include budgets, personnel, machines and equipment, and land. They are often fixed in the short-term. Policy constraints guide the development of plans by forcing the plan to adhere to either organizational goals or regulatory restrictions. Organizational policies, such as maintaining habitat suitability above a certain level, producing a certain revenue, or maintaining costs within a budget, are very common. Regulatory policies could include clearcut size restrictions (in states or provinces with forest practice laws), restrictions on sediment production from harvesting activities, as well as many others imposed by forces outside the organization.

A. Resource Constraints

We will first deal with resource constraints in the context of the management problem we have been building since Section III. In the problem that has been developed so far, the only type of resource we have introduced has been the land itself. We described the landbase (20 stands, 1,062.5 acres) in Table 7.1, and we suggested that the main decisions are to assign harvests to acres within each stand. It should seem obvious that since the time horizon of the plan being developed is relatively short (15 years), each stand should be harvested only once. Taking this a little further, as a planner, we should not schedule more acres for harvest than we have available. Thus the resource

constraints that we need to develop for this problem should indicate that the sum of the area assigned for harvest over the time horizon should not exceed the total size of each stand. These types of constraints need to be developed for each stand. The resource constraint for stand 1, for example, would be:

$$S1P1 + S1P2 + S1P3 \leq 12.3$$

In other words, the area scheduled for harvest during time period 1 (*S*1*P*1), plus the area scheduled for harvest during time period 2 (*S*1*P*2), plus the area scheduled for harvest during period 3 (*S*1*P*3) should not exceed the total size of the stand (12.3 acres, or the resource). The amount of area scheduled for harvest in a stand can certainly be less than the total size of the stand; in fact, no land area could be scheduled for harvest in a stand. Listed here is the full set of resource constraints for this management problem:

```
S1P1 + S1P2 + S1P3 <= 12.3
S2P1 + S2P2 + S2P3 <= 35.6
S3P1 + S3P2 + S3P3 <= 34.6
S4P1 + S4P2 + S4P3 <= 85.4
S5P1 + S5P2 + S5P3 <= 65.4
S6P1 + S6P2 + S6P3 <= 69.1
S7P1 + S7P2 + S7P3 <= 78.3
S8P1 + S8P2 + S8P3 <= 71.0
S9P1 + S9P2 + S9P3 <= 19.6
S10P1 + S10P2 + S10P3 <= 34.8
S11P1 + S11P2 + S11P3 <= 81.6
S12P1 + S12P2 + S12P3 <= 90.2
S13P1 + S13P2 + S13P3 <= 45.6
S14P1 + S14P2 + S14P3 <= 67.2
S15P1 + S15P2 + S15P3 <= 38.6
S16P1 + S16P2 + S16P3 <= 49.1
S17P1 + S17P2 + S17P3 <= 58.3
S18P1 + S18P2 + S18P3 <= 26.8
S19P1 + S19P2 + S19P3 <= 53.0
S20P1 + S20P2 + S20P3 <= 46.0
```

B. Policy Constraints

Any of the variables we have introduced so far, the decision variables or the variables created in conjunction with the accounting rows (from Section IV), can be used to constrain a solution. We will discuss briefly three of the common types of policy constraints, and provide some examples that relate to the management problem that has been developed in this chapter.

1. Constraints on Harvested Areas

Harvest area constraints are very common in natural resource management problems. If left unconstrained, the harvested area that is scheduled during each time period could fluctuate widely. These types of constraints are generally fairly simple to formulate. For instance, if we were interested in specifying a minimum harvest area of 300 acres during time period 1, then a policy constraint can be designed as this:

$$AC1 \geq 300$$

If we were interested in a maximum harvest area of 500 acres during time period 1, then another policy constraint would become:

$$AC1 \leq 500$$

Policy constraints do not necessarily have to include inequalities (i.e., less than or equal to, greater than or equal to). We could, as a policy, indicate that the harvest area during time period 1 needs to be some very specific value, such as 350 acres:

$$AC1 = 350$$

If we were interested in having scheduled harvest areas be the same from one time period to the next, then the relationships could be described with the following:

$$AC1 = AC2$$
$$AC2 = AC3$$

Of course, here we would need to shift all the variables to the left-hand side of each equation for these to be of use in a linear programming model:

$$AC1 - AC2 = 0$$
$$AC2 - AC3 = 0$$

We also developed accounting rows to accumulate the area of older forest harvested in each time period. We could use these variables (*OF*1, *OF*2, and *OF*3) to regulate how much of the older forest is scheduled for harvest during each time period. Constraints such as the following could be developed to accomplish this task:

$$OF1 \leq 100$$
$$OF2 \leq 100$$
$$OF3 \leq 100$$

2. Constraints on Harvested Volume

The volume scheduled for harvest also commonly is constrained in many natural resource management plans. Examples include minimum and maximum levels of harvest volume for various forest products, nonmarket goods, or rangeland resources. As these pertain to the management problem we have been building in this chapter, we could define minimum and maximum harvest levels:

$$HV1 \geq 30{,}000$$
$$HV1 \leq 40{,}000$$

Alternatively, we could ensure that an even-flow of timber harvest volume is scheduled. What this implies is that the scheduled harvest volume is the same from one time period to the next as their differences are equal to zero.

$$HV1 - HV2 = 0$$
$$HV2 - HV3 = 0$$

Example

A short-term harvest schedule was illustrated by Macmillan and Fairweather (1988) for an industrial forest ownership in northwestern Pennsylvania. The forest types included black cherry (*Prunus serotina*), red maple (*Acer rubrum*), sugar maple (*Acer saccharum*), and northern red oak (*Quercus rubra*). The time horizon was very short (5 years) and the objective was to maximize the NPV of the harvest over this time frame. Stumpage values, an estimate of the percent growth of stands, and depletion values were used to determine the potential NPV of harvesting each stand during each of the 5 years. The decision variables were assumed to represent the amount of land area (acres) in each stand that could be scheduled for harvest during each year, and as a result, they were assumed to be assigned continuous numeric values. The annual demand of a local mill was used as a policy constraint on the minimum and maximum volume to be harvested each year. A set of resource constraints limited the amount of area scheduled for harvest in each stand to the size of each stand. The problem formulation consisted of the following:

$$\text{Maximize} \left(\frac{\text{Net value per acre}_{X,Y}}{(1+i)^{Y-1}} \right)$$

where:

X = a stand
Y = a year
i = interest rate

subject to:

1. Resource constraints:

$$\sum_{Y=1}^{5} \text{Scheduled area}_{X,Y} \leq \text{Area}_X \quad \forall X$$

2. Wood flow constraints:

$$\sum_{X=1}^{N} \text{Scheduled area}_{X,Y} \text{ Harvest volume}_{X,Y}$$
$$\geq \text{Minimum harvest target} \quad \forall Y$$

$$\sum_{X=1}^{N} \text{Scheduled area}_{X,Y} \text{ Harvest volume}_{X,Y}$$
$$\leq \text{Maximum harvest target} \quad \forall Y$$

The notation $\forall Y$ simply means *for every Y* (in this case for every year), and $\forall X$ simply means *for each stand*. Thus the minimum and maximum harvest volume constraints were in effect each year, and each stand is associated with a resource constraint. Several sensitivity analyses were performed by adjusting the costs and growth rates assumed. These alternative plans helped to understand how expectations might change if economic or ecological conditions in the future change. When stand-level growth rates were assumed to increase, for example, some stands were scheduled later in the plan's time horizon rather than earlier under the initial assumptions, because the value growth rate of certain stands had become greater than the alternative rate of return. Further, when prices changed, the product mix scheduled for harvest over time changed. An increase in one product price created conditions that allowed the scheduled harvests of stands earlier than when they normally would have been scheduled.

3. Constraints on Habitat Availability

Wildlife habitat that is developed for any species of interest could be used to control the development of a plan of action. For example, using the example we have been building upon in this chapter, our wildlife biologist could indicate that one policy might be to maintain HSI levels above 0.600 for the entire 1,062.5 acre forest. Since we developed variables earlier to represent the overall HSI during each time period ($HSI1$, $HSI2$, and $HSI3$), these can be incorporated into constraints to guide the development of a plan.

$$HSI1 \geq 0.600$$
$$HSI2 \geq 0.600$$
$$HSI3 \geq 0.600$$

VI. DETACHED COEFFICIENT MATRIX

Another way of viewing a linear programming problem is to represent it as a matrix in what is called a *detached coefficient form*. A detached coefficient matrix is a tableau form of a problem, which is an alternative method to represent how you would arrange the variables and coefficients when solving the problem in a computer spreadsheet. This format appeals to different people. Here, we use a matrix where one column is reserved for each variable in the model, and one row is reserved for each equation (whether an objective function, constraint, or accounting row). Contained in the matrix are the technical coefficients associated with the use of each decision variable in each equation. To illustrate the use of a detached coefficient matrix, we will follow the development of a problem first proposed by Johnson and Stuart (1987) for a portion of the Brush Mountain area of the Jefferson National Forest in Virginia. The planning problem involves determining the assignment of land to various land allocations, which in turn reflects the

need to produce timber, wildlife habitat, cattle forage, and wilderness. The example allocation scheduling problem utilizes nine variables to represent six allocation decisions and three periodic harvest levels.

Y_{AE1} = Proportion of zone A assigned to timing choice 1 of the timber/birds allocation scheduling choice
Y_{AE2} = Proportion of zone A assigned to timing choice 2 of the timber/birds allocation scheduling choice
Y_{AF1} = Proportion of zone A assigned to timing choice 1 of the wilderness/birds allocation scheduling choice
Y_{BG1} = Proportion of zone B assigned to timing choice 1 of the timber/forage allocation scheduling choice
Y_{BG2} = Proportion of zone B assigned to timing choice 2 of the timber/forage allocation scheduling choice
Y_{BP1} = Proportion of zone B assigned to timing choice 1 of the wilderness/forage allocation scheduling choice
Q_1 = Timber harvest volume scheduled for time period 1
Q_2 = Timber harvest volume scheduled for time period 2
Q_3 = Timber harvest volume scheduled for time period 3

The Brush Mountain management problem was designed to maximize the NPV of a forest plan over a three-period time horizon.

$$\text{Maximize } 147\, Y_{AE1} + 143\, Y_{AE2} + 68\, Y_{AF1}$$
$$+ 85\, Y_{BG1} + 45\, Y_{BG2} + 40\, Y_{BP1}$$

The coefficients in the objective function represent the total NPV for each allocation of land in each management zone (in thousands of dollars), and included proposed timber harvests, costs associated with roads, trails, and other recreation-related revenues or expenses. The decision variables represent the percentage of the management zone assigned to each allocation decision. In the case of the two wilderness-related decision variables, the NPV represents a positive economic return to recreational experiences. Two land accounting rows were created to ensure that the proportion of the area assigned to each choice within a zone would equal, yet not exceed, 100%.

$$Y_{AE1} + Y_{AE2} + Y_{AF1} = 1$$
$$Y_{BG1} + Y_{BG2} + Y_{BP1} = 1$$

Three timber volume accounting rows were developed to accumulate the harvest volume from each of the four nonwilderness options.

$$621\, Y_{AE1} + 600\, Y_{AE2} + 300\, Y_{BG1} + 150\, Y_{BG2} - Q_1 = 0$$
$$594\, Y_{AE1} + 672\, Y_{AE2} + 330\, Y_{BG1} + 330\, Y_{BG2} - Q_2 = 0$$
$$708\, Y_{AE1} + 618\, Y_{AE2} + 520\, Y_{BG1} + 350\, Y_{BG2} - Q_3 = 0$$

The coefficients here represent values in 1,000 ft^3 that are available from each management zone during each time period. Once harvest volumes are accumulated into the variables Q_1, Q_2, and Q_3, a nondeclining harvest volume can be controlled with two constraints:

$$Q_2 - Q_1 \geq 0$$
$$Q_3 - Q_2 \geq 0$$

A sediment constraint was added to the problem to limit the sediment units (tons) produced as a result of management activities over background levels naturally occurring in the area.

$$3038\, Y_{AE1} + 2703\, Y_{AE2} + 25\, Y_{AF1} + 2105\, Y_{BG1}$$
$$+ 1305\, Y_{BG2} + 17\, Y_{BP1} \leq 50,000$$

Finally, at least one of the two management zones had to be assigned to a wilderness designation. As a result, the two variables related to a wilderness designation need to be assigned binary integer values to represent a yes/no allocation decision. The constraint that ensures the assignment of one of the two to a wilderness designation is:

$$Y_{AF1} + Y_{BP1} = 1$$

With this last constraint, it should be obvious that once a wilderness designation has been assigned to a management zone, the other options for that zone are moot, since the land accounting rows presented earlier will preclude the assignment of any additional portion of the zone to the timber/birds or timber/forage land allocations.

For this problem, the detached coefficient matrix shown in Table 7.3 presents the decision variables as column headers, and a description of the type of equation in the far left-hand column. The values within the matrix represent the coefficients related to each variable, as they are necessary within each linear equation. What you should notice in the matrix is that (1) there are no values for row-column intersections involving decision variables that are not present in the associated equation; in fact the majority of intersections are empty; and (2) the value "1" implicitly represents those row-column intersections where variables in associated equations have no coefficient value to modify the equation.

VII. MODEL I, II, AND III LINEAR PROGRAMMING PROBLEMS

The method for defining decision variables in a linear programming problem can fall into three classes: Model I, Model II, or Model III. A Model I linear programming problem utilizes decision variables that track the history of a stand or strata over the entire time horizon of the forest plan. For uneven-aged stands, a single decision variable may represent the structural condition of a stand over time, given the management prescriptions periodically scheduled for that stand. A decision variable such as $S1R1$ may represent stand 1, management regime 1, and include outcomes and associated conditions throughout the plan's time horizon if areas of the stand are assigned this management regime. In even-aged stands, the decision variable will often include the timing of the final harvests and associated outcomes and conditions prior to and after the harvest. In either case, a number of decision variables may be required to represent the various series of actions (regimes) that can be applied to a stand or strata over the entire time horizon of a plan, even if two different series of actions involve the same specific activity scheduled during the same specific time period.

TABLE 7.3 A Detached Coefficient Matrix for the Brush Mountain Planning Problem

	Y_{AE1}	Y_{AE2}	Y_{AF1}	Y_{BG1}	Y_{BG2}	Y_{BP1}	Q_1	Q_2	Q_3	RHS
Maximize	147	143	68	85	45	40				
Land resource constraint, zone A	1	1	1							= 1
Land resource constraint, zone B				1	1	1				= 1
Volume accounting row, period 1	621	600		300	150		−1			= 0
Volume accounting row, period 2	594	672		330	330			−1		= 0
Volume accounting row, period 3	708	618		520	350				−1	= 0
Nondeclining yield constraint, period 1							−1	1		≥ 0
Nondeclining yield constraint, period 2								−1	1	≥ 0
Sediment constraint	3,038	2,703	25	2,105	1,305	17				≤ 5,000
Wilderness constraint			1			1				= 1

Source: Johnson, K.N., Stuart, T.W., 1987. FORPLAN version 2: mathematical programmer's guide. US Department of Agriculture, Forest Service, Land Management Planning Systems Center, Washington, DC.

A Model II linear programming problem tracks the history of a stand only until a final harvest is scheduled. After the final harvest, the areas regenerated in a Model II problem are aggregated into a separate regenerated decision variable associated with a specific time period. Stated another way, at the point of regeneration, the history of the initial stand or strata is lost, and all regenerated areas are aggregated if they were scheduled for a final harvest during the same time period. As a result, Model II is best suited for even-aged management regimes. As compared to a Model I problem, fewer decision variables generally are required in a Model II problem. However, decision variables are required to track the initial stand or strata (1) until a final harvest occurs, (2) until the subsequent final harvest occurs, if the second stand is regenerated, and (3) through the end of the time horizon whether the initial stand or strata was scheduled for final harvest or not. To summarize the differences, Model I preserves the areas that form any stand or strata throughout the time horizon, and Model II aggregates areas after final harvest into regenerated classes. Both models require resource constraints to ensure that no more areas than are available can be assigned to the different management options. However, Model I requires only as many resource constraints as there are stands or strata, and Model II may require these constraints along with others to track the regenerated age class areas.

Johnson and Scheurman (1977) coined the terms Model I and Model II, and they remain widely used in natural resource management planning problems. Model III is less commonly used, however. This model aggregates stands or strata of the same age class at the beginning of a management planning analysis, and it is these age classes that are tracked through the time horizon of the management plan. Similar to Model II, once a final harvest occurs, the regenerated strata contains the amount of the area harvested. When using Remsoft's RSPS scheduling software (Remsoft, 2016), a Model III problem can be created when building the *Areas Module* from a shapefile, with age classes more than 1 year in length. Although Model II and Model III may seem to more efficiently represent a planning problem, these management problems can become cumbersome to develop when differences in site quality or forest type need to be recognized, for example. As a result, some very large linear programming models may arise, given the potential management pathways for land areas, and given the constraints necessary to control the movement of land (or other resource) through the network of options (Gunn, 2007).

VIII. INTERPRETATION OF RESULTS GENERATED FROM LINEAR PROGRAMMING PROBLEMS

A *problem formulation*, as it is known in operations research, is the mathematical expression of the management problem to be solved. Solving a problem requires using techniques such as the Simplex method (described in Appendix B) that are contained within computer programs ("solvers"). Although many different solvers are available to apply to linear programming models (we mention a few in Section XII) we focus here on input to, and output from LINGO (LINDO Systems, Inc., 2016), and the methods we would use to formulate the management problem we have been developing in this chapter.

As a first management scenario, assume using the example we have been developing in this chapter that we simply want to maximize the NPV of a management plan, and that the only constraints are the resource-related constraints. However, we will include several accounting rows as well, to quickly assess the number of acres scheduled for harvest in each time period, as well as the scheduled harvest volumes. LINGO requires that the objective function be stated first, followed by the words "subject to" (on a separate line), and then the constraints, each given a sequential number starting with the number 2. At the end of the list of constraints, the word "end" is placed. This work can be performed in any text editor program. In the following problem formulation, constraints 2 through 21 represent the resource constraints, and constraints 22 through 30 represent the accounting rows.

```
Maximize
        885.17   S1P1   +   1075.01   S1P2   +   1141.18   S1P3
  +   1416.27   S2P1   +   1508.48   S2P2   +   1412.89   S2P3
  +   2301.44   S3P1   +   2028.65   S3P2   +   1725.35   S3P3
  +   2257.18   S4P1   +   1993.97   S4P2   +   1711.77   S4P3
  +   1925.25   S5P1   +   1803.24   S5P2   +   1589.50   S5P3
  +    929.43   S6P1   +   1109.69   S6P2   +   1181.93   S6P3
  +   1438.40   S7P1   +   1543.16   S7P2   +   1440.06   S7P3
  +    442.59   S8P1   +    728.23   S8P2   +    869.47   S8P3
  +   1880.99   S9P1   +   1751.22   S9P2   +   1548.74   S9P3
  +   2235.05   S10P1  +   1976.63   S10P2  +   1698.18   S10P3
  +   1969.50   S11P1  +   1837.92   S11P2  +   1616.67   S11P3
  +   1372.01   S12P1  +   1473.80   S12P2  +   1372.13   S12P3
  +   2345.70   S13P1  +   2080.66   S13P2  +   1779.69   S13P3
  +    951.56   S14P1  +   1127.03   S14P2  +   1195.52   S14P3
```

```
+    486.84   S15P1   +    762.91   S15P2   +    910.22   S15P3
+    973.69   S16P1   +   1144.36   S16P2   +   1195.52   S16P3
+   1394.14   S17P1   +   1508.48   S17P2   +   1399.30   S17P3
+   2279.31   S18P1   +   2028.65   S18P2   +   1711.77   S18P3
+   1438.40   S19P1   +   1543.16   S19P2   +   1453.64   S19P3
+    398.33   S20P1   +    693.55   S20P2   +    828.71   S20P3
subject to
2) S1P1 + S1P2 + S1P3 <= 12.3
3) S2P1 + S2P2 + S2P3 <= 35.6
4) S3P1 + S3P2 + S3P3 <= 34.6
5) S4P1 + S4P2 + S4P3 <= 85.4
6) S5P1 + S5P2 + S5P3 <= 65.4
7) S6P1 + S6P2 + S6P3 <= 69.1
8) S7P1 + S7P2 + S7P3 <= 78.3
9) S8P1 + S8P2 + S8P3 <= 71.0
10) S9P1 + S9P2 + S9P3 <= 19.6
11) S10P1 + S10P2 + S10P3 <= 34.8
12) S11P1 + S11P2 + S11P3 <= 81.6
13) S12P1 + S12P2 + S12P3 <= 90.2
14) S13P1 + S13P2 + S13P3 <= 45.6
15) S14P1 + S14P2 + S14P3 <= 67.2
16) S15P1 + S15P2 + S15P3 <= 38.6
17) S16P1 + S16P2 + S16P3 <= 49.1
18) S17P1 + S17P2 + S17P3 <= 58.3
19) S18P1 + S18P2 + S18P3 <= 26.8
20) S19P1 + S19P2 + S19P3 <= 53.0
21) S20P1 + S20P2 + S20P3 <= 46.0
22) S1P1 + S2P1 + S3P1 + S4P1 + S5P1 + S6P1 + S7P1 + S8P1 + S9P1 + S10P1 + S11P1 + S12P1 + S13P1 + S14P1 + S15P1 +
    S16P1 + S17P1 + S18P1 + S19P1 + S20P1 - AC1 = 0
23) S1P2 + S2P2 + S3P2 + S4P2 + S5P2 + S6P2 + S7P2 + S8P2 + S9P2 + S10P2 + S11P2 + S12P2 + S13P2 + S14P2 + S15P2 +
    S16P2 + S17P2 + S18P2 + S19P2 + S20P2 - AC2 = 0
24) S1P3 + S2P3 + S3P3 + S4P3 + S5P3 + S6P3 + S7P3 + S8P3 + S9P3 + S10P3 + S11P3 + S12P3 + S13P3 + S14P3 + S15P3 +
    S16P3 + S17P3 + S18P3 + S19P3 + S20P3 - AC3 = 0
25) S3P1 + S4P1 + S10P1 + S13P1 + S18P1 - OF1 = 0
26) S3P2 + S4P2 + S5P2 + S9P2 + S10P2 + S11P2 + S13P2 + S18P2 - OF2 = 0
27) S2P3 + S3P3 + S4P3 + S5P3 + S7P3 + S9P3 + S10P3 + S11P3 + S12P3 + S13P3 + S17P3 + S18P3 + S19P3 - OF3 = 0
28) 40 S1P1+ 64 S2P1+ 104 S3P1+ 102 S4P1 + 87 S5P1 + 42 S6P1+ 65 S7P1 + 20 S8P1 + 85 S9P1 + 101 S10P1+ 89 S11P1 +
    62 S12P1 + 106 S13P1 + 43 S14P1 + 22 S15P1 + 44 S16P1 + 63 S17P1 + 103 S18P1 + 65 S19P1 + 18 S20P1 - HV1 = 0
29) 62 S1P2 + 87 S2P2 + 117 S3P2 + 115 S4P2 + 104 S5P2 + 64 S6P2 + 89 S7P2 + 42 S8P2 + 101 S9P2 + 114 S10P2 +
    106 S11P2 + 85 S12P2 + 120 S13P2 + 65 S14P2 + 44 S15P2 + 66 S16P2 + 87 S17P2 + 117 S18P2 + 89 S19P2 +
    40 S20P2 - HV2 = 0
30) 84 S1P3 + 104 S2P3 + 127 S3P3 + 126 S4P3 + 117 S5P3 + 87 S6P3 + 106 S7P3 + 64 S8P3 + 114 S9P3 + 125 S10P3 +
    119 S11P3 + 101 S12P3 + 131 S13P3 + 88 S14P3 + 67 S15P3 + 88 S16P3 + 103 S17P3 + 126 S18P3 + 107 S19P3 +
    61 S20P3 - HV3 = 0
end
```

A. Objective Function Value, Variable Values, and Reduced Costs

In solving this problem with LINGO, a lengthy report is generated, the size of which is a function of the number of decision variables and constraints included in the problem formulation. Just a portion of the report is shown here.

```
     OBJECTIVE FUNCTION VALUE
          1)      1688646.
VARIABLE         VALUE          REDUCED COST
  S1P1          0.000000         256.010010
  S1P2          0.000000          66.169991
  S1P3         12.300000           0.000000
```

S2P1	0.000000	92.209984
S2P2	35.599998	0.000000
S2P3	0.000000	95.589989
....		
....		
AC1	393.799988	0.000000
AC2	315.399994	0.000000
AC3	353.299988	0.000000
OF1	227.199997	0.000000
OF2	0.000000	0.000000
OF3	0.000000	0.000000
HV1	38036.199219	0.000000
HV2	27522.000000	0.000000
HV3	27215.500000	0.000000

We should begin interpreting the solution by first assessing the objective function value. What we find is that the maximum NPV possible, given the timber yields, prices, and discount rate assumed, is $1,688,646. We have to infer the units (dollars) based on our knowledge of how the objective function was developed. Examining the columns labeled VARIABLE and VALUE, we begin to understand the actual schedule of activities. For example, the optimal solution is to harvest 12.3 acres of stand 1 during time period 3, and 35.6 acres of stand 2 during time period 2, and so on.

The *reduced cost* represents the amount (in objective function value terms, which in this case is NPV) that the coefficient for each decision variable in the objective function must increase before that decision becomes competitive enough, given the other choices available, to enter the solution. For example, the reduced cost for variable $S1P1$ is $256.01001. What this implies, in dollars and cents, is that the NPV of harvesting stand 1 during time period 1 must increase $256.02 per acre, to $1,141.19 ($885.17+$256.02), before an acre (or more) of stand 1 is scheduled for harvest during time period 1. The reduced cost should be 0 for all variables that are currently "in the solution," or have a positive VALUE. And, if we look closely, as in the case of stand 1, the reduced cost is simply the difference between the option for stand 1 with the highest NPV (harvest during time period 3) and the option being considered (harvest during time period 1). In addition, the reduced cost could be viewed as a penalty imposed on the objective function value if a noncompetitive variable is made to be part of the solution space. For instance, if we made the solution use one unit (1 acre) of $S1P1$, then the objective function value would decrease by $256.02 to a total of $1,688,389.98.

We also find in this output the values of the variables that were introduced through the accounting rows. The variable $AC1$, for instance, was developed solely to hold the aggregate scheduled harvested area during time period 1 (393.8 acres). The value associated with the variable $OF1$ indicates that all the older forest areas are scheduled for harvest during time period 1, which is not surprising, since the NPV per unit area for the older stands declines as these stands get older. This also suggests that the rate of growth of the older stands is not as high as the interest rate that is assumed (5%). The scheduled wood flows can also be interpreted from this output, as the variables $HV1, HV2,$ and $HV3$ reveal that 38,036, 27,522, and 27,215 tons of wood are scheduled for harvest during time periods 1, 2, and 3, respectively.

B. Slack and Dual Prices

The remainder of the output report generated by LINGO for the forest management problem we developed in this chapter is shown below.

ROW	SLACK OR SURPLUS	DUAL PRICES
2)	0.000000	1141.180054
3)	0.000000	1508.479980
4)	0.000000	2301.439941
5)	0.000000	2257.179932
6)	0.000000	1925.250000
7)	0.000000	1181.930054
8)	0.000000	1543.160034
9)	0.000000	869.469971
10)	0.000000	1880.989990
11)	0.000000	2235.050049
12)	0.000000	1969.500000
13)	0.000000	1473.800049
14)	0.000000	2345.699951
15)	0.000000	1195.520020
16)	0.000000	910.219971
17)	0.000000	1195.520020
18)	0.000000	1508.479980
19)	0.000000	2279.310059
20)	0.000000	1543.160034
21)	0.000000	828.710022
22)	0.000000	0.000000
23)	0.000000	0.000000
24)	0.000000	0.000000
25)	0.000000	0.000000
26)	0.000000	0.000000
27)	0.000000	0.000000
28)	0.000000	0.000000
29)	0.000000	0.000000
30)	0.000000	0.000000

What we find in this section of the output report is some pertinent information related to the constraints. The ROW column refers to each constraint (or accounting row), and the number is consistent with the number of the constraint in the problem formulation that we developed at the beginning of this section. For example, ROW 2 suggests that the following constraint:

2) S1P1 + S1P2 + S1P3 <= 12.3

has a slack of 0, and a dual price of 1,141.18.

The *slack* associated with a constraint tells us how much of the RHS of each constraint is not being used in the solution that was provided. As we can see, the slack associated with constraints 2–21 is 0. This tells us that all of the land (acres) for each of the stands is being scheduled for harvest sometime during the time horizon of the plan. In constraint 2, we ensured that the sum of the scheduled harvested areas of stand 1 during time periods 1, 2, and 3 must be less than or equal to the size of the stand. If some part of the stand was not harvested during any of the three time periods, the slack associated with this constraint would be a positive number, indicating some acres were left unscheduled (uncut). If the slack associated with a constraint is 0, then the constraint is said to be *binding*. This implies that the constraint has some influence on the outcome of the solution (the management plan). Thus, we can often look at these values to determine if our formulation is correct.

Evaluation of the dual price or shadow price is one of the many techniques used in sensitivity analysis. The *dual price* associated with a constraint indicates how much the objective function value would increase if one more unit of the RHS of the constraint were available. For example, the dual price associated with constraint 2 is 1,141.18 ($1,141.18). If one more acre of stand 1 were available (making the stand 13.3 acres in size), then the objective function would increase by another $1,141.18, since this extra acre likely would be scheduled for harvest in time period 3. An important caveat is that you cannot keep adding more land (inputs) and continuously increase the objective function value while holding all other inputs constant. At some point, additional units of land will no longer increase the objective function value because other supporting inputs will become more constrained.

IX. ASSESSING ALTERNATIVE MANAGEMENT SCENARIOS

One thing that is obvious about the results of the scenario provided in Section VIII is that the scheduled harvest volumes vary widely from one period to the next. As a second management scenario, we will add two constraints to the problem formulation to ensure an even-flow of scheduled timber harvest volume.

```
31) HV1 − HV2 = 0
32) HV2 − HV3 = 0
```

After solving this problem with LINGO, the results for the optimal solution to this scenario show a scheduled harvest volume that is perfectly even throughout the time horizon (31,588.6 tons per 5-year period). This management scenario's scheduled harvest level is much lower during the first time period, and higher during the second and third time periods, as compared to the initial management scenario we modeled. The NPV of the even-flow scenario has decreased $12,758, however, to $1,675,888. Interestingly, the area scheduled for harvest during each time period is almost the reverse of the area scheduled for harvest in the initial scenario (320, 348, and 394 acres per period for time periods 1, 2, and 3, respectively). Finally, some older stands are scheduled for harvest in time periods 2 and 3, which was not the case in the initial scenario. One or more stands are scheduled for harvest in

more than one time period (i.e., stand 19), indicating that some acres are falling into the "old forest" class just before being harvested.

As a third management scenario, assume that due to site preparation equipment and personnel availability, you want to ensure that the areas scheduled for harvest are also equal from one time period to the next. Two constraints need to be added to the problem formulation we already have developed:

```
33) AC1 − AC2 = 0
34) AC2 − AC3 = 0
```

After solving the problem, the results for the third management scenario now suggest that the scheduled harvest volume is again perfectly even throughout the time horizon (30,965.7 tons per 5-year period), yet is lower than what we found in the second management scenario because we have added two more constraints to the problem. The NPV of this solution has decreased more significantly from the first scenario ($45,806), to $1,642,840. On the positive side of things, the area scheduled for harvest in each time period is even (354.2 acres), yet again some older forest is being scheduled for harvest in time periods 2 and 3.

As a fourth management scenario, assume that you now want to harvest an equal amount of older forest in each time period, to prevent the older forest from being liquidated in the first time period of the management plan. Two more constraints need to be added to the problem formulation we already have developed:

```
35) OF1 − OF2 = 0
36) OF2 − OF3 = 0
```

The results for this management scenario suggest that the scheduled harvest volume is again perfectly even throughout the time horizon, yet slightly lower than the previous scenario (30,726.6 tons per 5-year period). The NPV of this scenario has decreased even more significantly ($58,494) from the first scenario, to $1,630,152. The area scheduled for harvest in each time period is even (again 354.2 acres), and the amount of older forest scheduled for harvest in each time period is 113.6 acres.

As a fifth and final scenario, we add the 66 habitat-related accounting rows to the management problem, and add three constraints that force the solution to maintain an overall HSI of 0.250 during each time period:

```
37) S1P1 + N1P1 = 12.3
38) S1P1 + S1P2 + N1P2 = 12.3
39) S1P1 + S1P2 + S1P3 + N1P3 = 12.3
40) S2P1 + N2P1 = 35.6
41) S2P1 + S2P2 + N2P2 = 35.6
42) S2P1 + S2P2 + S2P3 + N2P3 = 35.6
43) S3P1 + N3P1 = 34.6
```

44) S3P1 + S3P2 + N3P2 = 34.6
45) S3P1 + S3P2 + S3P3 + N3P3 = 34.6
46) S4P1 + N4P1 = 85.4
47) S4P1 + S4P2 + N4P2 = 85.4
48) S4P1 + S4P2 + S4P3 + N4P3 = 85.4
49) S5P1 + N5P1 = 65.4
50) S5P1 + S5P2 + N5P2 = 65.4
51) S5P1 + S5P2 + S5P3 + N5P3 = 65.4
52) S6P1 + N6P1 = 69.1
53) S6P1 + S6P2 + N6P2 = 69.1
54) S6P1 + S6P2 + S6P3 + N6P3 = 69.1
55) S7P1 + N7P1 = 78.3
56) S7P1 + S7P2 + N7P2 = 78.3
57) S7P1 + S7P2 + S7P3 + N7P3 = 78.3
58) S8P1 + N8P1 = 71.0
59) S8P1 + S8P2 + N8P2 = 71.0
60) S8P1 + S8P2 + S8P3 + N8P3 = 71.0
61) S9P1 + N9P1 = 19.6
62) S9P1 + S9P2 + N9P2 = 19.6
63) S9P1 + S9P2 + S9P3 + N9P3 = 19.6
64) S10P1 + N10P1 = 34.8
65) S10P1 + S10P2 + N10P2 = 34.8
66) S10P1 + S10P2 + S10P3 + N10P3 = 34.8
67) S11P1 + N11P1 = 81.6
68) S11P1 + S11P2 + N11P2 = 81.6
69) S11P1 + S11P2 + S11P3 + N11P3 = 81.6
70) S12P1 + N12P1 = 90.2
71) S12P1 + S12P2 + N12P2 = 90.2
72) S12P1 + S12P2 + S12P3 + N12P3 = 90.2
73) S13P1 + N13P1 = 45.6
74) S13P1 + S13P2 + N13P2 = 45.6
75) S13P1 + S13P2 + S13P3 + N13P3 = 45.6
76) S14P1 + N14P1 = 67.2
77) S14P1 + S14P2 + N14P2 = 67.2
78) S14P1 + S14P2 + S14P3 + N14P3 = 67.2
79) S15P1 + N15P1 = 38.6
80) S15P1 + S15P2 + N15P2 = 38.6
81) S15P1 + S15P2 + S15P3 + N15P3 = 38.6
82) S16P1 + N16P1 = 49.1
83) S16P1 + S16P2 + N16P2 = 49.1
84) S16P1 + S16P2 + S16P3 + N16P3 = 49.1
85) S17P1 + N17P1 = 58.3
86) S17P1 + S17P2 + N17P2 = 58.3
87) S17P1 + S17P2 + S17P3 + N17P3 = 58.3
88) S18P1 + N18P1 = 26.8
89) S18P1 + S18P2 + N18P2 = 26.8
90) S18P1 + S18P2 + S18P3 + N18P3 = 26.8
91) S19P1 + N19P1 = 53.0
92) S19P1 + S19P2 + N192 = 53.0
93) S19P1 + S19P2 + S19P3 + N19P3 = 53.0
94) S20P1 + N20P1 = 46.0
95) S20P1 + S20P2 + N20P2 = 46.0
96) S20P1 + S20P2 + S20P3 + N20P3 = 46.0
97) 0.304 N1P1 + 0.482 N2P1 + 0.805 N3P1 + 0.765 N4P1 + 0.683 N5P1 + 0.377 N6P1 + 0.514 N7P1 + 0.181 N8P1 + 0.669 N9P1 + 0.788 N10P1 + 0.629 N11P1 + 0.447 N12P1 + 0.814 N13P1 + 0.397 N14P1 + 0.218 N15P1 + 0.361 N16P1 + 0.478 N17P1 + 0.799 N18P1 + 0.500 N19P1 + 0.141 N20P1 − HU1 = 0

98) 0.450 N1P2 + 0.663 N2P2 + 0.903 N3P2 + 0.882 N4P2 + 0.797 N5P2 + 0.526 N6P2 + 0.690 N7P2 + 0.352 N8P2 + 0.738 N9P2 + 0.869 N10P2 + 0.810 N11P2 + 0.609 N12P2 + 0.905 N13P2 + 0.505 N14P2 + 0.331 N15P2 + 0.474 N16P2 + 0.707 N17P2 + 0.872 N18P2 + 0.688 N19P2 + 0.363 N20P2 − HU2 = 0
99) 0.629 N1P3 + 0.768 N2P3 + 0.904 N3P3 + 0.972 N4P3 + 0.892 N5P3 + 0.692 N6P3 + 0.789 N7P3 + 0.501 N8P3 + 0.812 N9P3 + 0.903 N10P3 + 0.910 N11P3 + 0.779 N12P3 + 0.995 N13P3 + 0.654 N14P3 + 0.498 N15P3 + 0.649 N16P3 + 0.766 N17P3 + 0.962 N18P3 + 0.821 N19P3 + 0.505 N20P3 − HU3 = 0
100) 0.00094118 HU1 − HSI1 = 0
101) 0.00094118 HU2 − HSI2 = 0
102) 0.00094118 HU3 − HSI3 = 0
103) HSI1 >= 0.250
104) HSI2 >= 0.250
105) HSI3 >= 0.250

The results for this management scenario suggest that the scheduled harvest volume is again even throughout the time horizon (21,910.6 tons per 5-year period), yet significantly lower than the other scenarios due to the need to maintain older forest for wildlife habitat. The NPV of this solution has decreased even more significantly from the first scenario ($526,183), to $1,162,433. The area scheduled for harvest in each time period is even (217.5 acres), and the amount of older forest scheduled for harvest in each time period is 176.3 acres. Overall HSI levels range from 0.355 in the first time period to 0.250 (binding) in the third time period. The results from this scenario affect the economics of the management problem more significantly because some older stands must remain uncut for the overall HSI to be at least 0.250 in the final time period, as suggested by the constraint.

To be able to more easily compare the five scenarios presented here, the critical information related to each is presented in Table 7.4. In addition, managers may wish to know which stands are being scheduled for treatment in each time period, thus as a planner, you may need to examine the output and develop a harvest schedule for them.

X. CASE STUDY: WESTERN UNITED STATES FOREST

To study the development of a management problem with linear programming a little more closely, let's examine the western forest data that was developed for this text. The Lincoln Tract's stands GIS database contains 87 management units, each with an age, species, and potential volume for six 5-year time periods. To begin, assume that the landowner is interested in maximizing the volume of timber produced from this forest over the next 30 years. The constraints on the management problem are: (1) the landowner does not wish to harvest stands less than 35 years of age, and (2) the landowner wants an even volume from the harvest scheduled during each 5-year time period. As a result of this assessment of the landowner's

TABLE 7.4 A Comparison of the Five Management Scenarios for the 1,062.5 Acre Forest in South Georgia

Scenario	Net Present Value ($)	Harvest Volume (tons)	Area Harvested (acres)
Initial	1,688,646		
Time period 1		38,036.2	396.8
Time period 2		27,522.0	315.4
Time period 3		27,215.5	353.3
Even-flow	1,675,888		
Time period 1		31,588.6	319.9
Time period 2		31,588.6	348.4
Time period 3		31,588.6	394.2
Even-flow, even-acres	1,642,840		
Time periods 1–3		30,965.7	354.2
Even-flow, even-acres, equal older forest	1,630,152		
Time periods 1–3		30,726.6	354.2
Wildlife[a]	1,162,433		
Time periods 1–3		21,910.6	217.5

[a]*Includes even-flow, even-acres, equal older forest constraints.*

objectives, we need to develop an objective function that maximizes timber volume produced. We also need to develop constraints that restrict the harvest levels to an even amount each time period. To do this we may need to develop both accounting rows and policy constraints. In addition, we need to develop resource constraints so that we do not harvest more area than what is contained in each stand.

As an initial step in the process, the data for the 87 stands are extracted from the stand GIS database (Table 7.5). Decision variables will represent the number of acres of each stand to harvest in a given 5-year time period. To understand whether a stand is old enough to be harvested, we will determine the age at the beginning of each 5-year time period, and compare this to the landowner's desire not to harvest any stand below 35 years of age. The age noted in the GIS database is the age of each stand at the beginning of the first 5-year period, therefore stand 1, for example, will not be old enough to harvest until period 5 of the analysis, when it is 38 years old. In constructing the objective function, the contributions of stands too young for harvest should be omitted. The objective function might be designed as follows:

```
Maximize
     18.5 S1P5 +   25.3 S1P6 +   14.9 S2P6
 +   94.4 S4P1 +   98.4 S4P2 +  102.2 S4P3
 +  103.2 S4P4 +  104.0 S4P5 +  104.8 S4P6
 +   63.2 S5P1 +   68.7 S5P2 +   74.6 S5P3
 +   79.9 S5P4 +   84.9 S5P5 +   89.1 S5P6
 ....
 +   90.1 S85P1 +  93.7 S85P2 +  96.8 S85P3
 +   97.6 S85P4 +  98.5 S85P5 +  99.2 S85P6
 +   57.9 S86P1 +  64.2 S86P2 +  69.7 S86P3
```

TABLE 7.5 Data from the Lincoln Tract That Describe Each Age and Potential Volume of Each Stand over a 30-Year Time Horizon

Stand	Age[a]	Volume (MBF per acre)					
		Period 1	Period 2	Period 3	Period 4	Period 5	Period 6
1	18	0.4	2.2	6.1	11.7	18.5	25.3
2	10	0.0	0.0	0.7	3.3	8.4	14.9
3	7	0.0	0.0	0.3	1.5	4.9	10.0
4	100	94.4	98.4	102.2	103.2	104.0	104.8
5	70	63.2	68.7	74.6	79.9	84.9	89.1
...							
85	101	90.1	93.7	96.8	97.6	98.5	99.2
86	65	57.9	64.2	69.7	75.8	81.1	86.1
87	55	40.7	47.6	54.0	59.9	65.1	70.7

[a]*Age at the beginning of the first 5-year time period.*

```
+  75.8 S86P4 +  81.1 S86P5 +  86.1 S86P6
+  40.7 S87P1 +  47.6 S87P2 +  54.0 S87P3
+  59.9 S87P4 +  65.1 S87P5 +  70.7 S87P6
```

Notice in this abbreviated objective function that stand 3 is absent entirely. Stand 3 is only 7 years old at the beginning of the analysis. At the beginning of time period 6, stand 3 will be 32 years old, which is still too young to be considered for harvest.

The resource constraints indicate that the sum of the area assigned to be harvested during each 5-year time period must be equal to or less than the size of each stand. A subset of these constraints is as follows.

```
subject to
2) S1P5 + S1P6 <= 41.913
3) S2P6 <= 61.130
4) S4P1 + S4P2 + S4P3 + S4P4 + S4P5 + S4P6 <= 24.778
...
```

What you should observe here is the absence of resource constraints related to stand 3. Again, stand 3 is initially very young, and will never reach the minimum harvest age during the time horizon of the management plan. In addition, the resource constraints for stands 1 and 2 only reflect the feasible choices available for those stands. Stand 1, for example, can be harvested only in time periods 5 and 6 because of the initial age of the stand. To determine the volume scheduled for harvest, six accounting rows are needed to sum the volume harvested in each 5-year time period. They should resemble the following, which represent condensed versions of the accounting rows:

```
75) 94.4 S4P1 + 63.2 S5P1 .. + 40.7 S87P1 − H1 = 0
76) 98.4 S4P2 + 68.7 S5P2 .. + 47.6 S87P2 − H2 = 0
77) 102.2 S4P3 + 74.6 S5P3 .. + 54.0 S87P3 − H3 = 0
78) 103.2 S4P4 + 79.9 S5P4 .. + 59.9 S87P4 − H4 = 0
79) 18.5 S1P5 + 104.0 S4P5 + 84.9 S5P5 .. + 65.1
    S87P5 − H5 = 0
80) 25.3 S1P6 + 14.9 S2P6 + 104.8 S4P6 .. + 89.1 S5P6 +
    70.7 S87P6 − H6 = 0
```

Once the accounting rows have been developed, the policy constraints can be formulated. In the case of the Lincoln Tract, we are interested in an even amount of volume scheduled for harvest over the time horizon. As a consequence, constraints need to be designed to force the harvest volumes to be equal from one time period to the next. The constraints can be designed in a number of ways, such as ensuring volumes scheduled during all time periods 2−6 equal the first time period's harvest level. Alternatively, we could simply suggest that the volume scheduled in each subsequent time period equal the volume scheduled during the previous time period:

```
81) H1 − H2 = 0
82) H2 − H3 = 0
83) H3 − H4 = 0
84) H4 − H5 = 0
85) H5 − H6 = 0
```

After solving the problem using a linear programming solver such as LINGO, we find that the objective function value suggests 169,737.6 MBF of wood should be harvested over the next 30 years. The average harvest level for each of the 5-year time periods is 28,289.6 MBF, which, on an annual basis reduces to 5,657.92 MBF per year. If the timber were valued at $400 per MBF, the annual gross revenue arising from this plan would be $2,263,168. On average, the landowner would earn almost $500 per acre per year from the property as a whole. The location of the activities during the first 5-year time period is illustrated in Fig. 7.1.

XI. CASE STUDY: NORTHERN UNITED STATES HARDWOOD FOREST

As a final example of using linear programming, we will consider a selection harvest system for uneven-aged deciduous forests in northeastern North America. Ideally, these forests would be managed for shade-tolerant tree species such as sugar maple through the maintenance of a reverse J-shaped diameter distribution (see Chapter 2: Valuing and Characterizing Forest Conditions) that consists of trees in several age and size classes. This is one example of continuous cover management. The number of trees in larger diameter classes would likely decline as diameters increase, yet some larger trees would exist as dominant trees in the forest canopy to provide a continuous forest cover. An average basal area at the mid-point of a cutting cycle would be in the range of 80−100 ft^2 per acre (18.4−23.0 m^2 per hectare). Regular reestablishment of ingrowth is necessary to maintain the desired future conditions, therefore openings may need to periodically be developed so resources such as sunlight, water, and nutrients are available. For a nondeclining even-flow of scheduled harvest volumes, the volume harvested would be less than or equal to the growth, or the periodic increment, of the forest during the cutting cycle. We will assume here that the landowner will maximize the NPV of harvests associated with the management plan. In addition, a landowner may desire to extract a specific type of periodic volume from the management of the forest, and these goals can coincide in a course of action. Six cutting periods that are 5 years in length comprise the 30-year planning horizon.

The data used for this example consists of a mature northern forest, divided into four stands (Table 7.6). The potential management regimes for each stand are based on a 10-year recurring entry (cutting cycle), whereby a certain volume is extracted (Table 7.7). The harvest levels are conservative, and assume a residual stocking of around 9,000 board feet per acre. The volume projections suggest that the stands are in an uneven-aged steady state, or regulated, condition. The forest structure seems able to provide predictable harvest volumes where one can obtain the same amount of species distribution every 10 years.

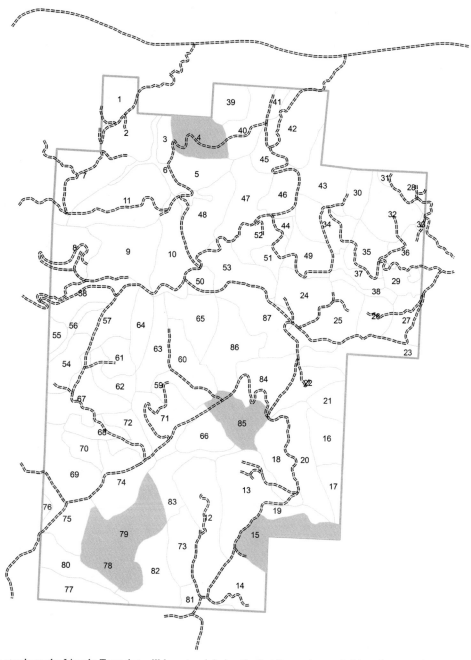

FIGURE 7.1 The stands on the Lincoln Tract that will be entered during the first 5-year time period in order to produce maximum timber volume, and to provide an even-flow of volume over the 30-year time horizon.

Future harvests are assumed to be performed in a manner that maintains the residual growing stock in roughly the same condition, emulating a reverse J-shaped tree diameter distribution (see Chapter 2: Valuing and Characterizing Forest Conditions).

Management regime 1 begins this cutting cycle in time period 1 (with entries again in periods 3 and 5), and management regime 2 begins this cutting cycle in time period 2 (with entries again in periods 4 and 6). To value the potential harvests, median stumpage prices, using the International 1/4″ rule for the Adirondack price

reporting region of New York, are employed (New York Department of Environmental Conservation, 2015). The prices assumed are: $175 per MBF for white ash (*Fraxinus americana*), $400 per MBF for sugar maple, $200 per MBF for yellow birch (*Betula alleghaniensis*), $75 per MBF for paper birch (*Betula papyrifera*), and $100 per MBF for all other tree species. A 4% discount rate is also assumed.

The NPVs associated with each management regime arise from the computation of a present value of a non-terminating periodic revenue (from Chapter 2: Valuing

TABLE 7.6 Data Associated With the Northern Hardwood Forest: Site Index (Sugar Maple (*Acer saccharum*), Base Age 50), Area, and Potential Net Present Value of Two Management Regimes

Stand	Site Index	Area (acres)	Management Regime	Net Present Value ($)
1	60	100	1	1,131.73
			2	930.20
2	60	130	1	1,102.39
			2	906.08
3	70	45	1	1,033.93
			2	849.81
4	70	60	1	1,135.22
			2	933.07

TABLE 7.7 Volume Extracted During Each Entry Into Each Stand, and Potential Nondiscounted Harvest Value

Stand	Volume Extracted (board feet per acre)					
	White Ash	Sugar Maple	Yellow Birch	Paper Birch	Other	Harvest Value ($ per acre)
1	100	700	300	300	250	405.00
2	90	650	350	250	300	394.50
3	600	150	300	800	850	370.00
4	700	200	250	850	900	406.25

TABLE 7.8 Net Revenue Computations for the Uneven-Aged Forest Harvest Entries Associated With the Northern United States Example

Basic Non-terminating Periodic Revenue Beginning in Year 10			Non-terminating Periodic Revenue Beginning in Year 2.5 With a 10-year Cycle		
Year	Revenue	Discounted Revenue	Year	Revenue	Discounted Revenue
10	405	273.60	2.5	405	367.17
20	405	184.84	12.5	405	248.05
30	405	124.87	22.5	405	167.57
40	405	84.36	32.5	405	113.21
50	405	56.99	42.5	405	76.48
60	405	38.50	52.5	405	51.67
70	405	26.01	62.5	405	34.90
80	405	17.57	72.5	405	23.58
90	405	11.87	82.5	405	15.93
100	405	8.02	92.5	405	10.76
110	405	5.42	102.5	405	7.27
120	405	3.66	112.5	405	4.91
130	405	2.47	122.5	405	3.32
140	405	1.67	132.5	405	2.24
150	405	1.13	142.5	405	1.51
160	405	0.76	152.5	405	1.02
170	405	0.51	162.5	405	0.69
180	405	0.35	172.5	405	0.47
190	405	0.24	182.5	405	0.32
200	405	0.16	192.5	405	0.21
210	405	0.11	202.5	405	0.14
220	405	0.07	212.5	405	0.10
230	405	0.05	222.5	405	0.07
240	405	0.03	232.5	405	0.04
250	405	0.02	242.5	405	0.03
Total		843.27[a]			1,131.67[b]

[a]If the timeline continued, the total should equal $843.32.
[b]If the timeline continued, the total should equal $1,131.73.

and Characterizing Forest Conditions). For stand 1, regime 1, without adjustment, this amounts to:

$$\text{Present value} = \left[\frac{\$405}{(1.04)^{10} - 1} \right] = \$843.32 \text{ per acre}$$

However, the result of the computation assumes that the revenue begins in year 10 (the cycle length), while our problem assumes that the revenue will begin in year 2.5 (the middle of time period 1) for regime 1. Therefore the present value of this opportunity is $1,131.73 when begun 7.5 years earlier. Since these are "non-terminating" net revenues, they are assumed to be obtainable every 10 years into perpetuity. Table 7.8 illustrates how these values can determined over a 250-year time frame in a more straight-forward manner. Another way to envision this transformation is to pull the $843.32 present value

(with harvests beginning in year 10) 7.5 years closer to *today*, using the 4% discount rate.

$$\$843.32 \times (1.04)^{7.5} = \$1,131.73$$

For regime 2, we assume that the revenue will begin in year 7.5 (middle of time period 2), therefore the present value of this opportunity is $930.20 when begun 2.5 years earlier.

The objective function of the problem might be designed as follows:

```
Maximize
1131.73 S1R1 + 930.20 S1R2 + 1102.39 S2R1 + 906.08
S2R2 + 1033.93 S3R1 + 849.81 S3R2 + 1135.22 S4R1 +
933.07 S4R2
```

In this problem, the decision variables are stated as $SxRy$, where x refers to the stand number, and y refers to the management regime. These decision variables are designed to contain the amount of land (acres) assigned to each management regime. The resource constraints therefore limit the amount of land available:

```
subject to
2) S1R1 + S1R2 <= 100
3) S2R1 + S2R2 <= 130
4) S3R1 + S3R2 <= 45
5) S4R1 + S4R2 <= 60
```

Since there are only two potential management regimes in this problem, the assignment of land to each regime is limited to the total size of each stand. These constraints also imply that land may not be assigned to either management regime. If a regime assignment must be made to each piece of land, then the inequality sign in the equations above should be changed to an equality sign.

Accounting rows might be developed to add up the total volume scheduled for harvest during each time period:

```
6) 1650 S1R1 + 1640 S2R1 + 2700 S3R1 + 2900 S4R1 -
   VS1 = 0
7) 1650 S1R2 + 1640 S2R2 + 2700 S3R2 + 2900 S4R2 -
   VS2 = 0
8) 1650 S1R1 + 1640 S2R1 + 2700 S3R1 + 2900 S4R1 -
   VS3 = 0
9) 1650 S1R2 + 1640 S2R2 + 2700 S3R2 + 2900 S4R2 -
   VS4 = 0
10) 1650 S1R1 + 1640 S2R1 + 2700 S3R1 + 2900 S4R1 -
    VS5 = 0
11) 1650 S1R2 + 1640 S2R2 + 2700 S3R2 + 2900 S4R2 -
    VS6 = 0
```

Here, VSz represents the total volume scheduled during time period z. Thus there are six accounting rows, one for each time period in the management plan. The volumes reflect the scheduled harvest for each regime during each time period. Observe, for example, that regime 1 decision variables are used for time periods 1, 3, and 5, while regime 2 decision variables are used for time periods 2, 4, and 6. A keen eye would also observe that only two of these constraints are really necessary (perhaps the period 1 and period 2 constraints). Given our assumption of removing the same amount of volume during each entry, each time a stand is entered every 10 years, the same total amount of volume will be removed. Therefore,

$VS1 = VS3 = VS5$, and $VS2 = VS4 = VS6$. This again suggests that the stands are in an uneven-aged steady state, or regulated, condition.

If one were interested in tracking the sugar maple volume scheduled for harvest, accounting rows can also be developed. Here, the variables $SMSz$ are used to contain total sugar maple volume scheduled for harvest during each time period (z).

```
12) 700 S1R1 + 650 S2R1 + 150 S3R1 + 200 S4R1 - SMS1
    = 0
13) 700 S1R2 + 650 S2R2 + 150 S3R2 + 200 S4R2 - SMS2
    = 0
14) 700 S1R1 + 650 S2R1 + 150 S3R1 + 200 S4R1 - SMS3
    = 0
15) 700 S1R2 + 650 S2R2 + 150 S3R2 + 200 S4R2 - SMS4
    = 0
16) 700 S1R1 + 650 S2R1 + 150 S3R1 + 200 S4R1 - SMS5
    = 0
17) 700 S1R2 + 650 S2R2 + 150 S3R2 + 200 S4R2 - SMS6
    = 0
```

As with the total volume accounting rows, technically only two of these constraints (perhaps numbers 12 and 13) are really necessary. Accounting rows for the area scheduled for harvest might also be developed:

```
18) S1R1 + S2R1 + S3R1 + S4R1 - AS1 = 0
19) S1R2 + S2R2 + S3R2 + S4R2 - AS2 = 0
20) S1R1 + S2R1 + S3R1 + S4R1 - AS3 = 0
21) S1R2 + S2R2 + S3R2 + S4R2 - AS4 = 0
22) S1R1 + S2R1 + S3R1 + S4R1 - AS5 = 0
23) S1R2 + S2R2 + S3R2 + S4R2 - AS6 = 0
```

The technical coefficients in these equations have a value of 1, and by convention are not included in the formulation (yet the value "1" will be included in the detached coefficient matrix for these six equations). Each of the decision variables will contain the areas scheduled for harvest. The $SxRy$ variables contain areas scheduled for harvest within each stand, while the ASy variables contain total areas scheduled for harvest in each year. Further, only two of these equations are technically necessary, given the 10-year reentry cycle.

After completing the development of the objective function, resource constraints, and accounting rows, policy constraints can be formulated. In this case, we might be interested in scheduling an even total volume for each time period of the 30-year plan. Constraints would then be designed to force the scheduled harvests to be equal during each time period. As in the Lincoln Tract example, we can simply suggest that the volume scheduled in each time period beyond the first be equal to the volume scheduled during the first time period:

```
24) VS1 - VS2 = 0
25) VS1 - VS3 = 0
26) VS1 - VS4 = 0
```

27) VS1 − VS5 = 0
28) VS1 − VS6 = 0

In conjunction with prior discussions, for the straight-forward problem we have described, if $VS1 = VS2$, and if $VS1 = VS3 = VS5$ and $VS2 = VS4 = VS6$, then the volume scheduled for all time periods would be equal if only constraint 24 were used.

After solving the problem, we find that the objective function suggests that the present value of the plan will be $345,759.70, and that the average total volume scheduled for harvest for each of the 5-year time periods is 336,850 board feet (336.85 MBF). In this scenario, every piece of land is assigned one of the two management regimes. Stands 3 and 4 are assigned regime 2. This makes sense for stand 4, as it has the highest present value for any regime 2 option. Since the same amount of total volume needs to scheduled in each time period, Stand 1 is assigned to regime 1, and stand 2 is split between regime 1 (104.8 acres) and regime 2 (25.2 acres). Areas scheduled for harvest range from 130.2 acres in time periods 2, 4, and 6 to 204.8 acres in time periods 1, 3, and 5. The scheduled sugar maple harvest volume ranges from 35,139 board feet (35.1 MBF) to 138,111 board feet (138.1 MBF) per period.

If we then wanted the areas scheduled for harvest to be the same during each time period, we might add the following five constraints to the problem formulation:

29) AS1 − AS2 = 0
30) AS1 − AS3 = 0
31) AS1 − AS4 = 0
32) AS1 − AS5 = 0
33) AS1 − AS6 = 0

After solving this second problem, we find that the objective function suggests that the present value of the plan will be $338,723.40, a decline of about $7,036. The average total volume scheduled for harvest for each of the 5-year time periods is again 336,850 board feet (336.85 MBF). As with the previous scenario, every piece of land is assigned one of the two management regimes, and areas scheduled for harvest are 167.5 acres in each time period. Stand 2 is assigned partly to regimes 1 (19 acres) and 2 (111 acres). Stand 4 is also assigned partly to regimes 1 (48.5 acres) and 2 (11.5 acres). Stand 1 is again assigned to regime 1, and stand 3 is assigned to regime 2. The scheduled sugar maple harvest volume ranges from 81,214 board feet (81.2 MBF) to 92,036 board feet (92 MBF) per period.

Finally, should we want the same amount of sugar maple volume to be produced during each time period, perhaps to supply a local mill, we might add the following five constraints to the problem formulation:

34) SMS1 − SMS2 = 0
35) SMS1 − SMS3 = 0
36) SMS1 − SMS4 = 0
37) SMS1 − SMS5 = 0
38) SMS1 − SMS6 = 0

After solving this third problem, we find that the objective function suggests that the present value of the plan will be $338,198.50, a decline of about $7,651 from the first scenario, and about $525 from the second. The average total volume scheduled for harvest for each of the 5-year time periods is again 336,850 board feet (336.85 MBF). As with the previous scenario, every piece of land is assigned one of the two management regimes, and areas scheduled for harvest are 167.5 acres in each time period. To provide an even amount of sugar maple during each time period, stand 1 is now assigned to regime 2. Stand 2 is assigned partly to regimes 1 (118.1 acres) and 2 (11.9 acres). Stand 2 is assigned partly to regimes 1 (0.4 acres) and 2 (44.6 acres). And stand 4 is assigned partly to regimes 1 (49 acres) and 2 (11 acres). The scheduled sugar maple harvest volume is 86,625 board feet (86.6 MBF) per period.

XII. SUMMARY

Linear programming is a mathematical programming method for locating the optimal solution to a set of linear equations with a linear objective function described in a problem formulation, and can be a powerful tool for forest management and planning. Linear programming has been used extensively in business applications and is used frequently to develop strategic forest management plans, especially for large industrial forest operations. Planning, scheduling, and assignment of activities to land classes can be accommodated relatively easily. The achievement of a landowner's objectives will be a function of the quantity of available land, the age class distribution of the forest, the structure of forests within each age class, and the constraints associated with managing the forest. The objective function for a linear programming model consists of a statement that describes the contribution of each decision variable to the achievement of the objective, whether it is an economic, an ecological, or a social objective. Constraints confine solutions to problems, by limiting the choices for the decision variables. Accounting rows simply sum values that may be useful in a constraint, or that may be of value in summarizing the results of a forest plan. Plans generated by models such as linear programming must be assessed to determine whether they are reasonable. Incorrect specification of the objective function or the constraints can lead to the development of forest plans that are either infeasible or inefficient.

Other limitations of linear programming, as Kidd et al. (1966) suggested over 50 years ago, include the capabilities of the software (size of the problem), the speed of the computer being used, and the ability of the planner to formulate the problem and interpret the results. Although we cannot control the first two of these, in this chapter we have given you some tools to address the latter (formulating the problem and interpreting results). Another limitation is in understanding how linear programming solves a problem. The Simplex method is one approach, and it assesses the quality of the corners of the feasible region of the solution space (as we described in Chapter 6: Graphical Solution Techniques for Two-Variable Linear Problems). For the braver students, the Simplex method is discussed in detail in Appendix B along with an example problem. One limitation of using linear models is in recognizing when decisions should no longer be modeled in a linear manner. For example, when the decisions need to be assigned discrete integer values (e.g., the building of a road), linear programming can accommodate these to a limited extent. However, mixed integer or integer programming approaches may be more appropriate. These advanced techniques are discussed in Chapter 8, Advanced Planning Techniques.

Linear programming solvers come in a variety of forms and use a variety of data formats. In addition to the format we described in this chapter for LINDO or LINGO, other common formats for designing problem formulations are the column-oriented MPS (mathematical programming system) format that can be used in conjunction with CPLEX and other solvers, and the Excel Solver format, which is essentially a detached coefficient matrix developed within an Excel spreadsheet. In each case, the objective function, accounting rows, and constraints must be designed appropriately to solve a management problem correctly. Additionally, large linear programming programs can be formulated and solved using advanced modeling languages such as the general algebraic modeling system (GAMS) or a mathematical programming language (AMPL). These problem-solving systems allow the user to formulate and solve large problems using a high level language that is similar to many other computer programming languages.

QUESTIONS

1. *Southern forest management.* Using the south Georgia model developed in this chapter (beginning in Section III), develop a scenario that maximizes NPV while harvesting only 300 acres in each time period. Solve the problem and compare and contrast these results to the scenarios that have already been developed (Table 7.4).
2. *Southern multiple use management.* If in the south Georgia model, the wildlife habitat constraints were used without the even-flow requirement, without the even-acres requirement, and without the equal older

forest harvest requirement, then how would the value of the plan compare to the other plans previously developed (NPV, acres harvested, volume harvested)?
3. *A small linear programming model.* Use the following data from western Oregon and the assumptions that are provided to develop a linear programming model that maximizes the NPV of a plan of action.

   Time horizon: 20 years
   Time periods: 4 (5 years long)
   Interest rate: 6%
   Stumpage price: $400 per MBF

Stand	Size (acres)	Period 1 volume (MBF/ac)	Period 2 volume (MBF/ac)	Period 3 volume (MBF/ac)	Period 4 volume (MBF/ac)
1	100.3	22.3	26.1	30.0	34.3
2	126.5	31.4	35.9	40.5	45.6
3	96.3	26.5	31.8	36.9	41.9
4	107.5	44.3	49.3	54.2	59.2
5	110.9	28.2	32.6	37.0	42.0
6	120.4	21.4	27.0	32.2	37.2
7	98.4	24.9	30.2	35.6	40.6
8	89.3	49.0	53.9	58.4	63.3
9	116.8	22.9	27.9	32.7	38.0
10	119.4	28.5	32.9	37.2	41.9

   a. What is the NPV of the plan if only resource constraints are applied?
   b. What is the NPV of the plan when an even-flow requirement is assumed?
   c. What is the NPV of the plan when an even-acres harvest is assumed, with no even-flow volume requirement?
4. *Your school forest plan.* For your school's forest, or some other familiar property, develop the stand-level data needed for a linear programming model (stand numbers, acres), and the growth projections necessary for a 20−30-year plan. Then assume some interest rate, and price for a forest product that is to be produced. Finally, develop the objective function, resource constraints, policy constraints, and accounting rows that are necessary to evaluate a reasonable management scenario for the property.
5. *Western forest example.* Develop and solve the even-flow harvest problem for the western forest, as described in Section X. Make sure that your answer is the same as what we have determined for the 30-year time horizon. Then remove the even-flow constraints and solve the problem again. How does the optimal

solution change? If the harvest volumes were allowed to deviate by 5% from one time period to the next, then how would you set up the problem? How would these results differ from the original problem?

REFERENCES

Bell, E.F., 1977. Mathematical programming in forestry. J. For. 75 (6), 317–319.

Curtis, F.H., 1962. Linear programming the management of a forest property. J. For. 60 (9), 611–616.

Gunn, E.A., 2007. Models for strategic forest management. In: Weintraub, A., Romero, C., Bjørndal, T., Epstein, R. (Eds.), Handbook of Operations Research in Natural Resources. Springer, New York, NY, pp. 317–341.

Gunter, J.E., 1987. Georgia Forest Landowner's Manual. Cooperative Extension Service, University of Georgia, Athens, GA, 70 p.

Johnson, K.N., Scheurman, H.L., 1977. Techniques for prescribing optimal timber harvest and investment under different objectives—discussion and synthesis. For. Sci. Monogr. 18, 29 p.

Johnson, K.N., Stuart, T.W., 1987. FORPLAN Version 2: Mathematical Programmer's Guide. US Department of Agriculture, Forest Service, Land Management Planning Systems Center, Washington, DC.

Kaya, A., Bettinger, P., Boston, K., Akbulut, R., Ucar, Z., Siry, J., et al., 2016. Optimisation in forest management. Curr. For. Rep. 2 (1), 1–17.

Kidd Jr., W.E., Thompson, E.F., Hoepner, P.H., 1966. Forest regulation by linear programming—a case study. J. For. 64 (9), 611–613.

LINDO Systems, Inc., 2016. LINGO, Version 15.0. Lindo Systems, Inc., Chicago, IL.

Macmillan, D.C., Fairweather, S.E., 1988. An application of linear programming for short-term harvest scheduling. Northern J. Appl. For. 5 (2), 145–148.

McTague, J.P., Oppenheimer, M.J., 2015. Rayonier, Inc., Southern United States of America. In: Siry, J.P., Bettinger, P., Merry, K., Grebner, D.L., Boston, K., Cieszewski, C. (Eds.), Forest Plans of North America. Academic Press, New York, NY, pp. 395–402.

New York Department of Environmental Conservation, 2015. Stumpage price report, winter 2015, #86. New York Department of Environmental Conservation, Albany, NY.

O'Connor, J.J., Robertson, E.F., 2003. George Dantzig. University of St. Andrews, St. Andrews, Scotland, UK. Available from: http://www-history.mcs.st-andrews.ac.uk/Biographies/Dantzig_George.html (Accessed 1/8/2016).

Papps, S., 2000. Harvest planning with uncertain prices and demands. In: Vasievich, J.M., Fried, J.S., Leefers, L.A. (Eds.), Seventh Symposium on Systems Analysis in Forest Resources. US Department of Agriculture, Forest Service, North Central Forest Experiment Station, St. Paul, MN, General Technical Report NC-205, pp. 241–248.

Remsoft, 2016. Remsoft Spatial Planning System Updates. Remsoft, Fredericton, NB. Available from: http://www.remsoft.com/downloads.php (Accessed 3/27/2016).

Siry, J.P., Bettinger, P., Merry, K., Grebner, D.L., Boston, K., Cieszewski, C. (Eds.), 2015. Forest Plans of North America. Academic Press, New York, NY, 458 p.

US Department of Agriculture, Forest Service, 1986. Service Forester's Handbook. US Department of Agriculture, Forest Service, Southern Region, State & Private Forestry, Atlanta, GA, Miscellaneous Report R8-MR11. 129 p.

Weintraub, A., Romero, C., 2006. Operations research models and the management of agricultural and forestry resources: a review and comparison. Interfaces 36 (5), 446–457.

Chapter 8

Advanced Planning Techniques

Objectives

Chapter 7, Linear Programming, that described linear programming techniques, we began our treatment of forest-level planning, where activities related to multiple stands need to be simultaneously evaluated. Linear programming, however, is limited in its ability to recognize and utilize nonlinear functional relationships among resources. In addition, as land managers you may find that the fractional solutions linear programming provides (due to the divisibility assumption) for activities assigned to stands are not suitable for your management needs. This suggests that in some cases, integer values should be assigned to decision variables rather than continuous real numbers. In this chapter, we introduce several alternative forest planning methods, and once completed, you should be able to:

1. Describe the differences between linear programming and mixed integer or integer programming methods for forest management and planning.
2. Compare and contrast the differences between linear and goal programming methods.
3. Understand how binary search can be used to develop a forest plan.
4. Describe how a heuristic method may be used to develop a forest plan.
5. Understand the variety of computer software programs that are readily available for addressing forest planning needs.

I. INTRODUCTION

A large portion of this book concerns the description and application of quantitative methods for solving problems. The problems involve the development of forest management plans. Our efforts would be relatively straightforward if each problem (plan) could be represented as a set of linear relationships. In practice, however, many functional relationships between potential actions (management activities) and potential outcomes are best represented in a nonlinear or discontinuous fashion. Alternatively, it may be of value to include more than one potential outcome in the objective of a problem, to explore the trade-offs associated with the simultaneous optimization of two or more goals. For these reasons

advanced planning techniques may be employed to assist in the development of management plans. This chapter therefore describes several of the common problem-solving methodologies that have been used in practice to assist land managers in their planning efforts.

II. EXTENSIONS TO LINEAR PROGRAMMING

As we suggested in Chapter 7, Linear Programming, linear programming is used widely in North America and many other parts of the world for assisting with the development of strategic forest plans. In doing so, decisions are based on land allocations, management areas, strata, and stands generally under a Model I or Model II format. These types of problem formulations usually are solved very quickly with a mathematical programming solver such as LINGO or Gurobi; however, not all management problems can be described with the continuous decision variables that are assumed in basic linear programming formulation. For example, when there is a need to control the placement of activities, perhaps for wildlife habitat or harvest opening size considerations, there may be an associated need to know exactly where the activity is placed. When continuous variables are used, unless they indicate 0 or 100% of a stand is treated in a specific manner, we do not necessarily know *where the activities are located within the stand*. As a result, some decision variables may need to be assigned integer variables to force the treatment of zero or 100% of a stand. In these cases, mixed integer or integer programming alternatives to basic linear programming methods would be useful. Another aspect of linear programming that is debatable among managers who cannot discern between two or more objectives is the fact that only one objective is included in the objective function (e.g., maximize habitat quality, maximize net present value, or minimize costs). There are methods, however, to incorporate multiple objectives into an objective function to accommodate the wishes of a landowner. In these cases, goal programming methods would be useful.

Forest Management and Planning. DOI: http://dx.doi.org/10.1016/B978-0-12-809476-1.00008-4

A. Mixed Integer Programming

As we have noted, linear programming assumes that all the decision variables, whose values are initially unknown, are required to be assigned continuous real numbers due to the divisibility assumption. If some, but not all, of the unknown variables are required to have integer values, then a problem is considered a mixed integer programming problem. The problem formulation for mixed integer programming is very similar to that of linear programming. An objective function is needed, and several resource or policy constraints, and perhaps some accounting rows, are required. What is different within the problem formulation is a formal specification of those decision variables that need to be assigned integer values. This formality also needs to address the type of integer values that will be assigned, either general integer (e.g., 0, 1, 2, 3, ...) or binary integer (0 or 1) values. Among the various methods that are used to solve mixed integer programming problems, the branch and bound and the cutting plane methods, rather than the simplex method, are two of the most widely applied. These can generate a solution and can, when fully implemented, generate an optimal solution. The mathematics behind each are presented more formally elsewhere. Here we simply provide a brief overview of these two methods.

The branch and bound process involves a systematic evaluation of the possible solutions to a problem. When a problem is being addressed by a solver (e.g., LINGO), a number of subproblems are developed and examined systematically by restricting the range of the integer values assigned to decision variables. When using binary values (0 or 1) to represent the types of decision variables, there are only two choices: set the state of the decision variable to 0 or to 1. A bound (upper or lower) generally is created by relaxing the need to assign an integer value to a decision variable. This results in a linear model that in turn provides an optimal solution to a relaxed problem. If all the decision variables that should be assigned integer values are not actually provided integer variables, then a tree-like search process is used to explore alternative solutions where a number of the values assigned to decision variables are fixed. If a subproblem at this stage of the search is infeasible, then no further expansion of the search from the point of the subproblem is performed. In this way, a number of alternative plans can be ignored by estimating upper and lower bounds on the quantity being optimized. If in a maximization problem one set of subproblems has an upper bound that is lower than the lower bound of a second set of subproblems (the estimated best solution of the first set is worse than the estimated worst solution of the second set), then the first set is discarded.

The cutting plane method uses a relaxed linear programming solution to a problem as a starting point.

This solution generally consists of a number of fractional results (decisions that we would like to have assigned integer values but which actually are assigned continuous values). The process then searches for an integer constraint that when added, will violate the feasibility of the linear problem, yet will reduce the size of the solution space. The addition of the integer constraint in effect reduces the size of the feasible region of the solution space, and is considered a cutting plane. This, in effect, subdivides the linear solution without removing any of the previously assigned integer decisions from the feasible region of the solution space. A subsequent linear solution is generated. If further fractional results are encountered, then the process continues, with the intent of locating solutions that are less fractional than before. When there are no further fractional results, an optimal integer solution has been located.

To illustrate the use of mixed integer programming problem formulations, assume that the landowner of the Putnam Tract is interested in producing the highest volume from timber harvests over the next 15 years from stands that are not located along the stream system. The minimum clearcut harvest age that the landowner would consider, regardless of the economics of the situation, is 25 years. Assume further that the landowner is interested in understanding which complete stands to harvest in the first 5 years. This suggests that the decision related to harvesting is *yes* or *no* during the first time period. Here, we might use a binary (0/1) integer value to represent the choice of harvesting each stand during time period 1. In this case, we could use a variable such as $S1P1$ to represent the opportunity to harvest stand 1 during time period 1, yet the value that will be assigned to the choice will only be 0 (which represents no harvest, and is the default) or 1 (which represents a harvest). In the second and third time periods, this decision is not as critical; therefore, although the decision variables are similar ($S1P2$, $S1P3$), the values that can be assigned to these variables are from a continuous range of numbers between 0 and 1. For example, if a value of 0.37 were assigned to the choice $S1P2$, then it would suggest that 37% of stand 1 is scheduled for harvest during the second time period. The differences in how choices for time period 1 and time periods 2−3 are handled are representative of a mixed integer planning problem.

Within the Putnam Tract stands GIS database, there are cordwood volumes for three time periods, the first 5 years, the second 5 years, and the final 5 years. After extracting this data from a GIS database, we can design an objective function that maximizes timber volume produced. The form of the objective function might be as follows:

$$\text{Maximize} \sum_{i=1}^{I} \sum_{t=1}^{T} a_i \, v_{it} \, SiPt$$

where:

i = a stand
I = the total number of stands
t = a time period
T = the total number of time periods
a_i = area of stand i
v_{it} = volume contained in stand i during time period t
$SiPt$ = a decision variable representing the harvest of stand i during time period t

For the Putnam Tract the objective function might resemble the following:

```
Maximize 2476.5825 S2P1 + 2738.61 S2P2 + 3025.995
S2P3 + 572.2315 S4P2 + 730.3065 S4P3 + 108.9329 S6P1
+ 135.6649 S6P2
```

You might notice that there is no reference to stand 1 in the objective function. The reason for this, as well as any other omissions, is that stands younger than 25 years at the beginning of each time period have been ignored. This is an implied constraint, since removing these options from problem formulation results in a constraint on the problem. The resource constraints for this problem would reflect that up to 100% of a stand can be harvested during the time periods in which each stand is old enough to harvest.

```
subject to
2) S2P1 + S2P2 + S2P3 <= 1
3) S4P2 + S4P3 <= 1
4) S6P1 + S6P2 + S6P3 <= 1
```

As before, any decisions that are moot (those related to the harvest of stands 1 or 3 or the harvest of stand 4 during time period 1) are not included in these resource constraints. If these were inadvertently included, then we run the risk of making infeasible decisions. In addition to the differences in the data type assigned to the decision variables, assume that the landowner also is interested in maintaining harvest levels within some reasonable bound from one period to the next. This might involve the use of accounting rows to accumulate the volume scheduled for harvest, such as the partial one shown here for time period 1:

```
51) 2476.583 S2P1 + 108.9329 S6P1 + 677.3607 S7P1
... + 1187.616 S76P1 + 966.8176 S79P1 + 2498.206
S81P1 - H1 = 0
```

Once each of the accounting rows have been developed for the three time periods, some control on the volume can be incorporated into the problem formulation. For example, assume that the landowner wanted a very consistent harvest scheduled for each of the three time periods, perhaps not varying by more than 100 cords per time period. Eqs. (54) and (55) suggest that if the harvest scheduled for time period 1 is greater than the harvest in time periods 2 and 3, then the difference can be no greater

than 100 cords. Eqs. (56) and (57) suggest the opposite, that if the harvests scheduled for time periods 2 and 3 are greater than the harvest scheduled for time period 1, then the differences can be no greater than 100 cords. As we discussed in Chapter 7, Linear Programming, these are considered wood flow policy constraints.

```
54) H1 - H2 <= 100
55) H1 - H3 <= 100
56) H2 - H1 <= 100
57) H3 - H1 <= 100
```

The objective function, resource constraints, accounting rows, and wood flow policy constraints form a mixed integer problem formulation only when the solver understands that the choices related to time period 1 have to be assigned binary integer values. Once this has been accomplished, a forest plan can be developed that explicitly determines the whole stands to harvest in the first 5 years (Fig. 8.1). The scheduled harvest volumes are 12,543 cords, 12,492 cords, and 12,643 cords for time periods 1 through 3, respectively. As you can see, these volumes do not deviate by more than 100 cords from the time period 1 volume.

B. Integer Programming

As we have just shown, when formulating a linear programming problem we might decide that some of the decision variables should be assigned integer values rather than continuous real number values. When some of the decision variables are handled in this manner, the solution process is called mixed integer programming. When nearly all the decision variables are considered to require an assignment of an integer value, the process is called integer programming. As we inferred earlier, some decisions in natural resource management are discrete, requiring a yes or no choice (as in the case of building a road or creating a snag), or requiring a choice that assumes a nonfractional value (number of trucks to purchase, number of stream pools to create). Integer programming problems are formulated much in the same manner as mixed integer programming problems, and are solved using the same types of mathematical methods (e.g., branch and bound, cutting plane). Integer programming generally assumes that the integer values range from 0 to ∞. Again, if the integer values can only be selected from the set of 0 and 1, then we are dealing with a binary integer programming problem. In this latter case, the value 0 usually represents *no*, and the value 1 represents *yes*.

To illustrate the application of integer programming to a natural resource management problem, assume that the landowner of the Putnam Tract was now interested in obtaining a forest plan where all the decision variables had to be assigned binary integer values. In addition, the landowner was not too keen on the idea that some of the

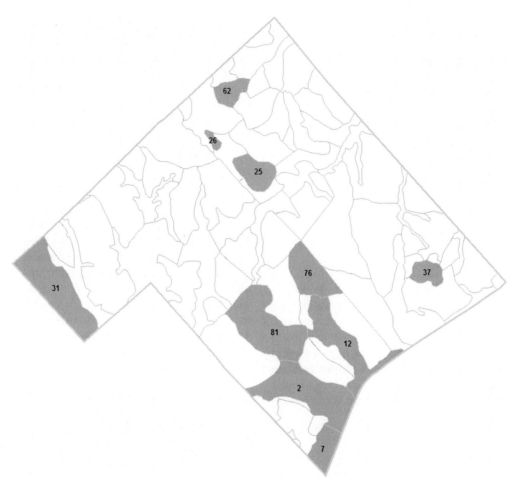

FIGURE 8.1 First period harvest units on the Putnam Tract using mixed integer programming methods.

harvests in the previous plan were placed so close to each other, which would have resulted in a very large cleared area. Two adjustments therefore are made to the mixed integer model described earlier: (1) all the stand-level harvest decision variables will be assigned a binary integer value (either 0 or 1), and (2) no two harvests can touch each other during a single time period. In making these changes, the problem formulation becomes basically an integer programming problem, where most of the decision variables will be assigned binary integers. The three periodic harvest level variables ($H1$, $H2$, and $H3$) are the only exceptions to this assumption. The clearcut harvest adjacency constraints can be handled in a number of different ways (see McDill et al. (2002) for a discussion on this subject), however, here we use simple pair-wise constraints to prevent two stands from being scheduled for harvest in a single time period. Pair-wise adjacency constraints take the form:

$$SiPt + SjPt \leq 1$$

which suggests that during time period t, either stand i or stand j can be scheduled for harvest, but not both of

them. In addition, it leaves open the possibility for neither stand to be harvested during the time period in question. Care must be taken to use only i, j pairs of stands that actually touch each other, if our goal is to prevent the harvest of stands that share a boundary. Otherwise, we may needlessly be constraining the harvest timing of stands that are not adjacent to each other. Almost 300 pairwise adjacency constraints are necessary for the Putnam Tract, a relatively small problem given the spatial arrangement of stands. A few of these constraints related to time period 1 include:

```
58) S2P1 + S6P1 <= 1
59) S2P1 + S7P1 <= 1
60) S2P1 + S12P1 <= 1
61) S2P1 + S81P1 <= 1
```

These constraints are associated with the timing of harvest of stand 2. As we can see, if stand 2 is scheduled for harvest during the first time period, then stands 6, 7, 12, and 81 have to be scheduled for harvest in one of the other two time periods (or not at all). Conversely, if any one of stands 6, 7, 12, and 81 are scheduled for harvest during time period 1, then stand 2 will have to be scheduled in a different time period (or not at all). Whether stands 6, 7, 12, and 81 can all be harvested during time period 1 will depend on their spatial juxtaposition on the landscape.

Once the integer programming problem formulation has been organized, a forest plan can be developed that explicitly determines the stands to harvest in each of the 5-year time periods, where no harvests physically touch each other within each time period. The spatial arrangement of harvests for time period 1 under this plan is shown in Fig. 8.2. As you can see, it is quite different than the previous plan we developed using mixed integer programming. The resulting scheduled harvest volumes, however, are 12,054 cords, 11,901 cords, and 12,124 cords for time periods 1 through 3, respectively. Although these scheduled volumes do not deviate by more than 100 cords from the time period 1 volume, they are significantly lower (4−6%) than those scheduled earlier using the mixed integer programming formulation, mainly due to the additional constraints that were included in the problem.

As we noted, there are other ways in which harvest adjacency can be modeled. In fact, our example represents a unit restriction process, where one stand (of any size) cannot be scheduled for harvest in a time period where an adjacent stand (of any size) is already scheduled. The problem with this is obvious - the sizes of the stands are not considered. An alternative is the area restriction process, where a collection of adjacent stands are prevented from being scheduled for harvest only when their total size exceeds some maximum size. For example, consider three stands that all share edges: A (50 acres), B (60 acres), and C (40 acres). Assume that the decision variables were binary (0 = not scheduled, 1 = scheduled for harvest) and there are no green-up allowances (time-based waiting periods for harvests). If it were of concern to prevent a final harvest opening from exceeding 120 acres, simple constraints for three time periods can be devised:

$$A1 + B1 + C1 \leq 2$$
$$A2 + B2 + C2 \leq 2$$
$$A3 + B3 + C3 \leq 2$$

FIGURE 8.2 First period harvest units on the Putnam Tract using integer programming methods and harvest adjacency constraints.

Any combination of two of these stands can be scheduled for harvest in any of the three periods, as alluded to in the equations above, but all three cannot be scheduled for harvest in the same time period. Final harvest openings can therefore be 40, 50, 60, 90, 100, or 110 acres in size, but no more. Readers should refer to Tóth et al. (2012) for more in-depth treatment of this subject.

C. Goal Programming

Linear programming in its traditional, conventional form requires a planner to specify a single objective for a management problem, and accompany it with a set of secondary requirements (constraints) imposed on the problem. Goal programming is a form of linear programming that could be of value for multiple-use management considerations, because in contrast to linear programming, multiple conflicting goals may be incorporated into the lone objective function (Field, 1973). There are two forms of goal programming, one sequentially best meets the goals (lexicographic goal programming), and the other that we describe here simultaneously minimizes the deviations from each of the goals. This type of goal programming problem usually is designed in such a way that deviations from individual goals are accumulated, and the optimal solution is the one that minimizes the sum of the deviations. Each goal can be weighted in the objective function, which would infer their importance to the landowner. However, defining weights may be difficult in some natural resource management situations (Dyer et al., 1979).

A significant difference between goal programming and linear programming is that a solution to a linear programming problem requires the attainment of constraint levels while maximizing (or minimizing) a single objective. As a result, in some instances we may only arrive at an infeasible solution. However, in a goal programming problem, a feasible solution is almost always assured. Here, we are simply minimizing deviations from goals. Since the important goals are incorporated into the objective function, feasible solutions are usually located because scheduling no activities can still result in a feasible solution (even though the deviations from the goals are great). The resulting values related to the goals may not be what we wanted, yet they ideally approach as closely as possible our desired levels. However, our desired level of goal attainment may not be met completely. Setting the priorities for each goal is important, and Field (1973) suggests the following when developing a goal programming problem formulation:

1. Determine what can be produced from the land currently and in the future, as well as the demand for the different products.

2. Estimate the physical capacity of the land to produce the various products.
3. Analyze the complementary and competitive relationships among the goals.
4. Determine the feasible set of desirable goals.
5. Express the goals as a single objective function, and design the problem formulation using the appropriate constraints.

In selecting the weights for the goals, the need to determine the intensity of desire (priority) for each goal is important (Dyer et al., 1979). This, in essence, represents the "satisficing" policy of the landowner, or the landowner's attempt to implement adequately a plan that meets multiple criteria rather than implementing a plan with a single objective. Often, when weights are difficult to determine, but precedence of the goals is more easily stated, lexicographic (priority-based) goal programming is recommended.

Even though goal programming could be used to address diverse considerations within the objective function to a problem, we will demonstrate it here using the problem framework we have been building in this chapter. In advancing the integer programming problem, assume that the owner of the Putnam Tract continues to be interested in producing timber volume, yet wants more in each time period than what was suggested could be produced with integer programming (about 12,000 cords per time period). In our problem, harvest volumes currently are being accumulated using accounting rows. After setting a target harvest level (12,500 cords, or approximately 30,000 m^3 if one assumes a cord (128 ft^3 of stacked wood) contains about 2.4 m^3 of stacked, not solid, wood), deviations from the target harvest levels are noted as additional constraints to the model.

$$H1 + NDevH1 - PDevH1 = 12,500$$
$$H2 + NDevH2 - PDevH2 = 12,500$$
$$H3 + NDevH3 - PDevH3 = 12,500$$

where:

$NDevHt$ = negative deviations from the right-hand side of the equation during time period t, in cases where the right-hand side volume of the equation cannot be scheduled

$PDevHt$ = positive deviations from the right-hand side of the equation during time period t, in cases where the right-hand side volume of the equation is exceeded

The objective function subsequently is designed to minimize the deviations among harvest levels, yet the deviations are given weights (w) that reflect their importance to the landowner.

Minimize $w_1 NDevH1 + w_2 PDevH1 + w_3 NDevH2 + w_4 PDevH2 + w_5 NDevH3 + w_6 PDevH3$

Assume that the landowner has thought about the relative importance of the goals, and has assigned weights to them as follows:

$w_1 = 0.35$
$w_2 = 0.15$
$w_3 = 0.20$
$w_4 = 0.10$
$w_5 = 0.15$
$w_6 = 0.05$

These weights suggest that the landowner is most concerned with a negative deviation from the harvest target during time period 1 (weight = 0.35), and least concerned about the positive deviation from the harvest target during time period 3 (weight = 0.05). Stated another way, the landowner is most concerned about not meeting the volume target in the first time period, and least concerned with exceeding the harvest target in the third time period.

In developing the problem formulation, the integer programming objective function was replaced with:

```
Minimize 0.35 NDevH1 + 0.15 PDevH1 + 0.20 NDevH2 +
0.10 PDevH2 + 0.15 NDevH3 + 0.05 PDevH3
```

In addition, the constraints that limited harvest volume deviations from one period to the next were replaced with:

```
347) H1 + NDevH1 - PDevH1 = 12500
348) H2 + NDevH2 - PDevH2 = 12500
349) H3 + NDevH3 - PDevH3 = 12500
```

The forest plan developed using this approach suggests that 12,493 cords can be obtained in time period 1, 12,408 cords in time period 2, and 11,971 cords in time period 3. The location of the scheduled harvests during time period 1 is illustrated in Fig. 8.3. If we were to readjust the weights so that they are basically opposite of the first set,

$w_1 = 0.05$
$w_2 = 0.15$
$w_3 = 0.10$
$w_4 = 0.20$
$w_5 = 0.15$
$w_6 = 0.35$

then we are emphasizing positive deviations in the third time period, and attempting to locate a plan that

FIGURE 8.3 First period harvest units on the Putnam Tract using goal programming methods.

meets or exceeds 12,500 cords in this time period. The resulting schedule of harvests suggests that 11,324 cords can be obtained during time period 1, 12,463 cords during time period 2, and 12,502 cords during time period 3.

Had we developed the relationships that accounted for changes in habitat suitability, recreational opportunities, or any other number of nontimber values, we could have used these in a goal programming framework to explore the trade-offs associated with a landowner's preferences. As you might imagine, the ability to insert several goals into the objective function may be more appealing to some landowners or managers than the traditional approach of using a single objective, and using resource or policy constraints to accommodate the attainment of other goals.

III. BINARY SEARCH

Binary search is a process that attempts to find a solution to a problem by making progressively better attempts at determining the optimal value of the objective function. For example, if we were interested in locating the forest plan that could produce the highest harvest level over time, we would use a binary search to determine the harvest level, and after doing so, either increase or decrease it, then try again. In other cases, a solution (harvest level) that is too high becomes the upper limit of the range of values to consider; and yet in other cases a solution that is too low becomes the lower limit of the range of values to consider. In the case of forest planning, the answer to the harvest level solution is used to determine whether to increase or decrease the proposed harvest level. In sum, binary search is a method for locating a value using a selection process that closes in on a solution by using targets that represent mid-points of classes. A basic binary search process works as follows:

1. Set a target value.
2. Determine the range of potential solution values in an ordered list.
3. Select a value representing the mid-point of the range.
4. Compare the selection with the target value, and determine whether it is greater than, less than, or equal to it.
5. Make a decision:
   a. If the selection is greater than the target, then reduce the target by one-half the range of values.
   b. If the selection is less than the target, then increase the target by one-half the range of values.
   c. If the selection equals the target, then stop and report the solution.
6. Return to step 1 if necessary.

As you can see, when searching through an ordered list for the correct solution, the range of solution values becomes progressively smaller due to the guesses being positioned at the mid-point of the range. In effect, with each guess, the range of solution values is reduced by one-half until the correct value has been located. As a result, the search process narrows the range of possible values by a factor of two with each choice made.

Example

Pick a whole number between 1 and 1,000. By using a binary search process, we can guess your number in less than 10 or 11 tries with a simple answer to a question associated with each guess—was the guess higher or lower than your chosen number? For example, assume you picked 732 as your number. The binary search process for locating your number would work as shown in Table 8.1. With each guess, the remaining solution space is divided in half, and the mid-point is selected in the subsequent attempt.

In natural resource planning, a binary search process that uses a simple harvest volume target and either stand-level volumes or strata-based volumes is relatively easy to perform. This type of planning model can be implemented within a spreadsheet environment if the attributes of the stands or strata that contribute to the objective function are available. Sorting the list and selecting the stands or strata for harvest is straightforward. This assumes that stands or strata can be assigned fractional values related to the harvesting decision, thus some are scheduled for harvest during more than one time period. When stands need to be modeled using integer decision variables, and when the adjacency of harvests must be recognized and accommodated, the use of binary search becomes more complex. In this case, the harvest levels will not be as even, and attention must also be paid to the relationship of each stand to others that have already been scheduled for harvest.

In forest management and planning, binary search traditionally has been used to determine the highest timber volume that can be produced over the assumed time horizon. However, a number of caveats are associated with the typical binary search process. These include modifications to the choice(s) made, the need for a sequential assessment by time period, and the need for a sorted list of choices based on a scheduling rule. For example, assume that the challenge is to determine the highest and most even timber volume possible each year, over a 5-year time horizon.

1. Since an individual stand will likely not produce as much volume as is necessary to achieve the highest even volume from a forest, stand-level harvest decisions are aggregated. It may take harvesting several stands in each year to produce the harvest level necessary.
2. The scheduling of individual stands begins with the first time period (in this case, year 1). Harvests are

TABLE 8.1 An Example Binary Search Process Locating a Randomly Selected Number Between 1 and 1,000

Guess Number	Range of Feasible Values	Guess	Correct?	Your Number Is	Decision
1	1–1,000	500	No	Higher	Range is 501–1,000
2	501–1,000	750	No	Lower	Range is 501–749
3	501–749	625	No	Higher	Range is 626–749
4	626–749	688	No	Higher	Range is 689–749
5	689–749	719	No	Higher	Range is 720–749
6	720–749	735	No	Lower	Range is 720–734
7	720–734	727	No	Higher	Range is 728–734
8	728–734	731	No	Higher	Range is 732–734
9	732–734	733	No	Lower	Range is 732–732
10	732–732	732	Yes	–	–

scheduled until the last one exceeds the harvest target for the year. Subsequently, harvests for the next time period are scheduled. This process continues until harvests have been scheduled for each time period, or we run out of harvesting options because no other stands are old enough to harvest.

3. The selection of harvest units is from a sorted list of harvesting options. If we need to schedule multiple harvests during each time period, then we need a rule that guides us in the selection of stands. For example, stands can be sorted by age, and the rule used could be stated as "oldest first." Alternatively, stands could be sorted by net present value, value growth percent, or volume per unit area. In the case of sorting by net present value or volume per unit area, the rule used could be "highest first." In the case of value growth percent, it might be "lowest first." In any event, as a planner, the sorting process and the selection rule should be considered carefully.

Once these decisions have been made, a binary search process for determining the highest and most even timber volume that can be produced over the time horizon can be developed. How this would work requires the following steps:

1. The target harvest volume is selected, perhaps in a subjective manner by the planner (e.g., 5,000 cords per year).

2. The selection of the initial increment also is made subjectively by the planner (e.g., increase or decrease the target volume 100 cords per year).

3. Stands are scheduled for harvest until the target volume has been reached. This may require scheduling for harvest a number of stands in each time period,

using the chosen rule for selecting stands from the sorted list of stands.

4. Once harvests have been scheduled for all time periods, sequentially from year 1 to year 5, an assessment of whether the target volume has been reached is made.

5. If the target harvest volume has been successfully scheduled for each time period within the time horizon, then the target volume increases, the schedule is eliminated, and the process begins anew. The increment may then need to be adjusted (divided in half).

6. If the target harvest volume has not been scheduled successfully in any one (or more) of the time periods within the time horizon, then the target volume decreases, the schedule is eliminated, and the process begins anew. As with step 5, the increment may then need to be adjusted (divided in half).

7. When the increment becomes too small, the binary search process stops, and the last feasible forest plan is reported.

One of the challenges to using this type of sequential harvest scheduling process is that the increment for increasing or reducing the target volume should not be divided in half until after a change in the direction of target harvest levels has been determined. Stated another way, if the target harvest volume is initially lower than the productive capacity of the forest, then the increment should not change in size until the scheduled timber volume is less than the target the first time (i.e., the target is not met for the first time). If the target is initially too high, then the increment should not change until the scheduled timber volume is greater than the target the first time (i.e., the target is met for the first time). After this point, the increment decreases by one-half of the previous level whether the target is met or not. Our desire is to search in this manner:

Attempt	Target	Target Met?	Increment
1	5,000	Yes	100
2	5,100	Yes	100
3	5,200	Yes	100
4	5,300	Yes	100
5	5,400	No	50
6	5,350	No	25
7	5,325	Yes	12
8	5,337	Yes	6
9	5,343	No	3
10	5,340	Yes	1

rather than in this manner:

Attempt	Target	Target Met?	Increment
1	5,000	Yes	100
2	5,100	Yes	50
3	5,150	Yes	25
4	5,175	Yes	12
5	5,187	Yes	6
6	5,193	Yes	3
7	5,196	Yes	1
8	5,197	Yes	–

In the first example, the target increases by 100 units through the first four attempts because the direction of the target harvest level did not change (it kept increasing because the target harvest level was met successfully in each of the first four attempts). Once the direction changed, at attempt 5, the increment began to decrease by one-half of the previous attempt's increment. In the second example, the increment changes by one-half of the previous attempt's increment from the start of the search process, even though the direction of the target harvest level did not change (as with the previous example, it kept increasing because the target harvest level was met). In doing so, this process leads to a poor solution to the planning problem.

To illustrate the use of binary search in a natural resource management problem, assume that the owner of the Putnam Tract has heard about binary search, and is interested in the technique since it can use a rule they have been using for some time: harvest the oldest stands first. If there are two or more stands with the same age,

then they will be further sorted by the amount of volume within each stand, and the one with the greatest amount of volume will be scheduled first. Since we found when using the mixed integer programming problem formulation that the harvest levels could be approximately 12,500 cords per time period, we will subjectively set the initial target a little lower, to a volume of 12,400 cords, and subjectively assume that the increment will be 500 cords. Initially, we will consider the scheduling of fractional portions of stands to illustrate the capability of the process. Complete stands will be assumed scheduled for harvest in a subsequent analysis. At the conclusion of scheduling each of the three time periods, a decision will be made: increase the target volume, or decrease the target. In addition, once the target volume needs to be decreased (i.e., it changes direction), the increment will be reduced by one-half after each iteration of the search regardless of whether we increase or decrease the target volume. Eventually the increment will become very small, say 10 or less cords, and we will decide to stop the search process and report the last feasible solution that was located.

When allowing the scheduling of fractional portions of stands, we attempted to schedule 12,400 cords per time period (Table 8.2), and found that we were successful in scheduling this volume for all three time periods. For the second attempt, the harvest schedule was first wiped clean, and then the target volume was raised 500 cords to 12,900 cords. Here, we found that we could not schedule enough volume during time period 3, and as a result, we lowered the target by one-half of the previous increment (250 cords). Since we know that the solution is somewhere between 12,400 cords and 12,900 cords, it makes sense next to attempt to schedule 12,650 cords, the midpoint of the current range of solution values. When attempting to develop a forest plan using 12,650 cords at the target, we find again that time period 3 falls short of volume, so we reduce the target by one-half of the previous increment (125 cords), to 12,525 cords per time period. This new target is the mid-point of the shrinking range of feasible solutions (now 12,400 cords to 12,650 cords). This process continues until the increment is very small, and we decide to stop. The best schedule, using fractional stands, involves a harvest target of 12,509 cords per time period.

When we use whole stands to develop the harvest schedule, we find that a target of 12,400 cords results in a volume scheduled for time period 3 that falls short of our target (Table 8.3). After reducing the target volume to 11,900 cords, we find that the target volume is less than what can be produced from the Putnam Tract, so the increment begins to be divided by one-half with each subsequent iteration, no matter whether the target can be achieved or not. As the increment gets smaller and

TABLE 8.2 Binary Search Results for the Putnam Tract Using Fractional Portions of Stands

Attempt	Target (cords)	Scheduled Volume (cords)			Target Met?	Increment
		Time Period 1	**Time Period 2**	**Time Period 3**		
1	12,400	12,400	12,400	12,400	Yes	500
2	12,900	12,900	12,900	11,595	No	250
3	12,650	12,650	12,650	12,155	No	125
4	12,525	12,525	12,525	12,483	No	64
5	12,461	12,461	12,461	12,461	Yes	32
6	12,493	12,493	12,493	12,493	Yes	16
7	12,509	12,509	12,509	12,509	Yes	8
8	12,517	12,517	12,517	12,499	No	—

TABLE 8.3 First Iteration of a Binary Search Process for the Putnam Tract, Scheduling Oldest Stands First

Stand	Age	Area (acres)	Available Volume (cords)			Scheduled Volume (cords)		
			Time Period 1	**Time Period 2**	**Time Period 3**	**Time Period 1**	**Time Period 2**	**Time Period 3**
12	47	62.21	1,947.24	2,121.43	2,332.95	1,947.24		
81	45	85.56	2,498.21	2,746.32	3,028.65	2,498.21		
2	45	84.53	2,476.58	2,738.61	3,026.00	2,476.58		
25	44	26.49	752.17	826.33	916.38	752.17		
7	44	23.28	677.36	768.14	842.63	677.36		
69	43	5.32	153.60	172.21	186.56	153.60		
76	40	42.72	1,187.62	1,358.50	1,503.74	1,187.62		
79	40	35.16	966.82	1,110.96	1,230.50	966.82		
26	40	4.18	112.00	128.30	141.67	112.00		
31	35	81.23	2,112.03	2,371.97	2,631.92	2,112.03		
62	35	17.28	428.52	489.00	539.10		489.00	
56	35	5.01	114.77	134.32	150.86		134.32	
58	34	29.25	532.40	628.94	719.62		628.94	
59	34	30.15	503.57	591.02	690.53		591.02	
57	34	20.27	350.60	411.40	474.22		411.40	
61	33	34.38	546.66	649.80	766.70		649.80	
60	33	18.03	295.68	351.57	409.26		351.57	
51	32	25.55	439.51	516.17	585.16		516.17	
74	30	59.24	1,321.10	1,617.31	1,795.03		1,617.31	
37	29	16.09	350.85	407.18	458.68		407.18	
30	28	109.84	2,251.68	2,756.93	3,075.46		2,756.93	

(Continued)

TABLE 8.3 (Continued)

Stand	Age	Area (acres)	Available Volume (cords)			Scheduled Volume (cords)		
			Time Period 1	Time Period 2	Time Period 3	Time Period 1	Time Period 2	Time Period 3
19	28	7.28	146.23	181.15	204.43		181.15	
29	27	19.94	386.84	478.56	532.40		478.56	
70	26	63.69	1,210.02	1,522.07	1,694.02		1,522.07	
27	26	41.80	764.94	969.76	1,095.16		969.76	
21	26	29.21	555.03	677.72	762.43		677.72	
71	26	25.72	473.23	599.25	676.41		599.25	
28	26	16.95	316.93	396.58	445.73			445.73
...	...	...	...	...	...			...
9	15	12.62	—	—	219.62			219.62
8	15	31.16	—	—	554.59			554.59
Total						12,883.63	12,982.14	11,589.64

TABLE 8.4 Binary Search Results for the Putnam Tract

Attempt	Target (cords)	Scheduled Volume (cords)			Target Met?	Increment
		Time Period 1	Time Period 2	Time Period 3		
1	12,400	12,884	12,982	11,590	No	500
2	11,900	12,884	12,383	12,266	Yes	250
3	12,150	12,884	12,383	12,266	Yes	125
4	12,275	12,884	12,383	12,266	No	64
5	12,211	12,884	12,383	12,266	Yes	32
6	12,243	12,884	12,383	12,266	Yes	16
7	12,259	12,884	12,383	12,266	Yes	8
8	12,267	12,884	12,383	12,266	No	

smaller, the scheduled volumes change very little, and the results are not too appealing (Table 8.4), since scheduling an entire stand during a single time period results in a significant shift in volumes. The location of the activities scheduled for time period 1 are illustrated in Fig. 8.4, which represents a plan somewhat different than when using the other scheduling techniques.

In practice, either whole stands are scheduled for harvest or fractional parts of stands, as in the case of linear programming. With the scheduling of fractional parts of stands, you will find that one stand will have its schedule split between each adjacent time period (time periods 1 and 2, and time periods 2 and 3 in our previous examples). This splitting of stands is necessary to meet the target volume exactly in each time period. Our rule of scheduling the oldest stand first could be modified so that the highest volume per unit area stands were harvested first, or so that the lowest value growth percent stands were harvested first. Our sub-rule determining which stands to schedule in the event of a tie also could have been modified to use the highest volume per unit area stands or the lowest value growth percent stands. Undoubtedly, a different plan of action would be generated using these alternatives.

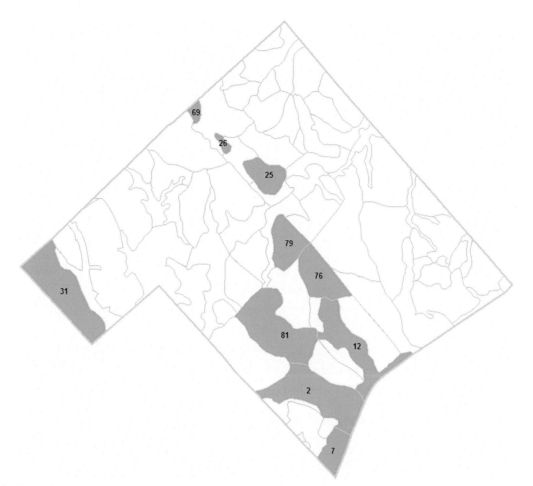

FIGURE 8.4 First period harvest units on the Putnam Tract using binary search methods.

IV. HEURISTIC METHODS

Heuristic solution methods use logic and rules of thumb to arrive at feasible and high-quality solutions to complex planning problems. There are generally two reasons why people would select a heuristic as their planning tool: (1) they desire to incorporate into the planning process quantitative relationships that are not easily described through linear equations (thus eliminating the use of linear programming to solve the problem), and (2) they want a solution to a complex problem to be generated very quickly. Heuristic methods have some unique qualities that sometimes draw the ire of pure mathematicians. First and foremost, there is no guarantee that the optimal solution to a problem will be located with a heuristic. Often there is no statement to the quality of the solution as the evaluation of the quality of results generated from any one heuristic generally requires a comparison to results generated from linear, mixed integer, or integer programming methods. Second, many of the relationships that can be quantified and recognized in a heuristic are difficult to describe with linear or even nonlinear equations. For example, the stream sediment evaluation rules that have been used in the Intermountain United States (US Department of Agriculture, Forest Service, 1981) involve a number of relationships that are best represented using computer programming logic (e.g., if-then-else statements, for-next loops, etc.) rather than equations. These and other types of complex natural resource evaluations can easily be incorporated into heuristics, but cannot easily be incorporated directly into linear programming models nor its extensions without simplifying the quantitative relationships.

A number of heuristic methods have been demonstrated for use in natural resource management, including Monte Carlo simulation, simulated annealing, threshold accepting, tabu search, and genetic algorithms. Some natural resource management organizations have hired planners specifically to design and implement these methods. Commercial or public domain software that use heuristic methods to address natural resource management and planning problem is limited to a few examples, some of which we provide toward the end of this chapter. One of the advantages of heuristic methods, that it can accommodate complex functional relationships, is also one of the reasons for the limited number of available software products. Heuristics commonly are developed to address an individual organization's planning problem, and to utilize their data format and structure. The bottom line is that many natural resource management organizations want a heuristic that can solve their specific problem. As a result, the applicability of the problem-solving technique to other natural resource management organization's problems can be challenging.

A. Monte Carlo Simulation

Monte Carlo simulation, which includes a large number of sampling techniques, relies on random samples from a population in the effort to develop a natural resource management plan. The name of this technique arose about 70 years ago in a reference to the famous gambling casino in Monaco. Monte Carlo simulation methods have been used in physics, finance, chemistry, and other fields where modeling complicated interactions among variables is necessary. To utilize a Monte Carlo heuristic in natural resource management planning, you would need to define an objective function (e.g., maximize net present value), then develop the set of choices from which you will pick to create a plan of action. These choices could be discrete management regimes that can be applied to individual stands. Stand 1, for instance, could have five different options (one being the stand-level optimum solution), whereas stand 2 could have seven. In any event, the choices are best described with integer decision variables. If the decision variables related to stand 1 were $S1R1$, $S1R2$, $S1R3$, $S1R4$, and $S1R5$, where $S1$ represents stand 1, and $R1$ represents regime 1, then a portion of a randomly defined feasible solution could have become:

$$S1R1 = 0$$
$$S1R2 = 0$$
$$S1R3 = 1$$
$$S1R4 = 0$$
$$S1R5 = 0$$

which indicates that management regime 3 has been chosen randomly for stand 1, and the other four regimes have not. This set of choices would likely have been constrained by the following relationship:

$$S1R1 + S1R2 + S1R3 + S1R4 + S1R5 = 1$$

Given this, should another choice for stand 1 be selected randomly, it would replace management regime 3 in the forest plan. As such, the relationship between management regimes indicates that only one of them can be chosen for a stand. The constraint could have utilized the logical operator less than or equal to ($\leq$) to suggest that none of the available management regimes could be scheduled for the stand. This null choice reflects doing

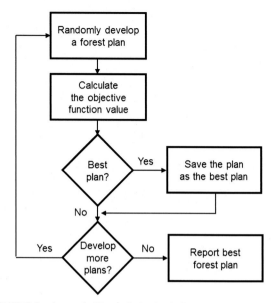

FIGURE 8.5 A generic Monte Carlo simulation process.

nothing in a stand during the time horizon, which may be a necessary alternative in heavily constrained planning problems that require some stands to be assigned no treatment to maintain feasibility.

Once the set of potential choices for each stand has been defined, a Monte Carlo heuristic would randomly select choices for the decision variables, and test their ability to provide a feasible solution (Fig. 8.5). Therefore, in this and other heuristics, some logic must be developed to test for violations of constraints. If an infeasible solution results with the inclusion of a random choice into a plan of action, then the choice generally is ignored, but placed back into the sample set of potential choices for possible selection later (if conditions among the variables change during the solution process). A number of randomly generated forest plans are assessed in this manner. The heuristic remembers the best forest plan developed during the search process (as measured by the associated objective function value), and stops after some predefined number of randomly developed forest plans. The quality of the best solution is gauged by the contributions of each randomly selected choice to the objective function value. Clements et al. (1990) and Nelson and Brodie (1990) were among the first to describe the use of Monte Carlo simulation in forest management and planning.

B. Simulated Annealing

Annealing is the process of the cooling of metal after it has been raised to a very high temperature. The hot material rearranges on its own and forms an optimal configuration as it cools (anneals) directly affecting the strength of the metal. The physical manner in which metal cools is the basis of the search process called "simulated annealing." The use of simulated annealing as a search process was suggested over 60 years ago (Metropolis et al., 1953), yet its potential has been explored by forest planners only within the last 30 years. A simulated annealing process used in natural resource management begins with a feasible forest plan, generated either randomly or using a deterministic method (e.g., schedule the best stand-level optimum decisions that do not violate broader forest-level goals). The simulated annealing search process then modifies the forest plan one aspect at a time. For example, a management regime scheduled for a single stand may be changed and the associated effects assessed.

Three parameters need to be defined prior to using a simulated annealing algorithm: how long to run the model, the initial temperature level, and the cooling schedule. The initial temperature is usually some relatively high value that is chosen based on initial trial runs of the model, so that most of the initial choices are made with almost a 100% probability of acceptance. The initial temperature is also usually case-specific, and determined through trial-and-error. The cooling schedule determines how the initial temperature will decline with each change to a solution. How long the model should be run can depend on how small the temperature becomes. In many instances, the search process stops when the temperature reduces to 1, 5, or 10 degrees. In other cases, the search process stops when it has run as long as the user had specified initially. As you may have gathered, arriving at the appropriate parameters for a simulated annealing search process is somewhat of an art.

For forest management purposes, generally a single stand is chosen at random (Fig. 8.6), and an alternative management regime is randomly assigned to the stand. How this change in the forest plan affects the overall quality of the forest plan is assessed by calculating the proposed objective function value. If the change in the forest plan results in a higher quality plan, then it is always acceptable. If the change in the forest plan results in a lower quality forest plan, then it, too, may be acceptable under one condition, if the result of:

$$EXP\ (-\ (proposed\ objective\ function\ value\ -\ best\ objective\ function\ value)/temperature)$$

is larger than a randomly drawn number between 0 and 1. As the temperature gets smaller, the result of the calculation of this acceptance criterion draws closer to zero. Therefore, the chance of accepting a change that results in a lower quality forest plan diminishes as the number of changes to the plan increase. This heuristic, as well as others, allows changes to a plan that decrease the value of the plan, to accommodate free movement around the solution space. In contrast to linear programming, these types of heuristics do not move around the corners

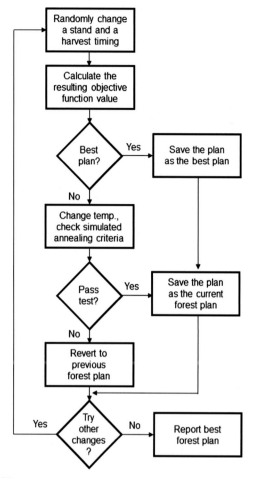

FIGURE 8.6 A generic simulated annealing process.

of the solution space. They simply test alternative solutions within the solution space, and hopefully move the solution toward the optimum. Though the association with the cooling of metal may be worrisome to some forest planners, this heuristic has been shown to provide very good solutions to complex forest planning problems (Bettinger et al., 2002; Öhman and Eriksson, 2002).

C. Threshold Accepting

Threshold accepting, introduced by Dueck and Scheuer (1990), is a heuristic search process that is similar to simulated annealing. An initial forest plan is developed (randomly or otherwise), and changes are made to the forest plan randomly as well. When a potential change results in a lower quality forest plan, the manner in which this heuristic evaluates whether to accept the lower quality solution represents the distinct difference between this heuristic and simulated annealing. In threshold accepting, any lower quality solution within a given threshold of the best solution is deemed acceptable during the search process. An initial threshold must be provided by the user, as well as a rate of change. In the case of maximizing net present value, for example, you may indicate that the initial threshold is $1,000. This suggests that at the beginning of the search process, any forest plan developed that is not as good as the best forest plan found thus far during the search, yet is within $1,000 of the best forest plan, will be deemed an acceptable move to make. Of course, like simulated annealing, the threshold gets tighter as the search process progresses, using the rate of change that is provided. At some point the threshold will become so small that only marginally worse solutions will be deemed acceptable moves to make. From this point forward, most of the alterations to a forest plan will involve changes that lead to higher quality solutions. This type of search process is more intuitive than simulated annealing; however, studies suggest that the two heuristics produce about the same high quality results to complex forest planning problems (Bettinger et al., 2002).

D. Tabu Search

Tabu search is a deterministic heuristic introduced by Glover (1989, 1990). In contrast to the previous heuristics, there are no random aspects within a general tabu search algorithm (Fig. 8.7). The name of the process infers that some of the decisions become taboo (off limits) during the development of a forest plan. Imagine a forest plan that initially was developed randomly or by some other means. Tabu search would evaluate all the potential changes to the plan (the different management regimes that could be assigned to each stand), then select the best choice from this set. To some, this may seem to be a more rational process for developing a plan of action. This examination of all potential changes also represents a distinct difference between tabu search and the other heuristics, where only a single potential change is considered. The selected change to the plan is made, and subsequently that same choice cannot be considered again until X number of other choices has been made. The value X is the length of time (number of choices) that a choice is off-limits (taboo). This tabu state must be defined by the user of the algorithm, generally after examining a few trial runs of the model. A short tabu state results in a cycling of forest plans and the revisiting of the same number of plans over and over again (not a desirable search process). Imagine a search rotating around the solution space, revisiting the same plans every X (or so) iterations of the model. Longer tabu states force the heuristic to explore more diverse options within the solution space, hopefully allowing the search to move toward the optimal solution.

There is only one caveat to considering a decision taboo: if a choice is unavailable (taboo), yet selecting it again before this distinction is removed will lead to a solution unlike (better) any other solution found thus far in a

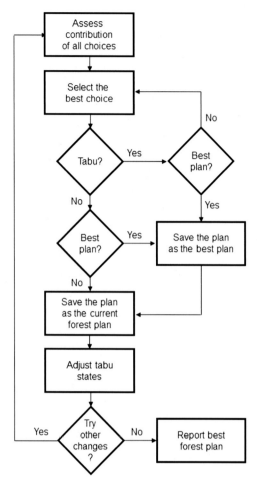

FIGURE 8.7 A generic tabu search process.

search, then the choice will be selected, effectively overriding the discriminatory factor. Standard tabu search is an effective heuristic for complex forest planning problems; however, enhancements may be necessary to allow the development of solutions that rival those developed by simulated annealing or threshold accepting. See Bettinger et al. (2002, 1997), Caro et al. (2003), and Richards and Gunn (2000) for a more detailed description of tabu search applications in forest management and planning.

E. Genetic Algorithms

Genetic algorithms initially were described by Holland (1975) as a way to search for an optimal solution to a problem in a manner similar to the way that the DNA of parents are combined to create children. A genetic algorithm heuristic search process (Fig. 8.8) begins with the development of a population of parents (forest plans). Imagine 200 randomly defined forest plans for a single property as the population of "parents." From this population two forest plans are selected, either randomly or based on their fitness (their objective function value in the most basic case). In some cases, one forest plan may

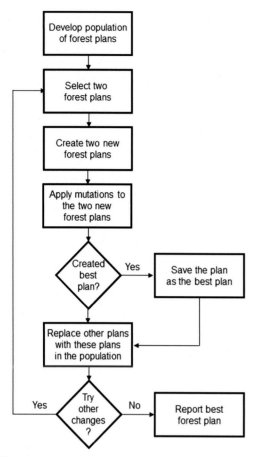

FIGURE 8.8 A general genetic algorithm search process.

Plan #2

Stand	Harvest Period
1	3
2	0
3	2
4	3
5	1
6	2
7	1
8	1

Each of these plans could represent the harvest timing for eight stands over a three-period planning horizon. A genetic algorithm can be designed to transfer some aspect of one plan to the other, and vice versa. For example, assume that a cross-over point was chosen randomly to be located just in front of stand 3 in the plans noted previously. At this point, the two plans are split, and recombined to form two "child" plans:

Child A

Stand	Harvest Period
1	1
2	2
3	2
4	3
5	1
6	2
7	1
8	1

be selected randomly and the other selected because it is the highest quality plan within the population of plans. In other cases, each parent may be selected based on a probability of selection that, in turn, is based on the fitness of each plan. Once the two parents are selected, some portion of their DNA is swapped. Inspect for a moment the two forest plans shown here:

Plan #1

Stand	Harvest Period
1	1
2	2
3	0
4	3
5	1
6	3
7	2
8	1

Child B

Stand	Harvest Period
1	3
2	0
3	0
4	3
5	1
6	3
7	2
8	1

The two children are then tested for feasibility, and their potential objective function values are assessed. At this point a number of options have been used in genetic algorithms to address the need to incorporate high quality children in the population of solutions. One or both of the two children, assuming they result in feasible solutions, could be placed back into the population. If only one child is assumed to survive (is feasible), then it is the higher quality of the two. One or more parents are then removed from the population, either randomly or based on their fitness (i.e., the lower quality forest plans may have a higher chance of being removed).

The forest plans within the population are then said to evolve with each iteration of the genetic algorithm search process. In theory, portions of a forest plan may be nearly optimal, and in swapping portions of plans, we might find a resulting combination that has a very good overall plan of action. To better resemble biological processes, the resulting children can also be mutated after they are created. Here, random changes are interjected into each child plan at a very low rate. One disadvantage of genetic algorithms is that splitting and recombining parents generally leads to a high number of infeasible solutions when clear-cut adjacency constraints are involved; therefore, some cases may warrant swapping only a small amount of DNA among parents. A number of biologists and ecologists prefer genetic algorithms over other heuristic search techniques, particularly for addressing reserve design problems. Falcão and Borges (2001) and Lu and Eriksson (2000) provide a more detailed description of genetic algorithm applications in forest management and planning.

F. Other Heuristics

As we have inferred, heuristics are useful for the development of forest plans that involve spatial objectives, or for other management planning problems that have nonlinear or complex evaluations that must be accommodated simultaneously with the scheduling of management activities. We have described the basic underpinnings of several basic heuristics that have been used in forest planning efforts, but a number of other heuristics exist as well, far too many to cover in this text. These include the HERO algorithm (Pukkala and Kangas, 1993), ant colony optimization (Zeng et al., 2007), the raindrop method (Bettinger and Zhu, 2006), and others. The various fields related to operations research have provided numerous other examples of heuristics in attempts to optimize production processes, streamline facilities and their management, and reduce costs associated with infrastructure development. Some advancements will likely filter over to natural resource management and planning as time moves forward.

V. FOREST PLANNING SOFTWARE

When developing linear, mixed integer, or integer programming forest planning problems, the problem formulation can be developed as a set of equations and solved using a number of commercial software programs (CPLEX, Gurobi, Lingo, etc.). The format of the problem can be designed as a set of equations, much like the examples we provided in this and earlier chapters, or can be designed using another format, such as the mathematical programming system format. For small to moderate planning problems, a matrix can be described in a spreadsheet, and a spreadsheet solver can be used to determine optimal solutions to linear problems. In many planning efforts, however, you may prefer to use an optimization software package designed specifically for natural resource management planning problems. Since natural resource management problems vary considerably in the scope of resources and goals recognized, and since the time and cost to develop such a model (and to keep it current given changes in computer technology) can be considerable, only a few of these are widely used in North America. We describe in the following sections four models that might be of value to forest management and planning efforts.

A. Spectrum

The Spectrum model was developed by the US Forest Service Ecosystem Management Analysis Center in Ft. Collins, Colorado, and the Rocky Mountain Forest and Range Experiment Station. Spectrum is an outgrowth of the FORPLAN model (Johnson et al., 1986) that was developed in the late 1970s, and used to assist with the formation of numerous US National Forest plans. Any number of resources (commodity or noncommodity related) can be modeled within Spectrum, as long as coefficients can be provided to link the response of the resources to management activities or to natural growth of forests. Spectrum includes linear programming, goal programming, and mixed integer programming modeling processes, as well as a simulation process for modeling natural disturbances. The objective function for a linear programming problem could involve maximizing or minimizing an outcome, or minimizing the over- or under-achievement of a goal. Goal programming formulations would involve minimizing the over- or under-achievement of several goals. A number of resource constraints, policy constraints, or accounting rows can be designed to control the scheduling of activities, or accumulate resource-related information. Spectrum is used primarily for strategic planning, although the incorporation of mixed integer programming techniques suggests that some tactical planning issues may be addressed as well.

These levels of planning processes are described in more detail in Chapter 13, Hierarchical System for Planning and Scheduling Management Activities. The management considerations that the model is well suited to address include analyses of land or resource allocation, the sustainability of resources, desired future conditions, and broad-scale natural resource management policies. Spectrum can be obtained free of cost through a US Forest Service Internet site.

B. Habplan

Habplan is a forest planning model that was developed by the National Council for Air and Stream Improvement (NCASI Statistics and Model Development Group, 2006), and is based on a process closely related to the simulated annealing heuristic. The planning model simulates forest development and management associated with both even-aged and uneven-aged stands, and a large number of management regimes for stands can be accommodated. The objective function within Habplan can consist of a single goal; however, generally the objective function is designed as multiobjective problem in much the same way as a goal programming problem. The model is built around flow components, which control the achievement of an output that is associated with the scheduling of a management action to a stand. Weights can be applied to the flows described in the objective function to allow one or more goals to have greater influence on the others. Harvest area sizes can be controlled by maintaining them within some upper and lower size limits. Achievement of harvest size goals can be incorporated into the objective function as a flow. Biological components can be specified to favor the assignment of management regimes for stands to best meet a landowner's objectives.

Habplan uses a Model I formulation, where stands are tracked throughout the length of the time horizon, and are not aggregated after harvest into a new age class or strata (as in Model II). In addition, Habplan contains a linear programming function that allows the development of a solution for a relaxed (nonspatial) problem with which to compare. And, a direct link to GIS is provided to allow the presentation of results. Habplan can be obtained free of cost through a National Council for Air and Stream Improvement Internet site.

C. Magis

The Multiple-resource Analysis and Geographic Information System (MAGIS) is a forest planning model developed by the US Forest Service, Rocky Mountain Research Station in Missoula, Montana, to accommodate the scheduling of management activities for forest stands and roads. The objective function could be designed to maximize or minimize a goal, and the plan developed could be constrained by a number of factors, such as those related to wood flows, areas harvested, types of management activities allowed, amount of nontimber products produced, or allocation of land into structural classes. Road management issues can be used to constrain scenarios developed in MAGIS. Stands are grown on "pathways," or transitions, which are related to natural succession, natural disturbances, or management activities.

The volume within a stand is modified using an annual growth percentage increment that is a function of the state of the stand and the pathway on which it travels. Scenarios can be designed to direct the structure of a forest toward a mixture of forest types while minimizing the cost of the plan. Fixed and variable costs for road maintenance are recognized given the likely path of logging trucks from each harvested stand. The express version of MAGIS uses a combination of simulated annealing and a network algorithm, and is meant primarily for tactical planning, where the focus is on timber harvests and access considerations. The professional version of MAGIS utilizes a mixed integer programming technique, and is meant for broader-scale analyses, where the incorporation of nontimber goals and the measurement of landscape-level effects are performed. MAGIS is linked directly to GIS to facilitate input processes (stand polygons and road networks), and to facilitate the display of outcomes related to each forest plan scenario. As with Spectrum, MAGIS can be obtained free of cost through a US Forest Service Internet site.

D. Remsoft Spatial Planning System (Woodstock/Stanley)

One widely used commercial forest management and planning software package is the Remsoft spatial planning system (RSPS) developed by Remsoft, of Fredericton, Canada. This software suite allows forest managers to develop user-defined detailed short- and long-term forest management plans. The RSPS software includes Woodstock, Spatial Woodstock, and Stanley. Woodstock allows a forest manager to build a problem formulation matrix that can be solved using simulation, optimization, or a combination of the approaches. In many natural resource management applications, a management situation is formulated in Woodstock as a linear programming problem. And in general, the problem formulation is represented by a Model I or Model II linear programming problem. An objective function can be designed either to maximize or minimize a quantity; however, an option to use goal programming is available. There is no restriction on the number of constraints that can be specified, and therefore, as with many linear programming problems, the greatest limitations may be associated with the capacity of the solver (number of rows and columns allowed in the detached coefficient matrix) and the computational power of the computer.

Woodstock is a detached coefficient matrix generator (as is Spectrum), and is organized into 12 modules (or sections) to enable a user to fully define and describe a management problem. More appropriate resources are available to facilitate delving deeply into the intricacies of each of the 12 sections. However, a few very brief descriptions seem necessary here to provide a taste of how the system is designed. The "landscape section" is perhaps the most important section of the system, where we would describe the forested landscape that is being evaluated. Themes are defined here, which are similar to land allocations, and current and future development types are described. A "development type" in Woodstock is a strata of the forest that is defined by a specific set of landscape theme attributes. A direct link between a development type and growth and yield information is needed. Woodstock allows the importation of growth and yield data from relevant forest vegetation growth simulators, or the use of specifically designed yield tables, to create deterministic transitions for the management actions identified in the model. The "lifespan section" is where the maximum age of a development type can be defined. The "control section" is where a forest manager can specify parameters such as the number of runs for a Monte Carlo simulation or the tolerance limits for binary search. The "transition section" describes the responses to management actions, which can either be deterministic or stochastic. Finally, any number of reports can be generated as long as they are designed in the "output section." One advantage of using Woodstock is that it allows forest managers to import GIS data containing the land classification themes (land allocations), and it automatically reads the attributes of the GIS database. One limitation is that a maximum of 50 landscape themes can be defined; however there is no limit on the number of attributes that these themes can contain.

Spatial Woodstock is a data management and spatial analysis tool that allows a user to manage the geographic data, and provides a link to GIS software. Spatial Woodstock allows a user to map the results of a forest plan, and in doing so forest managers can analyze and evaluate a plan's impact on the landscape. In addition, Spatial Woodstock allows users to locate adjacency violations, which facilitates an assessment of opening size or green-up violations.

Stanley is a tactical forest planning model that is based on a process similar to Monte Carlo simulation. Stanley acquires a Woodstock solution (generated using linear programming, perhaps), and attempts to allocate the area treated and products produced to specific ground

units (stands). As a result, Stanley modifies Woodstock results to incorporate and address adjacency and green-up constraints. Scenarios that explicitly address adjacency and green-up issues are directly compared with the Woodstock results. The goal is to locate a Stanley solution that minimizes the differences between it and the associated Woodstock results.

E. Tigermoth

Tigermoth (2016) is another commercial software package that incorporates the data entry, solver, and report writing processes into a single software package that does not require a user to purchase a separate license for the solver. The program is able to address and solve both linear and mixed integer forest planning problems. The program utilizes a data input process that resembles a spreadsheet, and utilizes Microsoft SQL database formats. Thus most analysts are able to quickly navigate formulation problems. Results are displayed in both standardized and customized reports that allow for sensitivity analysis of the solutions generated. Tigermoth is often used as part of a due diligence analysis to determine the value of forest plantation resources in Southeast Asia and Oceania.

VI. SUMMARY

Although linear programming continues to be widely used in forest management and planning throughout North America, a number of advanced problem-solving techniques are coming of age as well. The reasons for the adoption of advanced techniques range from the need to assign to management decisions integer values (yes or no, or 1, 2, 3, ... ∞), as well as the need to incorporate nonlinear or complex functional relationships into a planning process. The former requirement is necessary when developing a plan that recognizes the spatial juxtaposition of resources. For example, assigning integer values to decisions is necessary when there is a need to know where harvest units are in relation to one another, or a need to know where foraging and roosting habitat is in relation to each other. The latter requirement is problematic in forest management and planning. If functional relationships are very complex, and not easily transferable to linear equations, then they can be assessed prior to plan development and represented by rough coefficients in a linear model, incorporated directly into a forest plan using a heuristic method, or assessed after a plan has been developed. The first of these three options suggests some compromises will be made in the assessment of an effect, as complex relationships are reduced to rough coefficients. The second of the three options suggests the use of a nonstandard planning method. The third of the three options may be the most inefficient, since constraint violations may be assessed after the generation of a plan, rather than during the generation of a plan.

A variety of operations research methods have been used successfully in the development of forest plans due to the suitability of certain models to specific management situations and the need for efficient management of forest resources (Weintraub and Romero, 2006). Advances in computer software and hardware over the last 30 years also have facilitated the adaptation of modeling techniques to forest management problems. Although the choice of planning process may be guided by broader organizational issues, perhaps the most important factor for natural resource managers is the need for a basic understanding of how forest plans are created, and the advantages and disadvantages of each approach.

QUESTIONS

1. *Mixed integer programming forest management problem.* Acquire the Lincoln Tract data, and formulate and solve a mixed integer planning problem similar to the one described in Section II, Part A. Assume here that the landowner is interested in maximizing the net present value of the plan over the next 15 years (three time periods), yet the harvests scheduled for the first time period should be applied to whole stands. As a result, the decision variables associated with harvests in the first time period should be assigned binary integer values. In addition, assume a minimum harvest age of 35 years (as measured at the beginning of each time period). Solve the mixed integer programming problem, and provide in a memorandum to the landowner the objective function value, the schedule of harvest volumes, and a map that illustrates where the first time period harvests will be placed.

2. *Integer programming forest management problem.* Building on problem 1, now develop integer decision variables for all the harvesting decisions. In addition, constrain the harvests in the first time period such that stands scheduled for harvest do not touch one another. Develop another memorandum for the landowner that describes the objective function value, the schedule of harvest volumes, and a map that illustrates where the first time period harvests will be placed.

3. *Goal programming forest management problem.* Take the fifth scenario from Section IX of Chapter 7, Linear Programming, and rearrange the objective function to minimize the deviations from desired outcomes of net present value and downy woodpecker habitat quality. Use a single accounting row to add up the net present value of the forest plan, based on the previous objective function equation described in Section VIII of Chapter 7, Linear Programming. Then remove the HSI constraints (103−105). The goal for

the net present value is $1,688,646. The goal for the HSI values in each of the three time periods is 0.300. Since these are differently scaled values, each will be weighted differently in the objective function. In the new goal-oriented objective function, weight the positive and negative deviations of the net present value by 0.0001, and weight each of the positive and negative deviations of the HSI values by 25. The new objective function should resemble the following:

```
Minimize
0.00001 NDevNPV + 0.00001 PDevNPV + 25 NDevHSI1 +
25 PDevHSI1 + 25 NDevHSI2 + 25 PDevHSI2 +
25 NDevHSI3 + 25 PDevHSI3
```

The new goal programming constraints should resemble the following:

```
107) NPV + NDevNPV − PDevNPV = 1688646
108) HSI1 + NDevHSI1 − PDevHSI1 = 0.300
109) HSI2 + NDevHSI2 − PDevHSI2 = 0.300
110) HSI3 + NDevHSI3 − PDevHSI3 = 0.300
```

  a. How does the solution to this problem differ from the five scenarios described in Chapter 7, Linear Programming, in terms of net present value and harvest volumes?

  b. What are the average HSI values for each of the three time periods?

4. *Forest planning options.* You recently have been promoted to the regional office of a forestry company located in north Florida. The company traditionally has relied on linear programming methods for the development of forest plans. However, there is a feeling among upper-level management that clearcut size issues need to be addressed in forest plans, not only to provide better guidance to foresters in the field, but also to better assess the cost of additional organizational policies or regulatory restrictions. Develop a short report that discusses the options available to the company for moving beyond linear programming.

5. *Binary search forest planning problem.* Using the data provided in the following table, develop a harvesting plan using binary search. Assume that the landowner is situated in the Intermountain region of the United States, and that over the next 3 years they are interested in thinning 10 of their stands of trees. The uneven-aged stands are composed primarily of Engelmann spruce (*Picea engelmannii*) and subalpine fir (*Abies lasiocarpa*). The potential thinning volumes for each stand are illustrated in the table, and we need to assume that each stand will be thinned only once. The landowner wants to maximize the amount of volume harvested, and wants it to be approximately equal in each of the 3 years. To utilize binary search, assume that the harvesting rule involves entering the highest density stands first. To begin, develop a sorted

list of potential stand volumes (remember to multiply the volume per acre by the size of each stand). Then select a target volume and an increment. Assume that parts of stands can be harvested in different years. Develop a report for the landowner illustrating the potential harvest volumes in each year.

If the landowner insisted on harvesting whole stands each year, then how would the binary search solution change? See Section III for a refresher on the differences between harvesting whole stands within a time period and partial stands.

Potential Thinning Volume per Acre (MBF)

Stand	Relative Density[a]	Area (acres)	Period 1	Period 2	Period 3
1	74	10	3,450	4,010	4,570
2	85	15	3,780	4,330	4,880
3	72	25	3,680	4,220	4,760
4	65	10	3,550	4,110	4,670
5	62	20	3,640	4,190	4,740
6	87	30	3,910	4,450	4,990
7	76	15	3,820	4,380	4,940
8	68	20	3,460	4,010	4,560
9	69	10	3,950	4,510	5,070
10	66	15	3,560	4,120	4,680

[a]*Basal area per acre/(quadratic mean diameter).*$^{0.5}$

6. *Heuristic methods.* Assume for a moment that you need to develop a forest plan, and that your organization does not have the capability to use a mathematical programming technique such as linear programming. You could put a forest plan together based on your insight and knowledge of the property, and based on conversations with your co-workers. This ad-hoc method of developing a plan could arguably be called a heuristic technique, since logic and rules-of-thumb undoubtedly were used in the development of the plan. Given the search behavior of the standard heuristics that were described earlier in this chapter, how would you think they could improve upon the quality of the plan you developed?

REFERENCES

Bettinger, P., Zhu, J., 2006. A new heuristic for solving spatially constrained forest planning problems based on mitigation of infeasibilities radiating outward from a forced choice. Silva Fennica 40 (2), 315–333.

Bettinger, P., Sessions, J., Boston, K., 1997. Using Tabu search to schedule timber harvests subject to spatial wildlife goals for big game. Ecol. Modell. 94 (2–3), 111–123.

Bettinger, P., Graetz, D., Boston, K., Sessions, J., Chung, W., 2002. Eight heuristic planning techniques applied to three increasingly difficult wildlife planning problems. Silva Fennica 36 (2), 561–584.

Caro, F., Constantino, M., Martins, I., Weintraub, A., 2003. A 2-opt tabu search procedure for the multiperiod forest harvest scheduling problem with adjacency, greenup, old growth, and even flow constraints. For. Sci. 49 (5), 738–751.

Clements, S.E., Dallain, P.L., Jamnick, M.S., 1990. An operational, spatially constrained harvest scheduling model. Can. J. For. Res. 20 (9), 1438–1447.

Dueck, G., Scheuer, T., 1990. Threshold accepting: a general purpose optimization algorithm appearing superior to simulated annealing. J. Comput. Phys. 90 (1), 161–175.

Dyer, A.A., Hof, J.G., Kelly, J.W., Crim, S.A., Alward, G.S., 1979. Implications of goal programming in forest resource allocation. For. Sci. 25 (4), 535–543.

Falcão, A.O., Borges, J.G., 2001. Designing an evolution program solving integer forest management scheduling models: an application to Portugal. For. Sci. 47 (2), 158–168.

Field, D.B., 1973. Goal programming for forest management. J. For. 19 (2), 125–135.

Glover, F., 1989. Tabu search—part I. ORSA J. Comput. 1 (3), 190–206.

Glover, F., 1990. Tabu search—part II. ORSA J. Comput. 2 (1), 4–32.

Holland, J.H., 1975. Adaptation in Natural and Artificial Systems. University of Michigan Press, Ann Arbor, MI.

Johnson, K.N., Stuart, T.W., Crim, S.A., 1986. FORPLAN Version 2: An Overview. US Department of Agriculture, Forest Service, Land Management Planning Systems Section, Washington, DC.

Lu, F., Eriksson, L.O., 2000. Formulation of harvest units with genetic algorithms. For. Ecol. Manage. 130 (1–3), 57–67.

McDill, M.E., Rebain, S.A., Braze, J., 2002. Harvest scheduling with area-based adjacency constraints. For. Sci. 48 (4), 631–642.

Metropolis, N., Rosenbluth, A., Rosenbluth, M., Teller, A., Teller, E., 1953. Equation of state calculations by fast computing machines. J. Chem. Phys. 21 (6), 1087–1101.

NCASI Statistics and Model Development Group, 2006. Habplan User Manual. National Council for Air and Stream Improvement, Inc., Washington, DC, p. 74.

Nelson, J., Brodie, J.D., 1990. Comparison of a random search algorithm and mixed integer programming for solving area-based forest plans. Can. J. For. Res. 20 (7), 934–942.

Öhman, K., Eriksson, L.O., 2002. Allowing for spatial consideration in long-term forest planning by linking linear programming with simulated annealing. For. Ecol. Manage. 161 (1–3), 221–230.

Pukkala, T., Kangas, J., 1993. A heuristic optimization method for forest planning and decision making. Scandin. J. For. Res. 8 (1–4), 560–570.

Richards, E.W., Gunn, E.A., 2000. A model and tabu search method to optimize stand harvest and road construction schedules. For. Sci. 46 (2), 188–203.

Tigermoth, 2016. What Is Tigermoth? Tigermoth, Singapore. Available from: http://www.tigermoth.com/ (Accessed 3/30/2016).

Tóth, S.F., McDill, M.E., Könnyű, N., George, S., 2012. A strengthening procedure for the path formulation of the area-based adjacency problem in harvest scheduling models. Mathemat. Comput. For. Nat.-Resour. Sci. 4 (1), 27–49.

US Department of Agriculture, Forest Service, 1981. Guide for predicting sediment yields from forested watersheds. US Department of Agriculture, Forest Service, Soil and Water Management, Northern Region, Missoula, MT and Intermountain Region, Ogden, UT. 48 p.

Weintraub, A., Romero, C., 2006. Operations research models and the management of agricultural and forestry resources: a review and comparison. Interfaces 36 (5), 446–457.

Zeng, H., Pukkala, T., Peltola, H., Kellomäki, S., 2007. Application of ant colony optimization for the risk management of wind damage in forest planning. Silva Fennica 41 (2), 315–332.

Chapter 9

Forest and Natural Resource Sustainability

Objectives

Sustainability has long been a predominant theme in natural resource management, since as land managers we have an inherent need to provide current and future generations with the products and services that they desire. Natural resource managers have contemplated issues related to forest sustainability for centuries. Two of the earliest papers regarding the sustainability of forests and forest production were written in 1713 by von Carlowitz (Schmithüsen, 2013) and in 1826 by Hundeshagen (1826). Sustainability concepts, in fact, are regarded by many as successors to concepts related to the conservation of natural resources (Sample, 2004). However, arriving at a definition of sustainability on which everyone can agree and use is difficult, and the concept faces challenges posed by global social and economic pressures (Köpf, 2012). The journey toward sustainability involves determining what resources to sustain, determining the duration over which the resources will be sustained, and assessing the alternatives and associated costs for sustaining these and other resources (Wright et al., 2002). Understanding the history and intent of sustainability concepts and how they relate to forest management and planning is therefore an important aspect of natural resource management. At the conclusion of this chapter, you should be able to:

1. Understand the early inspiration for sustainable forests in Europe and the United States.
2. Explain how the concept of sustainability, as it relates to public land in the United States, has evolved over the past 100 years.
3. Describe how and when sustainability concepts might be applied as they relate to different landowners, different geographic regions, and different forest structural conditions.
4. Explain the current rationale for sustainability on federal lands in the United States.

I. INTRODUCTION

North American landscapes are quite diverse, ranging from the coniferous forests of the western United States and Canada, to the hardwood-dominated forests of the central United States, to the pine and hardwood forests of the southern United States, and the spruce/fir and northern hardwoods of the northeastern United States and Canada (Fig. 9.1). The land ownership pattern varies as well, partly due to the manner in which the continent was settled, thus there is a higher percentage of private forest ownership in the eastern portion of the continent as compared to the western portion, which has a relatively large portion of federal (public) ownership. It should be of no surprise that ownership intentions vary, from the traditional goal of sustaining the yield of commodities, to the desire to accommodate multiple uses, to the intent of sustaining ecosystems. The term *sustainability* often refers to the ability to maintain a resource indefinitely into the future, with no decline in quality or quantity, regardless of outside influences. Managing forests with sustainability in mind is therefore both an admirable and challenging endeavor.

Although a connection between timber production and sustainability of local economies can be made, more recently in natural resource management the notion of sustainability has come to include various aspects of plant nutrition, soil capacity, hydrologic stability, and other values (Farrell et al., 2000), or more collectively, *ecosystem services*. The notion of sustainability of any system is value-laden and subject to periodic change (Oliver, 2003). That is, our values reflect what we feel is important in managing natural resources (Xu et al., 1995). However, a society's values change over time, given the economic and social conditions that dominate an era. For example, in colonial times in the United States, sassafras (*Sassafras albidum*) and oak (*Quercus* spp.) were some of the most important natural resources, along with the American chestnut (*Castenea dentata*) (Kirwan and Bond, 2007). Some of these resources are not as socially or economically important today, due to changes in our values, the manner in which we consume forest resources, or the composition of forests. When it comes to sustainability, we might ask whether we should restore (e.g., chestnut) or maintain (e.g., sassafras) certain species or products, or restore and maintain the productive capacity of the forest as a whole. Alternatively, we might ask whether it is important to conserve various ecosystems, or to ensure

Forest Management and Planning. DOI: http://dx.doi.org/10.1016/B978-0-12-809476-1.00009-6

FIGURE 9.1 A hardwood-dominated forest in West Virginia. *Photo courtesy of Kelly A. Bettinger.*

that certain current or future forest-dependent human systems are not impaired. These are but a few of the issues that perplex land managers today.

Some broader examples of sustainability include balancing the long-term carbon storage levels, minimizing erosion, avoiding the depletion of soil nutrient stores, and keeping the chemical composition of soils and vegetation within the bounds of natural systems (Sverdrup et al., 2005). Some have suggested that sustainable forests can be maintained by keeping natural disturbances, such as fires, invasive species outbreaks, and insect and disease problems at bay with active management (Oliver, 2003). Others suggest the need for enhancing the self-organizing capability of a landscape by allowing natural disturbances to act on ecosystem functions and properties within a desired range of control. How these goals and objectives are addressed is a challenge for land managers, particularly where broader social, economic, or environmental concerns have an impact on management options. For example, in developing areas of the world, illegal logging, inadequate law enforcement, and economically nonviable management of nontraditional tree species are all barriers to the sustainable management of forests (Galloway and Stoian, 2007).

The concept of sustainability is admittedly ambiguous, but central to management of public and private forests and rangelands. Perhaps taken for granted by many within our field (natural resource management), it is a concept that has captured the hearts and minds of the public at large in a number of developed countries. Sustainability is a concept that should be thoughtfully considered by land managers today in their management plans. Management plans have been developed for about half of the world's forests (Food and Agriculture Organization of the United Nations, 2010), and a smaller portion of these are certified sustainable forests (Siry et al., 2005). However, while management plans and forest certification status may be

desirable of landowners, from a sustainability point of view they are not necessarily required to achieve a sustainable system if forests are managed appropriately. Although management of forests should focus on landowner objectives, if we were to consider the sustainability of an area of land to produce various products, some amount of planning seems necessary.

II. SUSTAINABILITY OF PRODUCTION

There was a time in the early development of North America that forest resources were considered an obstacle to expansion (agricultural, mainly), and viewed as an inexhaustible and unlimited supply of wood and income for local economies (Sample, 2004; Young, 1984). However, in the late 19th century and early 20th century growing concern over the possibility of a wood famine arose (Sample, 2004). The early work of European foresters in developing a sustainability philosophy for their forests reflects the desire to (1) supply dependable wood flows to urban industrial areas, (2) reduce the undesirable effects of changes in economic cycles, and (3) maintain watershed stability and the productivity of forests (Kennedy et al., 2001). Since most of the university-trained forest managers in the United States in the early 1900s were educated by faculty who themselves were influenced by early work of European foresters, these concepts naturally filtered down into our systems of natural resource management. Some contend that it took a change in people's attitude toward forest resources before any notion of sustaining the yield of timber products could take hold (Farrell et al., 2000). For example, in examining wood consumption patterns in the United States and the time required to regenerate cleared lands and return them to a productive state, Watson (1921) suggested the following:

It is evident that a severe timber shortage, covering a period of fifty to one hundred years, will surely occur.

as well as:

The outlook for adequate timber supplies of the near future is not very encouraging.

However, it wasn't long before critics arose, such as Baker (1933) who suggested:

The old threat of timber famine, which foresters used to arouse themselves and the Nation to active timber conservation, is so far from working out that the foresters, themselves, have lost faith in it.

And Boyce (1931) who offered the following:

... something has gone awry; our variously predicted timber famines have not occurred. Some of us feel duped.

The projected timber famine never materialized in the United States, and sustained production and management of forests now has evolved into a highly technical process of modeling growth, mortality, and risk to understand the level of wood products that can be derived from a forest over a long period of time (Sample, 2004). Today, a variety of private landowners in North America continue to embrace a sustained production philosophy, as do forest managers in other areas of the world. For example, the 52 million acres (21 million hectares) of state-owned forest land in Turkey are managed using plans that were developed with the objective of maximizing the sustained yield of timber production (Türker, 2007). Sustained production management will be the emphasis of forest management plans in South Korea as well, once the forest development phase has matured around 2030 (Park, 1990).

The early discussion of sustainability in the United States related mainly to public lands. The Organic Act of 1897 (16 USC § 475), in describing the purpose of the designation of National Forests, suggested the following:

No national forest shall be established, except to improve and protect the forest within the boundaries, or for the purpose of securing favorable conditions of water flows, and to furnish a continuous supply of timber for the use and necessities of citizens of the United States ...

As a result, an emphasis on the sustainability of production (timber and water resources) and protection was a broad management tenet that was associated with the US Forest Service at its inception. In fact, Gifford Pinchot (1905), head of the US Bureau of Forestry, underscored the sustained yield direction for federal lands with the following statement:

But whichever of the ways to using the forest may be chosen in any given case, the fundamental idea in forestry is that of the perpetuation by wise use; that is, of making the forest yield the best service possible at the present in such a way that its usefulness in the future will not be diminished, but rather increased.

This sustainability philosophy requires first identifying a desired level of production, then manipulating the environment to emphasize those products (Gale and Cordray, 1991). One of the requirements Pinchot (1905) notes for the "best service" of the forest is that a regular supply of trees will be available for harvest. Pinchot goes on to say that one of the central ideas of forestry should be that the amount of wood harvested nearly equals the amount grown. These two ideas suggest that the early direction for the US National Forests was indeed focused on the sustainability of wood resources. However, due to uneven age class distribution of many US National Forests, a standard of relating harvest levels to a long-term timber harvest capacity was adopted as the scheduling rule,

rather than relating harvest levels to the growth rate of forests (Clawson and Sedjo, 1984). In any event, organizations that are influenced by a policy of sustained yield of timber products have at their core the problem of determining the level of timber harvest that can be continuously harvested from a forest, or determining the level of harvest that could increase over time (Runyon, 1991).

One of the main arguments for a sustained yield of timber production was the belief that local communities need to be stabilized from a social and economic perspective. Another argument for a sustained yield of timber production concerned the supply of wood, and whether there would be enough for future generations. Through the predominant source of forestry literature in North America during the early 1900s (*Journal of Forestry*), we can locate a number of position statements proposing the adoption of a sustained yield policy, such as the following from Recknagel (1930):

There is only one possible solution of the problem of a permanent and sufficient supply of forest products in the United States — that is by the practice of sustained yield management.

The sustained production philosophy gained emphasis in the United States in the 1940s, and as a result, the even-flow of timber harvest volume concept was developed as a surrogate for the general philosophy of sustained production (Clawson and Sedjo, 1984). A nondeclining even-flow of wood products from National Forest land in the United States is one of the tenets of the National Forest Management Act (16 USC 1600 § 13(a)). In the United States, the Society of American Foresters (Winters, 1977) defines the sustained yield of timber production as:

The yield that a forest can produce continuously at a given intensity of management.

This implies that a planning process should be guided by the need to balance the growth of a forest with the harvest of forest products (Winters, 1977). The impression that timber yields need to be sustained has not been limited to the United States. Most provincial forest agencies in Canada for some time have utilized a policy of sustained yield in their forest plans (Runyon, 1991). In Ontario, for example, the Pulpwood Conservation Act of 1929 required all pulpwood companies to plan their future management on a sustained yield basis. Sustained yield forestry is one of the dominant themes in tropical forest management as well, although the complexity of these forests has limited our understanding of the growth factors related to the tree species found there. As a consequence, estimation of sustainable levels of harvests is tenuous in tropical forests since it is difficult to find any two forests that are similar in structure or form (Majid-Cooke, 1995).

Sustained yield of timber products has been loosely linked with the concept of the regulated forest (discussed in Chapter 10: Models of Desired Forest Structure), which generally implies an even-aged forest management structure is used. However, sustainability of timber products can be associated with uneven-aged forests as well. To be applied to uneven-aged forests, the residual live trees that remain after each entry (or cutting cycle) need to be of a certain density, species, and quality to provide a similar harvest in subsequent entries (Dauber et al., 2005). Managing the regeneration, and subsequent growth within an uneven-aged system is arguably more complex than managing the regeneration of an even-aged system, given the challenges related to survival rate of naturally seeded or sprouted trees under various environmental conditions.

Several issues need to be considered when using the sustained production philosophy in guiding the development of forest plans:

1. *The scale of the policy.* Planners and managers need to decide whether the sustained yield policy applies simply to a land ownership or some portion of a land ownership, to a watershed, a county, a province, a region, or a nation. The larger the geographical extent, the greater the sustained yield harvest volume. Larger areas provide a broader land base from which to draw wood production. In addition, the larger the extent, the greater amount of flexibility is provided to smooth out harvest allocation problems that may arise from the variable age class distributions of forest landowners.

2. *The intensity of forest management that is assumed.* Current and future stands of trees can be assigned a wide array of silvicultural treatments, which could affect the long-term timber harvest potential of an ownership or region. The assumption that all landowners will apply very intensive management practices may fail to recognize the economic or regulatory realities of public and private landowners. In this case, landowners either may not have a sufficient budget to explore intensive management treatments (e.g., mid-rotation fertilization), or may not have the regulatory latitude to implement certain management treatments (e.g., herbicides on some lands). The default assumption that all landowners will apply very general management practices may fail to recognize the desire and ability of industrial landowners to maximize economic values or wood production, and may lead to an underestimation of wood production when compared to the potential. In this case, we could fail to account for the use of genetically advanced seedlings, thinnings, fertilization, and other treatments commonly practiced on today's industrial lands.

3. *Market fluctuations.* Timber prices are affected by a variety of local, regional, national, and global forces. Planning for the sustainability of harvest volume generally ignores these forces, which subsequently implies that landowners are suboptimizing the economic value of their resource. For example, if prices are held constant in a plan, when timber prices rise (and this rise is not accounted for in the plan), the planned harvest is the same as when timber prices fall. As a result, too much wood would be produced in periods where prices were depressed, and not enough wood would be produced in periods when prices were high (Alston, 1992). A natural tendency may be to harvest more wood when prices are high and less wood when prices are low. This commonly occurs in real-world forest management situations, creating challenges for the planner to react and modify the sustained yield plan. Therefore, short-term deviations from a sustainable yield may be allowed in practice, yet may be inconsistent with the long-term goals.

4. *Application of the sustained production concept to nonregulated forests.* To manage a nonregulated forest using a sustained yield objective, you would need to take a forest described by a nonnormal age class distribution and convert it to a normal forest (as described in Chapter 10: Models of Desired Forest Structure). To do so may require assumptions about how the gaps in the age class distribution will be filled, and to what extent, perhaps, the older forest volume will be harvested as the forest transitions to a regulated state. We are focusing this point on even-aged forests, because as we noted, it is possible, but perhaps more difficult, to develop sustained yields from uneven-aged forests that are undefined by an age class distribution.

5. *Forest depletions from natural forces.* The supply of timber from a forest will be affected by harvesting activities as well as natural disturbances (insect outbreaks, fires, etc.). These reductions in forested area, growth, and quality will undoubtedly affect the sustained yield from a forest (Runyon, 1991). Although simulation models have been developed in the last 20 years to account for some natural disturbances, developing a sustainable forest plan that accounts for stochastic disturbances in an optimization model is an ongoing area of work.

Given these points, one of the main disadvantages of this philosophy for managing natural resources is that it does not necessarily guarantee stability. Controversy over employment reductions due to the habitat requirements of threatened and endangered species also has called into question whether land management should emphasize community stability (Gale and Cordray, 1991). In addition, a number of other uncontrollable forces, such as advances in technology and land use changes, suggest that the level of harvest calculated today may not be consistent with future trends in forestry, economic, or social conditions (Riihinen, 1992). Assumptions about genetic improvements and their impact on tree survival and growth could cause an increase in sustainable yields (see the discussion of the allowable cut effect in Chapter 11: Control Techniques for Commodity Production and Wildlife Objectives). Advancements in harvesting, milling, and equipment could increase the merchantability of products, and subsequently lower manufacturing costs could also have an effect on sustained yields. The extent of these changes generally is unknown (although might be predicted with a low level of accuracy) at the time a forest plan is being developed.

The implementation of the sustained production philosophy is complex, since it suggests that the biological rotation age is not sufficient for the sustainability of wood production because it is not responsive enough to the demand for wood. In addition, the biological rotation age, which has been used by many to represent the point in time in a stand where wood production is maximized, is inherently economically inefficient. Yet by taking an economic perspective, though markets for wood may guarantee sustainability through supply—demand relationships, the sustained yield policy may be compromised (Riihinen, 1992). Upon returning to one of the original questions of this chapter (What should the forests sustain?), the sustained production philosophy seems to become a constraint on forest management, rather than the goal itself (Alston, 1992).

III. SUSTAINABILITY OF MULTIPLE USES

In the early 20th century it became apparent to many people concerned about North America's natural resources that the viewpoint of an unlimited level of resource production may be unrealistic (Young, 1984). In fact, as we noted earlier, at various times during the 20th century there was a fear of a wood famine in some regions of North America. As we suggested in the previous section, many of the early thoughts regarding the sustainability of forests were centered on the need to produce the highest yield at the lowest cost, and the need to find the most effective way to meet the demands of the local and regional markets. However, in the middle of the 20th century, leisure time increased among residents of the United States, transportation systems became well-developed, and more people became interested in recreation, wildlife, and other noncommodity forest resources (Sample, 2004). In addition, the United States Congress formalized the nation's vision of sustainability on federal lands as the need to stabilize local communities *as well as provide multiple benefits to society*. In the Sustained Yield Forest

Management Act of 1944 (16 USC § 583), Congress attempted to encourage the development of sustained yield units, or collaborations of industry and federal land, as a means of maintaining communities dependent on timber harvests and other resource amenities. The purpose of the Act was to:

… promote the stability of forest industries, of employment, of communities, and of taxable forest wealth, through continuous supplies of timber; … to provide for a continuous and ample supply of forest products; and … to secure the benefits of forests in maintenance of water supply, regulation of stream flow, prevention of soil erosion, amelioration of climate, and preservation of wildlife. (16 USC § 583)

Although the Sustained Yield Forest Management Act may have suggested ("… secure the benefits …") that the sustainability of multiple resources is important, the Multiple-Use Sustained Yield Act of 1960 (Public Law 86-517) solidified the notion for federal lands:

… the national forests are established and shall be administered for outdoor recreation, range, timber, watershed, and wildlife and fish purposes. (16 USC § 528)

The Act also indicates that:

The Secretary of Agriculture is authorized and directed to develop and administer the renewable surface resources of the national forests for multiple use and sustained yield of the several products and services obtained therefrom. (16 USC § 529)

and defined both multiple use and sustained yield:

(a) *"Multiple use" means: The management of all the various renewable surface resources of the national forests so that they are utilized in the combination that will best meet the needs of the American people; making the most judicious use of the land for some or all of these resources or related services over areas large enough to provide sufficient latitude for periodic adjustments in use to conform to changing needs and conditions; that some land will be used for less than all of the resources; and harmonious and coordinated management of the various resources, each with the other, without impairment of the productivity of the land, with consideration being given to the relative values of the various resources, and not necessarily the combination of uses that will give the greatest dollar return or the greatest unit output.*

(b) *Sustained yield of the several products and services means the achievement and maintenance in perpetuity of a high level annual or regular periodic output of the various renewable resources of the national forests without impairment of the productivity of the land. (16 USC § 531)*

FIGURE 9.2 Forest and livestock use of federal land in central Oregon. *Photo courtesy of Kelly A. Bettinger.*

This philosophy of sustained production of multiple uses requires first identifying the desired set of diverse human-oriented outcomes (Fig. 9.2), then manipulating the environment to emphasize those products (Gale and Cordray, 1991). One of the main contributions of this approach to sustainability of natural resources is that the emphasis on multiple uses has encouraged the development and advancement of economic and operations research approaches in natural resource management (Xu et al., 1995). These approaches were made necessary as US National Forests were directed to value and integrate nontraditional products and services (i.e., recreation, water, wildlife, etc.) into forest plans. For example, the 1986 land management plan for the Prescott National Forest in Arizona (US Department of Agriculture, Forest Service, 1986) suggested that managers were guided by the need to respond to local and national demands for wood products, forage production for livestock, water yield, and a wide array of developed and dispersed recreational opportunities and wildlife-related uses. Their goal was to produce these outputs and provide these opportunities on a sustained basis while restoring or maintaining air, soil, and water resources to levels suggested by local, state, or federal standards. This multifunctional view of forests is not solely a North American idea, as areas of Europe have embraced this concept for several decades (Verbij et al., 2007). The integration of nontraditional products and services into forest plans continues to be valid and sometimes challenging today, as many natural resource organizations, including some industrial and state management organizations, have incorporated nontimber goals and constraints into forest plans.

Besides the classic objective of providing a continuous supply of timber products over time, the sustained yield of multiple uses has had many other meanings. For example,

according to Waggener (1982), we could view the objectives of managing forests and rangeland for sustained multiple uses as those that include provisions for the maintenance of forest productivity, of employment and other associated economic measures, of adequate levels of nontimber products, and interestingly, of political support and professional credibility. Multiple-use forest management has become the prevailing use on private forests in the United States (Sample, 2005), and, notably, continues to be the approach most preferred by students taking upper-level forest management courses at the University of Georgia (lead author, personal survey). Perhaps the main criticism of the approach is that each unit of land is not equally suitable for multiple-use management (although the definition of multiple use indicates that, "… some land will be used for less than all of the resources …"), and that some uses of the land (such as to conserve biological diversity) are not compatible with other uses of the land (Sample, 2005). Another is that many people feel that the balance struck between managing for commodities and managing for noncommodity uses is uneven (Sample, 2004). The crux of managing for the sustainability of multiple uses, however, is to optimize the use of human-valued products and services. This expands beyond simply commodity production, yet falls short of what is encompassed in our third and final view of sustainability.

IV. SUSTAINABILITY OF ECOSYSTEMS AND SOCIAL VALUES

Public attitudes concerning forests and natural resources in some areas of the world have evolved from viewing the resources as productive centers of local economies to viewing them as functioning ecosystems (Farrell et al., 2000). Some would argue that plans guided by a sustained yield of natural resources as a dominant goal ignore the "existence" values of forests. Recently in the United States, there has been a noticeable increase in the public's romantic, symbolic, and biological value of forests. In other words, we have seen an increase in values that emphasize ecological integrity over commodity production. These values span the fields of recreation, landscape amenities, and nongame wildlife, to name a few. As an example, recently developed United States national forest plans focus on the capability of ecosystems to provide a sustainable flow of beneficial goods and services through the achievement of desired forest conditions (social, economic, and ecological) (US Department of Agriculture and Forest Service, 2013, 2015). For forest managers who operate under this paradigm, nontimber products must be recognized and valued, and forest management activities need to be adjusted accordingly (Wiersum, 1995).

This perspective on forest and natural resource sustainability has been attached several labels over the past two decades, from "new perspectives," to "new forestry," to "ecosystem management." The central tenet of this approach is the need to maintain the integrity of entire forest ecosystems to provide a plethora of services. These ecosystem services refer to the benefits that ecosystems provide, and include provisioning services (energy, wood, food, water), regulating services (climate, erosion, flood, disease), cultural services (recreation, reflection, enrichment), and supporting services (photosynthesis, primary production, nutrient cycling, soil formation). It is believed with this philosophy that the flow of all products and services from forest depends on the maintenance and development of the functions and processes that sustain ecosystems (Xu et al., 1995). This philosophy suggests that the ecosystem should be able to sustain itself with minimum human assistance, and suggests low-level extraction of some forest products at best. Where ecosystems are unhealthy, this approach could permit a greater degree of active management intervention, however (Gale and Cordray, 1991). In contrast to the philosophy of sustained production, sustainability under this approach implies that it is a goal, rather than a constraint, and that the goal is related to the maintenance of long-term integrity of an ecosystem (Alston, 1992). Further, sustaining ecosystems does not necessarily imply that we are also sustaining multiple uses. For example, reserving an area from active management to allow an ecosystem to recover reduces timber harvest levels and may increase recreational activities, neither of which (lower harvest levels, increased recreational activities) may be viewed as sustainable.

If an organization or a society adopts this broader version of sustainability, it places new demands on the natural resources manager, who then has to expand the scope of the management plans to encompass a broader array of goods and services (Farrell et al., 2000). To derive the maximum contribution of forest, range, recreation, wildlife, fisheries, and other resources from the landscape requires the development of planning techniques that facilitate the multiresource facet of the current planning problem. Some concerns that could be incorporated into these analyses include land use changes, long-term wood supplies, and habitat development and maintenance. With this type of planning process, wood supplies can fluctuate according to social, environmental, and economic realities, and production would not necessarily be restricted to even-flows of products (Young, 1984).

Some may argue (Kennedy et al., 2001) that forest managers are in a transition period, where the technical and professional constructs learned in university coursework are being challenged by changing social conditions, as reflected in the need to collaborate with a diverse spectrum of colleagues and stakeholders.

It has been suggested that public land managers in the United States must now be more broadly focused on sustainable systems management rather than simply the technical components of forest, range, wildlife, or ecosystem management, where the "systems" include economic, ecological, and social dimensions (Kennedy et al., 2001). This may certainly be true in some areas of the world, such as on public lands in the United States; however, many other landowners continue to operate under the assumption of sustained yield of production (Runyon, 1991).

One of the arguments for the sustainability of production involved the need to stabilize communities and the economic development of local areas, which prompted the concept of nondeclining even-flow of wood products from national forests. The view of rural economies was that they were separated from others in time and space, that they were simple and less diverse than urban economies, and that they were at an economic disadvantage, thus requiring governmental intervention. Today, it is argued that communities should be viewed as adaptable, complex, and socioeconomically interrelated with other communities (Kennedy et al., 2001). Advocates of ecological sustainability doubt whether we can both use renewable resources (such as timber products) and maintain ecological integrity. Thus the belief that forests can be both managed and sustainable is not widely held within this group (Majid-Cooke, 1995).

Recently, the US Forest Service shifted its mission from multiple use management toward the sustainability of ecosystems. According to its current planning rule (36 CFR Part 219), National Forest managers must now plan with ecosystem sustainability in mind. The new planning process is strategic in nature, and allows each National Forest to develop broad goals for the forest (desired conditions, objectives, guidelines, suitability of areas for treatments) without being prescriptive, which implies that the implementation of the plan is left to the field managers. Plans developed under this rule must be consistent with the requirements of the previous Acts, such as the Multiple Use Sustained Yield Act of 1960 and the National Forest Management Act of 1976, yet they also require managers to develop a sustainable social, economic, and ecological framework that will guide the implementation of management activities.

Since the new direction places more emphasis on strategic planning, it consequently leaves the site-specific implementation of activities to the discretion of the land managers of the National Forests. Therefore, the overall forest planning cost is reduced, but not necessarily the implementation cost. According to a benefit/cost analysis for forest plan development, this paradigm shift will cost the US Forest Service more today (as compared to the forest plans developed in the 1980s) because of the need to (1) collaborate with stakeholders, (2) analyze multiple alternatives, (3) develop decisions, (4) document the processes involved, (5) assess sustainability requirements, and (6) monitor the implementation of the plan (US Department of Agriculture and Forest Service, 2005). However, as the rule suggests (yet in contrast to what we noted earlier), managing for sustainability in this manner may sufficiently provide for the maintenance of various resources without impairing the productivity of the land, which is a requirement of the Multiple Use Sustained Yield Act.

Governments in other parts of the world also have suggested through legislation that the sustainability of ecosystems and social values are important. For example, in Ontario, the Crown Forest Sustainability Act of 1994 (S.O. 1994, c. 25 (CFSA)) requires that management plans for public lands contain ecosystem sustainability provisions, as noted in the following guiding principles:

1. Large, healthy, diverse and productive Crown forests and their associated ecological processes and biological diversity should be conserved.
2. The long term health and vigor of Crown forests should be provided for by using forest practices that, within the limits of silvicultural requirements, emulate natural disturbances and landscape patterns while minimizing adverse effects on plant life, animal life, water, soil, air and social and economic values, including recreational values and heritage values. (CFSA 1994, c. 25, s. 2 (3))

Here, the forest resources on Crown land (public, federal) are meant to provide numerous forest-based values and to maintain ecosystem sustainability, which they define as long-term forest health (Ontario Ministry of Natural Resources, 2009).

Application of evolving sustainability concepts in developing regions of the world may be difficult, particularly where the wood resource is considered community property, and where wood is needed simply to meet the daily energy requirements of the people. For example, in many parts of Africa wood is the primary source of household energy, and is used daily for cooking and heating. If this is the only source of fuel for the community, and if people are allowed to extract wood from forests without restraint, then these short-term needs may outweigh the desire for the long-term sustainability of supply. These cause and effect issues may also be related to the land tenure situation in each community. In these cases, neither the sustainability of production nor the sustainability of associated environmental conditions is relevant, unless the products have some higher and better value (Clawson and Sedjo, 1984). Further, in other parts of the world, such as Venezuela, sustainability may still focus on the stability of local communities and the development of wood processing industries that can be sustainably supplied from public forests (Kammesheidt et al., 2001). Sustainable tourism has been advocated as a market-based mechanism for promoting both conservation and economic values, by providing income for local communities while urging higher conservation standards. However, this too has some potential pitfalls: profits may move out of local communities, and some forms of tourism may have a negative effect on the environment (Schloegel, 2007).

A goal for forest managers focusing on this concept of sustainability may be to develop forested ecosystems in a way to enable them to adapt to changing market and environmental conditions. Silvicultural decisions based on historical conditions are increasingly being considered as ways to increase the resiliency of forests to changes in climate or to natural disturbances (Churchill et al., 2013). These approaches have mainly been applied to public forests, and are closely related to the sustainability of ecosystems philosophy noted in this chapter. Feedback during the implementation stage of a forest plan seems essential in meeting the goals of policies aimed at the restoration of forests to historical conditions, therefore adaptive management plans with monitoring programs are necessary. Resilience can refer to the ability of an area to recover to pre-event ecosystem composition and structure, and relate to the tolerance of plant and animal species to change (climate or otherwise) (Pulla et al., 2015). The definition of pre-event conditions is an important aspect of measuring the achievement of sustainable, resilient forest systems. However, in some cases it may be a challenge to identify historical forest conditions undisturbed by man.

V. INCORPORATING MEASURES OF SUSTAINABILITY INTO FOREST PLANS

To incorporate quantitative measures of sustainability into forest and natural resource management plans, a number of approaches can be followed. Whether these quantitative measures are included in the objective function or represented by constraints of a plan is a matter of practicality for the planner and a matter of semantics for the other stakeholders involved in the planning process. Some biologists prefer ecological measures to be incorporated directly into the objective function rather than relegated to the constraints, even when the constraints may have more influence on the development of a forest plan. This might suggest the use of a goal programming approach to forest planning. If forest certification programs (one of the focuses of Chapter 15: Forest Certification and Carbon Sequestration) are engaged by landowners, then a number of the principles and criteria interact with forest management and planning processes, including:

- The need to comply with laws and regulations
- The need to develop a land management plan
- The need to monitor and assess forest conditions, yields, and perhaps the supply chain (the focus of Chapter 14: Forest Supply Chain Management)
- The need to maintain areas of high conservation values

In the examples we provide next, we illustrate a single output from a forest plan to show more clearly how a quantitative measure of sustainability might be accommodated.

1. Develop a schedule of activities that provides an even-flow level of output in each time period of the plan (Fig. 9.3). Using linear programming, which assumes that decision variables will be assigned continuous real numbers, developing a solution with an exact output level in each time period is relatively easy. However, some timber stands, range allotments, or other management units will likely be subdivided in the resulting schedule of activities. If the decision variables are assigned integer values (e.g., yes/no), then a plan requiring the exact same output from one time period to the next may be unobtainable. When using integer or mixed integer programming, and since land units (stands) are of varying sizes, developing a plan with exactly the same output from one period to the next is nearly impossible. This is because it is assumed in these situations that stands are considered to be treated entirely, or not at all. Examples include planning for the harvest of an entire timber stand or for the building of an entire road.

2. Develop a schedule of activities that provides outputs within some range (or bound) in each year of the plan (Fig. 9.4). This type of management plan allows deviations in outputs from one period to the next, and provides us with flexibility that may be needed, given the type of variables assumed. Alternatively, if the decision variables were, as in the case of linear programming, assigned continuous real numbers, then we could use this measure of sustainability to allow some deviation in output that responds to predicted changes in technical parameters (such as prices or costs), or to adjust outcomes given problems with the initial age class distribution of a forest.

3. Develop a schedule of activities that provides a nondeclining output of certain natural resource products (Fig. 9.5). This is one of the tenets of the National Forest Management Act (16 USC 1600 § 13(a)) that is applied to National Forest land in the United States, although it can be applied to any type of land ownership, where the sale of wood products is limited to quantities less than or equal to the quantity that can

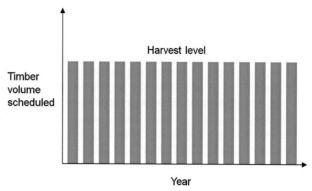

FIGURE 9.3 An exact, even output of timber products over time from a forest plan.

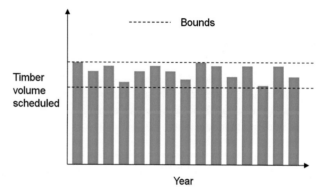

FIGURE 9.4 An output of timber products over time that is within some predefined bounds.

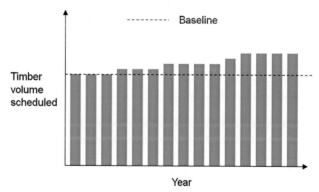

FIGURE 9.5 Output of timber products over time that is nondeclining.

be removed in perpetuity (Sample, 2004). For example, if the current state of a public or private forest will not provide a very high harvest level in the near-term, but due to expected growth and management action could provide a higher harvest level later in the time horizon of the management plan, then we might adopt this method. This indicates that near-term outcomes are sustainable, since longer-term outcomes can be realized at higher levels.

4. Develop a schedule of activities where harvest equals growth. In classical methods of forest regulation, foresters adopted a policy of harvest equaling growth. However, given the practical nature of land management, this policy could be modified to indicate that outcomes are equal to, or less than, the growth of the product (Fig. 9.6). Weather conditions, irregular age class distributions, damage to products during harvest, and other factors may all contribute, however, to harvests not equaling growth. The California Forest Practices Rules (14 CCR § 913.11, 933.11, 953.11) are one example of this policy (California Department of Forestry and Fire Protection, 2013). These rules govern all management actions on privately and state managed timberlands within the state, and are considered by many to be the most restrictive in the western United States. The rules define as the primary goal for timberlands the maximum sustained production of high quality timber products, and call for a balance between growth and harvest to the extent of harvest not exceeding the growth over a 10-year period. In addition, the rules call for the protection of soil, water, air, fish, and other wildlife species, and for making allowances for recreation, aesthetics, and regional economic vitality:

Consistent with the protection of soil, water, air, fish and wildlife resources a SYP [Sustained Yield Plan] shall clearly demonstrate how the submitter will achieve maximum sustained production of high quality timber products while giving consideration to regional economic vitality and employment at planned harvest levels during the planning horizon. The average annual projected harvest over any rolling 10-year period, or over appropriately longer time periods for ownerships which project harvesting at intervals less frequently than once every ten years, shall not exceed the long-term sustained yield estimate for a SYP submitter's ownership. (Section 1091.4.5, a)

5. Develop a schedule where the harvest of a product is less than the long-term sustainable harvest of that product (Fig. 9.7). This is a conservative measure of management, where we calculate the long-term sustainable harvest level first (perhaps with linear programming), then develop the actual harvest level such that it does not exceed the long-term level. The actual harvest level may either be planned (with integer or mixed integer programming) or the result of problems with implementation (i.e., weather, natural disasters, etc.).

Although these examples tend to focus on the sustainability of timber production, the concepts of sustainability need not be limited to commercial forest products when recognizing them in a natural resource management plan. For example, you could require that a plan of action

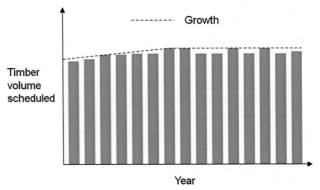

FIGURE 9.6 Output of timber products where harvest is less than or equal to the growth of a forest.

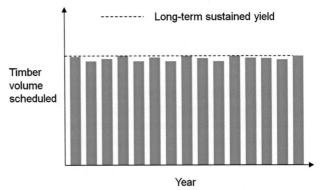

FIGURE 9.7 Output of timber products that is less than or equal to the long-term sustained yield.

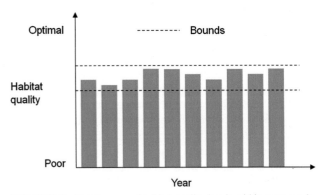

FIGURE 9.8 Maintenance of habitat quality levels within some predefined bounds.

maintain critical habitat levels within some range (or bound) that a biologist suggests is necessary to facilitate the breeding, nesting, or foraging requirements of the species (Fig. 9.8) or one where the habitat level does not decline over time (Fig. 9.9). Biodiversity, wildfire, carbon levels, aquatic concerns, and measures of forest insect spread and control all have been incorporated as quantitative measures in forest plans. By expanding the list of quantitative measures reported (and perhaps controlled with constraints), it is possible to develop

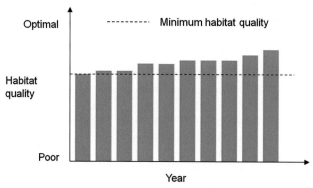

FIGURE 9.9 Maintenance of nondeclining levels of habitat quality over time.

management plans that address the sustainability of multiple uses, and to address some aspects related to the sustainability of ecosystems and social values. Generally speaking, the outputs measured need to be quantitatively derived. Qualitative measures (e.g., good or poor habitat, good or poor aesthetics) require numeric conversion to be of value here.

VI. SUSTAINABILITY BEYOND THE IMMEDIATE FOREST

In forest management and planning, sustainability concepts can be incorporated into management plans to ensure the development or maintenance of values deemed important to the landowner. In fact, in many cases, land managers consider sustainability to apply only to the forest or land that they manage. In reality, the management of individual forests has an effect on the resources and markets that reside in the surrounding community. Therefore, there may be instances where a wider geographic coverage or assessment of values and impacts may be needed to better describe the impact of management on broader economic, environmental, and social values. Sample (2005) suggests that intensively managed forest plantations and protected areas both may have a place within a comprehensive sustainability framework, regardless of who owns or manages these areas. However, one drawback is that although we might have very good information to describe the forest we manage, we could have limited information to describe the surrounding forests that other landowners manage. We are also often limited in our ability to collect information for land that others manage. And, although we may be intimately aware of our own management objectives and constraints, those that relate to other landowners may be elusive. A further challenge involves developing a process to encourage cooperation among landowners or land managers. These processes should avoid facilitating participants' engagement in antitrust behavior.

Cross-ownership challenges may confront natural resource managers in the future if over-arching landscape goals become more important. Should a cross-ownership plan or assessment be recommended, it suggests that landowners cooperate in the planning of their management activities. Irland (2003) proposed several options to accommodate the achievement of a long-term sustainable timber supply from multiple landowners distributed across the landscape, including letting the markets sort out the problem (balancing growth of forests and harvest), allowing politicians to persuade landowners to conform to a multi-landowner sustainability plan, banning exports, controlling local mill capacities, and controlling the harvest through regulation. For example, multiple landowner cooperation is suggested in the California rules, but not mandated. The political culture and institutional models currently in place may not be sufficient for multilandowner sustainability planning to be successful (Irland, 2003). Large-scale multiowner planning, however, has been attempted in other countries. In Australia, joint oversight and planning of public forest resources (Commonwealth government and State government resources) was initiated in the 1990s as a way to maintain ecological processes and biological diversity while also managing for the full range of environmental, economic, and social values. Unfortunately, conservation groups and other interests generally were dissatisfied with the outcome of the process (Ferguson, 2007). Even though the area conserved for reserves was increased, there was difficulty in adequately assessing the trade-offs between resource uses, and resource (timber) security for local industries was not realized. Therefore, the regional forest agreements in Australia now seek to ensure access and supply of timber products to local industries, ensure that forests are managed in an ecologically sustainable manner, and provide protection to the unique aspects of biodiversity found across the landscapes (Commonwealth of Australia, 2016).

VII. SUMMARY

In a recently published book, *Forest Plans of North America* (Siry et al., 2015), 48 different forest plans were described that guide the management of private and public lands of various forest ownership classes. Although sustainability concepts are often alluded to in the description of each plan, each chapter ends with a specific statement of how the plans address sustainability. If one were to review each of these chapters, one would find that the concepts of sustainability of timber production, multiple uses, and ecosystems are all actively integrated into forest management today. Each landowner or landowner group may view sustainability differently, yet the forethought and actions toward the well-being of future human, plant, and animal life reflects the concern forest managers place on the appropriate uses of land.

For natural resource managers, sustainability often refers to meeting the requirements of present generations without undermining the natural resource base and compromising the ability of future generations to use those same resources. This is an ethical issue regarding the need to maintain a forest system that will benefit future generations. The conundrum is that social values and environmental conditions evolve continuously, and as a result, future generations may have a different view of sustainability than the one(s) we use (Rametsteiner and Simula, 2003). Many private land managers (industrial, nonindustrial) are concerned about multiple use and ecosystem sustainability, along with the protection and maintenance of a long-term timber supply, yet the sustainability objectives of private land managers may not be readily apparent in the forest plans that they develop and implement. Public land managers, on the other hand, are guided by laws and regulations that make it apparent that sustainability measures are included in management plans. Forest certification programs (see Chapter 15: Forest Certification and Carbon Sequestration) have been argued as one method to promote the sustainability of forests and ecosystem biodiversity; however, evidence to support the assertion is limited, and some have argued that the effects of certification programs are highly variable and depend on the local circumstances surrounding the management of a property (Rametsteiner and Simula, 2003).

On a more technical note, to develop an estimate of the maximum contribution of forest, range, wildlife, recreation, and fisheries resources from an area requires the use of sophisticated planning techniques. In many cases, the necessary tools are not even developed at this point in time. We should keep in mind that models used, and the results generated, only should guide the management of natural resources. They are not a substitute for the daily negotiations that must take place in real-world management environments. In addition, they usually include generalized functional relationships between management activities and outcomes, and may even ignore some of the more elusive realities inherent in natural resource management (i.e., weather, market fluctuations, climate change, natural disasters).

QUESTIONS

1. *Sustainability philosophies for a managed landscape.* Assume you have been hired recently as a land manager for a company in the upper peninsula of Michigan. Your company traditionally has developed forest plans that emphasize the sustainability of wood production. However, other concerns (recreation, wildlife, biodiversity) guide the management of your forests. To communicate effectively with your peers and the public, you should be prepared to concisely put into perspective the various philosophies of management and how they relate to the land you manage. To help you understand these issues, develop a short report that describes the factors that influence the three views of sustainability of natural resources: timber production, multiple uses, and ecosystems.

2. *Your view of sustainability philosophies.* As a student in a natural resource management program, at this point in your career, which of the three views of sustainability is most influential and important to you? Why?

3. *Sustainability philosophies around the world.* Why would the governments of different countries promote and accept (by focusing on particular outputs) the different views on sustainability of natural resources?

4. *Sustainable management of private lands.* Assume that you are a consultant in South Carolina. You are advising a local dentist on the management of his land (2,500 acres). The dentist would like the land to produce some income, but he also uses the land for hunting and is concerned with improving the quality of streams and ponds on the land. In essence, he is not quite sure what philosophy should guide the management of his land. What types of trade-offs should this private landowner ponder when developing a management plan that provides:
   a. A sustainable flow of timber production
   b. A sustainable flow of multiple uses
   c. Functions and processes that lead to a sustainable ecosystem

5. *Sustainability across the broader landscape.* Assume you are a senior member of the forestry staff charged with the management of a large private forest in western Washington. You are assigned to a watershed planning panel that will be used to guide the development of landscape management scenarios for an area where your company owns a significant amount of land. What are some of the broader landscape issues related to the development of a landscape management plan that include provisions for "sustainability"?

REFERENCES

Alston, R.M., 1992. History of sustained yield in the United States (1937–1992). In: Gundermann, E. (Ed.), Sustained Yield, Historical, Economic and Political Aspects. Proceedings Berichte Comptes Rendus Actas. Danish Forest and Landscape Research Institute, Lyngby, pp. 19–30.

Baker, F.S., 1933. The timber problem in conservation. J. For. 31 (2), 167–171.

Boyce, C.W., 1931. Miracles. J. For. 29 (3), 391–397.

California Department of Forestry and Fire Protection, 2013. California Forest Practices Rules 2103. California Department of Forestry and Fire Protection, Resource Management, Forest Practice Program, Sacramento, CA.

Churchill, D.J., Larson, A.J., Dahlgreen, M.C., Franklin, J.F., Hessburg, P.F., Lutz, J.A., 2013. Restoring forest resilience: from reference spatial patterns to silvicultural prescriptions and monitoring. For. Ecol. Manage. 291, 442–457.

Clawson, M., Sedjo, R., 1984. History of sustained yield concept and its application to developing countries. In: Steen, H.K. (Ed.), History of Sustained-Yield Forestry: A Symposium. Forest History Society, Durham, NC, pp. 3–15.

Commonwealth of Australia, 2016. Regional Forest Agreements. Department of Agriculture, Fisheries and Forestry, Commonwealth of Australia, Canberra. Available from: http://www.agriculture.gov.au/forestry/policies/rfa (Accessed 5/7/2016).

Dauber, E., Fredericksen, T.S., Peña, M., 2005. Sustainability of timber harvesting in Bolivian tropical forests. For. Ecol. Manage. 214 (1–3), 294–304.

Farrell, E.P., Führer, E., Ryan, D., Andersson, F., Hüttl, R., Piussi, P., 2000. European forest ecosystems: building the future on the legacy of the past. For. Ecol. Manage. 132 (1), 5–20.

Ferguson, I., 2007. Integrating wood production within sustainable forest management: an Australian viewpoint. J. Sustain. For. 24 (1), 19–40.

Food and Agriculture Organization of the United Nations, 2010. Global Forest Resources Assessment 2010. Food and Agriculture Organization of the United Nations, Rome, FAO Forestry Paper 163.

Gale, R.P., Cordray, S.M., 1991. What should forests sustain? Eight answers. J. For. 89 (5), 31–36.

Galloway, G.E., Stoian, D., 2007. Barriers to sustainable forestry in Central America and promising initiatives to overcome them. J. Sustain. For. 24 (2/3), 189–207.

Hundeshagen, J.C., 1826. Die Forstabschätzung auf neuen wissenschaftlichen Grundlagen, Laupp, Tübingen, Germany. 428 p.

Irland, L.C., 2003. Capping the cut: preliminary analysis of alternative mechanisms. J. Sustain. For. 17 (4), 25–46.

Kammesheidt, L., Torres Lezama, A., Franco, W., Plonczak, M., 2001. History of logging and silvicultural treatments in the western Venezuelan plain forests and prospect for sustainable forest management. For. Ecol. Manage. 148 (1–3), 1–20.

Kennedy, J.J., Thomas, J.W., Glueck, P., 2001. Evolving forestry and rural development beliefs at midpoint and close of the 20th century. For. Policy Econ. 3 (1–2), 81–95.

Kirwan, J., Bond, B., 2007. Forests and forest products at the time of Jamestown. Virginia For. 63 (2), 4–7.

Köpf, E.U., 2012. Sustainable forestry in financial times. Leśne Prace Badawcze 73 (2), 175–181.

Majid-Cooke, F., 1995. The politics of sustained yield forest management in Malaysia: constructing the boundaries of time, control and consent. Geoforum 26 (4), 445–458.

Oliver, C.D., 2003. Sustainable forestry: what is it? How do we achieve it? J. For. 101 (5), 8–14.

Ontario Ministry of Natural Resources, 2009. Forest Management Planning Manual for Ontario's Crown Forests. Queen's Printer for Ontario, Toronto, ON, 447 p.

Park, T.-S., 1990. Republic of Korea. Forestry Resources Management, Report of the APO Symposium on Forestry Resources Management. Asian Productivity Organization, Tokyo, pp. 231–246.

Pinchot, G., 1905. A Primer of Forestry. Part II—Practical Forestry. US Department of Agriculture, Bureau of Forestry, Washington, DC, Bulletin No. 24. 88 p.

Pulla, S., Ramaswami, G., Mondal, N., Chitra-Tarak, R., Suresh, H.S., Dattaraja, H.S., et al., 2015. Assessing the resilience of global seasonally dry tropical forests. Int. For. Rev. 17 (S2), 91–113.

Rametsteiner, E., Simula, M., 2003. Forest certification—an instrument to promote sustainable forest management? J. Environ. Manage. 67 (1), 87–98.

Recknagel, A.B., 1930. Sustained yield for a permanent and sufficient supply of forest products. J. For. 28 (8), 1053–1056.

Riihinen, P., 1992. Sustained yield in a changing world. In: Gundermann, E. (Ed.), Sustained Yield, Historical, Economic and Political Aspects. Proceedings Berichte Comptes Rendus Actas. Danish Forest and Landscape Research Institute, Lyngby, pp. 47–56.

Runyon, K.L. 1991. Canada's Timber Supply: Current Status and Outlook. Forestry Canada, Maritimes Region, Fredericton, NB. Information Report E-X-45. 132 p.

Sample, V.A., 2004. Sustainability in Forestry: Origins, Evolution and Prospects. Pinchot Institute for Conservation, Washington, DC, Discussion Paper 6-04. 43 p.

Sample, V.A., 2005. Sustainable forestry and biodiversity conservation: toward a new consensus. J. Sustain. For. 21 (4), 137–150.

Schloegel, C., 2007. Sustainable tourism: sustaining biodiversity? J. Sustain. For. 25 (3/4), 247–264.

Schmithüsen, F., 2013. Three hundred years of applied sustainability in forestry. Unasylva 64 (1), 3–11.

Siry, J.P., Cubbage, F.W., Ahmed, M.R., 2005. Sustainable forest management: global trends and opportunities. For. Policy Econ. 7 (4), 551–561.

Siry, J.P., Bettinger, P., Merry, K., Grebner, D.L., Boston, K., Cieszewski, C. (Eds.), 2015. Forest Plans of North America. Academic Press, New York, NY, 458 p.

Sverdrup, H., Stjernquist, I., Thelin, G., Holmqvist, J., Wallman, P., Svensson, M., 2005. Application of natural, social, and economical sustainability limitations to forest management, based on Swedish experiences. J. Sustain. For. 21 (2/3), 147–176.

Türker, M.F., 2007. Importance of the Turkish forestry sector in the national economy: an input-output analysis. In: Dubé, Y.C., Schmithüsen, F. (Eds.), Cross-Sectoral Policy Developments in Forestry. CABI, Oxfordshire, pp. 190–194.

US Department of Agriculture, Forest Service, 1986. Prescott National Forest land and Resource Management Plan. US Department of Agriculture, Forest Service, Southwestern Region, Albuquerque, NM.

US Department of Agriculture, Forest Service, 2005. Cost-Benefit Analysis. The Final Rule (36 CFR 219) for National Forest Land Management Planning. US Department of Agriculture, Forest Service, Washington, DC. Available from: http://www.fs.fed.us/emc/nfma/includes/cba2.pdf (Accessed 12/31/2015).

US Department of Agriculture, Forest Service, 2013. Huron-Manistee National Forests, Land and Resource Management Plan (as Amended January 2012). US Department of Agriculture, Forest Service, Huron-Manistee National Forests, Cadillac, MI.

US Department of Agriculture, Forest Service, 2015. Land Management Plan, 2015 Revisions. Idaho Panhandle National Forests. US Department of Agriculture, Forest Service, Northern Region, Missoula, MT, 187 p.

Verbij, E., Turnhout, E., Schanz, H., 2007. Comparative analysis of framing the 'forest sector': case studies from Austria and The

Netherlands. In: Dubé, Y.C., Schmithüsen, F. (Eds.), Cross-Sectoral Policy Developments in Forestry. CABI, Oxfordshire, pp. 174–182.

Waggener, T.R., 1982. What are acceptable criteria for comparing sustained yield alternatives? In: LeMaster, D.C., Baumgartner, D.M., Adams, D. (Eds.), Sustained Yield. Washington State University, Cooperative Extension, Pullman, WA, pp. 131–139.

Watson, R., 1921. National needs and sustained annual yield of the nation. J. For. 19 (4), 390–393.

Wiersum, K.F., 1995. 200 years of sustainability in forestry: lessons from history. Environ. Manage. 19 (3), 321–329.

Winters, R.K. (Ed.), 1977. Terminology of Forest Science Technology Practice and Products. Society of American Foresters, Bethesda, MD.

Wright, P.A., Alward, G., Hoekstra, T.W., Tegler, B., Turner, M., 2002. Monitoring for Forest Management Unit Scale Sustainability: The Local Unit Criteria and Indicators Development (LUCID) Test. US Department of Agriculture, Forest Service, Inventory and Monitoring Institute, Ft. Collins, CO, Report No. 4. 370 p.

Xu, Z., Bradley, D.P., Jakes, P.J., 1995. Measuring forest ecosystem sustainability: a resource accounting approach. Environ. Manage. 19 (5), 685–692.

Young, W., 1984. Development of sustained yield forest management in British Columbia. In: Steen, H.K. (Ed.), History of Sustained-Yield Forestry: A Symposium. Forest History Society, Durham, NC, pp. 220–225.

Chapter 10

Models of Desired Forest Structure

Objectives

The control of forest structure and other natural resources is an aspiration of many management plans, especially those plans being developed for governmental organizations; however, as forest certification expands many private companies are including desired future conditions in their forest plans. See Chapter 15, Forest Certification and Carbon Sequestration, for more details on forest certification. Forest structure classes can be defined by stand density, size, or age characteristics. Stand age classes commonly are used to visualize forest structure, thus most of our examples in this chapter are based on the arrangement of age classes within an ownership. The desired forest structure of a landowner or land manager acts as a guide to many of the ensuing management decisions. After reviewing the discussion provided in this chapter, you should be able to understand:

1. The concept of the normal forest structure and why it was (or still is) important.
2. The potential shortcomings of a normal forest structure.
3. Why, and under what conditions, we can argue that a normal forest is at equilibrium.
4. The concept and use of normal yield tables.
5. How the regulated forest is different than the normal forest.
6. How a forest managed using the natural range of variability of forest types might be different from a forest managed to regulate timber harvests.
7. How the normal, regulated, or historical range of variability forest concepts might apply differently to different landowner groups.

I. INTRODUCTION

Every landowner or land manager has a vision of what their forest should look like (character, species, etc.) in the future. This vision often is used as a guide to manage the land. These desired future conditions can be formalized in a forest management plan and used as outcomes that are assumed achievable through the implementation of forest management activities. While we emphasize the condition of the forest resource in this chapter, some organizations have carried the desired future condition concept further to reflect their vision for other resources such as recreation areas, wildlife habitat, and aquatic resources. In addition to specific desired future conditions noted in their management plan for each management area, the Huron-Manistee National Forests in Michigan (US Department of Agriculture and Forest Service, 2013) described a number of desired conditions as part of their forest-wide management direction. A few of these include:

> The total of early successional habitat less than or equal to 15 years, and open-land habitat, such as agricultural, urban development and roads, should generally not exceed 66 percent of the area within any 6th level watershed on the forests.
>
> Areas with unique character are protected.
>
> Habitat needs of riparian-dependent species are met and that habitat is maintained, especially habitat for threatened, endangered and sensitive species.
>
> The cumulative amount of streamside stabilization over time does not exceed five percent of the total shoreline length of a river system within National Forest System boundaries.

The detail and extent of the vision for future forest structure will vary among landowners, based on the sustainability philosophy that guides forest management, the objectives for managing land, and other external influences, such as policies and regulations that influence management efforts. For example, it may be important for landowners seeking a sustainable timber supply to develop forest plans that move the managed forest estate to a fully regulated and sustainable system through a variety of management activities. What follows is a discussion of some classical thoughts on models of desired future conditions, as well as some recent alternative thoughts, all of which can influence the structure and composition of the future forest.

II. THE NORMAL FOREST

Ensuring regularity in forest production has long been discussed; in fact Hundeshagen (1826) is credited as the first to describe a "normal forest" in the early 1800s as a way to ensure the sustainability of forest production in

Forest Management and Planning. DOI: http://dx.doi.org/10.1016/B978-0-12-809476-1.00010-2

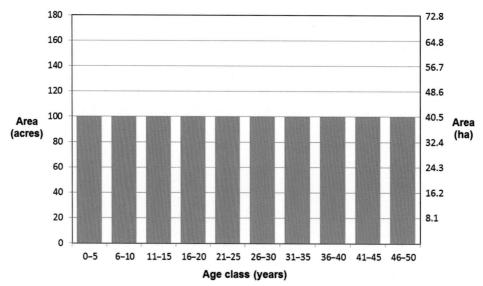

FIGURE 10.1 One example of the age class distribution of a normal forest.

Europe. The normal forest model is considered suitable primarily for even-aged forests where clearcutting is the main regeneration technique. The basic tenet is that there are equal areas of forest land in each age class (Fig. 10.1). The oldest age class corresponds to the desired rotation age. In a normal forest, no stand is allowed to grow in age beyond the desired rotation age. However, in the adjustment period between a nonnormal forest and a normal forest, some stands may be harvested at ages beyond the desired rotation age. Given the tidiness of these assumptions, we might consider this course of management to represent the outcome of a rational decision-making process. Chapman (1950) stated that a normal forest is the "ultimate form of the forest," developed and maintained by regulating timber production through regeneration practices.

In the course of managing a landbase as a normal forest, a land manager has to make one basic assumption—the rotation age. The rotation age sometimes is chosen as the one that produces the maximum sustainable timber harvest volume per unit of land over time (i.e., perfect sustainability of timber harvest volume), which is the point at which mean annual increment is maximized, often referred to as the *culmination of mean annual increment*. Other times, it is chosen based on an economic analysis (maximization of net present value or internal rate of return) or based on local or regional markets and practices (concepts discussed in Chapter 5: Optimization of Tree- and Stand-Level Objectives). Why is this important? Because the land manager then must adjust the structure of the forest such that there is an equal amount of area of forest land in each age class ranging from zero (recently clearcut) to the desired rotation age.

To determine how much area should be contained within each age class, a simple relationship is used:

$$\text{Area in each age class} = \left(\frac{\text{Total area of the forest}}{\text{Desired future rotation age}} \right)$$
$$\times (\text{Length of a planning period})$$

This assumes, of course, that the age class, the desired future rotation age, and length of the planning period all are measured in the same units (e.g., years).

Example

A landowner in Mississippi manages 27,000 acres of land, 16,500 acres of which are suitable for growing pine plantations. If the landowner wanted to create a normal forest on the area suitable for pine plantations, and the landowner has decided to use a 26-year rotation age for pine plantations as it maximizes the discounted net revenue, then how many acres would the landowner need in each 1-year age class? Regardless of the initial forest structure the landowner would need to have about 635 acres in each 1-year age class to obtain a normal forest.

$$\text{Area in each age class} = \left(\frac{16,500 \text{ acres}}{26 \text{ years}} \right)(1 \text{ year})$$

$$= 634.6 \text{ acres}$$

If a landowner managed a forest with the intent of it becoming a normal forest, and the landowner intended as well to maximize the mean annual increment of the forest, then the normal forest would contain and carry from one time period to the next the maximum timber volume possible. This level or amount of timber volume available is called the "normal growing stock." Associated with this is a "normal increment," where all stands have the same growth or productivity rate. In other words, all stands are assumed to have the same site index. In addition, the normal increment also assumes that the forest is healthy and fully stocked.

Regardless of whether the mean annual increment is being maximized, when a landowner is planning for the management of a normal forest, a number of other land and forest assumptions are incorporated into the planning process. For example, as we mentioned, regardless of the diversity of the actual forest, site index values are assumed constant across the entire area. Since the area harvested within a normal forest will be constant (same amount of land each year), this assumption prevents major deviations in projected harvest levels due to differences in stand productivity. In addition, when managing for a normal forest, we assume that the tree species diversity is consistent from one stand to the next (e.g., same percentage of Douglas-fir (*Pseudotsuga menziesii*) and western hemlock (*Tsuga heterophylla*) in each stand). Therefore, stands contained in any one age class are assumed to be identical in tree species composition as they grow into that age class. Similarly, stands are assumed to have the same stocking, density, and structural conditions within an age class. In addition to these biological assumptions, three operational assumptions usually are associated with the development and maintenance of a normal forest:

- When a normal forest has been created, none of the stands in the managed area are at ages above the desired rotation age
- Final harvest (clearcut) decisions are made by choosing the oldest stands first
- The same set of silvicultural management prescriptions are implemented in each stand

The final assumption prevents the occurrence of inconsistencies that may arise in stand structure as a result of implementing more- or less-intensive management practices over time.

One important point about the normal forest that is worth investigating further relates to forest growth and harvest removals. Simply put, the stock (volume) of the oldest age class in the age class distribution of a normal forest is equal to the periodic growth of the forest. Here, the period length and the length (or width if you prefer) of the age classes are exactly the same. For example, if there are *n* age classes being recognized, then the volume (V_n) in age class *n* (the oldest age class) is equal to the sum of the periodic growth of all the other age classes.

TABLE 10.1 Volume Per Acre for Loblolly Pine Stands

Age Class	Volume (cords/ac)
0–5	0
6–10	4
11–15	12
16–20	20
21–25	28
26–30	34
31–35	39
36–40	43
41–45	46
46–50	48

Example

Assume a landowner owns 1,000 acres of loblolly pine (*Pinus taeda*) forestland in Alabama, which is represented by a normal forest consisting of 10 age classes of equal area (100 acres each). Assume also that the 10 age classes are 5-year groupings: 0–5, 6–10, ..., 45–50. And assume that the volumes per acre, based roughly on site 65 (base age 25) yields from the Service Forester's Handbook (US Department of Agriculture and Forest Service, 1986), in the middle of each age class, are reflected by those shown in Table 10.1. If over a 5-year period the landowner were to harvest the oldest age class (46–50 year old stands), then they would generate about 4,800 cords of volume (48 cords per acre × 100 acres). The sum of the periodic growth, when the normal forest is allowed to grow 5 years into the future, would be reflected by the change in volume of each age class, as illustrated in Table 10.2. The sum of the periodic growth is therefore 4,800 cords, which is equal to the amount of the previous harvest, and also equal to the amount of the stock (volume) in the oldest age class (i.e., the new volume of new age class 46–50).

When these conditions exist, a normal forest is considered to be in equilibrium (from a timber production point of view), where the periodic growth of the forest equals the periodic timber removals. Another way to think about this transition through time is that the annual timber harvest level gets replaced each time period by the periodic growth of the forest.

The idea of the normal forest persists today as a management paradigm within which a landowner might operate if he or she were concerned about the sustainability of timber harvest volume, revenue, jobs, and other socioeconomic realities. The main advantages of maintaining a normal forest are that the size of the

TABLE 10.2 Periodic (5-year) Growth of a Loblolly Pine Stand

Old Age Class	New Age Class	Old Volume per Acre (cords)	New Volume per Acre (cords)	Difference (cords)	Growth[a] (cords)
0–5	6–10	0	4	4	400
6–10	11–15	4	12	8	800
11–15	16–20	12	20	8	800
16–20	21–25	20	28	8	800
21–25	26–30	28	34	6	600
26–30	31–35	34	39	5	500
31–35	36–40	39	43	4	400
36–40	41–45	43	46	3	300
41–45	46–50	46	48	2	200
46–50	0–5	48	0	–	0
Total				48	4,800

[a] 100 acres times the difference in growth.

periodic harvest area is constant, as is the size of the periodic harvest volume. This, in theory, could facilitate stabilization of the local economies that depend on the forest products produced, or those processing facilities that depend on consistent flow of raw materials. However, it may ignore the impact of market volatility of prices. In addition, managing under a normal forest paradigm could facilitate a smooth budgeting and planning process, since there would be little or no variance in the land areas that need to be treated (thinned, fertilized, etc.) from 1 year to the next. The desire to produce an even-flow of wood continues to be a goal of many large, integrated organizations. However, given the globalization of the timber (or forest) industries, lumber and pulp mills may encounter other market forces that influence how much timber can be processed. Because of this, the argument that local communities would be sustainable today simply because landowners develop normal forests may be unsupported.

The concept of the normal forest has other drawbacks as well. For example, in today's management environment ecological goals may have equal or greater importance than do economic and social goals, and these ecological goals may not be addressed when managing land under the assumptions of a normal forest, as it may include no reserve areas. In the normal forest model, the structural and biological diversity commonly found in natural forests may be lacking, given the need to assume that each age class contains a uniform stand of trees of uniform stocking. Along these lines, some critics (Kant, 2003) argue that

the normal forest concept is flawed because of its production orientation, and therefore forests managed in this manner may not be sustainable from a multiple-use or ecosystem process point of view (see Chapter 9: Forest and Natural Resource Sustainability).

From an operational perspective, there are a few problems as well with the normal forest. For example, the requirement that the oldest stands be harvested first each time period may not be operationally feasible. To minimize logging costs, a land manager may group younger stands with older stands in a harvest block, creating areas of contiguous or nearby harvests that make economic sense. Thus, the move-in cost for the logging equipment is associated with the larger area. This recognizes an economy of scale related to logging operations, where small stands may be economically inefficient to harvest alone due to the cost of moving equipment. In addition, deviations from a harvest plan may be related to weather conditions or road management issues, particularly where roads need to be built to provide access. Thus, the spatial arrangement of the age class distribution is important in determining whether the oldest-first harvest rule can be implemented operationally. In addition, fluctuating annual budgets may preclude the use of consistent management practices (e.g., fertilization or herbicide treatments) on every piece of land in a normal forest. Alternatively, as new developments in management technology (e.g., site preparation, genetics) arise, they could be utilized incrementally in portions of a normal forest, which would affect the growth rates of the trees in those areas, and effectively change the site index and associated productivity levels.

Why would it be difficult to find a normal forest in nature? In addition to the previously discussed limitations associated with the normal forest, a number of other external factors may prevent a landowner from maintaining a perfectly even age class distribution composed of stands with density and stocking levels that are consistent from one age class to the next. Natural disturbances (e.g., hurricanes, fires, insect outbreaks) and unwanted human-caused disturbances (e.g., arson) can affect the age class distribution (Fig. 10.2). As we mentioned earlier, market forces may influence management decisions that result in a deviation from the normal forest concept. For example, when timber prices are high at a mill, a landowner may tend to schedule harvests on more land area than that which is suggested by this model (Fig. 10.3). Alternatively, as timber prices decline, lower levels of harvests may result. Although the normal forest is a management paradigm that some landowners may continue to desire, the reordering of their inherently heterogeneous forest structures (Fig. 10.4) to produce a normal forest may require a considerable amount of effort.

A number of timber volume or yield tables used in forest management over the last 90 years are based on the

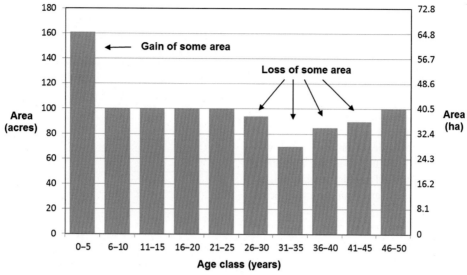

FIGURE 10.2   Impact of natural disturbances on the normal forest age class distribution.

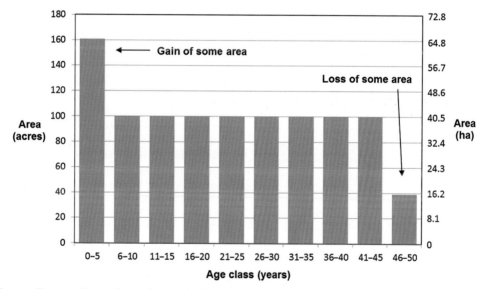

FIGURE 10.3   Impact of increased harvest levels, due to rising forest products prices, on the age class distribution of a normal forest.

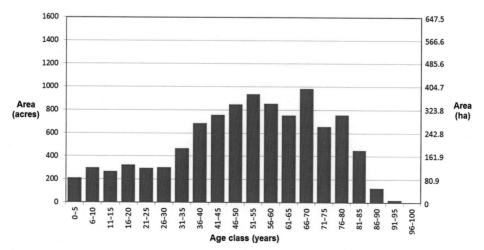

FIGURE 10.4   A heterogeneous older forest age class structure.

concept of the normal forest. Standard procedures for the development of *normal yield tables* were developed in the 1920s, and although the idea may be outdated, it is still used in some organizations, and thus worth exploring further. Normal yield tables are based on stands that are considered fully (100%) stocked. The tables present average yields for stands of "normal" stocking, and have been developed for numerous combinations of tree species, site indices, and ages. In contrast to this, *empirical yield tables* have been developed since to provide users with average volumes per unit area under actual conditions (i.e., not necessarily fully stocked stands), thus empirical yields per unit area are usually less than the corresponding normal yields.

To be able to construct a normal yield table, we would need to measure plots in forests where full stocking of trees is assumed. This usually implies the areas being measured did not undergo any of the intermediate stand treatments that are common today (e.g., thinnings), which could reduce tree stocking levels below full stocking for some period of time. Thus the use of normal yield tables has tended to fade as foresters and land managers moved from one-cut-per-rotation management of natural stands to intensive silvicultural management with one or more intermediate density-management treatments, such as precommercial or commercial thinnings (Curtis and Marshall, 2004). However, normal yield tables are still useful today for (1) serving as a reality check on other growth and yield estimates, and (2) estimating timber volumes across geographic regions that lack other readily available data (Curtis and Marshall, 2004).

To estimate timber volumes for a stand of trees using a normal yield table, we need to know three things about the stand: the stand age, the site index, and the tree stocking relative to a normal stand (i.e., 100% stocking). Age can be obtained from increment cores or stand history records. Site index can be obtained from height and age measurements in conjunction with site index equations. Stocking can be estimated from a stocking guide, which generally is based on trees per acre, basal area, or average diameter. Stocking can also be estimated simply by comparing basal area at the stand's age to the basal area of a normal forest at the same age. After determining the stocking level, we need to estimate stocking relative to a normal stand. So, for example if the stand of interest has 90 ft^2 per acre of basal area, and a normal stand should have 100 ft^2 per acre at the same age, then the stand of interest is said to be 90% stocked (90 ft^2 per acre/100 ft^2 per acre) relative to the normal stand.

Example

Assume you are using the normal yield table for shortleaf pine (*Pinus echinata*) shown in Table 10.3, and your

TABLE 10.3 Normal Yield Table for Second-Growth Shortleaf Pine Stands 4 Inches DBH and Over, in Cords Per Acre

Age	Site Index (base age 50)		
	70	**80**	**90**
20	18	25	30
25	31	38	43
30	41	48	54
35	49	57	64
40	56	65	73
45	61	71	80
50	66	77	87

Source: US Department of Agriculture, 1929. Volume, Yield, and Stand Tables for Second-Growth Pines. US Department of Agriculture, Washington, DC. Miscellaneous Publication No. 50. 202 p.

stand of interest is 40 years old, located on site index 80 (base age 50) land, and is 90% stocked relative to a normal stand. What would be your estimate of the cords per acre for your stand? If your stand is 90% stocked relative to a normal stand, then you could estimate the volume to be 58.5 cords per acre (0.9 × 65 cords per acre). We could also use the normal yield table to estimate future yields, based on the notion that the stand of interest will grow proportionally to a normal stand. So 5 years from now, your stand of interest would be expected to yield about 63.9 cords per acre (0.9 × 71 cords per acre).

Although the intervals of site indices and stand ages may be fairly broad in a normal yield table (10 ft and 5 years in the previous example), it is possible to interpolate between the values for stands that do not exactly fit the values provided. For example, assume the site index of our stand of interest was 85 (base age 50), which is not listed in the table. Also assume that the stand is 30 years old, and is 80% stocked relative to a normal stand. The difference between site index 80 and site index 90 at age 30 is 6 cords per acre (54−48 cords per acre). Since site index 85 is half-way between site index 80 and site index 90, and since half of the difference is 3 cords per acre, we could use this volume to develop (using a linear interpolation of the data) an estimate of the normal volume (100% stocking) at site index 85 (51 cords per acre). Our stand, however, is 80% stocked relative to normal stocking, so 80% of 51 cords per acre provides us with a volume estimate (40.8 cords per acre) for our stand of interest.

Example

Assume that you manage a shortleaf pine stand that is 33 years old, has a site index (base age 50) of 70, and is 90% stocked relative to a normal forest. What would be your estimate of the volume per acre using the normal yield table presented in Table 10.3? To begin, the difference between the volume per acre at ages 35 and 30 is 8 cords (49−41 cords per acre). Stand age 33 can be described as being 60% of the way (3/5 years) between ages 30 and 35; therefore, 60% of the difference in volume at those ages (8 cords per acre) is 4.8 cords per acre. Adding this to the volume of a normal stand at age 30 results in a normal stand (100% stocked) volume of 45.8 cords per acre (41 cords per acre + 4.8 cords per acre). Finally, 90% of 45.8 cords per acre provides us with an estimate (41.2 cords per acre) of the volume per acre for a 33-year-old stand that is 90% stocked relative to normal stocking.

As we noted earlier, we must keep in mind that normal yield tables are based on stands that are assumed fully stocked. These stands are difficult to find in nature. However, to develop the normal yield tables some subjective judgment was used to locate these stands, and the corresponding caveats were provided, as is noted in a US Department of Agriculture (1929) publication:

In selecting plots some leeway as to the meaning of full stocking was necessary in order that a sufficient number might be found without unreasonable expense. Since the tables give the average figures for the plots studied, they do not, strictly speaking, represent maximum possible volume. The additional fact that the great majority of the stands chosen had at one time or another been burned over (few unburned areas are known) and have developed naturally, should also assure higher yields than are shown in the tables, when protection and management are introduced.

Finally, some normal yield tables may have been developed for very broad areas, such as the extent of the southern pine region in the southern United States (Virginia to Texas), or the extent of the coastal Douglas-fir region in the Pacific Northwest United States (Washington to California). Since a representative sample of measurements covering all parts of a broad region may have been used to develop a normal yield table, the information contained within the table may need to be adjusted for local conditions. In addition, some adjustment to the normal yield tables may be necessary for differences in utilization standards. For example, if a normal table provides estimates of a board foot volume to a 4-inch top in a tree, and the merchantability standard for your working area (or mill) was a 6-inch top, then the yield you would be interested in using would be more or less than what might be indicated in the table, depending on the relationship between your merchantability standard and the standard utilized to develop the table.

III. THE REGULATED FOREST

A regulated forest is similar to a normal forest in that the goal of management is to produce a predictable and sustainable harvest of timber. However, when managing land using a regulated forest paradigm, some of the assumptions associated with the planning of the normal forest are relaxed. For example, the level of sustainable harvests (and hence rotation age assumed) could vary slightly within a forest, and can be sustained using a range of different management intensities (Sample, 2004). As with the normal forest, a landowner would need to define the desired rotation age, perhaps using an economic or biological criteria. If the objective of the management of the forest was to maximize the sustainable timber harvest volume, then the length of rotation generally is defined according to a biological rule, where mean annual increment is at its maximum (Sample, 2004).

Ultimately, if we want to develop a plan for an area of land that will lead to a regulated forest, then the aim is to approximate the normal forest. However, in a regulated forest, full stocking is not assumed to occur all the time on every unit of land, thus the size and quality of the timber produced can fluctuate over time, but remain about the same. As a consequence, there exists a stable relationship between inventory, harvest, and growth of the forest within a regulated forest (Beuter, 1982). Another difference between the regulated and normal forests is that site index values are allowed to vary across the landscape in a regulated forest, and are not necessarily assumed to be constant, as in the case of a normal forest. We can also assume with a regulated forest that there is some diversity in tree species across the landscape, and that a different mix, or percentage, of species may be present in each stand. Therefore, stands in any one age class are not identical to others as they grow into the next age class, and we would need to acknowledge the potential differences in stocking, density, and structural conditions. The three operational assumptions associated with the normal forest still hold, however:

- None of the stands are at ages beyond the desired rotation age
- Final harvest (clearcut) decisions are made by choosing the oldest stands first
- The same set of management prescriptions are used in each stand

In addition, the biological and operational caveats to these assumptions that were discussed earlier still apply.

Why would a landowner or an organization want to move an unregulated forest, one that violates these assumptions, toward a regulated state? The main reason often argued is that there is a need for more certainty in the outcomes available in the future (the structure of the forest, the timber yields, etc.) and more simplicity in the decision-making process. As we noted in our discussion of a normal forest, this argument tends to ignore or assume irrelevant many environmental, social, and economic uncertainties that could occur over time (e.g., fires, changes in markets). In any event, one of the main emphases of many of the US National Forest plans developed in the 1980s and early 1990s was the development of a fully regulated forest. For example, the forest plan for the Umpqua National Forest (US Department of Agriculture and Forest Service, 1990) in Oregon noted the following:

Areas with programmed timber harvest will be[come] a mosaic of stands of various sizes and ages. The desired condition of this available commercial forest land is that of a regulated forest where the stands exist in age and size class proportions and grow at rates such that a high level of timber yield can be sustained.

Of course, goals related to numerous other natural resources (wildlife, recreation, fisheries, etc.) were included in these National Forest plans; thus the regulated forest that was suggested as a desired future condition pertained to only a portion of the National Forest that was allocated to commercial timber harvesting.

The regulated forest concept has guided the development of forest plans throughout the world, although as Beuter (1982) notes, it may be more of an ideal that guides forest management decisions rather than the actual objective of forest management. For example, we could develop a forest plan with the objective of maximizing net present value, and the resulting forest plan may not lead to a regulated state (depending on the initial age class distribution of the forest). However, add to the forest plan policy constraints for controlling periodic wood flows, cash flows, and areas treated, and we might be able to develop a plan that could lead to a regulated state (Beuter, 1982).

In recent times, the regulated forest concept has come under scrutiny concurrent with the rise of importance of ecological and social values associated with forest management. Unfortunately, managers who utilize and adhere to the regulated forest concept are said to view the forest only for the commodity benefits they provide (which is not necessarily true), and are thought to ignore other intrinsic values (e.g., existence values of old growth, wilderness areas, and ecosystems). Further, the measures of stability that are used to promote the use of the regulated forest concept (income, jobs, etc.) may be affected more by forces outside the local area, such as the globalization of markets, the economic diversification of the timber industry, and changes in interest rates (Lee, 1983), than by scheduled timber harvest volumes.

IV. IRREGULAR FOREST STRUCTURES

Though the concepts of the normal or regulated forests may have influenced the development of many forest

FIGURE 10.5 An irregular forest structure illustrated in an aerial photograph of an area in western Washington.

plans, in actuality, most managed forests are composed of an irregular age class structure, which may be obvious in a graph of the age class distribution or when the landscape is viewed from above (Fig. 10.5). Periodic land sales, land purchases, land trades, natural disturbances, and increases or decreases in harvest rates all contribute to departures from management plans designed to lead to a regulated state. The main issues facing landowners and land management organizations concern whether their age class distribution of forests is manageable given their objectives, whether the current age class distribution simply represents a transition to some other (perhaps normal or regulated) state, or whether some other action (land sale or land purchase) is necessary to better achieve their objectives.

Without belaboring the numerous variations of irregular forest structures, examining three representative cases should be sufficient to describe this catch-all category of forest conditions. The first example (Fig. 10.6) represents the age class distribution of a young forest, where the area by age class is biased toward nonmerchantable, younger stands. A forest that may be over-cut could be described in this manner, as could some parcels of forest that recently have been sold to private landowners by industrial companies who are in the process of reducing the size of their land base. In addition, adding these forests to a landbase that is currently somewhat regular in structure effectively would make it irregular in structure. Natural disasters (e.g., the Mt. St. Helens eruption in 1980, Hurricane Katrina in 2005, or the widespread fires in Alberta in 2016) could also change the composition

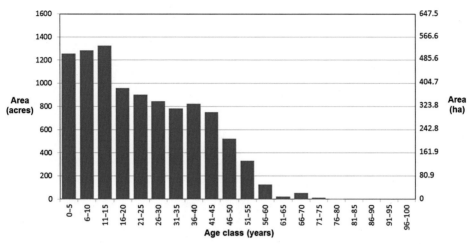

FIGURE 10.6   A young forest age class structure.

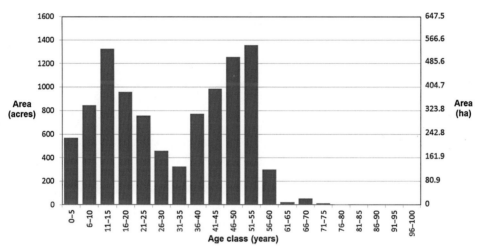

FIGURE 10.7   A bimodal forest age class structure.

of forests on a widespread basis and effectively create younger, irregular forest structures.

The second example, which was presented earlier (Fig. 10.4) represents an older forest, with the area by age class biased toward older stands. Today, we might find these types of forest structures being managed by nonindustrial private landowners, some state and federal agencies, and even some military bases, where a limited amount of harvesting has taken place since acquiring or purchasing the property. In addition, these forests have grown for quite some time without experiencing a major natural disturbance that effectively would transition the forest to a younger age structure. Finally, as a third example, we might find a bimodal distribution of age classes commonly managed by some landowners (Fig. 10.7). A variety of events may have led to this, from land sales and purchases (disrupting a more normal distribution), to natural disturbances (shifting the distribution of age classes), to changes in markets (leading to variable

harvest levels from year to year). The gap in the age class structure may eventually be a concern, if economic or commodity production goals are important.

Example

Using the data provided for the Lincoln Tract, and assuming 5-year age classes, which of the models described so far does the current forest age class structure resemble? As you might conclude from viewing Fig. 10.8, the age class distribution of the Lincoln Tract, which is composed mainly of even-aged stands, represents neither a normal, nor a regular forest structure. In fact, it is very irregular. With the exception of the first two classes and a few gaps (26–30, 41–45, 56–60), there seems to be a downward-sloping relationship between the amount of area by age class and the age class itself. Between 11 and 25 years prior to current time, almost 1,800 acres (almost 40% of the Tract) were regenerated. In addition to the minor gaps in the age class distribution and this overweighting

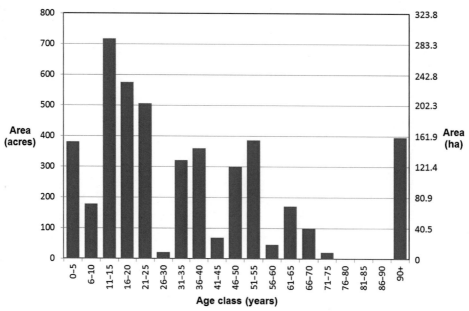

FIGURE 10.8 The age class distribution of the Lincoln Tract.

(perhaps) of younger age classes, there is almost a 20-year age class gap between the older forest structure (90+ years of age, and about 9% of the area) and the next lowest age class. As a result, moving this Tract toward a regulated forest structure, if that were the landowner's desire, may take quite some time.

V. STRUCTURES GUIDED BY A HISTORICAL RANGE OF VARIABILITY

Using the natural range of variability to guide the distribution of forest classes is one theme that permeates US National Forest planning today. It is hard to argue with the fact that over 100 years of forest management in the United States has guided some ecosystems outside of their natural range of variability. However, what is debatable is whether future management policy should be designed to move ecosystem processes back within their natural range of variability. If we were to adopt this perspective as a guiding paradigm for a management plan, then we may need to integrate natural and human disturbances into the planning process, for example, to adjust the structure of a forest to a range of variability that is acceptable. In developing a plan that emulates the natural range of variability, two planning elements are needed, (1) successional rules, which describe the potential changes of a stand or forest from one state (or condition) to the next, and (2) transition probabilities, which are used to simulate stand or forest changes over time.

The key to using this system of management is to determine the "natural range" of variability for the

resource in question, or the natural fluctuation in resource levels that existed prior to extensive human influence. This information can be derived by using:

- Dendrochronology to determine age structures and growth patterns
- Old land surveys and the information surveyors recorded about the landscape to infer something about variability
- Palynology (carbon or pollen stores in lakes, ponds, or wetlands) to understand changes in forest structure
- Information about the natural range of variability of forests in other regions to guide management actions (Committee of Scientists, 1999; Seymour et al., 2002).

Depending on the scale and size of the system being managed, acquiring actual measurements of variability may be impossible. For example, estimating the natural range of variability of southern United States forest structure, using current forests as examples, may fail to recognize types of forests that were present 200 or more years ago. In addition, dendrochronology and palynology require actual field measurements, and may require extensive sampling, and therefore may be quite expensive. Scale also matters, as the variability inherent in a small watershed may not be similar to the variability inherent across a broader landscape or region. In addition, the concept of managing landscapes and forests for the natural range of variability is not limited solely to forest structure itself. For example, in British Columbia there are regulations that tie the management of the forest and range vegetation resource on government (Crown) land to the natural range of variability of stream temperatures

(Government of British Columbia, 2004). Given the amount of information necessary to make the appropriate estimates of a natural range of variability, following this course of action may be difficult.

However, using a natural range of variability for broader ecosystem processes has been proposed to guide the development of plans for US National Forests. Many National Forest plans currently are being revised (or recently have been revised), and each may contain objectives to maintain, protect, or (in some cases) restore landscapes such that they contain a structure that is representative of the natural ranges of variability that occur—or once occurred—there. The Committee of Scientists (1999) make a number of compelling points regarding management of National Forests and Grasslands within their natural range of variability:

> [the natural range of variability] is best applied to coarse attributes of the landscape: the condition of streams; the distribution among seral stages of different forest types; the amount and distribution of large dead trees; and the size, frequency, and intensity of disturbances.
>
> Some dimensions of [the natural range of variability] are difficult to reestablish within some landscapes. As an example, the forests of the Western Cascades in Oregon and Washington will not be managed for the large, infrequent, high-intensity burns that created them. It is just not socially acceptable. Such burns may occur, but not through purposeful public policy.
>
> By many measures, much of our current standard of living is based on converting landscapes to conditions outside the [the natural range of variability]. The cities and farmlands of much of America are examples. Much nonfederal land around national forests and grasslands is also outside of [the natural range of variability]. Given that we wish to retain our native species, though, maintaining at least a significant portion of the landscape within [the natural range of variability] would seem prudent.

Other land management groups have adopted the natural range of variability paradigm for forested areas, such as the fire management plan for the Bandelier National Monument in New Mexico (Rodgers, 2005). Here, the plan indicates that one management objective is to change their forest conditions by defining:

> ... levels of fire use to restore and perpetuate natural processes given current understanding of the complex relationships in natural ecosystems.

Some stakeholder groups also have sought to encourage nearby federal land management organizations to adopt the natural range of variability concept as a driving force behind resource management. In restoration recommendations to public land managers in the southwestern United States, a reintroduction and enhancement of natural processes (e.g., fire) that once helped shape forest structure prior to widespread settlement of the western United States has been suggested (Allen et al., 2002). This philosophy is based on the belief that there is a need to restore local forest structures and natural processes so that they resemble stands created and maintained within the natural range of variability.

Using the natural range of variability paradigm to influence the development of a management plan has been suggested as a way to create and maintain species composition and stocking levels that will improve the resiliency of forests to disease, drought, and wildfire. The paradigm is based on the notion that the environmental conditions that are most likely to help conserve native species and habitats are those from which they evolved (Committee of Scientists, 1999). One disadvantage of using this approach to managing the structure of a forest is that it may fail to work as planned. The fact that human interaction may be needed to move a landscape back to within a range of natural variability is at the heart of the matter. Likely, there will be a high level of active management needed to align the forest structure to meet the desired conditions, as many areas have transitioned far from historical states. Under this approach, commodities can be produced as a byproduct of the active management to physically emulate natural disturbances. However, some argue that if natural disturbances and processes created a pattern that should be emulated, then allowing natural disturbances and processes to take the landscape back to these conditions is the best course. This effectively precludes the production of commodities, and significantly reduces the financial value of the management plan. Another disadvantage, as we noted earlier, is the fact that in many cases the natural range of variability of forests, ecosystems, or disturbance agents is unknown. As a result, a number of current research opportunities for inspired graduate students focus on estimating the natural range of variability for forest, aquatic, and rangeland ecosystems.

VI. STRUCTURES NOT EASILY CLASSIFIED

Some forested areas are not specifically managed using even-aged management systems, thus cannot be described simply by differences in age classes. These include the uneven-aged coniferous forests of the interior western United States, the deciduous forests of the northeastern United States and eastern Canada, and other forests across North America managed specifically with an uneven-aged forest structure. Placing an average age on an uneven-aged stand might be possible, but the variation in ages among trees within an uneven-aged stand could be high, and thus categorizing each unit of land within an uneven-aged forest with a single age class may be inappropriate. However, regulation of uneven-aged stands minimizes the risk of overcutting during any one cutting entry, and ensures desirable outcomes during future periods of time, barring any unforeseen problems such as natural disasters. Guldin (1991) suggests that the key to utilizing uneven-aged silviculture is in finding a way to ensure a long-term sustainable harvest.

Another example of a forest structure not easily classified is the management plan for the State of Oregon's northwest forests (Oregon Department of Forestry, 2010). Although generally guided by even-aged management principles, the landscape management concepts and strategies found in the management plan for these 615,000 acres uses a structure-based management approach. This approach is similar in concept to the natural range of variability paradigm, however, silvicultural techniques are planned in such a way as to develop and maintain a collection of forest stand structures that best meets the social, economic, and ecological direction for these lands. The forest plan includes goals related to developing and maintaining a high level of sustainable timber volume, a diversity of wildlife habitats, and diverse recreational opportunities, among others. To achieve these goals, the forests are managed so that the collection of forests will have the following array of forest structures:

Regeneration areas	15—25% of the landscape
Closed, single canopy forests	5—15% of the landscape
Understory initiation in single canopy forests	30—40% of the landscape
Layered forests	15—25% of the landscape
Older forest structure	15—25% of the landscape

The desired forest structure for the State of Oregon's northwest forests was designed to emulate the diversity of forests historically found in the mountains of western Oregon, while recognizing that the actual quantity and distribution of forest types was highly variable. Exact percentages of each class are not provided in the plan, therefore allowing managers some flexibility during the implementation of the plan. Although the Oregon Department of Forestry acknowledges that objective and subjective processes were used to determine the percent of land desired in each class, a number of factors were considered, including:

- The current state of the forest conditions on state forest lands
- The available data on historical distributions of older forest structure
- The type of wildlife habitat necessary to contribute to the conservation of native wildlife species

- The collection of forest types that simultaneously could meet habitat, biodiversity, and timber management goals
- The management intentions of the other landowners in the region

This desired forest structure thus contains some guidance provided by knowledge of the natural range of variability of western Oregon forest types, but the desired structure is adjusted to reflect the social and economic factors that influence the management of the state's forests.

The Chattahoochee-Oconee National Forest (Georgia) utilizes a similar approach in their most recent forest plan (US Department of Agriculture and Forest Service, 2004):

Objective 8.A.1-01 Manage forest successional stages to maintain a minimum of 50 percent of forested acres in mid- to late-successional forest, including old-growth; a minimum of 20 percent of forested acres in late-successional forest, including old-growth; and 4 to 10 percent per decade in early-successional forest.

Here, the supply of wood derived from the objectives of the forest plan is a coproduct of the need to achieve a number of nontimber objectives. Therefore, as in the Oregon case, the silvicultural techniques are planned in such a way as to develop and maintain a collection of forest stand structures that best meets the social, economic, and ecological direction for the National Forest.

VII. SUMMARY

One of the guiding premises for management plans developed for private, state, and federal lands is the desired forest structure of the landowner. Many of the desired forest structure models are more appropriate for even-aged forests, since uneven-aged forest structure is not as easy to characterize beyond the stand level. The normal and regulated forest models provide regularity in land area treated and timber volume produced. Normal forests are the ideal, and regulated forests relax many of the assumptions to reflect variation across the landscape. Reaching this state of management requires a number of planning assumptions that may not be realistic in today's dynamic management environment. However, management plans can reflect these desired outcomes by incorporating policy constraints that ensure evenness in area treated or volume produced. Irregular forest structures are prevalent throughout North America. Land sales, land purchases, market fluctuations, natural disturbances, and the various needs of landowners all influence the development of a forest structure that may not provide the continuity of harvest suggested by a regulated forest.

Some forest plans are now being guided by desired structures that emulate a natural range of variability. Other forest plans are guided by desired forest structures that cannot be easily classified. How these plans are implemented and how effective they are in meeting a landowner's objectives remains to be seen, yet their premise (e.g., reflect natural processes) is one that is attractive to a number of land managers and stakeholders.

QUESTIONS

1. *Using normal yield tables.* You recently have been hired by a small consulting firm in southeast Virginia. You have been asked to develop an estimate of the timber volume contained in a tract you recently cruised. You noticed during the cruise that the stand had been thinned in the last 5 years. You are tempted to use a normal yield table for loblolly pine to develop a quick estimate of the volume. However, what factors should you keep in mind as you develop this estimate?

2. *School forest structure.* Using your school or college forest as an example, develop a graphical description of the age class structure of the forest resources that they manage. Develop a second, hypothetical forest structure (same number of total acres or hectares) for a period of time 50 years from the present, guided by one of the concepts described in this chapter. What management controls might be used to direct the current forest structure to the desired forest structure?

3. *Normal and regulated forests.* You have been hired recently by the Lolo National Forest in Montana. Your supervisor is participating in a forest plan revision, and has had some discussions with other people on the interdisciplinary planning team about the differences between a regulated forest structure and a normal forest structure. From a planning perspective, develop a short memorandum for your supervisor that compares and contrasts the two approaches.

4. *Natural range of variability.* You work for the Coconino National Forest in Arizona. Your team is participating in a forest plan revision, and the desire to manage a portion of the forest using a natural range of variability paradigm for wildfires has been proposed. From a planning perspective, develop a short memorandum that addresses the issues that the team should consider in association with using this management paradigm.

5. *Normal yields.* Using the normal yield table for shortleaf pine (Table 10.3), estimate the volume per acre of a 47-year-old stand, that has a site index (base age 50) of 75, and is 110% stocked relative to a normal forest.

6. *Forest structure of the Putnam Tract.* The Putnam Tract consists of a mixture of pine, hardwood, and mixed pine-hardwood stands. The hardwood and mixed pine-hardwood stands may be better associated with uneven-aged stands, thus attaching an age to these is tenuous. However, the pine stands, for the most part, are even-aged stands. Develop an age class distribution for the pine stands, and describe the type of pine forest structure currently in place on the Putnam Tract. Are the pine stands within the Putnam Tract representative of a normal forest, a regulated forest, or some other type? Explain how you came to this conclusion.

REFERENCES

Allen, C.D., Savage, M., Falk, D.A., Suckling, K.F., Swetnam, T.W., Schulke, T., et al., 2002. Ecological restoration of southwestern ponderosa pine ecosystems: a broad perspective. Ecol. Appl. 12 (5), 1418–1433.

Beuter, J.H., 1982. The economic assumptions and implications of the regulated forest. In: LeMaster, D.C., Baumgartner, D.M., Adams, D. (Eds.), Sustained Yield. Washington State University, Cooperative Extension, Pullman, WA, pp. 37–43.

Chapman, H.H., 1950. Forest Management. The Hildreth Press, Bristol, CT, 582 p.

Committee of Scientists, 1999. Sustaining the People's Lands: Recommendations for Stewardship of the National Forests and Grasslands into the Next Century. US Department of Agriculture, Washington, DC.

Curtis, R.O., Marshall, D.D., 2004. Douglas-fir growth and yield: research 1909–1960. Western J. Appl. For. 19 (1), 66–68.

Government of British Columbia, 2004. Forest and Range Practices Act: Range Planning and Practices Regulation. Queen's Printer, Victoria, British Columbia, SBC 2002, Chapter 69.

Guldin, J.M., 1991. Uneven-aged DBq regulation of Sierra Nevada mixed conifers. Western J. Appl. For. 6 (2), 27–32.

Hundeshagen, J.C., 1826. Die Forstabschätzung auf neuen wissenschaftlichen Grundlagen, Laupp, Tübingen, Germany.

Kant, S., 2003. Extending the boundaries of forest economics. For. Policy Econ. 5 (1), 39–56.

Lee, R.G., 1983. Sustained-yield and social order. In: Steen, H.K. (Ed.), History of Sustained-Yield Forestry: A Symposium. Forest History Society, Inc., Durham, NC, pp. 90–100.

Oregon Department of Forestry, 2010. Northwest Oregon State Forests Management Plan, Revised Plan April 2010. Oregon Department of Forestry, Salem, OR.

Rodgers, M., 2005. Bandelier National Monument Forest Management Plan 2005. US Department of the Interior, National Park Service, Bandelier National Monument, Los Alamos, NM.

Sample, V.A., 2004. Sustainability in US Forest Policy: Origins, Evolution and Prospects. Pinchot Institute for Conservation, Washington, DC, Discussion Paper 6-04. 43 p.

Seymour, R.S., White, A.S., deMaynadier, P.G., 2002. Natural disturbance regimes in northeastern North America—evaluating silvicultural systems using natural scales and frequencies. For. Ecol. Manage. 155 (1–3), 357–367.

US Department of Agriculture, 1929. Volume, Yield, and Stand Tables for Second-Growth Pines. US Department of Agriculture, Washington, DC, Miscellaneous Publication No. 50. 202 p.

US Department of Agriculture, Forest Service, 1986. Service Forester's Handbook. US Department of Agriculture, Forest Service, Southern Region, State & Private Forestry, Atlanta, GA, Miscellaneous Report R8-MR11. 129 p.

US Department of Agriculture, Forest Service, 1990. Land and Resource Management Plan, Umpqua National Forest. US Department of Agriculture, Forest Service, Pacific Northwest Region, Portland, OR.

US Department of Agriculture, Forest Service, 2004. Land and Resource Management Plan, Chattahoochee-Oconee National Forests. US Department of Agriculture, Forest Service, Southern Region, Atlanta, GA, Management Bulletin R8-MB113A.

US Department of Agriculture, Forest Service, 2013. Huron-Manistee National Forests land and resource management plan (as amended January 2012). US Department of Agriculture, Forest Service, Eastern Region, Milwaukee, WI.

Chapter 11

Control Techniques for Commodity Production and Wildlife Objectives

Objectives

As a developer of a natural resource management plan, if you were to claim that your plan would ensure that a landowner's objectives will be met, you may need to justify how you arrived at the measures used to guide the forest into the future. For example, if wood flow constraints are incorporated into a plan, then how would you have determined the appropriate harvest levels to use as goals? It is the efficiency of the plan that is at stake, as Recknagel (1913) noted over 100 years ago:

> The regulation of yield in wood-lots must conform primarily to the wishes and desires of the owner, but it can usually accomplish these without the waste incident to haphazard management. . . .

In this chapter, we describe several classic methods for determining the appropriate measures (area scheduled for treatment and volume scheduled for harvest) to use in policy constraints. Most of these approaches were designed as ways to guide the development of a forest to a regulated state, even though regulated forests are seldom found in practice. However, several of the approaches continue to be used as general guidelines for forest plan development. An extension of these methods to the determination of wildlife habitat development is also proposed. Upon completing this chapter, you should be able to:

1. Describe how area control can be used in the development of a forest plan, why it might be used, and the outcomes we can expect.
2. Describe how various volume control methods can be used in the development of a forest plan, why each might be used, and the outcomes we can expect.
3. Understand how to determine the appropriate harvest volume targets for wood flow policy constraints.
4. Explain how one or more of the volume control approaches can be extended to other natural resources.

I. CONTROLLING THE AREA SCHEDULED

The control of the amount of harvest and habitat areas through limitations placed on their size and extent is relatively easy and straightforward with most forest planning

models. Within linear programming, minimum or maximum harvest areas can be controlled through policy constraints. Assume that the variable $AH1$ was designed to represent the area harvested during time period 1 of a forest plan. Policy constraints could be configured to require any reasonable level of harvest area to be controlled. Two examples are:

$$AH1 \leq 1,000$$
$$AH1 \geq 500$$

English units are used frequently in this chapter to support the examples, however we periodically (but not always) provide metric equivalents to offer international readers a more familiar context. Therefore, if the areas harvested were measured in acres, inequalities such as these will allow the area scheduled for harvest during a single time period to range between 500 and 1,000 acres (202−405 hectares). If the basic planning units were strata of land, or aggregations of stands, then some necessary flexibility should be allowed during the development of the plan. Further, if integer variables are used to represent the harvest of entire stands, then inequality constraints such as these would be necessary to avoid over-constraining the problem. For example, if the area constraints were represented by an equation such as:

$$AH1 = 750$$

the type of data assigned to the decision variables would likely need to be continuous real number values rather than integer values. The decisions when using continuous real numbers would consist of determining *how much* of a stand or strata to harvest, rather than *whether or not* to harvest the stand or strata. The main reason for using a continuous real number data type to represent harvest decisions is that if scheduled harvests were limited to entire stands or strata, then an equality constraint such as this could rather easily be violated. In other words, very few (if any) entire stands or strata, when scheduled for harvest, would produce exactly the value suggested on the

Forest Management and Planning. DOI: http://dx.doi.org/10.1016/B978-0-12-809476-1.00011-4

right-hand side of the constraint. If fact, we recommend that no equality constraints be used except for those used in accounting rows to prevent over-constraining the problem and causing infeasibility.

If you are uncertain about the allowable area to schedule for harvest during each time period, then *area control*, a harvest scheduling control technique, could be used. However, area control is a method used to help develop a regulated forest within the timeframe of one rotation of an even-aged stand of trees, and as a result, using this method assumes that a regulated forest is desired. When using area control, we would be interested in scheduling for harvest equal areas of land during each time period. By the end of one rotation of using this method, we should have a forest with equal areas in each age class. In effect, area control stabilizes the area harvested during the conversion period (one rotation). One disadvantage of this approach is that the volumes scheduled for harvest may fluctuate widely during the conversion period. Volumes scheduled for harvest after the conversion period, however, should be relatively stable. This approach has been used to develop maximum harvest rates, even for small properties (Kallesser, 2015). Area control has also been used to assist in the development of desired wildlife habitat conditions within a national wildlife refuge (LaPointe et al., 2015) and sustainable harvest levels for public forests (Barkley et al., 2015).

In a regulated or normal forest, the decision regarding how much volume to harvest each year is based on the area of harvest. Since in a regulated forest there should be the same number of units of land in each forested age class, once a landowner has defined the rotation age (*R*), determining the amount of land to schedule harvests upon each year is relatively straightforward:

$$\text{Area to harvest} = \left(\frac{\text{Area of forest}}{\text{Desired rotation age}}\right)$$

If a landowner decides to use natural regeneration after harvesting to start the next rotation of trees, then this method may be too rigid, since control over regeneration processes is not as certain for all tree species with natural regeneration as it may be when tree planting activities are assumed. In essence, while a stand may be clearcut today, in 10 years it may be considered only an 8- or 9-year old stand (or less) if natural regeneration was delayed. The area control method also is restricted to forests with relatively uniform conditions; therefore, it is best used either for even-aged plantations or for well-developed uneven-aged forests (Recknagel, 1913). The application of this approach to uneven-aged forests would not involve determining the amount of land to schedule for clearcut harvest entries, but would involve determining the amount of land to schedule for periodic partial cutting entries.

Example

Assume that you are working for a large landowner in north Florida, and your organization has determined that the desired future rotation age of slash pine (*Pinus elliottii*) plantations is 22 years. One may consider this as the rotation age that maximized the bare land value, as is often used in these types of problems. Assume that the landowner wants to eventually move the forest to a regulated state and that area control is the tool that will be used. Given the age class structure of pine plantations shown in Table 11.1, and the desire to use area control to schedule the harvests, what is the planned amount of land to schedule for clearcut harvest each year?

$$\text{Area to harvest} = \left(\frac{21{,}735 \text{ acres}}{22 \text{ years}}\right)$$

$$= 988 \text{ acres per year}$$

(or 400 hectares per year)

What harvest volume would you expect to produce with this plan of action during the first year? For the first year of this plan, assuming an *oldest age class first* harvesting rule, all of age class 27 would be scheduled for harvest (14,647 tons), as would 859 acres (95.3%) of age class 26, or about 93,823 tons. In sum, 108,470 tons would be scheduled for harvest during the first year of the plan. What harvest volume would you expect during the second year of the plan? Using the volume per acre that was assumed for each age class, the remaining 42 acres of the original age class 26 would move up to age class 27, resulting in 4,769 tons. Only 946 acres of the original age class 25 (now age class 26) are needed to schedule 988 acres for harvest. These 946 acres would produce 103,325 tons, and the total harvest would be 108,094 tons.

As you can tell, the area scheduled for harvest may be perfectly even, when using the area control method, but the volume scheduled for harvest during the conversion period (one rotation) will vary. The extent of the variation in harvest volumes will depend on the original age class distribution of the forest, and the condition of forests within each age class. With a very irregular age class distribution as a starting point, the volumes scheduled for harvest may vary considerably, which could be a cause for concern for the landowner. As we suggested, area control has been proposed as a guiding principle for regulating forests. The method moves a forest age class distribution toward that of a regulated forest as quickly as possible in an effort to both maximize the yield from a forest and to provide a steady harvest of timber, although ultimately these may be incompatible. Some form of control on areas treated or areas in habitat development and maintenance is regularly incorporated into forest plans being developed today; however, pure area control is seldom practiced without some modification of the harvest scheduling rules (Chapman, 1950).

TABLE 11.1 Area and Merchantable Volume for a North Florida Landowner Growing Slash Pine Plantations

Age	Acres	Merchantable Volume (tons)
0	253	0
1	1,242	0
2	965	0
3	234	0
4	1,843	0
5	785	0
6	560	0
7	290	0
8	1,206	0
9	893	0
10	502	8,520
11	206	4,732
12	1,205	35,304
13	503	18,004
14	587	24,851
15	729	35,604
16	308	17,016
17	1,952	120,070
18	942	63,687
19	1,092	80,282
20	295	23,372
21	792	67,110
22	386	34,753
23	931	88,565
24	1,022	102,222
25	982	102,833
26	901	98,410
27	129	14,647
Total	**21,735**	**939,982**

II. CONTROLLING THE VOLUME SCHEDULED

Volume control techniques involve processes where the schedule of harvests is forced to (1) represent a specific level of volume during each time period, (2) be within a range of volumes during each time period, or (3) be relatively constant over some period of time. The first two examples are relatively easy to implement within a linear programming model, since they are assumptions made by the land manager. The latter requires a determination of the appropriate level of scheduled volume, and in effect is not as easy to implement in practice. In addition, when requiring a relatively constant scheduled volume over time, the area harvested may fluctuate from year to year based on the productivity and age of the stands being harvested. Further complicating the situation, the lack of control on area harvested makes it more difficult to move a forest toward a regulated state. In practice, attempts have been made to adjust areas harvested so that the total volume produced would be similar in each planning period, yet obtainable from different timber size classes to meet local needs, essentially softening some of the assumptions of strict volume control (Corral-Rivas et al., 2015).

In the linear programming examples from previous chapters, we included in the problem formulation policy constraints, similar to those shown here, that can be used to define a specific amount of timber volume desired during each time period.

$$H1 = 30,000$$
$$H2 = 28,000$$
$$H3 = 33,000$$

One question a landowner or land manager might ask is how we arrived at the level of volume desired (30,000 units during time period 1, 28,000 units during time period 2, and 33,000 units during time period 3, in this case), and we will describe some of these methods soon. We could also define a range of desired volumes during a single time period with constraints such as these that place bounds on harvest levels:

$$H1 \geq 29,500$$
$$H1 \leq 30,500$$

Controls on scheduled volume can be developed for linear programming problems much in the same way as controls on area treated are developed. Assume for example that you developed two accounting rows, one to accumulate the volume harvested during time period 1, the other to accumulate the volume harvested during time period 2. Two variables, $H1$ and $H2$, then could be associated with these accounting rows to represent the scheduled harvest volume for the two time periods. To summarize, volume control, or wood flow, policy constraints could involve the following:

1. An exact scheduled harvest level.

```
H1 = 20,000
H2 = 22,000
```

Here, the question again becomes one of how the volume targets were defined. Some guidelines for developing the volume target using classic methods are described in the sections that follow.

2. A percentage fluctuation in scheduled harvest levels from one period to the next, such as allowing period 1 to be ± 5% of the period 2 volume:

H1 − 1.05 H2 ≤ 0
H1 − 0.95 H2 ≥ 0

For example, assume in each case that the volume harvested in time period 2 is 1,000 cords. The volume harvested in time period 1 cannot exceed 1,050 cords (105% of the time period 2 volume), nor drop below 950 cords (95% of the time period 2 volume). A version of this type of constraint was used in the development of forest management options for a planning unit in Turkey, where timber production, water production, and carbon sequestration were deemed important goals (Başkent and Küçüker, 2010).

3. A percentage departure from the average schedule harvest volumes over a period of time. Assume first that an equation was developed to compute the average volume for two time periods:

0.5 H1 + 0.5 H2 − AvgVol = 0

Policy constraints similar to the ones described in the previous section can then be developed to control the variation of harvests as they relate to the average volume.

H1 − 1.05 AvgVol ≤ 0
H1 − 0.95 AvgVol ≥ 0

4. An equal scheduled harvest volume during each time period. Here, an even-flow harvest volume could be ensured with policy constraints such as:

H1 − H2 = 0

As we have suggested in previous chapters of this book, even-flow policy constraints are the most challenging (almost impossible) to adhere to when the decision variables are assigned integer values. However, when the decision variables are assigned continuous real number values, even-flow policy constraints can ensure that the same volume is scheduled in each of the affected time periods.

Some harvest policies may be tied directly to the needs of a wood processing facility, and in these cases the desired harvest levels may be relatively easy to determine. However, if this were not the case, then other methodology must be used to determine the right-hand side value of the policy constraint. These methods might include:

- A harvest equals growth policy
- A "best judgment" policy
- A trial and error policy
- A determination of allowable harvest through classical volume control methods

The first of these suggests that a sustainable harvest volume is one where the scheduled harvest volume equals the growth of the forest. If we were to maximize the productivity of a forest in terms of wood volume produced, then this would imply that the rotation age is one where mean annual increment is maximized, and removals (harvests) would equal the increment of the forest (the growth) in each time period. However, when a forest age class distribution is not normal or regulated, or when the site productivity is highly variable across the forest, this policy could lead to ineffective management of the area. For example, assume at one end of the spectrum a forest is composed mainly of stands of old-growth timber that produce little new volume each year, and thus where the annual growth rate is relatively low. No matter what the desired rotation age, under this policy only an amount equal to the growth can be removed during each time period, and consequently, the forest will likely remain older (for better or worse). On the other end of the spectrum, assume that a forest is composed mainly of younger stands of trees, with relatively high annual growth rates. Using a policy of harvest equaling growth, the implied harvest volume may be so high relative to the current growing stock (standing volume) that the suggested harvest volume could not be sustained over the long-run. As a consequence, the landowner would eventually deplete the merchantable harvest volume.

The "best judgment" policy may be commonly practiced today, but might also leave a landowner with a level of uncertainty that may be unacceptable. When making a best judgment periodic harvest volume recommendation to a landowner, what would you base it on? Would the recommended harvest level be, in turn, defensible? Some best judgment policies may be palatable to a landowner when the objective of forest management is not necessarily contingent on commodity production or economic goals. Conservative harvest policies may in fact be consistent with a landowner's objectives, and as a result the selection of a harvest policy may be better made using the best judgment of the forest manager. Recknagel (1913) described a method of volume control (the Hufnagl method) similar to this that is based solely on the increment of a forest, and can be applied to even-aged forests as well as uneven-aged forests. At its root, the Hufnagl method suggests that harvest equals growth, and that the age-class distribution of the forest must approach a regulated state for the method to work effectively. However, where it differs from other methods is in the loose adjustments suggested to the sustained harvest level. For example, if the growing stock within a forest seems excessive, then the Hufnagl method suggests a harvest greater than the increment can be scheduled. Alternatively, if the growing stock is deficient, then the method suggests a harvest less than the increment can be scheduled. The upward or downward adjustment to the harvest level is left to the forest manager to determine. This is a very basic method for determining the sustained annual harvest; therefore, the best use of it may be for comparison against other methods. The Hufnagl method is similar to the Austrian formula (described later), except that the Austrian formula contains a rational and quantifiable approach for adjusting the harvest when basing it on the increment, whereas the Hufnagl method suggests a subjective adjustment upward or downward at the discretion of the forest manager. As with other methods that are based on the growth rate of a stand or forest, the increment (growth) of a forest needs to be reassessed every few years for these methods to work effectively (Recknagel, 1913).

Trial-and-error policies for determining the sustainable harvest level suggest the use of techniques such as binary search. Binary search, in fact, is one form of volume control. As we mentioned in Chapter 8, Advanced Planning Techniques, it is a technique for meeting the objectives of a landowner by repeated simulations of harvest levels until a feasible maximum or minimum harvest volume is located. The approach is usually specific to the landowner, the forest conditions, and the assumptions regarding the selection of harvest areas. However, binary search is an efficient trial-and-error method for identifying upper and lower bounds on sustainable harvest volumes within a specific time horizon.

Other classic methods for determining the appropriate harvest volume are described next. These include the Hanzlik formula, the Austrian formula, and others. As with area control, volume control techniques could be used to develop a regulated forest. Volume control techniques use forest growth and age class distributions to determine the conversion period allowable harvest levels. When using volume control methods, the scheduled volumes should be stable throughout the conversion period, yet the area harvested may fluctuate widely.

A. The Hanzlik Formula for Volume Control

A graduate of the University of Washington forestry program, Edward Hanzlik worked as a forester for many years in the Pacific Northwest of the United States. He is best known for his method of determining the sustainable annual harvest level from virgin forests (Hanzlik, 1922). The Hanzlik formula, as it has come to be known, was used extensively in the northwestern United States and British Columbia in the middle part of the 20th century, in an effort to ensure that the supply of wood would meet the demands of society at that time as well as demands assumed to occur in the future. The Hanzlik formula for determining the sustained annual yield centered on the concern over transitioning virgin forests to normal forests (which was described in Chapter 10: Models of Desired Forest Structure). The calculation of the near-term sustained annual yield is for even-aged stands, since it is a function of the amount of wood volume over a predefined

rotation age (determined by the landowner), as well as the growth rate of the stands within a forest that are younger than the desired future rotation age:

Sustained annual yield

$$= \left(\frac{\text{Mature timber volume above rotation age}}{\text{Years in the rotation}} \right)$$

$$+ \text{MAI of immature timber}$$

To shorten the formula and make it consistent with the others that follow, assume that the volume of mature timber above rotation age is represented by the variable V_m, and that the mean annual increment of immature timber is represented by I. The years in the rotation age can be represented simply by the rotation age (R). The equation then becomes:

$$\text{Sustained annual yield} = \left(\frac{V_m}{R} \right) + I$$

Hanzlik argued that to move a forest structure toward that of a normal forest, the removal of surplus mature growing stock needed to be regulated through an adjustment of the annual harvest so that the removal of wood in stands with ages above the desired rotation age coincided with the growth of the immature forests. If we remove (V_m/R) from the formula, then we have simply a policy of harvest equals growth (I). The incorporation of (V_m/R) represents the removal of surplus growing stock over the length of the desired rotation. The process is not perfect, because V_m could grow or decline over the period R, and these additions or subtractions are not accounted for in the determination of sustained annual yield. In effect, the goal of using the Hanzlik formula is to ration the harvest of mature wood until some point in time when the immature wood can replace it in the harvest schedule. We indicate that the result is a near-term sustained yield because the allowable annual harvests were determined in many cases for shorter periods of time than R; in other cases experts suggest periodically reevaluating the sustained yield computation given changes in the biological and political environment. In 1945, Sloan (1945) described the use of a process similar to the Hanzlik formula for estimating the allowable cut in British Columbia, but cautioned that he did "not consider it a safe guide for any greater period than for the next ten years." Over longer periods of time, changes to the land and forest base due to land sales, land purchases, access problems and opportunities, regeneration delays, and losses from natural disturbances require periodic remeasurement of the allowable harvest.

The Hanzlik formula can be applied at the stand or forest level. To demonstrate the application of the approach on a small ownership, assume that a landowner in Virginia owns 30 acres (12.1 hectares) of land; 20 acres (8.1 hectares) are comprised of older

TABLE 11.2 Average Yield for a Loblolly Pine Stand

Age	Volume (ft^3 per acre)
20	1,750
30	3,440
40	4,290
50	4,740
60	5,070
70	5,330
80	5,560

TABLE 11.3 First Decade Harvest for a Loblolly Pine Stand Using Harvest Volumes Estimated Using the Hanzlik Formula

Stand Age	Acres	Before Harvest		Harvest Area (acres)	Harvest Volume (ft^3)
		Volume per Acre (ft^3)	Total Volume (ft^3)		
60	20	5,070	101,400	8.44	42,800
20	10	1,750	17,500	—	—
Total				8.44	42,800

(55 years) loblolly pine (*Pinus taeda*), and 10 acres (4 hectares) of younger-growth (15 years) pine. The yields expected from this area are described in Table 11.2. Assume that the landowner is interested in knowing the volume to harvest over three 10-year time periods, and given a desired rotation age of 30 years. Assume further that the harvests are scheduled for the middle of each time period. Using the Hanzlik formula, we find that the volume (growing stock) in stands older than the desired rotation age is:

$$V_m = (5,070 \text{ ft}^3 \text{ per acre})(20 \text{ acres}) = 101,400 \text{ ft}^3$$

If we assume that the MAI of immature timber is 30 ft^3 per acre per year (2.1 m^3 per hectare per year), then:

$$I = (30 \text{ ft}^3 \text{ per acre per year})(30 \text{ acres}) = 900 \text{ ft}^3 \text{ per year}$$

As a result, the sustained annual yield using the Hanzlik formula is:

$$\text{Sustained annual yield} = \left(\frac{101,400 \text{ ft}^3}{30 \text{ years}} \right) + 900 \text{ ft}^3$$

$$\text{Sustained annual yield} = 4,280 \text{ ft}^3 \text{ per year}$$

Over an entire decade, the harvest should be 42,800 ft^3. If this harvest policy were instituted, then the harvested area would be 8.44 acres in the first decade (Table 11.3), 8.03 acres in the second decade (Table 11.4), and 8.93 acres in the third decade (Table 11.5). The forest, after three decades, seems to be moving to a regulated state, as the age classes are moving downward toward the desired future rotation age. However, the area harvested is not equal from one decade to the next, and unless there are differential growth rates in the various age classes, the forest may never reach a regulated state. As a consequence, the sustainable harvest rate should probably be reassessed each decade.

TABLE 11.4 Second Decade Harvest for a Loblolly Pine Stand Using Harvest Volumes Estimated Using the Hanzlik Formula

Stand Age	Acres	Before Harvest		Harvest Area (acres)	Harvest Volume (ft^3)
		Volume per Acre (ft^3)	Total Volume (ft^3)		
70	11.56	5,330	61,615	8.03	42,800
30	10.00	3,440	34,400	—	—
10	8.44	0	17,500	—	—
Total				8.03	42,800

TABLE 11.5 Third Decade Harvest for a Loblolly Pine Stand Using Harvest Volumes Estimated Using the Hanzlik Formula

Stand Age	Acres	Before Harvest		Harvest Area (acres)	Harvest Volume (ft^3)
		Volume per Acre (ft^3)	Total Volume (ft^3)		
80	3.53	5,560	19,627	3.53	19,627
40	10.00	4,290	42,900	5.40	23,173
20	8.44	1,750	14,770	—	—
10	8.03	0	0	—	—
Total				8.93	42,800

Example

Assume that you are working for a large landowner in north Florida, and your organization has determined that the desired future rotation age of slash pine plantations is 22 years. Assume that the mean annual increment of the immature forests is 90,700 tons per year. Given the age class structure of pine plantations shown in Table 11.1, and the desire to use volume control to schedule the harvests, what is the planned volume to harvest using the Hanzlik formula? With an inspection of the data provided in the table, we find that $V_m = 406,677$ tons. This is the total volume of the stands with ages 23−27 years, or all stands older than the assumed rotation age. Once this is known, we can estimate the sustained annual yield using the Hanzlik formula:

$$\text{Sustained annual yield} = \left(\frac{406,677 \text{ tons}}{22 \text{ years}} \right)$$
$$+ (90,700 \text{ tons per year})$$
$$\text{Sustained annual yield} = (18,485 \text{ tons per year})$$
$$+ (90,700 \text{ tons per year})$$
$$\text{Sustained annual yield} = 109,185 \text{ tons per year}$$

Two areas of concern have been expressed about the Hanzlik formula, the first of which also applies to other methods that also involve using the forest increment (I) to determine sustainable harvest levels. Because the forest increment is associated with the estimated harvest level, concern must be placed on how the increment measurement was developed. The mean annual increment at the rotation age could be used as a proxy for the increment, or the increment could be calculated for each age class up to the desired future rotation age, or perhaps beyond, then accumulated. These and other methods for determining the increment could lead to wide variations in scheduled harvest levels. Further, some suggest that the Hanzlik formula be applied only in cases where there exists a vast amount of volume in stands older than the desired rotation age, that were once common in the forests of the western United States in first half of the 20th century. If this were the case, and if the amount of mature forest volume were very large, then the mature forest volume may dwarf the desired increment, and a volume of (V_m/R) would dominate the scheduling process (Davis, 1954).

B. The Von Mantel Formula for Volume Control

The Von Mantel formula has been found useful in making quick estimates of the allowable harvest levels for even-aged stands. However, it assumes that the age-class distribution of the forest currently approximates a regulated forest. The Von Mantel formula is based entirely on the volume of standing growing stock, and assumes that the volume by age class is roughly a triangular relationship (Fig. 11.1).

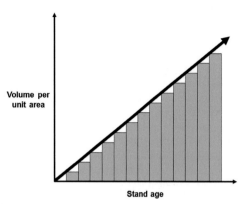

FIGURE 11.1 Triangular relationship assumed by the Von Mantel formula.

$$\text{Sustained annual yield} = \left(\frac{2(\text{Growing stock volume})}{\text{Rotation age}} \right)$$

Using the terminology we presented earlier, this becomes:

$$\text{Sustained annual yield} = \left(\frac{2 G_a}{R} \right)$$

Since the Von Mantel formula suggests a triangular representation of volume by age, applying the formula to an irregular, older, or younger forest age class structure would not seem appropriate. The nonlinear relationship that generally describes the accumulation of standing volume over the life of a stand also suggests that the accuracy of this method could be questionable when applied to a single stand. When measuring volume with solid units (e.g., cubic feet, cubic meters), the formula could overestimate the harvest, due to the fact that volume growth is nonlinear. When measuring volume, using merchantable units (e.g., board feet), the formula could underestimate the harvest. In the latter case, merchantable volume does not begin to appear until after some minimum stand age that effectively ignores the nonmerchantable growing stock of younger trees. In these cases, an adjustment (R-a) generally is assumed to be of value, where a represents the first year at which the merchantable volume is recognized.

$$\text{Sustained annual yield} = \left(\frac{2 G_a}{R-a} \right)$$

One advantage of both the original and adjusted Von Mantel formulas is that they are relatively simple to employ, since all we need to understand is the total current growing stock (standing volume) on a forest or in a stand. As a rough first estimate of the appropriate level of harvest, the Von Mantel formula may prove useful to some land managers. When applied to a forest using merchantable volumes, it is considered a conservative approach to the estimate of annual harvest volumes. However, other methods are available that account for the fact that most forests are not regulated.

Example

Assume you manage 1,000 acres (404.7 hectares) of land that contains a regulated forest of Douglas-fir (*Pseudotsuga menziesii*) plantations. Assume that the maximum rotation age for this landowner is 50 years. Therefore, there are 20 acres (8.1 hectares) in each 1-year age class. Using the data provided in Appendix A for an example Douglas-fir stand, we can plot the total cubic foot and board foot volume for the regulated forest (Fig. 11.2). The growing stock volume can be estimated through linear interpolation of the per-acre volumes for the 5-year age classes provided in Appendix A. In doing so, we might find that the forest contains 3,057,820 ft^3 or 10,241 MBF (thousand board feet). The sustained annual yield using the uncorrected Von Mantel formula would be:

$$\text{Sustained annual yield} = \left(\frac{2(3,057,820 \text{ ft}^3)}{50 \text{ years}} \right)$$

$$= 122,312.8 \text{ ft}^3 \text{ per year}$$

or

$$\text{Sustained annual yield} = \left(\frac{2(10,241 \text{ MBF})}{50 \text{ years}} \right)$$

$$= 409.6 \text{ MBF per year}$$

These estimated harvest levels are below the volume contained in the last age class, because the relationships represented by real data in Fig. 11.3 are not perfectly triangular. For example, the 50-year age class, prior to harvest, would contain 188,940 ft^3 (9,447 ft^3 per acre × 20 acres) and 745.4 MBF (37.27 MBF per acre × 20 acres). The estimated sustained harvest levels are 65% and 55% of the volume contained in the 50-year age class, for the cubic foot and board foot volumes, respectively. The differences are greater for the estimated merchantable harvest volume because the merchantable volume is not recognized until the stand is nearly 20 years of age. Had we incorporated the adjustment factor (*a*) of 20 years into the Von Mantel formula, we would find that the estimated harvest levels are now:

$$\text{Sustained annual yield} = \left(\frac{2(3,057,820 \text{ ft}^3)}{50 \text{ years} - 20 \text{ years}} \right)$$

$$= 203,854.7 \text{ ft}^3 \text{ per year}$$

or

$$\text{Sustained annual yield} = \left(\frac{2(10,241 \text{ MBF})}{50 \text{ years} - 20 \text{ years}} \right)$$

$$= 682.7 \text{ MBF per year}$$

With this adjustment, we now find that our estimated cubic foot harvest volume is slightly higher (8%) than the volume contained in the 50-year age class, and that the estimated merchantable harvest volume is slightly lower (92% of the volume contained in the 50-year-age class). Although informative, the use of the Von Mantel formula is best restricted to situations where there is a need to determine a provisional sustained annual harvest while other information about the forest is being collected (Recknagel, 1913).

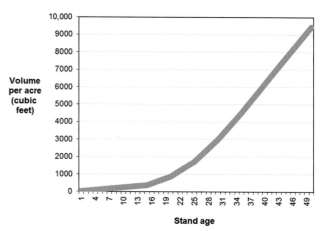

FIGURE 11.2 Cubic foot volume per acre for a Douglas-fir stand.

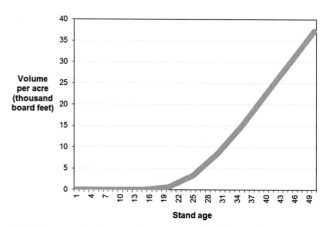

FIGURE 11.3 Board foot volume per acre for a Douglas-fir stand.

C. The Austrian Formula for Volume Control

The Austrian formula is a method for determining the appropriate volume to harvest given the annual growth of a forest, the growing stock volume that a landowner wants to maintain over time from a regulated forest, and the current growing stock (standing volume) of the forest. The method was developed around 1788, as a result of a decree by the Vienna Hofkammer (court chamber or official having jurisdiction over revenue matters). The goal of regulating the yield is to allow the forest to approach a normal growing stock and normal increment (Recknagel, 1913). An adjustment period also is assumed to deal with the difference in volumes between the desired and actual growing stock volumes. The annual growth, or increment of a forest was defined earlier (I). To describe the relationships in the Austrian formula, variables representing the current growing stock volume (G_a), the desired growing stock volume of a regulated forest (G_r), and the adjustment period (*a*) are hence defined. As Recknagel (1913) points out, technically the Austrian formula should use the mean annual increment of the forest to represent the increment (I). However, the current annual increment has been applied by others in lieu of the mean annual increment. In addition, the adjustment period technically was meant to represent an entire forest rotation rather than a period of time defined by the landowner. In the latter case, Recknagel (1913) called this *Karl's method*, which also employed the current annual increment rather than the mean annual increment. Although the Austrian formula can be applied to both even-aged and uneven-aged stands, this approach of using an adjustment period defined by the landowner or forest manager is necessary for application of the method to uneven-aged stands, where no rotation length is assumed. However, reentry cycles are assumed in uneven-aged management, and in estimating the appropriate harvest volume levels, the Austrian formula utilizes the difference between the current and desired growing stock volume as an addition or subtraction from the annual increment.

$$\text{Sustained annual yield} = \left(\frac{G_a - G_r}{a} \right) + I$$

Example

Assume that a forest you manage has an annual increment of 139,000 ft^3. Assume also that the current growing stock volume is 5,730,000 ft^3, and that the landowner wants a future growing stock volume of 6,323,000 ft^3. This assumes that the forest needs to accumulate growing stock volume 593,000 ft^3 over some period of time to arrive at the desired growing stock volume. If the adjustment period to arrive at the desired growing stock volume is 20 years, then the sustainable harvest volume becomes:

$$\text{Sustained annual yield} = \left(\frac{5,730,000 \text{ ft}^3 - 6,323,000 \text{ ft}^3}{20 \text{ years}} \right)$$
$$+ 139,000 \text{ ft}^3$$

Sustained annual yield $= -29,650$ ft^3 $+ 139,000$ ft^3

Sustained annual yield $= 109,350$ ft^3

Had the adjustment period been 5 years or less, the harvest level effectively would have been no volume at all. Although the Austrian formula would have indicated a negative sustainable harvest volume, it simply implies that no harvest could have occurred if the landowner wanted the growing stock to build up to a desired level within the time frame that they indicated. If the desired future growing stock volume of a regulated forest were equal to the current growing stock volume, then the sustainable annual harvest volume using the Austrian formula would simply become the increment. This suggests a policy of harvest equaling growth of the forest. If the current growing stock volume, say 10,000,000 ft^3, were much greater than the desired future growing stock volume, then the sustainable harvest level, in this case 322,850 ft^3, would overshadow the current increment. This suggests that the adjustment is to lower the current growing stock volume to the desired volume by spreading the surplus harvest volume over the years in the adjustment period.

Caution must be taken when using these volume estimation rules. Besides problems encountered in estimating the increment, the growth rate of the mature growing stock is not taken into account as the plan develops. As a result G_a may be higher than expected when the conversion period ends. Therefore, we can expect that the annual growth of the forest should equal the scheduled harvest after the conversion period, the current growing stock may be greater than what was expected in a regulated forest.

Example

Assume again that you are working for a large landowner in north Florida, and your organization has determined that the desired future rotation age of slash pine plantations is 22 years. The mean annual increment assumption here, of the immature forests, is 90,700 tons per year. Given the age class structure of pine plantations shown in Table 11.1, and the desire to use volume control to schedule the harvests, what is the planned volume to harvest using the Austrian formula? Assume that the desired growing stock level is 600,000 tons and that the adjustment period is 20 years.

Here, we find that $G_a = 939,982$ tons, $G_r = 600,000$ tons, and $a = 20$ years. Therefore,

$$\text{Sustained annual yield} = \left(\frac{939,982 \text{ tons} - 600,000 \text{ tons}}{20 \text{ years}} \right)$$
$$+ 90,700 \text{ tons}$$

Sustained annual yield $= 16,999$ tons $+ 90,700$ tons

Sustained annual yield $= 107,699$ tons

Example

Assume you manage a fairly large stand of uneven-aged mixed conifers in eastern Washington, in the Pacific Northwest of the United States, where the average current growing stock volume is 9,500 board feet per acre and the average growth rate is 300 board feet per acre per year. If the landowner wanted a reserve growing stock of 11,000 board feet per acre, and was amenable to a 15-year adjustment period, then what would be the average annual harvest using the Austrian formula?

$$\text{Sustained annual yield} = \left(\frac{9,500 \text{ board feet} - 11,000 \text{ board feet}}{15 \text{ years}} \right)$$
$$+ 300 \text{ board feet per acre}$$
$$\text{Sustained annual yield} = \left(\frac{-1,500 \text{ board feet}}{15 \text{ years}} \right)$$
$$+ 300 \text{ board feet per acre}$$

Sustained annual yield $= 200$ board feet per acre

In light of the fact that the current average growing stock is below the desired growing stock, the landowner could sustain an annual harvest of about 200 board feet per acre per year for the next 15 years and still reach their goal of a reserve growing stock of 11,000 board feet per acre.

Methods for regulating harvests based on volume can be adapted to irregular or uneven-aged forests. However, forest managers using the Austrian formula, as well as other formulas that are based on the increment and the current level of growing stock, should be aware that the increment may change over time, as may the current growing stock volume. As a result, the desired harvest volume needs to be recalculated every 5 or 10 years to ensure that the appropriate course of action is being taken.

D. The Hundeshagen Formula for Volume Control

The Hundeshagen formula was developed in the early 1800s as a method for estimating the sustained annual harvest level, and is based on the relationship between the current standing growing stock and the desired future growing stock level. Once the growth (increment) of a fully regulated forest has been estimated, the ratio of the current and desired growing stock levels is used to adjust upward (in the case of current growing stock being greater than desired) or downward (in the case of current growing stock being less than desired) the suggested annual harvest level. The Hundeshagen formula is:

$$\text{Sustained annual yield} = \left(\frac{\text{Current growing stock volume}}{\text{Desired future growing stock volume}} \right)$$
$$\times (\text{Regulated forest annual increment})$$

If we assume the variable I_r to represent the annual increment of a fully regulated forest, and we use the terms we have defined earlier, then the equation becomes:

$$\text{Sustained annual yield} = \left(\frac{G_a}{G_r} \right) (I_r)$$

Example

Assume that you manage a longleaf pine (*Pinus palustris*) forest in south Alabama that consists of 1,500 acres (607 hectares) of various age classes of stands. The current growing stock volume of the forest is 4,687,000 ft³, or about 3,125 ft³ per acre. If the desired growing stock level were 4,310,156 ft³, and the increment of a normal forest of this size were 108,750 ft³ per year, then what would become your estimate of the sustained annual yield for this forest?

$$\text{Sustained annual yield} = \left(\frac{4,687,000 \text{ ft}^3}{4,310,000 \text{ ft}^3} \right) (108,750 \text{ ft}^3)$$
$$= 118,262 \text{ ft}^3 \text{ per year}$$

The Hundeshagen formula is a straightforward method for approximating the sustained annual yield for a forest,

and provides a mechanism for building up or drawing down the current growing stock volume based on its relationship to the desired growing stock volume. What length of time may be required to arrive at the regulated forest is another question entirely.

Example

Assume, as in the case of the Austrian formula, that you manage a fairly large stand of uneven-aged mixed conifers in eastern Washington, where the average current growing stock volume is 9,500 board feet per acre, and that the regulated forest annual increment is 300 board feet per acre per year. If the landowner wanted a reserve growing stock of 11,000 board feet per acre, and was amenable to a 15-year adjustment period, then what would be the average annual harvest using the Hundeshagen formula?

$$\text{Sustained annual yield} = \left(\frac{9,500 \text{ board feet per acre}}{11,000 \text{ board feet per acre}} \right)$$
$$\times (300 \text{ board feet per acre per year})$$

The sustained annual yield in this case is about 259 board feet per acre per year for the uneven-aged mixed conifer stand.

It should seem obvious in these examples that as the growing stock is being built up or drawn down, the sustained annual harvest level should be reassessed. One of the drawbacks of using the Hundeshagen formula, as in the case of other methods, is that it does not take into account whether the growing stock itself may be unacceptable and therefore in need of adjustment (Davis, 1954). Ultimately, the suggested sustained harvest level will be nearly equal to the annual increment of a fully regulated forest. However, overmature forests can require a lengthy period of time to be adjusted to a regulated state, and immature forests can provide little harvest if the growing stock levels are deficient (Recknagel, 1913).

E. The Meyer Amortization Method for Volume Control

In the mid-20th century vast areas of forestland in the United States were characterized by immature, understocked forests as a result of the use of silvicultural operations that tended to harvest only the mature trees (Meyer, 1952). Often, these forests contain cohorts of regeneration of various sizes, along with remnants of older trees that were perhaps too small to remove during previous harvest entries. To understand the sustained annual harvest that can be provided from forests characterized in this manner requires a dynamic assessment since the growth rate and growing stock situation could constantly change as a result of management activities or natural disturbances. Under these conditions, a reassessment of the harvest potential at various time points would seem necessary. Here, the desired rotation age and the annual sustained harvest levels are not as important as the intermediate entry cycle employed and the choice of silvicultural system necessary to build up the growing stock to a desirable level.

Meyer (1952) suggested an amortization schedule of harvests for areas such as these, where the harvest will be distributed equally over all the tree size classes. Excluding the growth rate of ingrowth into the forest, the method for determining the sustainable harvest level over the near term is determined using the following equation:

$$\text{Annual harvest} = \text{Growth rate}\left(\frac{G_a(1+\text{Growth rate})^n - G_r}{(1+\text{Growth rate})^n - 1}\right)$$

Example

Assume for a moment that the growth rate of a large area of immature, understocked pine-hardwood forest in Tennessee is 5.5% per year. If the current standing growing stock is 3,500,000 ft^3 and the desired future growing stock volume at the end of a 10-year time horizon is 3,900,000 ft^3, using this method, what would be your estimate of the allowable annual harvest level?

Annual harvest

$$= 0.055\left(\frac{3,500,000\text{ ft}^3(1.055)^{10} - 3,900,000\text{ ft}^3}{(1.055)^{10} - 1}\right)$$

$$= 161,432.9\text{ ft}^3\text{ per year}$$

To prove that the method works, examine Table 11.6. Here we assume that currently we have a forest that contains 3,500,000 ft^3 of immature, understocked forests. Some patches of larger trees remain from previous harvests to provide the growing stock volume represented. At the end of year 1 we show growth of the growing stock (3,500,000 × 1.055) less the scheduled harvest, or:

$$3,500,000\text{ ft}^3(1.055) - 161,432.9\text{ ft}^3 = 3,531,067.1\text{ ft}^3$$

At the end of year 2, we take the growing stock available at the beginning of the year (same as the growing stock available at the end of year 1), apply the growth rate, and subtract the harvest. This continues for all 10 years, where we find at the end of the 10th year a residual growing stock volume of about 3,900,000 ft^3. Hopefully, the harvests were designed in such a way as to reduce the problem of maintaining immature and understocked forests. It should be obvious that in light of the harvests that are scheduled, since the desired growing stock increased, the average stocking level should have increased. The sustained annual harvest therefore should be recalculated every 5 or 10 years as growth rates of the forest change.

TABLE 11.6 Growth and Annual Harvest of an Immature, Understocked Forest in Tennessee

Time	Harvest Volume (ft^3)	Growing Stock (ft^3)
Now	—	3,500,000.0
End of year 1	161,432.9	3,531,067.1
End of year 2	161,432.9	3,563,842.9
End of year 3	161,432.9	3,598,421.3
End of year 4	161,432.9	3,634,901.6
End of year 5	161,432.9	3,673,388.3
End of year 6	161,432.9	3,713,991.8
End of year 7	161,432.9	3,756,828.4
End of year 8	161,432.9	3,802,021.1
End of year 9	161,432.9	3,849,699.3
End of year 10	161,432.9	3,899,999.9

Example

Assume as in previous cases that you manage a fairly large stand of uneven-aged mixed conifers in eastern Washington, where the average current growing stock volume is 9,500 board feet per acre, and that the regulated forest annual increment is 300 board feet per acre per year. Assume also the growth rate is about 3.2%. If the landowner wanted a reserve growing stock of 11,000 board feet per acre, what would be the average annual harvest using the Meyer formula over a 15-year time horizon?

Annual harvest

$$= 0.032\left(\frac{9,500\text{ board feet}(1.032)^{15} - 11,000\text{ board feet}}{(1.032)^{15} - 1}\right)$$

In this case, the sustained annual harvest is estimated to be about 225 board feet per acre per year, which represents a level of harvest between what was estimated with the Austrian and Hundeshagen formulas for this same problem.

F. The Heyer Method for Volume Control

The Heyer method, as with the previous methods, is based on the assumption that the landowner wants to move an irregular forest toward a normal forest state. The Heyer method assumes that the level of sustained harvest is equal to the increment as long as a normal growing stock is maintained, a normal increment is realized, and the age classes are all distributed in a normal fashion (Recknagel, 1913). Should the actual increment (growth rate) fall below the normal increment, the harvest levels are assumed to be represented by the actual increment, and if the actual growing stock is not equal to the normal growing stock, then it can be brought to normality through adjustments in the harvest levels. As with the Austrian formula, a period of adjustment (a) is needed to spread out over time the excess harvest (in the case of actual growing stock being greater than the normal growing stock), or spread out over time the reduction in harvest that is necessary (in the case of actual growing stock being less than normal). The formula developed by Heyer is:

$$\text{Sustained harvest level} = \left(\frac{G_a + (I)(a) - G_r}{a} \right)$$

Example

Assume that you manage a longleaf pine forest in south Alabama that consists of 1,500 acres (607 hectares) of various age classes of stands. The current growing stock volume of the forest is 4,687,000 ft³. The desired growing stock level is 4,310,156 ft³, and the increment of a normal stand is assumed to be 108,750 ft³ per year. If the adjustment period were 20 years, then what would be your estimate of the sustained annual yield for this forest?

Sustained harvest level

$$= \left(\frac{4,687,000\,\text{ft}^3 + (108,750\,\text{ft}^3)(20\,\text{years}) - 4,310,156\,\text{ft}^3}{20\,\text{years}} \right)$$

$$= 127,592.2\,\text{ft}^3$$

The Heyer formula is effective for both even-aged and uneven-aged forests. As with other volume control methods, the central goal is to move a forest toward a regulated state by increasing or decreasing the allowable harvest given the state of the system (growth rate and current growing stock volume). Recknagel (1913) suggests that it is one of the better volume control methods in situations where the current growing stock levels are greater than the desired future growing stock. However, when rearranged, the Heyer formula becomes the Austrian formula:

$$\text{Sustained harvest level} = \left(\frac{G_a - G_r}{a} \right) + I$$

$$\text{Sustained harvest level} = \left(\frac{4,687,000\,\text{ft}^3 - 4,310,156\,\text{ft}^3}{20\,\text{years}} \right)$$

$$+ 108,750\,\text{ft}^3 = 127,592.2\,\text{ft}^3$$

The difference between the two is that the Austrian formula technically should use the mean annual increment of a fully regulated forest to represent the increment (I), whereas the Heyer formula could use any expression of increment over the adjustment period.

G. Structural Methods for Volume Control

When using volume control methods for uneven-aged stands, an allowable cut is determined using the periodic increment of the stand, and a guiding diameter limit generally is used as the basis from which the allowable cut will be obtained (Guldin, 1991). What this suggests is that any tree over the minimum diameter assumed can contribute to the allowable cut. This also suggests two limitations about this type of management approach, that the trees in smaller diameter classes are not necessarily managed, and that little guidance is provided for addressing regeneration, other than to suggest that larger gaps created through group selection patches may favor the shade intolerant tree species.

The structural regulation of uneven-aged stands involves the negative exponential distribution we described in Chapter 4, Estimation and Projection of Stand and Forest Conditions. Leak (1964) suggested using the *BDq method* for stand-level regulation to determine the desired after-harvest diameter distribution of an uneven-aged stand. The *BDq* method depends on three factors: the desired residual, after-harvest basal area of the stand (B), the maximum diameter in which trees will be retained (D), and the slope of the reverse J-shaped diameter distribution (q). The residual basal area can be selected by drawing values from commercial thinning operations that are structurally similar to those the landowner wants, or by selecting a value that represents a conservative harvest given the initial stand conditions. Determining the maximum diameter and the q factor may require some thought. For example, setting the maximum diameter to a low DBH value suggests that a high level of volume may be produced, which may lead to considerable residual stand damage from the logging operations. A high q factor suggests that residual stands will contain a large amount of smaller trees relative to the larger trees in the diameter distribution. A smaller q factor (around 1.1) suggests that residual stands will contain a lower amount of smaller trees relative to the larger trees in the diameter distribution, and thus may favor the production of sawtimber products. The allowable cut for uneven-aged stands is then determined by comparing the diameter distributions of the before-harvest stand and the projected after-harvest stand. The volume associated with the removal of trees in each diameter class is accumulated to arrive at a per-acre harvest target. As Guldin (1991) suggests, adjustments to the schedule of harvests are more than likely necessary given the damage likely to the smaller trees as larger trees are felled and processed.

III. APPLICATION OF AREA AND VOLUME CONTROL TO THE PUTNAM TRACT

The Putnam Tract is composed of hardwood, pine (planted and natural regeneration), and mixed pine-hardwood stands. If we simply consider the pine stands (planted and natural regeneration), then there are 2,107 acres (852.7 hectares) of land that comprise an irregular age class distribution. Here we will assume that the land-owner is interested in regulating the forest, and is curious about the various area control and volume control options for regulating a forest. A few assumptions are necessary prior to beginning the analysis.

1. The desired future rotation age is 30 years.
2. The growth rate of the normal forest is 4.5% per year.
3. The mean annual increment of the stands less than 30 years of age results in a forest increment of 1,300 cords per year.
4. The mean annual increment of the future regulated forest is 1,467 cords per year.
5. The volume of mature timber 30 years and older is 16,720 cords.
6. The current growing stock (standing volume) is 31,573 cords.
7. The desired future growing stock is 18,870 cords.
8. The adjustment period, where necessary, is 15 years.

A. Area Control

The amount of land that should be scheduled for final harvest on the Putnam Tract in each year of a forest plan that seeks to move the pine part of the forest toward a regulated state should be

$$\text{Area to harvest} = \left(\frac{2,107 \text{ acres}}{30 \text{ years}} \right) = 70.2 \text{ acres per year}$$

If an *oldest stand first* harvesting rule were employed, then this would entail scheduling stand 12 for harvest in the first year, removing about 1,947 cords (62.2 acres × 31.3 cords per acre). Eight acres from stand 2 then are scheduled, resulting in a harvest of 234 cords. Therefore the total harvest during the first year of the plan is 2,181 cords.

B. Volume Control—Hanzlik Formula

When using the Hanzlik formula for volume control, we find that the annual yield should be about 1,857 cords per year given the growth rate of the immature timber and the amount of volume in stands considered overmature.

$$\text{Sustained annual yield} = \left(\frac{16,720 \text{ cords}}{30 \text{ years}} \right)$$
$$+ 1,300 \text{ cords per year}$$
$$= 1,857 \text{ cords per year}$$

Using an oldest-stand-first harvesting rule, this would require the harvest of most (95%) of stand 12 (a little over 59 acres). No other stands would need to be scheduled for harvest during the first year to reach the volume target.

C. Volume Control—Von Mantel Formula

The Von Mantel formula requires the use of the current growing stock volume. Here we find that the sustained annual yield is 2,105 cords per year.

$$\text{Sustained annual yield} = \left(\frac{2(31,573 \text{ cords})}{30 \text{ years}} \right)$$
$$= 2,105 \text{ cords per year}$$

This plan would require harvesting all of stand 12, and 6.4% of stand 2, or 5.4 acres (resulting in a harvest of 158 cords from stand 2).

D. Volume Control—Austrian Formula

With the Austrian formula, we take the difference between the current growing stock volume and the desired future growing stock volume, and spread this out over the adjustment period. The mean annual increment of the future regulated forest is then added to this surplus to determine the annual harvest level over the near term.

$$\text{Sustained annual yield} = \left(\frac{31,573 \text{ cords} - 18,870 \text{ cords}}{15 \text{ years}} \right)$$
$$+ 1,467 \text{ cords per year}$$
$$= 2,314 \text{ cords per year}$$

This plan involves harvesting all of stand 12 in the first year, along with 14.8% (12.5 acres) of stand 2. Since the adjustment period is rather long, and since the growth of the stands over the desired future rotation age is not factored into the current growing stock volume, this estimate of the annual harvest volume should be revisited periodically.

E. Volume Control—Hundeshagen Formula

Using the Hudeshagen formula, where the relationship between the current and desired future growing stock volumes is used, we find that the estimate of the sustained annual harvest is higher than the other estimates thus far.

$$\text{Sustained annual yield} = \left(\frac{31,573 \text{ cords}}{18,870 \text{ cords}} \right)$$
$$\times (1,467 \text{ cords per year})$$
$$= 2,455 \text{ cords per year}$$

Here, all of stand 12 would be scheduled for harvest during the first year, as well as 20.5% (17.3 acres) of stand 2.

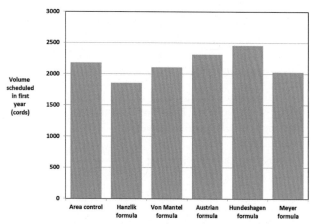

FIGURE 11.4 A comparison of volume control approached when applied to the Putnam Tract.

F. Volume Control—Meyer Formula

The Meyer formula for volume control takes into account the growth rate of the forest along with the current and desired future growing stock levels.

Annual harvest

$$= 0.045 \left(\frac{31{,}573 \text{ cords} (1.045)^{15} - 18{,}870 \text{ cords}}{(1.045)^{15} - 1} \right)$$

$$= 2{,}032 \text{ cords per year}$$

We find when using this approach that the volume suggested to be scheduled for harvest is closer to the Von Mantel or area control approaches (Fig. 11.4). Under this plan, all of stand 12 would be harvested in the first year, along with a minor portion (3.4%) of stand 2.

IV. AREA—VOLUME CHECK

When developing a regulated forest, area control does not ensure an even harvest over time, and volume control methods do not ensure that an even amount of land is treated over time. Thus the goal of the area-volume check method is to adjust the plan being developed to provide an even area treated and an even volume harvested. Chappelle (1966) described area-volume check as a process containing six steps:

1. Derive a first approximation of the sustainable harvest, and consider it a trial scheduled harvest.
2. Develop a list of areas that can be harvested, and sort them by order of harvesting priority.
3. Determine the time required to harvest each stand or age class, using the first approximation of the sustainable harvest level. Calculate the average age of the stands when they are harvested.
4. Determine the per unit area yield of the stands when they are harvested.

5. Check to see whether the number of years needed to harvest all the areas that can be harvested is equal to the desired future rotation age or the desired conversion period.
6. If the criteria for area and volume harvested are not met, then change the approximated harvest level and work through the process again until the results are acceptable.

This area-volume check is necessary since some of the assumptions behind the use of volume control methods are not necessarily valid the further we move out into the time horizon of the plan. Incorporating the additional increment associated with commercial thinnings, salvage operations, and the growth of regenerated stands, then subsequently smoothing the corrected increment over time is the central theme of area-volume check. The area-volume check process is applicable in both even-aged and uneven-aged stands. In even-aged stands, it is meant to provide regularity in management by means of maintenance of age classes. In uneven-aged stands, it is meant to provide regularity of management by controlling the cycle of harvest entries and by maintaining the optimum reserve growing stock level (Davis, 1954). With this method, a land manager would develop a schedule of activities using both area control and volume control, then other issues (silvicultural, organizational) are considered, and after adjusting the schedule, a decision is made. This process is repeated at the beginning of each planning period, and thus it represents somewhat of a rational approach, since more information affects the management decision than when simply using area or volume control methods in isolation. It may seem obvious to many that the area-volume check is very similar to the binary search technique for forest planning, which we discussed in Chapter 10, Models of Desired Forest Structure. However, binary search focuses on the volume scheduled for harvest, whereas area-volume check also would consider the area treated. Although more complicated to employ than area control and volume control methods, the area-volume check has been used extensively in practice, and continues to be used today to assess timber supply issues (Silvacom, Ltd., 2001).

V. WILDLIFE HABITAT CONTROL

The traditional control techniques described in this chapter have focused on commodity production goals; however, the use of controls on wildlife habitat and other noncommodity resources frequently have been incorporated into forest plans. For example, minimum desired habitat levels, either described as a weighted index for a management area, or as a total number of "suitable" habitat land areas, have been used in many US National Forest plans. In many cases, however, a minimum desired habitat level may not be achievable for some time. Therefore, there is a need to reasonably establish the level of habitat production over time. This estimate can then be used as a control in forest planning exercises. Of the approaches we described earlier, area control has the most direct association with the desire to maintain habitat at some predefined level. However, we are still left to wonder how much area should be controlled.

Application of the relationships represented in the volume control formulas to the management of habitat is more problematic. Some volume control formulas require a growth rate (increment) of habitat, which could be difficult to determine, since a stand or forest either is or is not suitable habitat depending on the underlying structural conditions (see Chapter 4: Estimation and Projection of Stand and Forest Conditions, for a discussion of this issue). However, if an average amount of growth (increment) of land into suitable habitat conditions can be estimated, and if a landowner or land manager can estimate the current and desired future amount of suitable habitat, modified versions of the Austrian or Heyer formulas can be employed. These would be used to determine how much habitat we would need to create, and how much we reasonably expect to build through natural processes, to a desired amount over an adjustment period. First, assume that G_r represents the desired habitat level, and G_a represents the current habitat level. Rather than subtract the surplus (or deficit) from the actual (since we are not harvesting products) we would need to subtract the actual from the desired. And since the increment (I) will occur naturally, this too would be subtracted from the computation, and the resulting equation would become:

$$\text{Annual development of habitat} = \left(\frac{G_r - G_a}{a}\right) - I$$

Example

Assume for a medium-sized forest the suitable habitat for pileated woodpeckers (*Dryocopus pileatus*) is building at a rate of 10 acres (4.0 hectares) per year, given the structural growth of the stands within the forest. Assume also that currently there are 1,000 acres (404.7 hectares) of suitable habitat, but 1,500 acres (607 hectares) are desired at the end of the decade. Using the previous modified equation, we would find that we need to create 40 acres (16.2 hectares) per year through active management and silvicultural methods to enable us to reach our goal of having 1,500 acres of suitable woodpecker habitat at the end of the decade.

$$\text{Annual development of habitat} = \left(\frac{1,500\text{ acres} - 1,000\text{ acres}}{10\text{ years}}\right)$$
$$- 10\text{ acres per year}$$
$$= 40\text{ acres per year}$$

Extensions of these methods can also be made to the development of carbon stocks, continuous crown cover, and other aspects of forests that are meaningfully related to the vast social, environmental, or economic objectives of the landowner.

VI. THE ALLOWABLE CUT EFFECT

In the early 1970s, a concept was proposed that seemed to enhance the attractiveness of investments in intensive silviculture. The concept became known as the *allowable cut effect*, because it suggested that an investment in a young stand today could result in increased harvests of older stands today. The allowable cut effect was first proposed as a way to justify increased harvests on US National Forest lands. Soon afterward, the concept became an issue on all types of land ownerships where management activities were being considered that would increase growth rates of young stands. Lundgren (1973) argued that if we were to assess the value of an investment in a forest, rather than the value of an investment in each individual stand, then we would need to place the stand-level investment into context with the impacts on broader forest-wide goals. Specifically, if an additional management action *in one area* increased the potential harvest sometime out in the future (through increased growth rates), then land managers could reap some of that benefit in the near-term *in another area* today through an increase in sustained yields. In addition, it has been argued that the additional, near-term income that would result from the higher harvest levels should be included in the economic analysis of the additional stand-level management activity.

In a standard stand-level analysis, we would determine the value of an additional management activity applied to a stand by assessing the costs and revenues associated with the activity. It is not unreasonable to assume, and silvicultural research often confirms, that intensive management activities applied to young stands in the near-term may result in increased stand-level harvests in the future. These activities may include improved genetic breeding of trees, herbaceous weed control, release treatments, fertilization treatments, and others.

Example

Assume, using a standard stand-level analysis, that within a recent clearcut in Georgia, we can plant standard nursery-grown seedlings or genetically improved seedlings. The additional cost incurred for the genetically improved seedlings is $25 per acre. If we planted the genetically improved seedlings, and we planned to clearcut the stand in 25 years, the added revenue is expected to be $200 per acre. The rate of return on this investment would be about 8.7% (invest $25 today at an 8.7% interest rate, and in 25 years, we should have a little over $200).

This stand-level decision ignores the opportunities associated with the broader forest. The main stand-level question is, given an 8.7% return on the investment, would the management organization pursue it? In other words, we would need to rank the investment relative to the other potential investments that an organization could pursue, assuming they had the funds today to invest in genetically improved seedlings at a rate of $25 extra per acre in planting costs.

The allowable cut effect acts to transfer the anticipated increases in stand-level timber yields to decisions about forest-level scheduled harvests in the near-term. In one way, it allows a land manager to argue for an increase near-term harvest levels in anticipation of higher future yields. In another, it allows a land manager to justify, on economic grounds, investments in long-term timber production.

Example

Building on the previous example, assume that in a recent clearcut in Georgia we can plant genetically improved seedlings at an additional cost of $25 per acre, and in 25 years the additional revenue will be $200. Given the additional volume available in the future, we also assume that we can harvest slightly more volume today (and every year for the next 25 years) from older stands, resulting in $12 per acre additional net revenue. This, of course, would occur in other stands; however, we could argue that the actual cost of the genetically improved seedlings is really $13 per acre ($25 actual cost per acre—$12 additional net revenue from other harvests). The rate of return on this investment might arguably now be about 11.5%, making it more attractive in the eyes of the decision-makers.

In sum, the allowable cut is the amount of wood that may be harvested over a period of time to maintain sustained production. The allowable cut effect involves the allocation of anticipated future timber yields to the near-term, and the projected increase in future inventories due to changes in management practices, which will be harvested in equal amounts beginning in the near-term, and extending over a period equal to one rotation (Teeguarden, 1973). It is used mainly to argue for increased current harvest levels, especially when harvest levels are limited by even-flow or sustained yield constraints. In other words, the effect of the analysis is that an organization can argue for an immediate increase in scheduled harvests based on the amount of wood that they hope will be made available some time in the future. The allowable cut effect can also be thought of as an increase in near-term harvests from a forest as a result of changes in assumptions about the productivity of future forests.

Example

Assume that in a forest in Louisiana we will clearcut 100 acres (40.5 hectares) this year, then plant genetically improved loblolly pine seedlings. We hope to harvest this stand once again in 25 years, resulting in a gain of 25 tons per acre over stands that were planted using standard nursery-grown seedlings. The total gain in wood volume from this investment in genetically improved seedlings is 2,500 tons (100 acres × 25 tons per acre). If this expected gain in wood volume were spread equally over 25 years (from today until the wood volume gain actually is realized), the allowable cut effect suggests that the annual scheduled harvests could be increased by 100 tons per year (2,500 ton gain/25 years).

This example represented a relatively straightforward analysis of the allowable cut effect. However, a number of factors can influence the allowable cut effect, including the desired rotation age, the current and desired forest structure (age class distribution), and organizational wood flow policies. The allowable cut effect will be greater for forests containing a regulated age class distribution than for forests containing a larger amount of older forest structure, because in the former case more acres are needed to meet the volume requirements of the regulated forest, whereas in the latter case fewer acres of older forest are needed to meet the same volume requirement. Thus in the case of a regulated forest, more acres are placed in the alternative management treatments sooner, allowing a landowner to realize the productivity gains sooner. In the case of younger forests, increasing the annual harvest levels requires harvesting more acres of younger forests sooner, which could require shortening the minimum harvest age, if this is possible (Schweitzer et al., 1972).

Two main assumptions support the assessment of the allowable cut effect: (1) a reserve of mature timber stands must be available for harvest, and (2) there must be a policy constraint in the forest plan that controls the rate of harvest (Teeguarden, 1973). Others, however, have suggested that a wood flow policy constraint is not necessary (Schweitzer et al., 1972), yet the allowable cut must be based on some form of volume regulation.

A number of criticisms have been lodged against the use of the allowable cut effect in economic analyses. These include the following:

- The allowable cut effect is based on current and future scheduled timber volume, not necessarily on increases of value, although the increase in wood volume can suggest higher rates of return in the investments in young stands.
- It is the future forest inventory, rather than the near-term annual scheduled harvest that is influenced by alternative management practices (Teeguarden, 1973). It is therefore speculative.

- The rate of return on the investment in the young stand is influenced by the productivity gains of the alternative management practices, as well as organizational wood flow policies (Teeguarden, 1973).
- Incorrect management decisions may occur simply because the analysis suggests that the opportunity exists to harvest more timber in the near term (Teeguarden, 1973). Since we are planning several decades into the future, we can argue that a continued program of intensive silvicultural activities may not materialize due to changes in budgets, markets, and other factors. In addition, an intensive silvicultural program may not be implemented at the scale it was assumed in some forest planning analyses.
- An overestimation of the rate of return on investments may occur due to the joint product of investments in alternative practices and the ability to harvest sooner some surplus wood volume contained in older stands.
- The allowable cut effect is not applicable to alternative management practices that increase the quality of wood (such as pruning), rather than the quantity of wood (Teeguarden, 1973).

VII. SUMMARY

The control techniques we described in this chapter are not universally applicable to all types of forest conditions, and therefore must be used with care. One of the challenges facing the management of uneven-aged forests may be the poor distribution of diameter size classes within stands. Developing a plan that reduces the harvest, effectively lengthening the entry cycle for harvesting, could allow a satisfactory diameter distribution to be created, yet may create an economic strain on the landowner (Zillgitt, 1951). If within a broader forest ownership some land can be acquired to fill in the gaps in diameter distribution, then it may provide more flexibility in meeting the sustained annual harvest desired by some landowners. Some stands may be in such poor shape, however, that the initial entry may be used to remove the culled and deformed trees. Subsequent entries may be conducted in short time intervals to remove commercially desirable trees, but the initial entry might need to be devoted to removing those trees that would likely be lost in the short-term, or that are using resources that may be better suited for other more desirable trees.

These broader considerations are accentuated by the fact that an insufficient amount of land containing the tree species and sizes desired by local manufacturing facilities could present a problem for the sustained yield management of both even-aged and uneven-aged forests. Landowners are generally aware of the risks associated with committing funds to forest management. Given the uncertainty of some markets, one challenge for natural resource managers will be to convince landowners that sustained yield management is an investment worth considering (Zillgitt, 1951). In uneven-aged stands, where the reserve growing stock is at a level acceptable to the landowner, and where the distribution of species and diameter classes suggest that regulation may have been obtained, all that is needed to determine the annual harvest under a harvest-equals-growth scenario is to estimate the annual increment. However, broader concerns, such as transportation system limitations, business-related costs and issues, forest health concerns, wildlife habitat needs, and perhaps the need to manage for sustainable ecosystems, can affect the harvest levels estimated with each of the techniques presented in this chapter. Carter (1922) noted some of these issues as a comment on the introduction of Hanzlik's method.

Requiring a harvest level that equals the level of growth is a characteristic of a regulated forest, and a desire of many landowners throughout North America as one aspect of a sustainability framework. When using volume control methods, the estimate of volume to harvest each year is a function of either the standing volume of growing stock alone, or the standing volume of growing stock and the rate of growth of the forest (the annual or periodic increment). If used long enough, then management by controlling harvest levels will transition to management by controlling area treated. However, most forests currently are not regulated, nor will they ever become regulated. Other physical issues brought about as a result of irregular age class distributions, organizational concerns (budgetary and time constraints), or outside factors (changing markets, natural disturbances) suggest that a regulated forest policy may not be feasible in practice. In irregular or uneven-aged forests, for example the increment may not be representative of the growth rate of the ideal forest. In addition, when a desired future growing stock level is suggested for an uneven-aged forest, the level chosen is difficult to justify, unless some form of stand-level optimization also was performed to provide evidence of its appropriateness. However, the current increment may be of limited usefulness in determining the annual harvest level, since other management issues may be more important, such as the need to adjust irregular age-class distributions. In addition, growth conditions change periodically, and growth rates may decrease due to insect or disease problems.

The Swiss needle cast issue in coastal Douglas-fir forests is a good example of how the growth rate of seemingly healthy stands may unexpectedly change within a few years. Additional management actions, such as fertilization practices, may increase the growth rate of trees as well. In summary, although controlling forest plans through area treated or volume scheduled may be a desire of a landowner, factors associated with the entire management situation may preclude reaching a regulated state.

QUESTIONS

1. *Area control.* Assume that the owners of the Lincoln Tract are interested in area control and the possible plan that might be developed using the area control method for regulating a forest. Assuming no intermediate treatments that would produce harvest volumes, and given the size of the forest (4,550.3 acres) and a desired future rotation age of 45 years:
   a. How much land area would be scheduled for final harvest in each year?
   b. Which stands would be harvested during the first year of the plan?
   c. How much harvest volume would be produced?

2. *Area control.* Develop a memorandum addressed to the owners of the Lincoln Tract that discusses the advantages and limitations of using area control as a guide for developing a forest plan.

3. *Volume control.* Assume the following about the Lincoln Tract:
   - The desired future rotation age is 45 years.
   - The growth rate of the current forest is about 5% per year.
   - The mean annual increment of the stands less than 45 years of age is 1,703 MBF per year.
   - The mean annual increment of the future regulated forest is 2,953 MBF per year.
   - The volume of mature timber over 45 years of age is 83,065 MBF.
   - The current growing stock (standing volume) is 96,972 MBF.
   - The desired future growing stock volume is 34,082 MBF.
   - The adjustment period is 20 years.
   - The harvesting rule is oldest stand first.

   a. Using the Hanzlik formula, what is the estimate of the sustained annual yield? How much land would be required to meet this harvest target during the first year of the plan?
   b. Using the Von Mantel formula, what is the estimate of the sustained annual yield? How much land would be required to meet this harvest target during the first year of the plan?
   c. Using the Austrian formula, what is the estimate of the sustained annual yield? How much land would be required to meet this harvest target during the first year of the plan?
   d. Using the Hundeshagen formula, what is the estimate of the sustained annual yield? How much land would be required to meet this harvest target during the first year of the plan?
   e. Using the Meyer formula, what is the estimate of the sustained annual yield? How much land would be required to meet this harvest target during the first year of the plan?

4. *Volume Control.* In a memorandum addressed to the landowners of the Lincoln Tract, compare and contrast the various volume control options for the owners.

5. *Volume Control.* Discuss for the owners of the Lincoln Tract the advantages and limitations of using volume control as a guide for developing a forest plan.

6. *Habitat control.* Assume that currently there are 1,071 acres of high quality pileated woodpecker habitat on the Lincoln Tract. The owners of the Lincoln Tract are interested in increasing the high quality habitat to around half the area of the Tract within the next 20 years. If the natural growth rate (increment) of high quality habitat is around 20 acres per year, on how many other acres would management actions be needed to improve the habitat quality of forested stands to meet their objective? What relation would this objective have to the generation of harvest revenue from the Tract?

7. *Allowable cut effect.* Within a southern United States forest where the minimum harvest age is 23 years, we can implement intensive management activities today that effectively increase the site index (base age 25) from 65 to 75. Assume that the increase in volume at age 23 is 39.7 tons per acre, and the stumpage price assumed is $25 per ton. The additional regeneration cost that would be required to change the site index is $50 per acre. What is the rate of return on this investment?

   Assume further that the increase in timber volume one can obtain this year, after developing a forest plan that maximized the even-flow harvest volume when acknowledging the increased future yields, was 681 tons. This additional revenue ($17,025) was obtained from 869.46 acres, resulting in an average increased revenue per acre of $19.58. Given this allowable cut effect on the annual harvest volume, what would be your rate of return on the intensive site preparation investment?

REFERENCES

Barkley, J., Bodine, J., Hoffman, C., Koslowski, J., Stevens, L., Schwantes, J., et al., 2015. Bayfield County Forest, Wisconsin, United States of America. In: Siry, J.P., Bettinger, P., Merry, K., Grebner, D.L., Boston, K., Cieszewski, C. (Eds.), Forest Plans of North America. Academic Press, New York, NY, pp. 265–275.

Başkent, E.Z., Küçüker, D.M., 2010. Incorporating water production and carbon sequestration into forest management planning: a case study in Yalnızçam planning unit. For. Sys. 19 (1), 98–111.

Carter, E.E., 1922. Comment. J. For. 20 (6), 626.

Chapman, H.H., 1950. Forest Management. The Hildreth Press, Bristol, CT, 582 p.

Chappelle, D.E., 1966. A Computer Program for Calculating Allowable Cut Using the Area-Volume Check Method. US Department of Agriculture, Forest Service, Pacific Northwest Forest and Range Experiment Station, Portland, OR, Research Note PNW-44. 4 p.

Corral-Rivas, J.J., Hernández-Díaz, J.C., Sánchez, C.A.L., Soto, J.E.L., von Gadow, K., 2015. Ejido Borbollones, Durango, Mexico. In: Siry, J.P., Bettinger, P., Merry, K., Grebner, D.L., Boston, K., Cieszewski, C. (Eds.), Forest Plans of North America. Academic Press, New York, NY, pp. 61−68.

Davis, K.P., 1954. American Forest Management. McGraw-Hill Book Company, Inc., New York, NY, 482 p.

Guldin, J.M., 1991. Uneven-aged *DBq* regulation of Sierra Nevada mixed conifers. Western J. Appl. For. 6 (2), 27−32.

Hanzlik, E.J., 1922. Determination of the annual cut on a sustained yield basis for virgin American forests. J. For. 20 (6), 611−625.

Kallesser, S.W., 2015. Camp No-Be-Bo-Sco, New Jersey, United States of America. In: Siry, J.P., Bettinger, P., Merry, K., Grebner, D.L., Boston, K., Cieszewski, C. (Eds.), Forest Plans of North America. Academic Press, New York, NY, pp. 1−9.

LaPointe, T., Casey, P., Drew, I., Flint, S., 2015. Umbagog National Wildlife Refuge, New Hampshire, United States of America. In: Siry, J.P., Bettinger, P., Merry, K., Grebner, D.L., Boston, K., Cieszewski, C. (Eds.), Forest Plans of North America. Academic Press, New York, NY, pp. 235−244.

Leak, W.B., 1964. An expression of diameter distribution of unbalanced, uneven-aged stands and forests. For. Sci. 10 (1), 39−50.

Lundgren, A.L., 1973. The allowable cut effect: Some further extensions. J. For. 71 (6), 357−360.

Meyer, H.A., 1952. Structure, growth, and drain in balanced even-aged forests. J. For. 50 (2), 85−95.

Recknagel, A.B., 1913. The Theory and Practice of Working Plans (Forest Organization). John Wiley & Sons, New York, NY, 235 p.

Schweitzer, D.L., Sassaman, R.W., Schallau, C.H., 1972. The allowable cut effect: some physical and economic implications. J. For. 70 (7), 415−418.

Silvacom, Ltd, 2001. Supplemental Timber Supply Analysis Procedures & Results. Silvacom, Ltd., Edmonton, AB, Silvacom Reference # F-057.

Sloan, G.M., 1945. Report of the Commissioner Relating to the Forest Resources of British Columbia. C.F. Banfield, Victoria, BC, 195 p.

Teeguarden, D.E., 1973. The allowable cut effect: a comment. J. For. 71 (4), 224−226.

Zillgitt, W.M., 1951. Converting mature northern hardwood stands to sustained yield. J. For. 49 (7), 494−497.

Chapter 12

Spatial Restrictions and Considerations in Forest Planning

Objectives

Commodity production continues to be the emphasis of forest management activities of many landowners across North America. However, the sustainable management and planning of natural resources for many land areas uses approaches that increasingly accommodate broader economic, ecological, and social goals. As a result, many landowners and land managers are becoming increasingly interested in forest plans that acknowledge a wider set of complex functional relationships between the growth and harvest of trees and other natural resource values, a number of which are considered nonlinear or contain conditional relationships that seem convoluted on first inspection. As a result, many practical management problems are becoming too complex to be addressed with classical optimization techniques—such as linear programming—that use single objective functions and variables assigned continuous real number values. As the forest management planning environment continues to evolve in an increasingly complex regulatory and social context, we will likely see more use of spatial restrictions and nonlinear goals in forest management plans.

Spatial information can be used to facilitate the development of a management plan that helps landowners achieve goals that otherwise would need to be addressed in an ad-hoc manner at the time of implementation. Spatial relationships that involve forest structure can range from maintaining minimum stand sizes, to preventing harvests larger than a certain size, and to ensuring that adjacent stands with a certain type of forest structure that are larger than a certain minimum size are next to stands of another type of forest structure that are also larger than a certain minimum size. Further, road and trail management issues relate to the construction, access, connectivity, and removal of routes of access. In this chapter, we describe a number of these types of spatial restrictions and relationships that will allow students to become aware of contemporary natural resource management planning issues. Upon completing this chapter, you should be able to:

1. Understand how adjacency and green-up restrictions for clearcut harvests can be incorporated into natural resource management plans.
2. Understand how complex, nonlinear wildlife habitat relationships can be incorporated into natural resource management plans.
3. Conceptualize and solve road and trail development plans.

I. ADJACENCY AND GREEN-UP RULES AS THEY RELATE TO CLEARCUT HARVESTING

Rules that pertain to the adjacency of harvests and subsequent green-up of regenerated stands relate directly to the spatial and temporal juxtaposition of harvests, and are perhaps the most widely used spatial constraints in forest planning today (Bettinger and Zhu, 2006). Controlling the size of harvests has been viewed as a way to benefit wildlife, diversity, and aesthetics. However, there are unintended consequences of spreading out harvests over space and time, such as the associated increase in forest edge and fragmentation (Tarp and Helles, 1997). Adjacency and green-up rules arise from laws, regulations, voluntary certification programs, and organizational policies (Bettinger and Sessions, 2003). For example, in Sweden the maximum clearcut harvest area is 49.4 acres (20 hectares) (Dahlin and Sallnäs, 1993). In the United States, some states (e.g., Oregon, California, and Washington) have laws indicating the maximum clearcut size allowed. Some forestry companies have incorporated adjacency constraints into their forest planning efforts to adhere to forest certification standards (McTague and Oppenheimer, 2015). In addition, some US National Forests have placed a limit on the maximum size of clearcuts, such as the 40 acre (16.2 hectare) limit for the Chattahoochee-Oconee National Forest (US Department of Agriculture and Forest Service, 2004). In Europe, on sites where the shelterwood system is used to support natural regeneration of even-aged forests, harvest areas may be divided into strips that advance progressively across the land in one direction, perpendicular to the prevailing winds (Konoshima et al., 2011). The planning of these harvests can be handled as a spatially-constrained problem with adjacency considerations.

Two common misconceptions forest managers have concerning adjacency and green-up constraints is that their implementation in a planning effort is relatively straightforward, and that the effects (economic and otherwise) of

Forest Management and Planning. DOI: http://dx.doi.org/10.1016/B978-0-12-809476-1.00012-6

recognizing these constraints are clear. In fact, controlling the timing and placement of harvests in a forest plan may require an extensive set of adjacency and green-up constraints. Two conceptual models of adjacency and green-up commonly are used in forest planning: the unit restriction model (URM) and the area restriction model (ARM). Both of these models are described in detail by Murray (1999). The URM is used to control the placement of harvest activities by disallowing the scheduling of a harvest that might physically touch (or be in proximity of) another harvest that previously has been scheduled or implemented, no matter how large or small each of the two stands are, and no matter how large or small the combined clearcut area might become. The ARM controls the size of harvests, and allows adjacent harvests to be scheduled concurrently as long as the total size of the aggregate harvest area does not exceed some predefined limit. If forest stands (management units) are small relative to the maximum harvest area allowed, then using the URM constraints may significantly misrepresent the problem (Barrett and Gilless, 2000). In contrast, if forest stands are about the same size as the maximum harvest area allowed, using the URM constraints may be appropriate. This may be important when attempting to analyze the economic impact of different maximum clearcut sizes, since changing the average size of stands of trees maintained in a geographic information system is relatively difficult. The ARM is not constrained in this regard, and planners and managers can utilize the most disaggregate geographic data available (Murray and Weintraub, 2002).

One assumption that needs to be made in a forest planning problem that includes these types of relationships is the adjacency model that is recognized. The three general models of adjacency include: (1) stands that share an edge, (2) stands that share an edge or simply a corner, and (3) stands that are within some distance of each other (Fig. 12.1). The first two of these can rather easily be ascertained within a geographic information system by understanding which polygons share a common line (edge) or point. The latter is more computationally difficult since a proximity or buffering analysis must be performed to determine the set of polygons that are within a certain distance of each other. Each of these three cases has been used in practice.

Also inherent in the use of the URM and ARM constraint models is the green-up period, or exclusion period. This is the length of time, usually expressed in years, that must pass before harvesting activities are allowed in adjacent areas. Conceptually, the green-up period represents the amount of time a regenerated stand needs to "green-up," or allow the regenerated trees to grow to a certain height, before an adjacent harvest can be

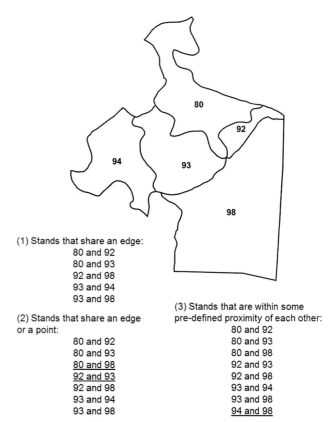

(1) Stands that share an edge:
   80 and 92
   80 and 93
   92 and 98
   93 and 94
   93 and 98

(2) Stands that share an edge or a point:

   80 and 92
   80 and 93
   80 and 98
   92 and 93
   92 and 98
   93 and 94
   93 and 98

(3) Stands that are within some pre-defined proximity of each other:
   80 and 92
   80 and 93
   80 and 98
   92 and 93
   92 and 98
   93 and 94
   93 and 98
   94 and 98

FIGURE 12.1   Three types of adjacency relationships.

scheduled. Typical green-up periods are 2−3 years in the southern United States, and 5−10 years in the western United States and Canada. Longer green-up periods have been proposed for US National Forests.

A number of methods have been assessed for developing constraints that will recognize and allow the control of adjacent harvest areas. Jones et al. (1991), Yoshimoto and Brodie (1994), Murray and Church (1995), McDill et al. (2002), Goycoolea et al. (2005), and Tóth et al. (2012) all describe several constraint formulations for representing adjacency relationships in forest planning problems. The goal of these efforts has been to increase the efficiency of integer programming problem-solving methods, since in general, reducing the number of constraints in a problem formulation also reduces the time required to solve the problem. Pair-wise adjacency constraints are the most simplistic, although perhaps not the most computationally efficient, formulations of adjacency constraints. Pair-wise adjacency constraints typically take the following form:

$$S1Y1 + S2Y1 \leq 1$$

Here, $S1Y1$ represents a binary integer decision variable that describes the potential harvest of stand 1 during year 1. *S2Y1* represents a similar binary integer decision variable describing the potential harvest of stand 2 during year 1. If these two stands somehow are considered adjacent, then the pair-wise constraint prevents them both from being scheduled for harvest in year 1. One or the other could be scheduled (or neither), but not both. Other constraints would be needed to reflect the fact that these two stands should not be scheduled for harvest in years 2-*n*. As you may have gathered, pair-wise constraints are useful for URM adjacency planning problems.

Example

Assume that we need to develop a 15-year (3-period) forest plan for the Lincoln Tract. The plan will be designed to control the placement and timing of harvests during the first time period using the URM process. Within the Lincoln Tract GIS database each stand is assigned a volume per unit area (thousand board feet per acre, or MBF per acre) for six time periods, in 5-year increments. After extracting this data from a GIS database, we can design an objective function that maximizes the discounted net revenue of harvests over the first three time periods. Assume that the minimum harvest age in this example is 50 years (i.e., a stand has to be at least 50 years old at the beginning of the time period). Assume also that the stumpage price is $450 per MBF, and that the discount rate assumed by the landowner is 5%. All revenues will be discounted from the middle of each time period (years 2.5, 7.5, and 12.5), and no costs will be assumed in the problem. In addition, we will assume that stands representing riparian areas (6, 38, 56) will not be scheduled for harvest. And, as a general wood flow constraint, the harvest volume per time period will be limited to 30,000 MBF or less. The form of the objective function might follow the following:

$$\text{Maximize} \sum_{i=1}^{I} \sum_{t=1}^{T} a_i \, v_{it} \, sp \, SiPt$$

where:
- i = a stand
- I = the total number of stands
- t = a time period
- T = the total number of time periods
- a_i = area of stand i
- v_{it} = volume contained in stand i during time period t
- sp = the stumpage price assumed
- $SiPt$ = a binary integer decision variable representing the harvest of stand i during time period t

For the Lincoln Tract the objective function would resemble the following:

```
Maximize 2302296.93 S4P1 + 1880346.75 S4P2
+ 1530196.64 S4P3 + ....+ 1437297.62 S87P1 +
1317081.82 S87P2 + 1170720.08 S87P3
```

You might notice that there is no reference to stands 1, 2, or 3 in the objective function. The reason for this, as well as any other omissions, is that stands younger than 50 years at the beginning of each time period have been ignored. The resource constraints for this problem would reflect that stands can be harvested only once, and each decision variable can be assigned a binary integer value.

```
subject to
2) S4P1 + S4P2 + S4P3 <= 1
3) S5P1 + S5P2 + S5P3 <= 1
4) S15P1 + S15P2 + S15P3 <= 1
```

As before, any decisions that are moot (e.g., those related to stands 1, 2, or 3) are not included in the resource constraints. If these decision variables were included inadvertently, then we run the risk of making infeasible decisions. The wood flow constraint might involve the use of accounting rows to accumulate the volume scheduled for harvest, such as the partial one shown here for time period 1:

```
30) 5779.92 S4P1 + 2966.99 S5P1 + .... +
    3608.34 S87P1 - H1 = 0
```

Once each of the wood-flow accounting rows have been developed for the three time periods, some control on the volume can be incorporated into the problem formulation. For example, when we assumed that the landowner wanted to harvest 30,000 MBF or less per time period, the following wood-flow constraints could be utilized:

```
33) H1 <= 30000
34) H2 <= 30000
35) H3 <= 30000
```

In addition to these constraints, the unit restriction harvest area constraints need to be incorporated into the problem formulation. These constraints indicate that the choice of harvesting each stand is dependent on the choice assigned to neighboring stands. Any choices that involve harvesting stands that touch one another are precluded from the forest plan. Some examples of these constraints include:

```
36) S4P1 + S5P1 <= 1
37) S4P2 + S5P2 <= 1
38) S4P3 + S5P3 <= 1
39) S23P1 + S84P1 <= 1
40) S23P2 + S84P2 <= 1
41) S23P3 + S84P3 <= 1
```

The objective function, resource constraints, accounting rows, and wood flow policy constraints form a mixed-integer problem formulation only when the solver understands that the choices related to time period 1 need to be assigned binary integer values. Once this has been accomplished, a forest plan can be developed that maximizes the discounted net revenue ($28,629,096) and explicitly determines the stands to harvest, such as those for time period 1 (Fig. 12.2). The scheduled harvest volumes are 29,990, 29,964, and 29,981 MBF for time periods 1 through 3,

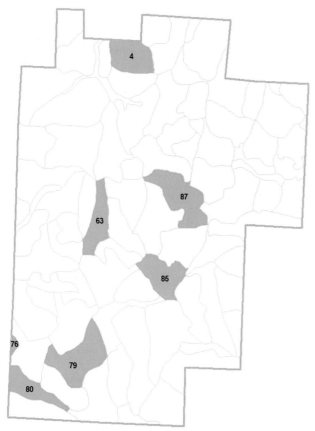

FIGURE 12.2 Unit restriction model of adjacency applied to the first-period harvests on the Lincoln Tract.

respectively. As you can see, these scheduled volumes are all very close to the desired maximum harvest level of 30,000 MBF.

Had the planning problem assumed that the green-up period was longer than one time period in the planning horizon, the adjacency constraints would have needed to acknowledge the fact that harvesting one stand in one time period affects the choices available for other stands in subsequent time periods. For example, assume that the green-up period lasted two time periods from the time of harvest. Rather than simply controlling the selection of either stand 4 or 5 during a single period, from Eqs. (36) to (38) above, we would need constraints such as the following:

36) S4P1 + S5P1 + S5P2 <= 1
37) S4P2 + S5P1 + S5P2 + S5P3 <= 1
38) S4P3 + S5P2 + S5P3 <= 1

Eq. (36) now indicates that if stand 4 is scheduled for harvest during the first time period, then stand 5 cannot be scheduled for harvest during time periods 1 or 2. Even worse, if stand 4 were scheduled for harvest during time period 2 (Eq. 37), then stand 5 could not be scheduled for harvest during any of the three time periods. Eq. (38) simply indicates that if stand 4 is scheduled for harvest during time period 3, stand 5 could not be scheduled for harvest during time periods 2 or 3.

Forest planning problems that utilize area-based (ARM) clearcut size restrictions can ultimately require a large number of constraints to provide control over the planning of adjacent harvests. In these cases, constraints are formulated to allow more than one adjacent stand to be harvested during each time period, as long as the total size of the harvest does not exceed the maximum limit. Assume, for example that stands 1, 2, and 3 are all 55 acres (22.3 hectares) in size, and that they all touch each other (stand 1 is adjacent to stand 2 and stand 3, and stand 2 is adjacent to stand 3 as well). Assume as well that the maximum clearcut size is 120 acres (48.6 hectares). It is obvious from this information that (at most) only two of the three stands can be scheduled for harvest during the same time period, since together, harvesting two of the stands will create a clearcut that is 110 acres (44.6 hectares) in size, which is below the maximum size allowed. To recognize this limitation, a constraint can be developed using the following form:

$$S1Y1 + S2Y1 + S3Y1 \leq 2$$

If the decision variables are assigned binary values, then this constraint allows, at most, two of the three stands to be scheduled for harvest during year 1. Again, other constraints would be needed to reflect the fact that these two stands should not be scheduled for harvest in time periods 2-n. These types of constraints, and other more efficient forms described in Yoshimoto and Brodie (1994), Murray and Church (1995), McDill et al. (2002), Goycoolea et al. (2005), and Tóth et al. (2012) are useful for ARM adjacency planning problems. We must keep in mind, however, that as the potential maximum clearcut size increases relative to the average size of a stand, the number of constraints and potential constraint redundancies will increase (Yoshimoto and Brodie, 1994; Crowe et al., 2003).

Example

Using the Lincoln Tract problem presented earlier, assume that the clearcut harvests can be as large as 120 acres during the first time period. In some cases, it is possible that pairs of neighboring stands can be scheduled for harvest during the same time period. For example, stands 63 and 71 are adjacent neighbors of stand 62. The combined area of stands 62 and 63 is about 86 acres, and the combined area of stands 62 and 71 is about 118 acres. However, the combined area of all three is about 171 acres; therefore, we can allow stands 62 and 63 to be scheduled for harvest during time period 1, or stands 62 and 71, but not all three stands. One way to handle this situation in a plan of action is to create constraints such as the following:

51) S62P1 + S63P1 <= 2
52) S62P1 + S71P1 <= 2
53) S63P1 + S71P1 <= 1

The first two equations allow the scheduling of adjacent stands, because we know that the combined areas will be less than our desired maximum scheduled area. The third of the three equations prevents the scheduling of the two stands that will result in a clearcut area larger than our maximum desired area. A more logical equation such as the following might have been considered:

54) S62P1 + S63P1 + S71P1 <= 2

However, this would still allow stands 63 and 71 to be scheduled for harvest during the same time period, violating our assumption of maintaining clearcut areas below 120 acres. Once all the ARM adjacency relationships have been accounted for in the problem formulation, a forest plan can be developed that maximizes the discounted net revenue ($28,633,724) and explicitly identifies the stands to harvest (Fig. 12.3). The scheduled harvest volumes in this scenario are 29,984, 29,980, and 29,989 MBF for time periods 1 through 3, respectively. As before, these scheduled volumes are all very close to our desired maximum harvest level of 30,000 MBF. However, the ARM formulation allows more flexibility in the placement and timing of activities, thus the discounted net revenue is higher here than in the previous case when we used the URM problem formulation.

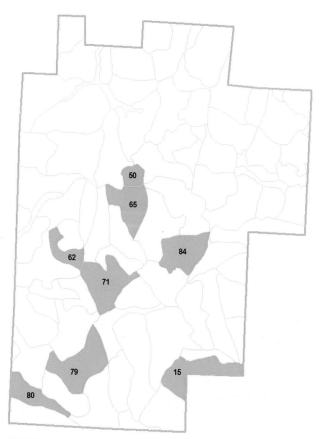

FIGURE 12.3 Area restriction model of adjacency applied to the first-period harvests on the Lincoln Tract.

Although linear programming has been, and continues to be, widely used in the development of forest management plans, the fractional results provided by linear programming (e.g., suggesting the harvest of only a portion of a stand) are a problem for planning efforts requiring the recognition of adjacency relationships. For example, if a plan indicates a harvest of 56% of a stand, then we do not necessarily know which part of the stand will be harvested, or which of the adjacency relationships should be assessed. As a result, binary choices (harvest/do not harvest) generally are used in situations that require the recognition of adjacency and green-up constraints. A number of mathematical techniques other than linear programming can be used to formulate and solve spatial forest planning problems that include binary decision choices, including mixed-integer programming and heuristic methods such as Monte Carlo simulation, tabu search, and simulated annealing. As we suggested in Chapter 8, Advanced Planning Techniques, mixed-integer programming is similar to linear programming (with the exception of the binary integer variables) and is considered an exact approach to problem solving. In general, if we can solve a problem with mixed-integer programming, then we can assume that the resulting solution is the optimal solution to the problem. The two examples presented thus far have been mixed-integer programming problems. We clearly stated in Chapter 8, that heuristic methods cannot guarantee that the optimal solution to a management problem will be found; however, high quality solutions to complex problems can be generated relatively quickly with these methods.

When using the URM constraints in forest planning, the size of the maximum harvest area is implied (size of largest single stand). When using ARM constraints in forest planning, the maximum harvest area is flexible, and stands are aggregated in the scheduling process up to the maximum area allowed. In general, forest plans that include URM constraints are lower in economic value than forest plans that include ARM constraints, because of the added flexibility that the ARM constraints provide in arranging the timing and placement of harvests.

Research has shown that the economic value of forest plans tends to decrease as the maximum clearcut size becomes smaller (Borges and Hoganson, 1999). Larger maximum clearcut sizes may provide forest managers more flexibility in the spatial and temporal arrangement of harvests, leading to higher economic and commodity production values. In addition, reductions in forest plan value may occur when a minimum harvest area size is assumed, which can force into a plan a number of harvest choices that may be suboptimal (Jamnick and Walters, 1993). The cost of these types of spatial restrictions generally ranges from 3% to 20% of the potential unconstrained forest plan value (Jamnick and Walters, 1993;

Nelson and Finn, 1991; Boston and Bettinger, 2001a; Boston and Bettinger, 2006). The impact, of course, depends on the type of forest modeled, the maximum clearcut size, and the length of the green-up period. However, in some cases a doubling of the green-up period can result in a 30–40% reduction in net present value. In other cases, the harvesting delay related to these constraints can result in a shift of products assumed to be produced, which may affect wood-flow needs (Boston and Bettinger, 2001b).

Since the timing and placement of harvesting activities across a property is dynamic, the availability of other forest characteristics should be viewed as dynamic as well. For example, older forest areas will increase in size as forests grow, yet may decrease in size as adjacent areas are scheduled for harvest (Boston and Bettinger, 2006). As a result, large, contiguous patches of older forest may become more limited when harvests are required to be spread out across the landscape. In fact, research results indicate that as the maximum clearcut size decreases, forest fragmentation actually can increase (Nelson and Finn, 1991). The resulting increase of forest edges and the subsequent decrease of forest patch sizes may have an effect on the quality of wildlife habitat, and may provide the impetus for other damage to the forest (e.g., increased windfall along the edges, from strong winds).

II. ADJACENCY AND GREEN-UP OF GROUP SELECTION PATCH HARVESTS

In stands designated for group selection harvesting activities, the harvest openings (patches) are generally small (0.5–4 acres, or 0.2–1.6 hectares) compared to the size of the stand (50+ acres or 20.2+ hectares). Group selection patches are designed and delineated on maps by taking into account the existing road system, the topography, and the potential array of logging options. In most instances, group selection patches are logically and consistently placed within the forest management framework. In many group selection management situations, the timing of entry and the type of activity assigned to each patch must be spatially and temporally coordinated. For example, in a stand where 0.5 acre patches are to be created, the timing of harvest entry into those patches might be coordinated such that (1) adjacent patches are not clearcut within a certain time window, creating openings larger than what was desired, and (2) other patches that are not clearcut are applied a simultaneous management action, such as a thinning. Additionally, the order that patches are clearcut may be restricted to prevent logging through the advanced regeneration, which can cause unacceptable damage. This suggests that the timing of

treatments (clearcuts and thinnings) might be synchronized to facilitate an efficient harvesting operation.

Example

Bettinger et al. (2003) describe a management situation where the objective function for the forest planning problem was to maximize the net revenue from timber harvests, subject to a number of constraints related to group selection harvests. First, the land managers required a minimum net revenue from each entry into a stand to make the system operationally feasible. Second, the age of the trees in the group selection patches that were clearcut harvested needed to be above a minimum harvest age. Third, the timing of clearcuts in the group selection patches needed to be synchronized with the timing of the thinnings in other group selection patches within the stand. Fourth, unit restriction adjacency constraints needed to be used to disallow the placement of group selection clearcut patches next to other group selection clearcut patches during the appropriate green-up time period. To further complicate the problem, as you can see in Fig. 12.4, the patches may be made up of several smaller pieces of stands. For patches within stands designed for group selection harvesting activities, the green-up time period could have been as long as 30 years. To control the level of group selection harvest activities within a stand, only a certain percentage of patches in a stand should be clearcut in each entry period. For example, the assumption that only 20% of the potential group selection patches within a stand could be clearcut during an entry was used in this work. When the reentry interval is assumed to be 10 years, this implies a 50-year rotation, where after 50 years, all the patches would have been clearcut harvested, and a second round of patch clearcut harvests would have begun.

FIGURE 12.4 Group selection patch harvests where only 20% of a stand is entered, and patches do not share an edge.

III. HABITAT QUALITY CONSIDERATIONS

In many cases, planning for the development and maintenance of habitat patches requires that we view a planning problem from a spatial perspective. Simply assuring that a suitable amount of land of a certain type of habitat will be available throughout the lifespan of a forest plan can be relatively straightforward and can be accommodated using linear programming methods. However, assuring that patches of a certain size are available through time is not as easy to accommodate, nor is the ability to assure managers and landowners that patches of a certain size will be positioned next to other types of habitat, or within some distance of other types of habitat. These goals generally require the use of binary integer decision variables that will enable the use of constraints that are similar to the adjacency constraints described earlier. Applications in this area of planning and analysis are numerous, and to further illustrate the vast array of possibilities, the remainder of this section of the chapter provides four case studies on the integration of spatial wildlife goals into forest planning efforts.

A. Case 1: Elk Habitat Quality

Objectives or goals for ungulates (elk and deer) can be expressed as the need to maintain a desired percentage of a management area in adequately sized blocks of forage and another desired percentage of the area in adequately sized stands of trees that act as cover. These two types of habitat areas serve the purpose of providing food and protection for elk and deer. As an example of an actual elk habitat quality goal, the 1990 Wallowa-Whitman National Forest plan (US Department of Agriculture and Forest Service, 1990) contained direction for managing areas that were considered Roosevelt elk (*Cervus elaphus*) summer range areas. The rules for arranging forage and cover across the landscape were twofold:

1. Maintain at least 80% of the forage area within 600 ft (183 m) of a patch of cover that was at least 6 acres (2.4 hectares) in size.
2. Maintain at least 80% of the forage area within 900 ft (274 m) of a patch of cover that was at least 40 acres (16.2 hectares) in size.

These conditions present a rather complex spatial planning problem. Obviously some sort of forest management activity would be needed to create and maintain forage areas. These could be represented on the landscape by managed food plots (fields containing agricultural vegetation), or more than likely by recently clearcut areas. The clearcuts, however, would be able to support only vegetation for adequate foraging for a certain amount of time, thus the spatial position of forage areas would need to move around the landscape over time. The cover areas would need to consist of overstory trees and understory vegetation that had a structure sufficient to provide elk and deer with areas both to hide and maintain warmth.

The functional relationships that describe the attainment and maintenance of elk summer range habitat are nonlinear (a management action does not necessarily lead to a linear and predictable response), and would therefore be difficult, if not impossible to incorporate within a strategic plan that utilized a linear programming framework. The difficulties arise with the scheduling of each activity (e.g., harvest), where a planning model would need to assess the impact of each activity on (1) the distance to forested cover, (2) the resulting size of the cover, if the activity affected the size of the cover, and (3) how much of the resulting forage (prior to the activity and created by the activity) is within a minimum distance of the resulting cover. Since these concerns could not be incorporated directly into the development of a strategic forest plan, more than likely they would be assessed at the tactical or operational planning levels (at or near the time of activity implementation) or after a plan of action was developed. Other nontraditional planning methods (heuristics) can be developed to simultaneously schedule activities and handle complex, spatial relationships such as these, and although we provided an introduction in Chapter 8, Advanced Planning Techniques, these methods are beyond the scope of this book. For further reading on this subject, Bettinger et al. (1997) suggested a tabu search method for incorporating these rules into a forest plan.

B. Case 2: Bird Species Habitat Considerations

To illustrate the incorporation of spatial considerations in forest planning when bird habitat is considered, we briefly describe two approaches introduced in (Bettinger et al., 2002), which cover minimum patch size goals and complementary patch goals. In these two cases, it is assumed that the decision variables related to activities are represented by binary (0,1) numbers, which indicate that an activity (harvest) assigned to a stand is assigned to the entire stand, not some portion of it less than 100%. In this work, two activities were considered: either clearcut harvest or no harvest. Three overarching constraints were assumed: (1) a minimum harvest volume was required to be produced during each time period, (2) stands could be scheduled for harvest only once, and (3) stands had to be above a minimum age before they could be harvested. The time horizon was 50 years, and thus 10 5-year time periods were used. The planning problem was developed in such a way as to maximize the area of "habitat":

Maximize

$$\left(\sum_{t=1}^{T} \sum_{i=1}^{I} \sum_{k=1}^{K} A_i \, H_{i,t,k} \right) \Bigg/ \left(\sum_{i=1}^{I} A_i \right)$$

where:

i = a stand
I = the total number of stands
t = a time period
T = the total number of time periods
k = a wildlife species
K = total number of wildlife species
A_i = area of stand i
$H_{i,t,k}$ = a binary variable indicating whether (1) or not (0) stand i is considered habitat for species k during time period t

To assess a minimum patch size goal for bird habitat, Bettinger et al. (2002) used a generalization of the habitat requirements for three western United States forest birds, varied thrush (*Ixoreus naevius*), winter wren (*Troglodytes troglodytes*), and Hammond's flycatcher (*Empidonax hammondii*), each of which were assumed to need intact stands of mature or old-growth stands greater than 49.4 acres (20 hectares) in size. They further assumed that mature or old-growth forests were stands greater than 80 years of age. To account for the fact that more than one stand could be joined together to create a patch, the ARM (Murray, 1999) was used to determine the size of individual older forest patches (those consisting of stands greater than 80 years of age). If a stand was part of one of these patches in any one time period, then it was considered to have $H_{i,t,k} = 1$ during that time period.

A planning goal that seeks to achieve the most area in complementary, adjacent patches is one order of magnitude more complex than the minimum patch size goal. Here, we would assume that a type of wildlife habitat (such as a patch of older forest of a certain size, perhaps for nesting and roosting purposes) should be located adjacent to another type of wildlife habitat (such as a patch of young forest of a certain size, perhaps for foraging purposes), and that this situation would be most beneficial to a particular wildlife species. One example used by Bettinger et al. (2002) was for the great gray owl (*Strix nebulosa*), which seems to prefer younger forests for foraging and mature or old-growth forests for roosting and nesting. To incorporate these concerns into a forest plan, it was assumed that older forest stands were those with an age greater than 80 years, and that younger forest stands were those with an age equal to or less than 10 years. In addition, they assumed that the size of the older forests had to be 49.4 acres (20 hectares) or greater,

and that the size of the adjacent younger forests had to be 24.7 acres (10 hectares) or greater.

To accommodate these goals, two recursive functions using the ARM process were used to evaluate the size of patches that consist of forests that are greater than or equal to 80 years old, and forests that are less than or equal to 10 years old. Then a process using the URM approach was used to determine whether any of the older forest patches were touching any of the younger forest patches. Therefore, stands that were greater than or equal to 80 years of age, or less than or equal to 10 years of age were not necessarily assigned a value $H_{i,t,k} = 1.0$ unless the patches that they belonged to were touching during a given time period.

These two goals are complex and nonlinear in nature, and as a result, incorporating these considerations into linear programming would be difficult, if not impossible. As an alternative, heuristic methods have been developed to allow planners to create forest plans that both mathematically assess these types of goals, and either constrain problems such that the goals are met, or incorporate them into the objective function to maximize or minimize some value. As in the previous examples, the only spatial information required for the assessment of these types of bird habitat relationships is the adjacency relationships among stands of trees.

C. Case 3: Red-Cockaded Woodpecker Habitat Considerations

The red-cockaded woodpecker (*Picoides borealis*) was listed as an endangered species in the United States in 1970 due to severe population declines and losses of habitat (US Department of the Interior and Fish and Wildlife Service, 2003). The bird species lives in open, mature, and old-growth pine forests in the southeastern states. Currently, less than 3% of estimated pre-European settlement population of these woodpeckers remains (US Department of the Interior and Fish and Wildlife Service, 2003). Red-cockaded woodpeckers are a cooperatively breeding bird species that live in family groups consisting of a breeding pair of birds and one or two male helpers. At the finest scale, one of the critical resources for the species are cavities excavated in live pines, a task that usually takes the birds several years to complete. Red-cockaded woodpeckers exploit the ability of live pines to produce large amounts of resin, and after cavities are created the resin acts as a barrier against predators such as climbing snakes. Longleaf pine (*Pinus palustris*) is the preferred species because it produces more resin than other southern pines, and produces it over a longer period of time (US Department of the Interior and Fish and Wildlife Service, 2003). At a broader scale, the woodpeckers need open pine woodlands and savannas for nesting and roosting habitat. Nesting habitat consists of open pine stands that contain little or no hardwood midstory or overstory structure. Encroachment by hardwoods as a result of fire suppression is a well-known cause of habitat abandonment. Foraging habitat consists of mature pine stands with an open canopy, a low density of small pine trees, little or no hardwood or pine midstory structure, little or no overstory hardwood trees, and ground cover that consists of native bunchgrass and forbs (US Department of the Interior and Fish and Wildlife Service, 2003). Fragmentation of habitat conditions can limit the number of breeding groups and isolate different groups. In general, public lands in the southern coastal United States are managed in such a way as to increase the population, whereas private lands may be managed in such a way as to stabilize the population.

Some of the red-cockaded woodpecker management guidelines encourage maintenance of low density stands, where the ideal state is somewhere around $40-80\,\text{ft}^2$ of basal area per acre ($9.2-18.4\,\text{m}^2$ per hectare) of pine trees. To maintain the desired open understory characteristic, some land managers use biennial, growing season prescribed burns. As a consequence of these frequent growing season burns, however, advanced regeneration that becomes established under mature stands may be destroyed before it can grow to a size that can survive such a prescribed fire management schedule. Thus an unintended result of this type of management may be to create forests that are comprised of mature and over-mature timber stands and do not have viable regeneration in place to replace the older trees when they ultimately die due to natural causes or other reasons. As a result, some questions managers have that relate to forest planning include:

- Do current management practices result in sustainable habitat over time?
- Do current management practices result in sustainable timber production?
- How should wildlife guidelines and timber management practices be modified to help insure the sustainability of both concerns?

The fitness of woodpecker habitat is a function of the structure of the forest resource, the character of the environment (openness), and the condition of the ground cover. As an illustration of the type of data needed, the following planning guidelines were acquired from the red-cockaded woodpecker recovery standard (US Department of the Interior and Fish and Wildlife Service, 2003). The recovery standard varies by the productivity of each forested site. Medium and high productivity sites (those with a pine $SI_{25} \geq 60$) require 120 acres of "good quality" habitat (defined as follows) within 0.5 miles (0.8 km) of the center of the cluster, and 50% or more of this within 0.25 miles

(0.4 km) of the center, *for each group of woodpeckers*. Low productivity sites (those with a pine $SI_{25} < 60$) require more area ($200-300$ acres, or $80.9-121.4$ hectares) of good quality habitat within 0.5 miles of the center of the cluster, and 50% or more of this within 0.25 miles of the center, for each group of woodpeckers. Good quality habitat includes areas with the following characteristics:

- Stands with at least 18 trees per acre (TPA) (44.5 trees per hectare) of pines ≥ 60 years of age, and ≥ 14 inches (35.6 cm) diameter at breast height (DBH). The minimum basal area for these stands is $20\,\text{ft}^2$ per acre ($4.6\,\text{m}^2$ per hectare).

  In these same stands,
- The basal area of pines $10-14$ inches ($25.4-35.6$ cm) DBH is between 0 and $40\,\text{ft}^2$ per acre ($9.2\,\text{m}^2$ per hectare).
- The basal area of pines <10 inches (25.4 cm) DBH is less than $10\,\text{ft}^2$ per acre ($2.3\,\text{m}^2$ per hectare), and below 20 TPA (49.4 trees per hectare).
- The basal area of pines ≥ 10 inches (25.4 cm) DBH is at least $40\,\text{ft}^2$ per acre ($9.2\,\text{m}^2$ per hectare).

The recovery standard is not clear when it comes to the maximum basal area for longleaf, shortleaf (*Pinus echinata*), or slash pine (*Pinus elliottii*) stands, yet a maximum is probably assumed by natural resource managers. The maximum basal area for loblolly pine (*Pinus taeda*) stands is $80\,\text{ft}^2$ per acre ($18.4\,\text{m}^2$ per hectare). In addition, the following forest conditions are used to further define good quality habitat:

- Bunchgrass and other native, fire-tolerant, fire-dependent herbs account for at least 40% of the ground and midstory plant community, and these are dense enough to carry a growing season fire at least once every 5 years.
- The hardwood midstory, if present, is sparse, and less than 7 ft (2.1 m) in height.
- Hardwoods in the canopy comprise less than 10% of the TPA in longleaf pine stands, and less than 30% of the TPA in loblolly pine and shortleaf pine stands.
- Foraging habitat is not separated from the cluster by more than 200 ft (61 m) of nonforaging areas, which include hardwood forests, young pine stands less than 30 years of age, cleared areas, paved roads, rights-of-way, and bodies of water.

Although population-specific foraging guidelines may be developed by various natural resource managers, a planner would need to understand whether any of these exist prior to modeling the habitat conditions. Other silvicultural recommendations of the recovery plan include the following:

- For two-aged pine stands, (1) rotation ages are at least 120 years for longleaf and shortleaf pine, and 100 years for loblolly pine, slash pine, and pond pine (*Pinus serotina*); (2) regeneration cuts are limited to 25 acres (10.1 hectares) in woodpecker habitat areas with less than 100 breeding groups and 40 acres (16.2 hectares) in areas with 100 or more breeding groups; (3) 6–10 pines per acre (14.8–24.7 trees per hectare) are left as residual live trees; and (4) all flat-topped, turpentine, or relict pines are retained.

- For uneven-aged pine stands, (1) 5 TPA (12.4 trees per hectare) of the oldest live pines are retained during each entry; and (2) all flat-topped, turpentine, or relict pines are retained.

If stands are more than one mile (1.6 km) from an active or recruitment cluster, then even-aged, two-aged, and uneven-aged management systems can be used to restore areas to native pine species. Regeneration harvests of up to 80 acres (32.4 hectares) in size are acceptable in these circumstances. Some of these guidelines are not only nonlinear, but they also include spatial relationships. As a result, these guidelines would be very difficult (although not impossible) to incorporate into a forest planning system. We could rank habitat with a binary quality (good habitat/not good habitat) and map it, or we could develop a ranking procedure to differentiate levels of "goodness." Afterward, a simulation or optimization model can be developed to project through time how habitat conditions will change, and how management actions might be used to maintain or increase habitat levels. If a quantitative objective for the management of an area can be ascertained, then we also could develop an optimization procedure to maximize the objective while also maintaining or increasing habitat levels (or alternatively, minimizing the loss of habitat). Boston and Bettinger (2001b) made headway into this area 15 years ago. If successful in these endeavors, a planner could provide broad-scale insights into the potential effects of current and alternative forest management policies by comparing the joint production of commodity and noncommodity products from a forested landscape.

D. Case 4: Spotted Owl Habitat Quality

One timely opportunity for the integration of economic and ecological goals in forest planning involves the development and maintenance of northern spotted owl (*Strix occidentalis caurina*) habitat in the Pacific Northwest. Although some regulations related to the owl are rather straightforward (e.g., do not harvest within a certain distance of owl nesting areas), the assessment of habitat quality across the landscape is generally more computationally difficult, depending on how it is accomplished. For example, the habitat capability index (HCI) model developed by McComb et al. (2002)

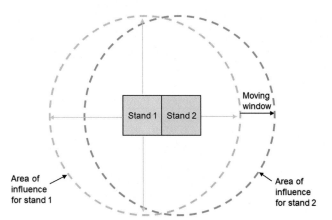

FIGURE 12.5 A moving window approach to the assessment of landscape conditions.

contains a spatial component that requires assessing both within-stand nesting characteristics and landscape-level home-range characteristics. This habitat model requires assessing the structure of all stands within three buffer distances from each stand of interest, one of them being 1.5 miles (2.4 km) wide. Therefore any scheduled activity within 1.5 miles of a stand can affect the estimated habitat quality of that stand. To assess the habitat quality of all stands across a landscape then would require some sort of moving window approach (Fig. 12.5).

One aspect of these types of habitat models is that a single habitat value will be assigned to each stand during each time period, assuming data are available to assess habitat quality across all periods of time. Another important aspect from a visual perspective is that we can assess how habitat values change across the landscape by mapping them. With spatial habitat models, mapping the values using a geographic information system can help planners and managers better understand how management may affect (positively or negatively) habitat quality. Further, the stand-level habitat quality values can be aggregated up to a landscape-level weighted average (weighted by the area of each stand for the entire landscape, for watersheds, for management areas, etc.) to arrive at a single habitat value for the area in question. These weighted landscape values can then be constrained in a forest planning model with equations that require a minimum level attainment, prevent large deviations in habitat quality from one period to the next, or prevent declines in habitat quality across time.

To explore this further, in the McComb et al. (2002) model, HCI levels are estimated by calculating nest stand and landscape capability for providing habitat, which then are combined into a single HCI value:

$$HCI_{it} = (NCI_{it}^2 LCI_{it})^{(1/3)}$$

where:

NCI_{it} = nest stand capability component for stand i during time period t

LCI_{it} = landscape capability component for stand i during time period t

Developing projected stand conditions for each potential management prescription, using a growth and yield model, is a necessary step in estimating habitat quality, since NCI_{it} is a function of the density of trees that are 4−10 inches (10−25 cm) DBH, 10−20 inches (25−50 cm) DBH, and 30+ inches (75+ cm) DBH. The landscape-level portion of the index, LCI_{it}, is a function of the percentage of large and medium sized trees within three buffer zones around each stand: 984 ft, 2,625 ft, and 1.5 miles (300, 800, and 2,400 m).

Methods for incorporating this habitat quality assessment within a forest planning model that had a time horizon of 40 years, using planning periods that were 5 years long, have been demonstrated (Bettinger and Boston, 2008). The main goal of the problem was to emulate current broad-level activity in northwestern forests by maximizing the even-flow of timber harvest volume subject to several constraints (adjacency of clearcuts, green-up period, minimum harvest age, and minimum habitat levels). To demonstrate the utility of the approach, management activities were limited to two decisions: clearcut and replant, or no harvest. The initial characteristics of each stand were provided as lists of trees, since tree characteristics where needed to estimate wildlife habitat. Modeling the growth of the trees was accomplished using the ORGANON growth and yield model (Hann et al., 1997). A heuristic forest planning model was developed to schedule potential activities across the landscape, since the habitat model could not be reduced easily to linear equations, and since the number of stands modeled (39,500) was quite large.

The ARM of adjacency was accommodated using a recursive function that could determine the actual size of all proposed clearcuts. In doing so, the function senses the size of potential clearcuts from neighbors of proposed sales, and their neighbors, and so on.

$$x_{it}A_i + \sum_{z \in Ni \cup Si} x_{zt} A_z \le MCS$$

where:

A_i = the area of stand i

t = a time period

z = a stand either adjacent to stand i or adjacent to a stand that is adjacent to stand i

Ni = the set of all stands adjacent to stand i

Si = the set of all stands adjacent to those stands adjacent to stand i

MCS = the maximum clearcut size assumed

x_{it} = a binary decision variable representing harvest (1) or no harvest (0) of stand i during time period t

This spatial model for forecasting the quality of owl habitat is, of course, a simplification of a much more complex biological and ecological system. Achieving reliability in a such a simulation process is difficult, given the potential for error in the underlying data, the potential for error in the projection of forest stand data into the future, and the potential for the habitat model to score some areas low (or high) when they actually are being used (or not used) by owls. Although these potential errors seem daunting, in practice we would assume that they provide a reasonable representation of the vegetative structure and wildlife habitat quality. As a result, the landscape model (Bettinger and Boston, 2008) can be viewed as a modeling structure that facilitates the evaluation of trade-offs among forest policies. In doing so, it recognizes economic and ecological goals, and models their attainment across a large area and over a long period of time.

Example

To illustrate the incorporation of a simple spatial wildlife habitat relationship into a forest plan, assume we are using the URM problem from earlier in the chapter, where we attempted to maximize the discounted net revenue while ensuring harvest areas were not touching each other during the first time period. In addition to this, assume that the two oldest stands (78 and 79) represent owl habitat and should not be scheduled for harvest in any of the time periods. The adjacent stands to these (73, 74, 75, 80, 82, and 83) are to remain uncut as well during the first time period. To accommodate these goals, several constraints can be developed that simply indicate the choices that are now unavailable:

```
133) S78P1 = 0
134) S78P2 = 0
135) S78P3 = 0
136) S79P1 = 0
137) S79P2 = 0
138) S79P3 = 0
139) S73P1 = 0
140) S74P1 = 0
141) S75P1 = 0
142) S80P1 = 0
143) S82P1 = 0
144) S83P1 = 0
```

Equations 133 through 138 indicate that neither stand 78 nor stand 79 should be scheduled for harvest during our three-period time horizon. Equations 139 through 144 control the scheduling of the adjacent buffer stands during the first time period of the analysis. Given the manner in which we have developed these constraints, these buffer stands may be available for harvest in subsequent time periods. However, the URM harvest constraints still apply.

In addition to these owl habitat constraints, assume that the landowner is interested in maintaining elk cover structure in stand 85. Here we will assume that the stand should remain uncut within the time horizon of the forest plan, and that the landowner desires to complement this habitat with elk forage areas that are directly adjacent to this stand. As a result, either stand 59, 71, 84, or 86 must be harvested during time period 1, since they are the only adjacent stands to stand 85 that are above the landowner's minimum harvest age. The constraints that might be developed to accommodate these concerns include the following:

    145) S85P1 = 0
    146) S85P2 = 0
    147) S85P3 = 0
    148) S59P1 + S71P1 + S84P1 + S86P1 = 1

Eqs. (145)−(147) suggest that stand 85 will not be scheduled for harvest during the time horizon of the plan, but Eq. (148) suggests that one of its neighbors must be scheduled for harvest in the first time period. After solving the problem we find that the objective function value is $23,798,826, or about $4,830,270 lower than the solution to the URM problem, mainly because of the delay or unavailability of harvests. The scheduled harvest volumes, in fact, do not necessarily approach the limit of 30,000 MBF per time period. Scheduled harvests for time periods 1−3, respectively, are 27,625, 24,595, and 20,933 MBF. However, the wildlife habitat goals are maintained (Fig. 12.6), and

the results should be of interest to the landowner in understanding the trade-offs between commodity production and multiple use management.

IV. ROAD AND TRAIL MAINTENANCE AND CONSTRUCTION

The objective of many road system management problems is to minimize the cost to the landowner for the maintenance, construction, and removal of roads. There are, however a number of ways we can look at these types of problems, which includes viewing them simply as a transportation problem where products are produced at different places on the landscape and need to be moved to various markets. Other ways include viewing these problems as a minimum cost problem where we need to determine the lowest cost for routing products or people across a network to a specific endpoint, or viewing them as a shortest path problem where we need to determine the shortest route from one place to another. Each of these types of problems has to be developed in conjunction with the need to provide access to areas of a property necessary to fulfill the objectives of the landowner. In terms of a problem formulation, the objective function could then reflect the total discounted cost of road construction, maintenance, and removal over the entire planning horizon. In addition, a budget constraint could be used to ensure that the money allocated to construction, maintenance, and removal of roads during each time period does not exceed what would reasonably be assumed to be available during those periods of time. Trail management problems can be viewed in much the same manner as well. What makes these problems different from the others we have described thus far in this chapter is that road and trail problems are represented using a network structure that is embedded into the larger forest planning problem that utilizes as spatial features lines (arcs) and intersections or endpoints (nodes) (Fig. 12.7).

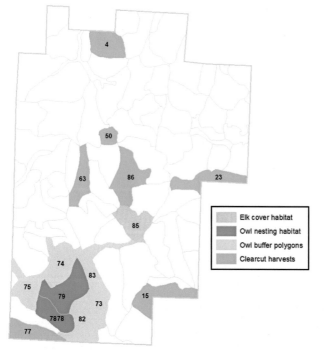

FIGURE 12.6 Wildlife habitat restrictions applied to the Lincoln Tract forest plan.

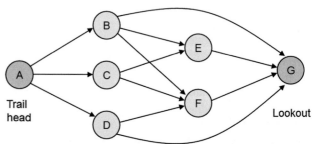

FIGURE 12.7 An example network problem beginning with a trail head (A) and illustrating several routes (through B−F) that lead to a lookout (G).

A. Case 1: Road Management Problem

To begin exploring a basic road management problem, examine Fig. 12.8. Here we find four stands of various sizes and standing volumes. Each stand has at least one access point to a potential road network, and one stand (Stand 2) has two entry points. The segments in the figure are actually vectors representing proposed roads, and each has to be developed at a cost of $20,000 per mile. A maintenance cost is assumed, at $0.05 per ton of wood transported over each road, per mile of road. The objective of the problem is to determine the optimal road system that minimizes the transportation costs associated with moving the harvested wood from each stand to the county public road system (accessed by nodes 4 and 5). We can define the decision variables perhaps as these:

$SEGx$ = a binary integer variable indicating the construction (or not) of road segment x

$VUyNz$ = the amount of volume (tons) moving from unit (stand) y to node z

$VNyNz$ = the amount of volume (tons) moving from node y to node z

The objective function can then be designed, taking into account the length of each potential road and the fixed and variable costs associated with each road, and can be formulated as follows.

```
Minimize 8000 SEG1 + 12000 SEG2 + 10000 SEG3 + 10000
SEG4 + 14000 SEG5 + 12000 SEG6 + 14000 SEG7 + 14000
SEG8 + 12000 SEG9 + 0.02 VU1N1 + 0.03 VU2N1 + 0.025
VU2N2 + 0.025 VU3N3 + 0.035 VU4N3 + 0.03 VN1N5 +
0.035 VN2N5 + 0.035 VN3N2 + 0.03 VN3N4
```

Several of the variables in the objective function represent the construction of roads. The coefficients for these decision variables are developed by multiplying the cost ($20,000 per mile) by the length of each segment. For example, the coefficient for $SEG1$ is $20,000 times the length of segment 1 (0.4 miles). The other variables in the objective function relate to the variable cost of transporting wood over each road segment. For example, the coefficient for $VU2N1$ is developed by multiplying the cost per ton per mile ($0.05) by the length of each associated route (in this case 0.6 miles). Some constraints need to be developed for the network to track the flow of volume harvested. What these constraints imply is that volume must originate from somewhere, and must arrive somewhere else, thus they are considered conservation of flow constraints. Further, it is assumed that none of the volume is lost along the road network. As they relate to this problem, the conservation of flow constraints might include the following:

```
2) VU1N1 = 2500
3) VU2N1 + VU2N2 = 3190
4) VU3N3 = 3240
5) VU4N3 = 3630
6) VU1N1 + VU2N1 - VN1N5 = 0
7) VU2N2 + VN3N2 - VN2N5 = 0
8) VU3N3 + VU4N3 - VN3N2 - VN3N4 = 0
9) VN3N4 + VN1N5 + VN2N5 = 12560
```

Here,

- Eq. (2) represents the amount of volume originating in stand 1, and transported to node 1 in the network.
- Eq. (3) represents the amount of volume originating in stand 2, and transported to either node 1 or node 2 in the network.
- Eq. (4) represents the amount of volume originating in stand 3, and transported to node 3 in the network.
- Eq. (5) represents the amount of volume originating in stand 4, and transported to node 3 in the network.
- Eq. (6) represents the amount of volume moved from stands 1 and 2 to node 1, then transported to node 5 in the network.
- Eq. (7) represents the amount of volume moved from stand 2 to node 2, and potentially from stands 3 and 4 (through node 3), then transported to node 5 in the network.
- Eq. (8) represents the amount of volume moved from stands 3 and 4 to node 3, then either transported to node 4 in the network, or moved through to node 2.
- Eq. (9) represents the amount of volume arriving onto the county road from segments 6, 7, or 9 in the network.

To ensure that some of these roads are actually built, we will use what is termed the "big M" method from the

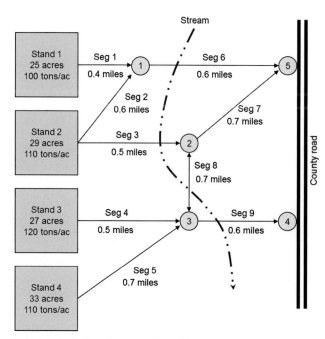

FIGURE 12.8 A road construction problem.

field of operations research. These equations will force a road to be built if wood volume is scheduled to be transported across it. The big M method equations use the following form:

$$X1 - MY1 \leq 0$$

This equation suggests that if $Y1$ is a binary integer, and if $X1$ represents a continuous real number, then whenever $X1$ is positive, $Y1$ has to be equal to 1. However, this works only when the value of M (a coefficient) is greater than or equal to whatever value $X1$ can become. As a result, the M coefficient has to be at least as large as $X1$. When the maximum value of $X1$ is unknown, M is set to some arbitrarily large number (hence the "big M"), and the less than or equal to inequality is used. When maximum value of $X1$ is known, M is known, and the equality sign can be used in the equation. These types of constraints can afford us the ability to designate roads for construction (the Y variables here) if some volume is scheduled to move across the road (the X variable in this brief example). For our problem, we know just how large M should be for some segments of the road system, therefore we can develop the road construction equations to resemble the following:

```
10) SEG1 = 1
11) VU2N1 - 3190 SEG2 = 0
12) VU2N2 - 3190 SEG3 = 0
13) SEG4 = 1
14) SEG5 = 1
15) SEG6 = 1
16) VN2N5 - 10060 SEG7 <= 0
17) VN3N2 - 6870 SEG8 <= 0
18) VN3N4 - 6870 SEG9 <= 0
```

Here we find the following:

- Eq. (10) suggests that road segment 1 must be built— it represents the only access to unit 1.
- Eq. (11) suggests that if any volume is moved from stand 2 onto road segment 2, then the variable $SEG2$ has to be given a value of 1, indicating that the road segment needs to be built.
- Eq. (12) suggests that if any volume is moved from stand 2 onto road segment 3, then the variable $SEG3$ has to be given a value of 1, indicating that the segment needs to be built.
- Eq. (13) suggests that road segment 4 must be built— it represents the only access to stand 3.
- Eq. (14) suggests that road segment 5 must be built— it represents the only access to stand 4.
- Eq. (15) suggests that road segment 6 must be built— it represents the only access to stand 1.
- Eq. (16) suggests that if any volume is moved from node 2 to node 5, then the variable $SEG7$ has to be given a value of 1, indicating that the road segment

needs to be built. The value 10,060 represents the maximum amount of volume that can be transported along the road segment.

- Eq. (17) suggests that if any volume is moved from node 3 to node 2, then the variable $SEG8$ has to be given a value of 1, indicating that the road segment needs to be built. The value 6870 represents the maximum amount of volume that can be transported along the road segment.
- Eq. (18) suggests that if any volume is moved from node 3 to node 4, then the variable $SEG9$ has to be given a value of 1, indicating that the road segment needs to be built. The value 6870 again represents the maximum amount of volume that can be transported along the road segment.

The optimum, or minimum cost solution to the problem indicates that road segments 1, 2, 4, 5, 6, and 9 must be built to harvest all four stands. The minimum cost solution requires a budget of \$68,730.55. Harvested volume from stands 1 and 2 will move through node 1 then node 5, and onto the county road. Harvested volume from stands 3 and 4 will move through node 3, then node 4, then onto the county road.

To increase the realism of these types of problems, the unused amount of the budget for any 1 year could be inflated, allowing the interest to accrue, and transferred to the next year. This suggests that the amount of budget available in any 1 year is the annual budgeted amount, plus any balance that is unused in previous years, plus the interest accrued on the unused balances. Further, for truck routing problems, methods that account for the slope of each road, the maximum speed allowed (or possible), and the cost of the trucking system would enable the problem to represent more closely the actual management situation.

B. Case 2: Trail Development Problem

Assume that the managers of the Lincoln Tract are considering the development of a trail system (Fig. 12.9). A budget has been developed (\$100,000 maximum), and funds must be spent in the current year. Ideally, the managers would like a trail system that is close to the highway, yet provides users with a challenging experience for both hiking and mountain biking. A parking area has been designated, as have several potential trail segments. As a start, one route from the parking area (A) to the turn-around point (B) is desired. After performing a reconnaissance of the area, you determine the costs related to developing and maintaining each segment of trail (Table 12.1). On average, the managers of the Lincoln Tract expect 10,000 visitors per year on the trail system. Since the cost of the trail system is limited, which

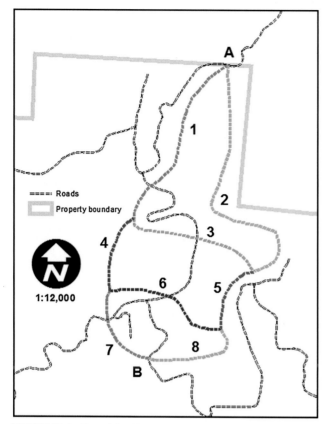

FIGURE 12.9 Potential trails system for the Lincoln Tract.

TABLE 12.1 Costs Associated With the Development of a Trail System on the Lincoln Tract

Trail Segment	Length (miles)	Construction Cost ($/mile)	Maintenance Cost ($/mile/ person)
1	0.65	48,000	0.75
2	0.94	40,000	0.70
3	0.49	47,000	0.75
4	0.30	62,000	0.85
5	0.25	43,000	0.70
6	0.48	46,000	0.75
7	0.35	54,000	0.65
8	0.45	41,000	0.70

trails would you recommend that the managers construct? What is your estimate of the cost of the trail system? Using the assumptions laid out here, what would be the projected annual maintenance costs?

To solve this problem, we can visualize it as a complex network (Fig. 12.10) and assign variables to the fixed and

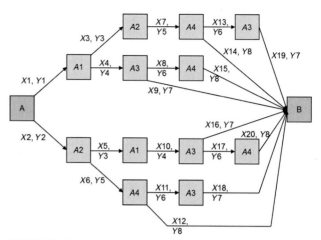

FIGURE 12.10 Network flow diagram for the potential trails system.

variable cost elements. We know that A represents the potential location of parking lot, and that B represents the potential location of a turn-around point. The question is, how do we most efficiently develop a trail system from point A to point B? Just for illustrative purposes, let's define the intermediate nodes of the network, where:

$A1$ = the junction of trails 1, 3, and 4
$A2$ = the junction of trails 2, 3, and 5
$A3$ = the junction of trails 4, 6, and 7
$A4$ = the junction of trails 5, 6, and 8

We must represent other aspects of the problem with binary integer variables that consider the construction costs:

$Y1$ = a binary variable indicating whether or not trail 1 was built
$Y2$ = a binary variable indicating whether or not trail 2 was built
$Y3$ = a binary variable indicating whether or not trail 3 was built
$Y4$ = a binary variable indicating whether or not trail 4 was built
$Y5$ = a binary variable indicating whether or not trail 5 was built
$Y6$ = a binary variable indicating whether or not trail 6 was built
$Y7$ = a binary variable indicating whether or not trail 7 was built
$Y8$ = a binary variable indicating whether or not trail 8 was built

The variable costs associated with the number of people using each trail will be assigned to the following variables:

$X1$ = the number of people using trail 1 (Parking lot to $A1$)
$X2$ = the number of people using trail 2 (Parking lot to $A2$)

$X3$ = the number of people using trail 3 ($A1-A2$)
$X4$ = the number of people using trail 4 ($A1-A3$)
$X5$ = the number of people using trail 3 ($A2-A1$)
$X6$ = the number of people using trail 5 ($A2-A4$)
$X7$ = the number of people using trail 5 ($A2-A4$)
$X8$ = the number of people using trail 6 ($A3$ to $A4$)
$X9$ = the number of people using trail 7 ($A3$ to turn-around point)
$X10$ = the number of people using trail 4 ($A1-A3$)
$X11$ = the number of people using trail 6 ($A4-A3$)
$X12$ = the number of people using trail 8 ($A4$ to turn-around point)
$X13$ = the number of people using trail 6 ($A4-A3$)
$X14$ = the number of people using trail 8 ($A4$ to turn-around point)
$X15$ = the number of people using trail 8 ($A4$ to turn-around point)
$X16$ = the number of people using trail 7 ($A3$ to turn-around point)
$X17$ = the number of people using trail 6 ($A3-A4$)
$X18$ = the number of people using trail 7 ($A3$ to turn-around point)
$X19$ = the number of people using trail 7 ($A3$ to turn-around point)
$X20$ = the number of people using trail 8 ($A4$ to turn-around point)

Shown here is the problem formulation for the trail development problem.

```
Minimize 0.49 X1 + 0.66 X2 + 0.37 X3 + 0.26 X4 + 0.37
X5 + 0.18 X6 + 0.18 X7 + 0.36 X8 + 0.23 X9 + 0.26 X10 +
0.36 X11 + 0.32 X12 + 0.36 X13 + 0.32 X14 + 0.32 X15 +
0.23 X16 + 0.36 X17 + 0.23 X18 + 0.23 X19 + 0.32 X20 +
31200 Y1 + 37600 Y2 + 23030 Y3 + 18600 Y4 + 10750 Y5 +
22080 Y6 + 18900 Y7 + 18450 Y8

subject to
2) X1 + X2 = 10000
3) −X1 + X3 + X4 = 0
4) −X2 + X5 + X6 = 0
5) −X3 + X7 = 0
6) −X4 + X8 + X9 = 0
7) −X5 + X10 = 0
8) −X6 + X11 + X12 = 0
9) −X7 + X13 + X14 = 0
10) −X8 + X15 = 0
11) −X10 + X16 + X17 = 0
12) −X11 + X18 = 0
13) −X13 + X19 = 0
14) −X17 + X20 = 0
15) X19 + X14 + X15 + X9 + X16 + X20 + X18 + X12
    = 10000
16) X1 − 99999 Y1 <= 0
17) X2 − 99999 Y2 <= 0
18) X3 + X5 − 99999 Y3 <= 0
19) X4 + X10 − 99999 Y4 <= 0
20) X6 + X7 − 99999 Y5 <= 0
```

```
21) X8 + X11 + X13 + X17 − 99999 Y6 <= 0
22) X9 + X16 + X18 + X19 − 99999 Y7 <= 0
23) X12 + X14 + X15 + X20 − 99999 Y8 <= 0
end
```

In this problem we sought to minimize the cost of construction (the "Y" variables) and maintenance (the "X" variables). The coefficients for $Y1$ were calculated using the length of each trail and the associated cost. For example, trail section 1 will cost \$48,000 per mile $\times$ 0.65 miles, or \$31,200. The variable cost coefficients (Xs) were calculated using the cost per person per mile and the length of each trail. For example, the variable cost associated with traveling between nodes $A2$ and $A4$ is \$0.70 per person per mile $\times$ 0.25 miles = \$0.175 per person. Other coefficients were calculated in a similar fashion. Constraint rows 2 through 15 account for the flow of people along the paths from the proposed parking lot to the turn-around point. Rows 16 through 23 ensure that the problem knows that if some people are expected to have traveled a trail, that the trail is built. The solution to the problem is to build trail sections 2, 5, and 8 (parking area to node $A2$, node $A4$, then to the turn-around point). The cost of the plan is \$78,400, well below the budget. The annual maintenance cost is projected to be \$11,600, and the initial trail construction cost is estimated to be \$66,800.

V. SUMMARY

In some areas of the world, advances in research and public concerns about aesthetics, biodiversity, and sustainability have resulted in the incorporation of spatial relationships into forest plans. Concerns over the spatial position of management activities affect both the timing and placement of activities across the landscape. Unfortunately, accommodating these restrictions into forest management plans may result in plans that include higher levels of forest fragmentation, lower wood flows, and lower revenues. In addition, accommodating adjacency and green-up constraints or wildlife habitat relationships may require significantly more work on the part of planners to develop the problem formulations, and subsequently the plans of action, that recognize the restrictions. The effect of spatial restrictions on the quality of a forest plan may vary based on the initial age class structure of a forest; for example many dimensions of the problem can be affected by the introduction of adjacency constraints (Kašpar et al., 2014). While complex planning problems may currently be difficult to solve with classical techniques, as computer software and hardware evolve issues associated with the computation time required may eventually be moot for large combinatorial problems. In the meantime, more reliance may be placed on simulation and optimization with heuristics.

QUESTIONS

1. *Adjacency and Green-up Constraints.* Your supervisor is interested in understanding the various aspects of "green-up" requirements related to clearcut harvests. Prepare a short memorandum for them that describes the following:
   a. The various ways they can measure adjacency of harvest units.
   b. The general idea of a green-up period.
   c. The general ways in which adjacency and green-up requirements might be incorporated into a forest plan.
   d. The differences between the URM and the ARM.

2. *Trail Management.* The managers of the Lincoln Tract are considering the development of a trail system (see Fig. 12.9). A revised estimate of the construction costs has been developed (see the following).

Trail	Length (miles)	Construction cost ($/mile)	Maintenance cost ($/mile/person)
1	0.65	75,000	0.75
2	0.94	40,000	0.70
3	0.49	35,000	0.75
4	0.30	55,000	0.85
5	0.25	75,000	0.70
6	0.48	35,000	0.75
7	0.35	56,000	0.65
8	0.45	59,000	0.70

On average, the managers of the Lincoln Tract now expect 15,000 visitors per year. Since the cost of the trail system is limited, which trails would you recommend that the managers construct? What would be the cost of the trail system to develop? Using the assumptions laid out here, what would be the projected annual maintenance costs?

3. *Unit Restriction Model.* Develop the problem formulation and solve a three-period URM for the Putnam Tract assuming the following:
   a. The minimum harvest age is 25 years.
   b. The landowner wishes to maximize the net revenue associated with the harvests.
   c. The stumpage price is $25 per cord.
   d. The landowner uses a discount rate of 6%.
   e. Only pine stands can be harvested.
   f. The time periods are 5 years long.
   g. Harvests occur during the mid-point of the time period.
   h. A wood-flow policy constraint suggests a maximum of 11,000 cords should be harvested per time period.

   i. Use the age at the beginning of each time period as the age of each stand. What is the discounted net revenue for this plan? Which stands would be harvested during the first time period? What wood flows should be expected using this plan?

4. *Area Restriction Model.* Develop the problem formulation and solve a three-period forest plan for the Putnam Tract assuming the following:
   a. The minimum harvest age is 25 years.
   b. The landowner wishes to maximize the net revenue associated with the harvests.
   c. The stumpage price is $25 per cord.
   d. The landowner uses a discount rate of 6%.
   e. Only pine stands can be harvested.
   f. The time periods are 5 years long.
   g. Harvests occur during the mid-point of the time period.
   h. A wood-flow policy constraint suggests a maximum of 11,000 cords should be harvested per time period.
   i. The landowner wants to control the size of the harvested areas in the first period, and wants to use the URM for the remaining two periods.
   j. The maximum clearcut size is 120 acres.
   k. Use the age at the beginning of each time period as the age of each stand. What is the discounted net revenue for this plan? Which stands would be harvested during the first time period? What wood flows should be expected using this plan? How does this plan compare to the plan developed using the previous question's assumptions?

5. *Wildlife Habitat Plan.* Using the problem formulation from Question 3, enhance it to account for and control the maintenance of wildlife habitat. Assume that wildlife habitat for a certain species is reflected by the following relationship:

$$\text{Habitat quality} = 0.1 \text{ Age} - 0.0025 \text{ Age}^2$$

Some assumptions that are necessary include the following: (1) use the stand ages at the beginning of each time period as the "age" of each stand, (2) if habitat quality falls below zero, then it is zero.
   a. Develop accounting rows to accumulate habitat units (Habitat quality * nonharvested acres) for each of the three time periods.
   b. Constrain the problem to produce 100 habitat units in each time period of the plan.

   What is the discounted net revenue for this plan? Which stands would be harvested during the first time period? What wood flows should be expected using this plan? How does this plan compare to the plan developed using the previous question's assumptions?

6. *Road Management Plan with Sediment.* Assume for a moment that the tonnage of sediment produced per ton of wood transported across each road section for each section of potential road described in the road management problem (Section IV.A.) that crossed the stream was:

VN1N5	0.001
VU2N2	0.002
VN3N2	0.003
VN3N4	0.004

a. Develop an accounting row to accumulate the sediment produced.
b. How much sediment was produced?
c. Constrain the problem to require a 10% reduction in sediment produced.
d. Did the optimal solution to the road management plan change? If so, then how?

REFERENCES

Barrett, T.M., Gilless, J.K., 2000. Even-aged restrictions with sub-graph adjacency. Ann. Oper. Res. 95 (1–4), 159–175.

Bettinger, P., Boston, K., 2008. Habitat and commodity production trade-offs in coastal Oregon. Socioecon. Plann. Sci. 42 (2), 112–128.

Bettinger, P., Sessions, J., 2003. Spatial forest planning: to adopt, or not to adopt? J. For. 101 (2), 24–29.

Bettinger, P., Zhu, J., 2006. A new heuristic method for solving spatially constrained forest planning problems based on mitigation of infeasibilities radiating outward from a forced choice. Silva Fennica. 40 (2), 315–333.

Bettinger, P., Sessions, J., Boston, K., 1997. Using Tabu search to schedule timber harvests subject to spatial wildlife goals for big game. Ecol. Modell. 94 (2–3), 111–123.

Bettinger, P., Graetz, D., Boston, K., Sessions, J., Chung, W., 2002. Eight heuristic planning techniques applied to three increasingly difficult wildlife planning problems. Silva Fennica. 36 (2), 561–584.

Bettinger, P., Johnson, D.L., Johnson, K.N., 2003. Spatial forest plan development with ecological and economic goals. Ecol. Modell. 169 (2–3), 215–236.

Borges, J.G., Hoganson, H.M., 1999. Assessing the impact of management unit design and adjacency constraints on forestwide spatial conditions and timber revenues. Can. J. For. Res. 29 (11), 1764–1774.

Boston, K., Bettinger, P., 2001a. The economic impact of green-up constraints in the southeastern United States. For. Ecol. Manage. 145 (3), 191–202.

Boston, K., Bettinger, P., 2001b. Development of spatially feasible forest plans: a comparison of two modeling approaches. Silva Fennica. 35 (4), 425–435.

Boston, K., Bettinger, P., 2006. An economic and landscape evaluation of the green-up rules for California, Oregon, and Washington (USA). For. Policy Econ. 8 (3), 251–266.

Crowe, K., Nelson, J., Boyland, M., 2003. Solving the area-restricted harvest-scheduling model using the branch and bound algorithm. Can. J. For. Res. 33 (9), 1804–1814.

Dahlin, B., Sallnäs, O., 1993. Harvest scheduling under adjacency constraints—a case study from the Swedish sub-alpine region. Scand. J. For. Res. 8, 281–290.

Goycoolea, M., Murray, A.T., Barahona, F., Epstein, R., Weintraub, A., 2005. Harvest scheduling subject to maximum area restrictions: exploring exact approaches. Oper. Res. 53 (3), 490–500.

Hann, D.W., Hester, A.S., Olson, C.L., 1997. ORGANON User's Manual Department of Forest Resources. Oregon State University, Corvallis.

Jamnick, M.S., Walters, K.R., 1993. Spatial and temporal allocation of stratum-based harvest schedules. Can. J. For. Res. 23 (3), 402–413.

Jones, J.G., Meneghin, B.J., Kirby, M.W., 1991. Formulating adjacency constraints in linear optimization models for scheduling projects in tactical planning. For. Sci. 37 (5), 1283–1297.

Kašpar, J., Marušák, R., Sedmák, R., 2014. Spatial and non-spatial harvest scheduling versus conventional timber indicator in over-mature forests. Lesnícky Časopis—For. J. 60 (2), 81–87.

Konoshima, M., Marašák, R., Yoshimoto, A., 2011. Spatially constrained harvest scheduling for strip allocation under Moore and Neumann neighbourhood adjacency. J. For. Sci. 57 (2), 70–77.

McComb, W.C., McGrath, M.T., Spies, T.A., Vesely, D., 2002. Models for mapping potential habitat at landscape scales: an example using northern spotted owls. For. Sci. 48 (2), 203–216.

McDill, M.E., Rebain, S.A., Braze, J., 2002. Harvest scheduling with area-based adjacency constraints. For. Sci. 48 (4), 631–642.

McTague, J.P., Oppenheimer, M.J., 2015. Rayonier, Inc., Southern United States of America. In: Siry, J.P., Bettinger, P., Merry, K., Grebner, D.L., Boston, K., Cieszewski, C. (Eds.), Forest Plans of North America. Academic Press, New York, NY, pp. 395–402.

Murray, A.T., 1999. Spatial restrictions in harvest scheduling. For. Sci. 45 (1), 45–52.

Murray, A.T., Church, R.L., 1995. Measuring the efficacy of adjacency constraint structure in forest planning models. Can. J. For. Res. 25 (9), 1416–1424.

Murray, A.T., Weintraub, A., 2002. Scale and unit specification influences in harvest scheduling with maximum area restrictions. For. Sci. 48 (4), 779–789.

Nelson, J.D., Finn, S.T., 1991. The influence of cut-block size and adjacency rules on harvest levels and road networks. Can. J. For. Res. 21 (5), 595–600.

Tarp, P., Helles, F., 1997. Spatial optimization by simulated annealing and linear programming. Scand. J. For. Res. 12 (4), 390–402.

Tóth, S.F., McDill, M.E., Könnyü, N., George, S., 2012. A strengthening procedure for the path formulation of the area-based adjacency problem in harvest scheduling models. Math. Comput. For. Nat.-Resour. Sci. 4 (1), 27–49.

US Department of Agriculture, Forest Service, 1990. Land and Resource Management Plan, Wallowa-Whitman National Forest. US Department of Agriculture, Forest Service, Pacific Northwest Region, Portland, OR.

US Department of Agriculture, Forest Service, 2004. Land and Resource Management Plan, Chattahoochee-Oconee National Forests. US Department of Agriculture, Forest Service, Southern Region, Atlanta, GA, Management Bulletin R8-MB 113 A.

US Department of the Interior, Fish and Wildlife Service, 2003. Recovery Plan for the Red-Cockaded Woodpecker (*Picoides borealis*).

US Department of the Interior, Fish and Wildlife Service, Washington, DC. Available from: http://www.fws.gov/rcwrecovery/finalrecoveryplan.pdf (Accessed 3/11/2016).

Yoshimoto, A., Brodie, J.D., 1994. Comparative analysis of algorithms to generate adjacency constraints. Can. J. For. Res. 24 (6), 1277−1288.

Chapter 13

Hierarchical System for Planning and Scheduling Management Activities

Objectives

Forest planning processes guide natural resource management, and allow land managers and landowners to compare the consequences of alternative courses of action. Planning involves methods for determining the best course of action to meet the goals of the landowner. Under most circumstances, and depending on the detail involved, plans should help a landowner understand the economic, ecological, and social consequences of a set of management activities, and allow them to understand the associated risks and uncertainty. This is important because in many cases, landowners seek plans that involve low levels of risk and uncertainty. The risks and uncertainties may be obvious in the presentation of the outcomes, determined through an analysis of alternatives, or may be inherent given the objectives (e.g., maximize even-flow, minimize environmental problems) or constraints (minimum harvest levels, minimum environmental values) that are recognized. In many instances, planning allows a landowner to further prepare for the changes that may arise financially and otherwise (changing vistas, perhaps), and enables them to develop ways to mitigate difficult circumstances.

Organizational processes, for land held and managed by corporations, investment organizations, states, provinces, and federal governments, vary in breadth and complexity. Therefore, formal planning may occur at different points in time, cover different time periods, and focus on different scales or resolutions of resources. For example, many if not all the US National Forest plans are very broad in scope, providing strategic direction for field managers who implement management activities across broad areas. During the implementation phase, lower levels of planning may be involved to ensure the higher-level concerns are consistently accommodated for activities proposed in specific places. This planning environment is somewhat similar in other areas of the world as well; however, the extent of the area covered by plans may vary depending on the socioeconomic atmosphere of each country. For example, the phases of planning for forests in Korea consist of basic plans, regional forest plans, and unit forest plans (Park, 1990). The basic forest plan is developed for *all forests* within the country, whether publicly or privately owned, and provides a strategic management perspective. Regional forest plans deal with public and private forests located within each district within a province. Unit forest

plans are concerned with the activities associated with specific forests in individual management areas. These three levels of planning are basically strategic, tactical, and operational in nature. Although variation exists in the content and complexity of these three types of planning efforts, at the conclusion of this chapter you should be able to:

1. Describe the basic characteristics of strategic, tactical, and operational planning as they relate to natural resource management.
2. Compare and contrast the general differences between strategic, tactical, and operational planning.
3. Understand the level of planning required for various types of natural resource management decisions that must be made.

I. STRATEGIC PLANNING

The term *strategic forest planning* sometimes is used as a synonym for long-term forest management and planning, although it is debatable whether some forest management strategies are meant to examine long-term consequences and impacts. In theory, strategic planning involves long-term forecasts of the economic, ecological, and social consequences of selective courses of action. Traditionally, strategic planning in forestry concentrated on the interaction between management decisions and timber sustainability or economic viability; however, a number of ecological and social concerns (habitat, sustained yields, etc.) may now be incorporated into a strategic plan. For example, in Chapter 1, Management of Forests and Other Natural Resources, we described several strategic objectives of the McPhail Tree Farm, which included a desire to practice sustainable forest management, with an emphasis on commercial timber production and associated revenue, and desires to promote the development and maintenance of wildlife habitat, aesthetic, and recreational qualities of the property (Straka and Cushing, 2015).

Forest Management and Planning. DOI: http://dx.doi.org/10.1016/B978-0-12-809476-1.00013-8

The City of San Francisco urban forest plan is also strategic in nature in that it describes how the city should encourage tree planting activities to support local wildlife, promote social equity, and advance sustainability policies and other programs of interest to the city (Swae, 2015).

Broad-scale production levels and other landscape-level issues are addressed during this planning process. Among the most widely used quantitative methods for facilitating strategic forest planning are linear programming and simulation models. Linear programming has been used mainly to allocate land and resources to various broad-scale concerns of a landowner. The Spectrum and Woodstock planning models (briefly discussed in Chapter 8: Advanced Planning Techniques) are examples of computer programs that can perform the analyses required to complete strategic planning efforts. Simulation has been used mainly as a way to evaluate the long-term productivity of forests, with a focus on forest growth dynamics. The growth and yield models we discussed in Chapter 4: Estimation and Projection of Stand and Forest Conditions, are one type of simulation model. Almost every large company and governmental agency in North America has used linear programming and simulation models (perhaps in concert) to develop strategic forest plans. However, the focus is generally on land availability and broader landscape-level questions.

Strategic plans are designed to illustrate a trajectory of the forest through time, using decisions that are meant to lead to long-term sustainable production levels. These plans may help an organization assess long-term sustainability, or decide how to best compete for long-term survival in its marketplace. Strata-based geography and generalized growth and yield information typically are used to address these issues. As we noted in earlier chapters, strata are aggregations of stands with similar conditions. Strata may be based on the age, species composition, site index, stand density, or combinations of these and other attributes. Obviously the more subdivisions of strata, the more complex the modeling system becomes. However, representing the various forest structural changes that may occur on different sites, or with different species compositions, can be more accurately portrayed when the description of the forest more closely represents the actual conditions on the ground (as opposed to using broader averages of a wide range of conditions). As a result, there is a greater emphasis here on the activities and outcomes of forest plans than on the details of the spatial arrangement of activities (Church, 2007).

The most common objectives used during strategic planning include the maximization of net present value, cash flow, net revenue, and wood flow of timber harvests, or the minimization of habitat and other ecological value degradation. Constraints typically are related to the allocation of land to various resource objectives, budgetary issues, wood flows, and the extent of land assigned to forest management activities. Measures of outcomes include revenue, habitat, jobs, commodities, water yields, recreational opportunities, carbon sequestered, and many others. The combination of these that provide the strategy for the management of a forest generally arises from upper-level managers and landowners, yet the implementation of the strategy is left to the mid-level managers and field personnel, who must also attend to tactical and operational issues (Gunn, 2007). Depending on the interaction among people interested in the management of a forest, the development of a strategic plan may require a significant effort and a significant amount of time, as we noted in the Chattahoochee-Oconee National Forest example forest plan presented in Chapter 1, Management of Forests and Other Natural Resources (Bettinger et al., 2015).

II. TACTICAL PLANNING

Implementing forest plans without regard to the spatial proximity of management activities to other resources or other activities can result in practical problems in the field, as well as result in the inability to evaluate many of the potential consequences of a plan of action. Tactical forest plans take into account the spatial relationships between management activities, and cover periods of time ranging from 1 year to perhaps 20 years. These are plans that use stand-based geography and fine-scale growth and yield information to characterize the spatial and temporal distribution of forest conditions and proposed management activities. Most forest companies, in fact, develop 1–3-year harvest plans that could be considered tactical plans. These plans suggest where and when to implement various natural resource management activities. In addition, spatial analysis of the impact of activities is incorporated here rather than in strategic forest plans. For example, in the Rayonier, Inc. forest plan summary presented in Chapter 1, Management of Forests and Other Natural Resources, it was suggested that harvest plans of this type would address the size, shape, and placement of final harvests for visual quality purposes, and support long-term sustainable harvest levels (McTague and Oppenheimer, 2015).

Although numeric coefficients can be developed that relate activities to habitat development, the assessment of habitat quality in relation to the location of planned harvests, as we described in Chapter 12, Spatial Restrictions and Considerations in Forest Planning, is one example of spatial analysis integrated into tactical planning. In Chapter 12, we also discussed variations of the harvest placement and timing problem that are common to many forest companies in North America. Green-up and adjacency rules, for example require spatial information and are accommodated in tactical forest plans to allow land managers and landowners to recognize these issues and to use them to control the development of forest plans.

Issues in tactical planning generally revolve around how to implement a strategic plan, and incorporate aspects of management that were not recognized in the strategic plan. The outcomes reported from strategic plans are spatially disaggregated here, in an attempt to locate the landscape position of activities, habitats, and forest conditions suggested. For example, harvest area size or shape, riparian management areas, and wildlife habitat patches may all be recognized spatially at this level of planning (Sessions et al., 2007). Tactical planning may be critical for understanding future landscape patterns, since what is planned may have a lasting impact on the pattern of vegetation. One of the purposes of a tactical plan is to analyze the cumulative effects of management activities, since at this scale spatial concerns are first recognized and can assist in the assessment of nontraditional outcomes. However, nontimber values may be difficult to address even in a tactical plan. These include concerns about water quality, aesthetics, recreational values, and certain aspects of wildlife habitat. It is left to the operational plan to address the final issues associated with the implementation of management activities.

Similar to strategic planning efforts, common objectives used during tactical planning include the maximization of net present value, cash flow, net revenue, and even-flow of timber harvests, or minimization of habitat quality and other ecological value degradation. Constraints again typically involve the location and timing of management activities, the location and timing of habitat development, budgetary issues, wood flows (by product perhaps), and the extent of land assigned to forest management activities. Measures of outcomes include revenue, habitat, commodities, water yields, recreational opportunities, carbon sequestered, and others. Although some of these arise from discussions with upper-level managers and landowners, others (adjacency and green-up) may be suggested through regulations or voluntary certification guidelines.

III. OPERATIONAL PLANNING

Operational planning involves determining the specific courses of action and allocation of resources that are needed to achieve the higher-level goals. This may involve daily, weekly, or monthly budgeting or resource allocation activities, or it may involve project-specific logistics. Like tactical plans, operational plans also include spatially explicit information. However, tactical plans do not lead directly to actual harvesting activities being implemented, they only generally indicate *where* and *when* the activities should occur, and not *how* the activities will be carried out (Epstein et al., 2007). Road and trail building and scheduling, two problems we discussed in Chapter 12,

Spatial Restrictions and Considerations in Forest Planning, are examples of operational planning processes. The emphasis here is usually on plans of action that cover less than a year, and the timing of activities within the year is important.

Other examples related to harvesting operations include:

- Optimal log bucking decisions
- Location and use of harvesting machinery and personnel
- Primary transportation routes of wood from the stump to a landing
- Secondary transportation routes of wood from the landing to a mill or wood yard

Operational concerns can also be extended to multidisciplinary activities such as these:

- Location and use of pruning or precommercial thinning crews
- Location and timing of tree planting activities
- Location of specific resources (e.g., trees) to be manipulated to create wildlife habitat (e.g., red-cockaded woodpecker nest trees)
- Optimal trail system route decisions
- Optimal decisions concerning areas for mushroom harvesting
- Optimal decisions regarding the set of stream segments that need enhancement (logs or boulders)

Although linear programming and dynamic programming have been shown to be useful at this level, heuristic techniques are becoming more commonly used to solve these problems (Sessions et al., 2007). In addition, benefit-cost analysis, internal rate of return, and other economic criteria have been used to evaluate management alternatives at this scale. Common objectives used during this level of planning include the minimization of costs, determination of optimal paths, maximization of net present value, or minimization of habitat quality values and other quantifiable ecological values associated with landscape management. Constraints typically involve the location and timing of management activities, the location and timing of habitat development, budgetary issues, wood flows by product, and other logistical concerns. Measures of outcomes include revenue, commodities, and elements of habitat quality and other ecological values. As with tactical planning, some of these will arise from discussions with upper-level managers and landowners, and others (some logistical concerns) may be suggested through regulations or voluntary certification guidelines. The Chico Mendes Extractive Reserve forest plan described in Chapter 1, Management of Forests and Other Natural Resources, may be considered an operational plan to some extent, as tree felling techniques, road design principles, and safety guidelines are all described.

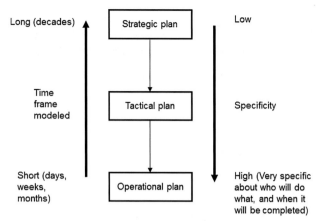

FIGURE 13.1 Differences in time frame modeled and specificity of information for the three forest planning processes.

If these concerns are associated specifically with planned activities, one can be assured that the plan contains the appropriate amount of detail for it to be considered *operational*.

IV. VERTICAL INTEGRATION OF PLANNING PROCESSES

To summarize the previous discussion, strategic plans generally are ones where comprehensive options are assessed using single or multiple objective processes along with broad-scale constraints on land and resource allocations. The time frame of the strategic planning analysis generally extends multiple decades or multiple rotations into the future (Fig. 13.1). Tactical plans generally involve more detailed spatial information and concern the location and timing of activities over a decade or two, but usually about one even-aged forest rotation. Operational plans utilize the most detailed and timely information to make management decisions related to activities that will be implemented within weeks or months, or within a year. Granted, every natural resource management organization is different. As we have suggested, each may adopt and implement a different type of planning process. In fact, in many cases the lines become blurred between performing strategic and tactical planning exercises, and between performing tactical and operational planning exercises.

Example

Millar Western Forest Products Ltd. developed a set of ground rules for the implementation of forest management plans for lands that they manage in a sustained yield unit in Alberta (Alberta Sustainable Resource Development, 2008). The strategic plan identified the stands eligible for harvest, yet basically used a nonspatial analysis to

develop a 20-year allocation of stands to harvest. A subset of these stands is used in a general development plan that covers a 5-year time period. Stands available for harvests are identified, and road development (tactical) plans are made for this period of time. An annual operating plan describes in more detail the operating schedule and final harvest plans. This operational planning phase provides greater detail about the planned activities, as compared to those suggested in the tactical plan. For example, the operational plan describes the actual harvest design and the detailed road management plan. However, in the set of Millar Western Forest Products plans, each level of planning incorporates the same set of general planning steps:

1. Identify current resource values.
2. Develop goals and objectives for management.
3. Evaluate impacts of alternative management strategies.
4. Select and apply the plan that achieves the management objectives and minimizes negative environmental impacts.
5. Monitor the effectiveness of the plan.

Example

When dealing with state-managed forest roads, the Oregon Department of Forestry (2000) conducts three levels of planning that are consistent with the strategic, tactical, and operational plans we have suggested.

Level I. This involves planning at a broad level, and across a long time frame. Established here are long-term goals and strategies that are consistent with legal requirements as well as the overall objectives of forest management. Specific activities are not identified in this type of plan.

Level II. This involves mid-level planning, and across a moderate time frame (one decade). Here, the current condition of the road system is described, and the general manner in which the system moves toward the envisioned road system is identified. Major road management activities such as road construction and maintenance are identified in this planning effort. This level of planning is consistent with the goals and strategies from Level I, and coincides with the development of implementation plans.

Level III. This involves the most detailed level of planning, and covers a relatively short time frame (2 years or less). Activities planned here must be consistent with the goals and objectives from Level II planning. In this effort, activities are site-specific and their exact locations are identified on the ground. The road management activities that are designed are also usually associated with, or in support of, other specific management activities.

In general, depending on the political environment of the country within which forests are managed, the levels of planning should be either evident within an organization (if forests are privately managed), or evident across broader organizational scales (if forests are publicly managed). For example, in areas of the world where local citizens have input into and influence on the management of forest resources, planning may occur at the national level (strategic), district level (tactical), and local level (operational). Whether the planning is coordinated top-down or bottom-up is also a function of the political environment. A bottom-up approach can be used to encourage land users to operate in ways that increase the productivity of a forest in a sustainable manner, yet may require more effort on behalf of the planning team. At the national level, planning processes concern the establishment of broad goals and priorities, such as the need to balance competing demands for land, or to allocate resources for development. District level planning processes involve understanding the diversity of the resources, and involve translating national priorities into local plans. Activities planned here include determining the general location of forest plantations, water supplies, and roads. Local planning processes involve small areas, such as a watershed. Using local knowledge of the resources, this level of planning concerns implementation of specific activities (where, when, and who will be responsible) (Food and Agricultural Organization of the United Nations, 1996). In other situations, where community involvement is not a necessary condition of management, each level of planning may be contained within the organization itself.

Vertical integration could also involve the scale of planning as it moves from the individual tree to the landscape. As we described in Chapter 5, Optimization of Tree- and Stand-Level Objectives, some decisions can be made at the tree level, using economic criteria (value growth rate of trees), biological criteria (susceptibility to disease), and other rules. These qualities of the collection of trees within a stand may inform the development of stand-level decisions, such as whether to let the stand continue to grow, to thin the stand, to apply an improvement harvest, or to apply a final harvest. We have suggested that stand-level optimal decisions are useful, yet at the forest-level other alternatives may be selected for individual stands, perhaps in response to wood flow concerns or to harvest adjacency and green-up concerns. Finally, when placed within the context of the larger landscape, the forest-level plan may be adjusted to address broader concerns, such as the impact of activities on a viewshed or on water quality.

V. BLENDED, COMBINED, AND ADAPTIVE APPROACHES

Spatially explicit forest planning processes (tactical planning) provide an opportunity to link strategic and operational planning through either top-down or bottom-up approaches to management. In a top-down approach, strategic goals inform the development of tactical forest plans. For example, the goals provided by the linear programming model Woodstock to the spatial planning model Stanley are one example of a top-down approach. In a bottom-up approach, operational feasibility can be incorporated into a tactical plan, which then informs the strategic plan about the alternatives available for the management of a forest. When detailed operational plans are condensed into a spatially explicit forest plan, planning is being directed from the bottom-up, and may lead to a strategic plan that is both operationally realistic, and that reflects reasonable management decisions made by landowner (Brown, 2001). However, the bottom-up approach is not suitable in all management situations, particularly when strategic direction is provided prior to the development of tactical or operational plans.

Adjustments to plans as they are being implemented at the operational stage are common. Many argue that field personnel should have the flexibility to adjust plans as is necessary given unexpected weather conditions and market fluctuations. If numerous changes to a plan are made, however, mainly due to the planning process' failure to recognize certain attributes or limitations of the operational system, then some consideration should be given to planning methods that better represent these factors. These changes may be inherent in the updated data provided by the field personnel at the end of a planning period, typically the end of a year. However, more frequent notification of deviations from a plan will likely be of value to an organization. And with these changes comes the possibility of strategic drift, where the objectives set at a higher-level are modified and changed at the lower level, resulting in a different plan being implemented. Purposefully changing a plan in reaction to new information, as we suggested in Chapter 1, Management of Forests and Other Natural Resources, might be considered adaptive management. Some land management organizations utilize an adaptive management planning approach, where feedback from activities that have been implemented is used to periodically adjust plans. One such approach is the adaptive management planning process used by Weyerhaeuser (Weyerhaeuser Company, Ltd, 2005) for the sustained yield unit it manages in Alberta (Fig. 13.2). Here, tactical forest planning processes are conspicuously absent, and more than likely incorporated into the strategic or operational planning

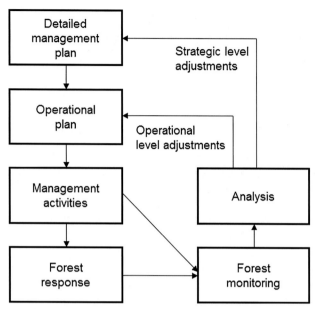

FIGURE 13.2 Adaptive management planning process used by Weyerhaeuser (Weyerhaeuser Company, Ltd, 2005) for land managed in Drayton Valley, Alberta.

FIGURE 13.3 Conceptual model of synergies related to forest planning processes.

processes. In fact, as computer and database management technologies advance, the differences between strategic and tactical planning become less obvious. It is not unreasonable to suggest that today forest plans can be developed using spatially explicit goals and analysis, yet cover long periods of time while addressing broad-scale resource allocation concerns.

In many cases, the differences between the three levels of planning are not clear, either because (1) the levels of planning address similar concerns, (2) the scope of planning differs among organizations or agencies, or (3) one or more levels is absent. For example, the blending of the strategic and tactical approaches is becoming very common (Sessions and Bettinger, 2001), however blended strategic/tactical approaches require assessing more information at one time than when simply performing a straightforward strategic planning exercise. When using a blended approach, one might be able to avoid strategies that are not feasible given tactical goals and constraints. Synergies that take advantage of staff collaboration and normal work flows may be realized in certain land management organizations. Benefits related to the sharing of information, models, technology, and personnel may enable more efficient planning processes to be conducted. In these cases, the hierarchy may be viewed as overlapping processes rather than a linear system of events and outcomes (Fig. 13.3).

Within the US Forest Service, programmatic decisions (strategic plans) cover the broadest land area, and describe management activities to be implemented in a very general manner. The land and resource management

plans for each US National Forest are examples of these types of strategic plans. Although these are considered long-range plans, because the analysis generally looks several decades into the future, US National Forest plans are generally applicable only for about 15 years, and describe the objectives and desired conditions anticipated through implementation of the plan over this period of time. As a result, National Forest plans provide guidance and information only for lower-level planning activities. Before site-specific activities can be implemented, a site-specific project plan is developed. Examples of site-specific management activities include prescribed burning, timber harvesting, wildlife habitat improvement, and recreational area development. This level of planning may be considered a blend of the tactical and operational planning processes, since the operational details are most likely considered at this time. However, the association between site-specific activities and forest-level objectives is something field-level managers need to constantly monitor and evaluate.

Example

In Pennsylvania, a draft strategic plan was developed to address conservation concerns and the long-term sustainability of state forests in an effort to, as the plan (Pennsylvania Department of Conservation and Natural Resources, 2015) suggests,

Ensure the long-term health, viability, and productivity of the commonwealth's forests and to conserve native wild plants.

and will accomplish this by (among other actions):

Managing state forests under sound ecosystem management, to retain their wild character and maintain biological diversity while providing pure water, opportunities for low-density recreation, habitats for forest plants and animals, sustained yields of quality timber, and environmentally sound utilization of mineral resources.

The draft State Forest Resource Management Plan lists a number of goals that are associated with objectives that provide guidance for future management decisions and allocations of resources to site-specific projects. This plan is suggestive of a mid-level planning effort, and further suggests operational planning will occur as the activities are developed to meet specific objectives.

One can view the hierarchical process as either being developed from the top (strategic level) down (to the operational level) or vice versa, using a bottom-up approach. Others (Kangas et al., 2015) have suggested that a combination of the two approaches may better represent the selection of the better stand-level decisions that are possible within the framework of broader landscape and longer time frame goals. In any event, the complexity of the process may direct how the levels of the hierarchy interact. The available budget for planning, along with the data, computer processing software, and planning expertise may also influence the effectiveness of using a planning hierarchy to develop forest management plans.

VI. YOUR INVOLVEMENT IN FOREST PLANNING PROCESSES

After this brief discussion of the various levels of forest planning, you might be asking yourself where you would fit into the natural resource planning processes. As a relatively new natural resource manager you will likely be involved with the development of operational plans for the land that you manage. If you do not directly develop these plans, perhaps you will do so indirectly under the guidance of other seasoned professionals. As time goes by, you will likely be afforded the opportunity to develop these plans yourself and implement them as well. As you will find, in many situations the annual goals of a property and some of the means for achieving them typically are defined by the landowner or the upper-level managers. These include annual harvest and budget levels, silvicultural options, and ecological and social concerns, all of which may be passed down to the field level through a strategic plan. However, the data and knowledge that would be required to develop tactical or strategic plans will likely arise from the field offices (Fig. 13.4). As a result, you should be aware of how the management activities that you implement support the goals recommended in the tactical or strategic plans. Not only will you and your colleagues in the field offices

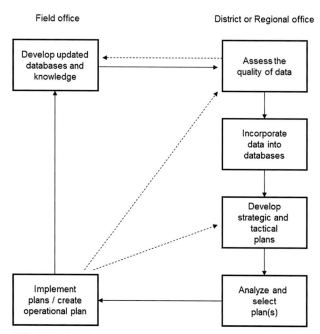

FIGURE 13.4 Flow of information in a typical timber company over the course of a year.

be the most familiar with the current condition of the natural resources being managed, but also you will be relied on to inform the planners and managers of the management prescriptions that are most likely to be implemented, and their associated costs. Although a set of optimal prescriptions for timber stands may be generated during one or more of the planning processes, understanding the current actual management implementation process is essential so that suggested courses of action are not outside the bounds of reasonable courses of action. At some point in your career as a natural resource manager, you will either be consulted on these types of issues during the planning process, or perhaps you will become the person responsible for developing tactical or strategic plans. Understanding the hierarchy of planning processes within your organization is therefore an important learning process and one that will allow you to work effectively with your peers.

VII. SUMMARY

Land managers and landowners want plans of action that provide guidance and assessments of the goals and objectives of a property. Some want an evaluation of strategies or general directions for managing resources and achieving long-term goals. Some want tactical plans to determine where activities should be planned over the next year or two. Others require operational plans to facilitate implementation of daily, weekly, or monthly management activities. Ideally, a single plan of action would include a geographic perspective that allows land managers and landowners to understand the extent of activities planned, the relationship between planned activities and the broader set of natural resources contained within the property (or landscape), and the potential courses of action for implementing the management activities. Through these plans, managers and landowners should be able to develop a sense for the short-term, intermediate, and long-term goals associated with the property. Further, an analysis of alternatives at each level allows the decision-makers to understand the trade-offs involved in selecting fine-scale and broader-scale courses of action. Understanding how you might contribute to each planning process may be one of the keys to a successful career.

QUESTIONS

1. *Operational planning related to commercial thinning operations.* You have recently been hired as a forester for a timber company in western Washington. Your main responsibility is to plan and implement all the commercial thinning activities across the land the company owns (200,000 acres). Higher-level plans call for a certain amount of volume to be thinned from a certain amount of land each year. A general thinning prescription is applied to stands that are 25−30 years old. Discuss the various aspects of your new job that might be included in a monthly operational plan.
2. *Operational planning related to wildlife habitat improvement.* As a natural resource manager for a military base in the southern United States, you are charged with developing a certain amount of red-cockaded woodpecker habitat each year. Based on the guidelines provided in Chapter 12, Spatial Restrictions and Considerations in Forest Planning, Section III, Part C, what operational planning issues would you expect to encounter over the course of a year?
3. *Forest planning processes.* The organization that you work for in central Maine utilizes a forest plan that describes for the land managers the areas to be harvested each year, the volume expected from each harvest, and the potential effects of the plan on wildlife habitat and water quality values. The plan provided to the field personnel is 1 or 2 years in length, but you are aware that it is modeled for 50 years to determine the effects of harvesting on long-term sustainable wood supplies. How would you describe this type of planning process? Why?
4. *Selecting a planning process.* The owner of a tract of land in Wisconsin is interested in having you develop a forest plan for their property. After a brief discussion with the landowner, you determine the

following: (1) they are interested in a long-term sustainable harvest from uneven-aged forests, (2) they want some flexibility in locating the harvests, therefore are not interested in a site-specific schedule, just a ball-park figure (volume and area) to work from, (3) though they desire a 50-year plan to ensure sustainable harvests, they are most interested in the costs and revenues over the next 10 years. Given sufficient data to develop the forest plan, which approach described in this chapter would you select? If the landowner later indicated that they are very concerned about deer habitat quality, specifically as measured using a spatial wildlife habitat quality model, would your choice of approach change? If so, then how would it change?
5. *The hierarchy of forest planning.* Assume you are a manager of a large area of land in central Missouri, and assume that you are interviewing recent graduates for an entry-level field forester position. During the interview, you describe for the candidates the types of management activities that the new employee will be performing. In addition to prescribed burning, timber inventory, and some harvest layout activities, the position includes supporting the strategic and tactical planning processes of your organization, and includes the direct use of operational planning processes. Since these planning processes may overlap somewhat, and since decisions at one level affect opportunities at another, the goals and objectives of the planning processes may be a little confusing to the students you are interviewing. To help clarify matters, develop a short, one-page memorandum that describes how the three processes differ, and how the new employee will eventually become involved.

REFERENCES

Alberta Sustainable Resource Development, 2008. 2002 Millar Western Forest Products Ltd. Ground Rules. Alberta Sustainable Resource Development, Edmonton, AB.

Bettinger, P., Merry, K., Mavity, E., Rightmyer, D., Stevens, R., 2015. Chattahoochee-Oconee National Forest, Georgia, United States of America. In: Siry, J.P., Bettinger, P., Merry, K., Grebner, D.L., Boston, K., Cieszewski, C. (Eds.), Forest Plans of North America. Academic Press, New York, NY, pp. 277−284.

Brown, C., 2001. ATLAS-SIMFOR Project Extension Report, Linking Operational and Strategic Planning (The RCFC Story). Silvatech Consulting, Ltd., Salmon Arm, BC, 12 p.

Church, R.L., 2007. Tactical-level forest management models. In: Weintraub, A., Romero, C., Bjørndal, T., Epstein, R. (Eds.), Handbook of Operations Research in Natural Resources. Springer, New York, NY, pp. 343−363.

Epstein, R., Karlsson, J., Rönnqvist, M., Weintraub, A., 2007. Harvest operational models in forestry. In: Weintraub, A., Romero, C., Bjørndal, T., Epstein, R. (Eds.), Handbook of Operations Research in Natural Resources. Springer, New York, NY, pp. 365−377.

Food and Agricultural Organization of the United Nations, 1996. Guidelines for Land-Use Planning. Soil Resources, Management and Conservation Service, Food and Agricultural Organization of the United Nations, Rome, FAO Development Series 1. Available from: http://www.fao.org/docrep/T0715E/t0715e00.HTM (Accessed 4/24/2016).

Gunn, E.A., 2007. Models for strategic forest management. In: Weintraub, A., Romero, C., Bjørndal, T., Epstein, R. (Eds.), Handbook of Operations Research in Natural Resources. Springer, New York, NY, pp. 317–341.

Kangas, A., Nurmi, M., Rasinmäki, J., 2015. From a strategic plan to a tactical forest management plan using a hierarchic optimization approach. Scandi. J. For. Res. 29 (Suppl. 1), 154–165.

McTague, J.P., Oppenheimer, M.J., 2015. Rayonier, Inc., Southern United States of America. In: Siry, J.P., Bettinger, P., Merry, K., Grebner, D.L., Boston, K., Cieszewski, C. (Eds.), Forest Plans of North America. Academic Press, New York, NY, pp. 395–402.

Oregon Department of Forestry, 2000. State Forests Program, Forest Roads Manual, Transportation Planning. Oregon Department of Forestry, Salem, OR, pp. 2-1 to 2-2.

Park, T.-S., 1990. Republic of Korea. Forestry Resources Management, Report of the APO Symposium on Forestry Resources Management. Asian Productivity Organization, Tokyo, pp. 231–246.

Pennsylvania Department of Conservation and Natural Resources, 2015. Draft 2015 State Forest Resource Management Plan. Department of Conservation and Natural Resources, Harrisburg, PA, 193 p.

Sessions, J., Bettinger, P., 2001. Hierarchical planning: pathway to the future? Proceedings of the First International Precision Forestry Cooperative Symposium. Institute of Forest Resources College of Forest Resources, University of Washington, Seattle, WA, pp. 185–190.

Sessions, J., Bettinger, P., Murphy, G., 2007. Heuristics in forest planning. In: Weintraub, A., Romero, C., Bjørndal, T., Epstein, R. (Eds.), Handbook of Operations Research in Natural Resources. Springer, New York, NY, pp. 432–448.

Straka, T.J., Cushing, T.L., 2015. McPhail Tree Farm, South Carolina, United States of America. In: Siry, J.P., Bettinger, P., Merry, K., Grebner, D.L., Boston, K., Cieszewski, C. (Eds.), Forest Plans of North America. Academic Press, New York, NY, pp. 87–96.

Swae, J., 2015. City of San Francisco, California, United States of America. In: Siry, J.P., Bettinger, P., Merry, K., Grebner, D.L., Boston, K., Cieszewski, C. (Eds.), Forest Plans of North America. Academic Press, New York, NY, pp. 285–292.

Weyerhaeuser Company, Ltd, 2005. SYU R12 Detailed Forest Management Plan, 2000–2015. Weyerhaeuser Company, Ltd., Drayton Valley, AB.

Chapter 14

Forest Supply Chain Management

Objectives

The forestry supply chain for wood products can be viewed as a system of operations that begins with the resource; moves through harvesting, processing, and sorting stages; and ends with the delivery of the product at one or more processing facilities (Mitchell, 1992). The wood manufacturing supply chain is broader, and can include the various steps involved in creating a product. The planning processes associated with the forestry supply chain are bounded by both environmental and economic considerations, and success is based on performance related to these considerations as well as other social concerns. Supply chains in general can be centralized, where all the decisions are made by a single decision-making organization. This system can occur in a vertically integrated timber company that owns forests, manufacturing facilities, and harvesting and hauling equipment, and that employs people to operate and manage these. Supply chains can also be decentralized, where each step of the supply chain involves one or more different decision-making organizations. In this case, there is often limited ability or willingness to share information across the organizational boundaries.

In this chapter, we will present an overview of the forestry supply chain as a hierarchical system that is similar to the hierarchical planning problems presented in the previous chapter. We will describe general solution methods that can be used to solve these problems with the mathematical techniques presented earlier in the text. Finally, we will conclude with a discussion of the uncertainty inherent in supply chain planning systems. Upon completion of this chapter, you should be able to:

1. Describe the components of a forestry supply chain management system.
2. Be able to suggest the appropriate mathematical technique to be applied to various aspects of the supply chain.
3. Understand the sources of variation in the supply chains.
4. Understand the potential benefits from managing the supply chain.

I. INTRODUCTION

Supply chains can be characterized as "push" systems, where production is emphasized and inventory is sold from the manufacturer to end-users generally through a variety of marketing advertisements. Supply chains can also be characterized as "pull" systems, where customer demands or requests are used to pull the products through the manufacturing process. An example of a pull system would be where a special order is requested by a customer, or a product is demanded in excess of supply. Most forestry supply chains are a combination of push and pull systems, with some products from the supply chain being in high demand, and others requiring marketing efforts to sell.

In general, competition for sources of material increases the efficiency of supply chains, but the degree of improvement depends on the structure of the system and the amount of competition and collaboration among the participants. The actors involved in a typical forestry supply chain can include the following:

- The landowner, cooperative, or concession holder
- The logging contractor
- An intermediary (perhaps) such as a wood yard operation or wood dealer
- The mills or processing centers
- Other downstream operations that use the by-products of the primary processing of wood

Supply chains are not perfect, and in many cases are inefficient. In fact, there has been little published research assessing the efficiency of supply chains (Perakis and Roels, 2007). In developing countries, supply chains typically involve numerous actors, and are tightly associated with long-standing social structures (Woods, 2004). In both developing and developed countries, transportation issues, stumpage costs, global competition, and the loss of local markets can be among the most concerning aspects of the supply chain. In one study of the Appalachian forest products industry (Buehlmann et al., 2007), most of the people surveyed suggested that the development of better relationships among potential up-stream customers in the supply chain was the most important response to globalization of the industry. Cooperation in the decision-making processes among the different organizations within a supply chain can significantly reduce the costs and increase the efficiency of the entire system (Hall and Potts, 2003). To improve performance in the supply chain, a number of

Forest Management and Planning. DOI: http://dx.doi.org/10.1016/B978-0-12-809476-1.00014-X

contracts (buybacks, rebates, revenue sharing, etc.) also have been proposed, however price-only contracts are the most prevalent (Woods, 2004).

The forest products industry's supply chain differs from most manufacturing businesses in that smaller products (lumber, paper, etc.) are generally decoupled from a larger item (a log) that is a decoupling of an even larger product (the tree). In other industries, products are assembled from a collection of smaller parts. As a result, in forest management, the demand for a single grade of log or lumber can result in the production of many other logs or lumber as coproducts of the production process. Therefore, the supply chain must not only account for the primary customer demand, but it must also include the delivery of products to other customers as a result of the joint production that arises from the decoupling process. This pull—push supply chain has both customer demand and production-driven components, whereas most other supply chains focus exclusively on only one of these components.

Supply chains are integrated, multifunctional networks in both space and time that facilitate the transformation of raw materials to finished products, which eventually are sold to customers or end-users. Although we described a few earlier, a wide number of manufacturing organizations

and products can be associated with a forestry supply chain (Fig. 14.1). Natural resource managers should be aware of how their decisions will influence the profitability of their own business as well as the other organizations that are dependent on the wood that is harvested. Individual planning without regard to supply chain partnerships may result in goods produced at noncompetitively high costs, and come with an associated poor level of customer service that can have an effect on long-term profitability. Zielke and Pohl (1996) developed the concept of the virtually vertically integrated firm as one that can easily cross organizational boundaries and result in efficient production. Here, the alignment of supply chains among the participants can facilitate an improvement in profitability among all the organizations involved. For example, a landowner that wants to increase his or her international log sales may not be able to achieve this goal if the port that they typically use does not have the excess loading capacity (or plans to expand this capacity) necessary. However, the landowner could work collaboratively with the port, after sharing its strategy of increasing sales of logs to international markets, and encourage the port to increase its capacity for the additional business. The end-result would be both an increase in business for the port, which may improve its profitability, and an increase in international sales for the landowner.

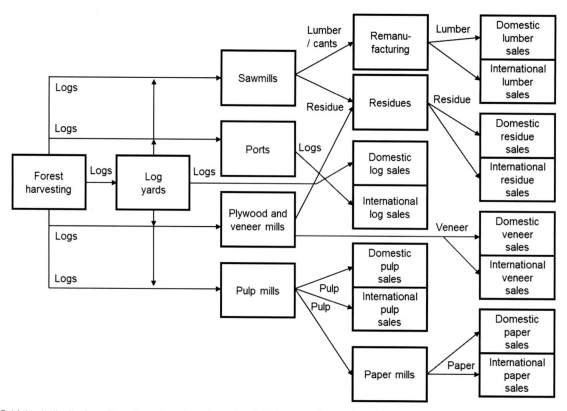

FIGURE 14.1 A distribution of logs through various processing facilities to markets. *Adapted from Weintraub, A., Epstein, R., 2002. The supply chain in the forest industry: models and linkages. In: Supply Chain Management: Models, Applications, and Research Directions (J. Geunes, P.M. Pardalos, and H.E. Romeijn, Eds.), Kluwer Academic Publishers, London. pp. 343—362.*

II. COMPONENTS OF A FORESTRY SUPPLY CHAIN

From a natural resource management perspective, the supply chain can be described by functions that involve data collection, demand estimation, planning, execution, and reporting. The functions (Fig. 14.2) are integrated within various phases of planning, from higher-level strategic planning processes to lower-level, time-critical, operational planning processes (Boston, 2005). The supply chain functions include the following:

1. *Supply planning*. This function involves the management of the availability and purchase of raw material from all potential sources, such as owned (fee) timberland, long-term log and stumpage agreements, short-term log sales, or gate-wood (delivered by other loggers or landowners). One source of data is the inventory associated with some of the forests being managed; another source is the projections obtained from growth and yield analyses.

2. *Forest planning*. This function involves the development of the strategic, tactical, and operational plans that are designed to address the management, harvesting, and transportation activities. In addition, this function results in the coordination of the logistical concerns related to aligning wood demand with wood supply in an effort to meet both end-user demand and the desired inventory levels of wholesalers or retailers.

3. *Demand planning*. This function involves the development of forecasts, through planned orders, of sales to customers. It evaluates and ranks the customer importance. This process can include rankings based on price or other less quantifiable criteria. Additionally, price-related information is managed during this function.

4. *Execution*. This function involves the implementation of the management plans, and may include the tracking of harvesting and transportation processes necessary for establishing a chain of custody.

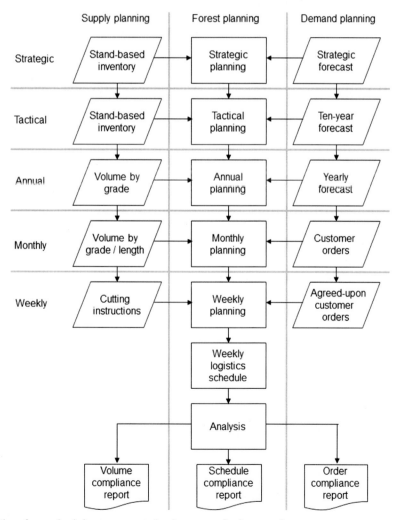

FIGURE 14.2 An illustration of a supply chain management planning process for forest products.

5. *Knowledge collection and reporting*. This function involves the collection of data regarding the performance of the supply chain. The information developed here allows an organization to analyze and recommend improvements throughout the supply chain system.

Execution of an efficient supply chain utilizes logistical systems that optimize the distribution of the materials from the forest to the end-users. One goal is to ensure that the end-users receive their orders during the time period desired. The ultimate goal of the supply chain is to execute the plans with the maximum efficiency while meeting all confirmed needs of each organization along the way. As we noted, a common element of the execution function within the supply chain may be to perform chain-of-custody auditing requirements, which are now part of forest certification programs that are described in greater detail in Chapter 15, Forest Certification and Carbon Sequestration. Tracking the chain-of-custody involves the recording and tracking of the possession of logs and lumber from the forest to the customers, and the tracking process often is associated with the certification of forest products. Since most certification systems require that forest landowners track wood products from the forest to the final customer, the chain-of-custody assessment ensures that customers will be able to purchase forest products originating from certified forests. Stated another way, chain-of-custody certification verifies the flow of certified forest products through the supply chain. The other common role of a chain-of-custody verification process is to prove ownership of the wood produced to reduce trade in illegally harvested wood products. Here, wood buyers, processing facilities, forest landowners, and conservationists all are involved in tracking wood through the wood supply system to better understand the efficiency of their supply chain performance.

Chain-of-custody issues are linked to forest planning, harvesting operations, transportation systems, and processing operations. Tracking wood through the system is not without some significant challenges, which can include systematic inhibitors to product quality throughout the supply chain, and a general lack of acceptance of technological improvements in the various wood processing sectors. Each of these is not unexpected when multiple organizations, each with perhaps different objectives and goals, need to interact to effectively track wood products.

A number of technologies have been applied to the methods for tracking wood products. The oldest and most common are log brands, which consist of simple patterns that are hammered into the end of logs. Nowadays plastic-coated paper barcodes or specially formulated paint with tracers that glow under various light sources commonly are used for international wood products. Some have suggested using microtag paint, which contains identification codes that can be viewed with a microscope. Radio frequency tags or aroma tags also have

been suggested for tracking wood products (Murphy and Franich, 2004). There are several issues that need to be considered when attempting to track forest products; one issue involves the number of tags that will be needed, as the number of logs can easily exceed millions from a large industrial forest products firm. Another issue involves the need for easy identification of a product along the supply chain. Therefore, these tools must be durable, and they must not interfere with subsequent processing systems. This may effectively eliminate from further consideration the use of metal or plastic tags, which may affect pulping or sawing processes.

The reporting function within the supply chain serves two roles, the first of which is to provide the information needed to address the general business practices of an organization. This includes invoicing customers and making payments to contractors and suppliers. The second role is equally important but not as well recognized, and that is to collect the information necessary to develop further an understanding of the business. Some typical questions that can be addressed with information generated during this function of the supply chain include:

1. Were the confirmed orders fulfilled at their agreed price?
2. Was the production of the crews at their predicted levels?
3. Did the grade distribution from each stand meet the anticipated goals?
4. Did log yard inventories remain within the desired range?
5. Were the harvest levels by strata, silvicultural regime, or forest type scheduled in the tactical and operational plans similar to those recommended by the strategic plan?
6. Were the stands included in the tactical plan used in the operational plan, or was the tactical plan allowed to drift without reason?
7. Was the accuracy of the long-term sales forecasts sufficient?
8. Was the accuracy of the shorter-term, weekly forecast sufficient?

Those items that have large variances between their planned performance levels and their actual performance levels should be targeted for further analysis, to determine the causes of divergent performance and to locate areas for improvement in the system.

III. ASSOCIATION WITH THE HIERARCHY OF FOREST PLANNING

The supply chain functions are integrated with strategic, tactical, and annual, monthly, and weekly operational forest plans (Fig. 14.2). The execution function also includes logistical scheduling, which is used to optimize the transportation resources as well as maintain a record of the chain-of-custody. For the natural resource manager, the bulk of the work will involve developing the supply estimates through the use of inventories in conjunction with growth and yield models. As we noted in Chapter 13, Hierarchical System for Planning and Scheduling Management Activities, strategic planning covers a broad time frame, perhaps a length of time equivalent to two to three typical rotations in an even-aged management system. Strategic plans for public (government-managed) lands may be developed once every 5–15 years, whereas strategic planning processes usually are conducted every 1 or 2 years by many forest companies. In the book *Forest Plans of North America* (Siry et al., 2015), an excellent example of a private company (Rayonier) needing to develop strategic supply estimates for their timberlands in the southeastern United States is provided (McTague and Oppenheimer, 2015). The objective of a strategic planning process is to develop a long-term vision that allows an organization to best define how it can adequately compete in current and future markets by integrating strategies where mutual gains can be produced across the entire supply chain.

In the forestry supply chain, one common goal is to align the volume and log types that can be produced from a forest with the capacities of the most logical manufacturing facilities. This is accomplished in the strategic planning process by constraining the use of various silvicultural prescriptions (for even-aged and uneven-aged stands) and by adjusting the forest rotation ages (for even-aged stands). The outcomes from a strategic forest plan can include the area to be treated by various silvicultural methods and logging systems, the desired rotation ages for stands containing various tree species and managed with various silvicultural regimes, and the projected harvest volumes by grade and species. Logging and transportation analysis examines the existing harvesting and transportation systems and considers the costs associated with changing the capacity of these systems.

Additionally, the strategic plan associated with a forest should consider the transportation infrastructure projects that may be necessary, such as the construction or maintenance of wharves or highway upgrades, each of which could influence the ability to meet the needs of potential customers or to take advantage of lower operating costs. In reaction to these outcomes, strategies can be developed for (1) the preferred location of new manufacturing facilities, (2) the disposal of properties that do not support future needs, and (3) the acquisition of properties that can support future needs. In addition, an organization can develop a strategic plan to acquire or dispose of long-term cutting agreements, and perhaps can develop a strategic plan that targets specific sources of externally procured wood volume.

Tactical planning is the second step in many forest management planning hierarchies, and as we suggested in Chapter 13, Hierarchical System for Planning and Scheduling Management Activities, one of the key differences between strategic and tactical planning is the type of data employed. In a strategic planning process, the decision variables generally are described by continuous real numbers that represent areas of forest types, silvicultural regimes, or otherwise delineated strata. However, in tactical planning processes and all the remaining levels in the supply chain hierarchy, the decision variables generally are represented by discrete integer values, such as whole timber stands or logging units. The time horizon associated with a tactical plan is also approximately one-half of one even-aged forest rotation length, or much shorter. Many tactical plans are prepared annually or every other year. Since the decisions are discrete, tactical planning is able to incorporate into plans spatial constraints, such as green-up and adjacency requirements. These constraints, and others related to wildlife habitat, may be incorporated into some state forest practices rules, into recovery plans for endangered species, or into voluntary certification program guidelines, and thereby can change supply estimates and impact supply chain performance (Boston and Bettinger, 2001, 2006).

The tactical forest planning process is where the combined harvest and transportation decisions can be accommodated, and where significant financial improvements in the efficiency of transportation system can be realized. In many parts of the world, roads may be built only during a 3- to 6-month period of the year, due to wet or frozen soils. In addition, roads in some areas of the world may be used for only a limited period of time, such as when the ground is frozen. Thus, landowners may have limited opportunities to change forest plans since many management activities are dependent on access to management units. As a result, many forestry supply chains may inherently contain limited near-term flexibility for accessing some stands, as the road infrastructure may require a longer period of time to change. Therefore, the current road system may have the most influence on where a landowner is able to harvest. Additionally, logging and hauling capacities can be included in tactical forest plans to better represent the management environment. Assumptions regarding the use of these systems are relatively robust over the near-term, as there may be limited opportunities to quickly change logging and hauling systems. Significant changes in these systems may require new equipment to be manufactured and marketed, but in the near-term their capacity is often fixed.

The objective function of a tactical forest plan may be to maximize net revenue subject to constraints associated with the guidance suggested by a strategic forest plan. Since individual stands or harvesting units (rather than coarser strata, silvicultural regimes, or forest types) are now the decision variables, there is an increase in the resolution of the supply and production data. The forecasted wood flow projected through a tactical plan similarly is refined with finer gradations in yields by grade classes, as compared to broader product classes typically recognized in strategic plans. The outcomes from tactical plans can include a set of stands to be treated by various silvicultural prescriptions and logging systems, and a set of road projects to be completed each year. In addition, with the outcomes from a tactical plan, land managers and landowners can confirm that the projected wood supply may be (or may not be) available to the manufacturing facilities considered in the strategic plan. Thus, land managers may be able to understand the composition of stands that can be used to best supply manufacturing processes, loggers may be able to match their equipment needs to stand conditions, and trucking capacity can be aligned to specific needs based on wood volume and distances being hauled. Therefore, the outcomes of a tactical plan can facilitate a number of efficiency gains in the operational capacity of a wood supply system.

As we noted in Chapter 13, Hierarchical System for Planning and Scheduling Management Activities, operational planning is the lowest level of forest planning, and involves the most direct contact between organizational goals and the forest itself. In fact, an organization may develop and use several operational plans, at the weekly, monthly, and annual time scales. Whereas tactical plans use virtual crews with average production rates, operational plans typically incorporate production and cost estimates for specific crews, information provided through knowledge of the contractors used for various management activities within an ownership. At this stage of planning, contractors can demonstrate that they have used the information provided by a tactical plan to become a competitive provider of the services associated with management activities. The inventory and forest structure data related to a forest typically are used to assign crews to harvest units and road construction and maintenance projects. In addition, the order in which the crews will complete these tasks to best satisfy the objectives of the landowner or the collective objectives of the supply chain partners can be determined, perhaps so that wood yard inventories do not fall short nor become overly excessive. Operational plans use very fine-scale information to assign work crews to harvest units, and these plans should be designed to ensure that all the management actions suggested by the tactical forest plan are completed. In response to this concern, a set of constraints can be designed for operational plans to force compliance with the tactical or strategic plan goals, thus the hierarchy of planning is enforced.

The wood supply data for operational plans generally arises from preharvest forest inventories that are collected a year prior to harvesting. This information is used to support log allocation decisions, since planners have the ability to create multiple log stock tables that represent alternative harvesting instructions. Additionally, supply managers may incorporate the volumes available from nonfee land or third-party timber sales to understand how to approach these opportunities based on their contribution to the wood supply mix. The outcomes of operational planning, as they relate to the forestry supply chain, can include:

- Weekly delivery schedules of wood volumes to key customers
- Work crew and harvest unit assignment instructions
- Estimates of the wood volume needed to be procured in weekly increments
- Reports of unsold, scheduled harvest volume in the landowner's inventory in weekly increments

Example

Carlsson and Rönnqvist (2005) described a supply chain management system used by Södra Cell AB in Sweden. Södra Cell AB is one of the largest producers of market pulp, and operates three pulp mills in Sweden and two pulp mills in Norway. Sales are made mostly to companies in Europe, particularly Sweden and Germany. The Södra Cell AB planning approach used a hierarchical system where strategic planning processes consider investments and improvements in pulp mills in light of the available wood supply. Strategic plans also were made to support the logistical capabilities of the supply chain, to increase its ability to supply future markets. A tactical planning process considered a 12-month rolling time horizon that matches the supply of wood to the demand for pulp based on knowledge of markets from key account managers and sales offices. An operational planning process was performed monthly, and consisted of decisions regarding the shift from one type of market pulp to another, based on production issues at each mill and the ability to transport the resulting products to various markets. Optimization models were used to allocate wood supplies to production facilities in an effort to minimize transportation costs.

Changes in wood markets can occur rapidly and may be difficult to accurately forecast, and as a result monthly planning processes allowed for these market changes to be recognized, and thus could be used to adjust harvesting and trucking capacity. A monthly plan could result in operational plans that actually look out over a 3-month time horizon, and these could be used to balance the needs of longer time horizons associated with international sales with the demands of shorter domestic wood sales cycles. These operational plans allowed a natural resource manager to manipulate the scheduled management activities and account for market changes, since one goal may be to confirm sales orders over the first 2 months. The third month was used primarily to identify future opportunities and to adjust the supply chain to improve revenue or customer service. This level of planning represents a significant change in the supply chain management process as changes in wood demand can be recognized and management reactions can be developed.

Whereas higher levels of planning suggest that any portion of logs produced can be sold to any customer, a monthly operational plan recognizes that logs can be sold using a variety of log allocation policies. Some landowners may allocate logs to customers based on those opportunities with the highest net revenue, and thus they treat each grade of log produced from their land independently of other grades. However, customer priority systems can be developed to provide better service to customers. For instance, common customer priority planning systems include:

- Acceptance of only a full order, thus the order is treated as a binary variable in planning efforts
- Acceptance of any part of an order, thus the order is treated as a continuous variable in planning efforts
- Acceptance of any amount of an order above some minimum volume, but with the caveat that the percentage of log grades must meet a desired target

In the latter case, the order is assumed to be represented by a continuous variable in planning efforts, but the percentage assigned is further constrained.

When developing monthly operational plans, log yard inventory levels are estimated to determine if they are in compliance with the established organizational policies. If the inventories are above some maximum acceptable level, then additional wood sales can be sought through price negotiations, or woods crews can be placed on work quotas (or eliminated) to reduce the production capacity. If the inventories are below the minimum acceptable stock levels, then either production can be increased with extended work weeks, or additional wood can be procured through increased short-term gate-wood purchases. In either event, once plans are in place to move inventory levels within a desired range, orders are confirmed and communicated with customers, and harvesting crews are notified of their next harvest unit. In summary, the results of a monthly operational plan include the assignment of wood flow orders to individual logging and transportation crews, the development of a specific wood procurement schedule, and an analysis of log yard inventories.

The development of a weekly operational plan is the final element in the planning and scheduling function of the supply chain analysis, and it is very similar to the master production schedule found in many generic factory scheduling systems. The production schedule assigns

crews a cutting schedule, and an order for wood volume by product, grade, and length is inherent. As actual orders can change from the forecasted order, a weekly operational plan attempts to accommodate these changes without sacrificing those orders that already have been confirmed through the monthly planning process. In sum, the weekly operational plan is the last attempt to manage market and customer changes.

Example

In an illustration of the challenges and opportunities facing the forestry supply chain in Chile, Weintraub and Epstein (2002) suggested that the supply chain begins with the harvesting of each tree, as the actions applied to a tree initiate the flow of material from the forests to the various manufacturing facilities. Here, given strategic goals, tactical decisions were supported by the program PLANEX (Epstein et al., 1999), which uses raster geographic information system databases, often with a spatial resolution of 10 m, to aid in the design of harvest unit boundaries. This analysis allowed the land managers to determine the harvesting system requirements for each stand. The PLANEX solution process, at this level of planning, used a heuristic planning technique. Operational planning was performed using the program OPTICORT (Epstein et al., 1999), which is a linear programming system that helps managers address the following questions with guidance from the tactical plan:

- What stands should be harvested?
- What cutting instructions need to be applied to each stand?
- What is the destination for the products produced from each stand?
- What are the harvesting system requirements?

This operational planning process led to a weekly and daily schedule that relied on manual adjustments to satisfy changing orders. Once logs were delivered to a landing, the logistics scheduling system ASICAM was used to assign trucks to landings and to assign customers to the delivered products (Epstein et al., 1999). The planning process developed a daily schedule, where truck travel times are considered along with the demand for the various log grades and the loading capabilities on each landing. Trucks could be assigned new routes to avoid high congestion at the landing or at a processing facility. Using this planning process, Epstein et al. (1999) estimated a gain of over $20 million per year in system efficiency. The use of the logistics scheduling system alone could result in a 15—35% reduction in transportation costs. This type of forestry supply chain planning system may be even more important today as some companies seek to minimize their carbon footprint. Thus when adopting a supply chain management approach to managing one's business, a manager must make an assessment of the potential profitability and compare this against the increased complexity in running the operation.

IV. MATHEMATICAL FORMULATIONS ASSOCIATED WITH FORESTRY SUPPLY CHAIN COMPONENTS

The mathematical processes that can be used to analyze supply chain issues already have been described previously in this book, and by others. Rönnqvist (2003) provided an assessment of processes that considers the application, the time horizon, the computational efficiency, and the solution quality of mathematical processes to address various stages of the supply chain problem (Table 14.1). When addressing supply chain problems, planning methods typically utilize a hierarchical approach that allows decisions from the higher levels to filter down to the lower levels as goals or constraints. Thus, solutions from a strategic forest plan will provide guidance to a tactical forest plan, which further influences decisions that are made when developing an operational plan. Although the techniques used to address these problems have been discussed previously, the difference when applied to supply chain analysis is in how these plans are developed. In previous chapters, it was assumed that a planner was able to generate all the data necessary for an analysis from within their organization. However, in a supply chain analysis, the data required might be needed from various organizations outside the one in which the planner works, since the goal is to develop collaborative networks that share data across organizational boundaries and increase the efficiency of the system. Thus, a supply chain planner may need to develop relationships with managers from outside their organization to obtain the necessary logging costs, production rates, future volumes, and so on to effectively develop plans of action.

TABLE 14.1 Planning Horizon, Solution Time, Solution Quality, and Model Type Used in a Supply Chain Planning Process

Application	Time Horizon	Solution Time	Solution Quality	Model Type
Board cutting	1 s	<0.1 s	Optimal	Dynamic programming
Truck dispatching	5 s	<1 s	High quality	Integer programming
Truck scheduling	1 day	20 min	Near optimal	Integer programming
Annual planning	1 year	1 h	Near optimal	Integer programming
Tactical planning (with road management)	5 years	>1 h	High quality	Integer programming
Strategic planning	100 years	20 min	Optimal	Linear programming

Source: Rönnqvist, M., 2003. Optimization in forestry. Mathematical Programming, Series B 97 (1–2), 267–284.

The longer term, strategic forest planning problems ideally are solved with linear programming techniques, where the supply data are highly aggregated into strata-based averages. Demand data also is aggregated into broad product classes where the individual firms involved in the supply chain may not yet be identified, although it may include both data from known customers as well as from econometric-based forecasts. A generic strategic forest planning problem formulation related to supply chain management might include the following:

Strategic Planning

Maximize discounted net revenue
  Subject to:
- Area constraints—minimum or maximum limits on productive available land in each time period, including land for purchase, sale, or trade.
- Volume flow constraints—minimum or maximum limits on the volume scheduled for harvest for the overall property plus those timber sales that are available for purchase in the surrounding area.
- Cash flow constraints—minimum or maximum levels related to the management of the overall property.
- Silvicultural limitations—minimum and maximum limits on areas to be treated with the various available silvicultural prescriptions.
- Habitat requirements—minimum or maximum limits on areas to be maintained in various conditions for habitat maintenance and protection.
- Volume flow by grade groups—minimum or maximum limits on volume scheduled for particularly important customers.

As we move down the hierarchy to tactical planning problems, the data becomes more refined and the decision variables are associated with individual harvest units and road projects, and are represented with discrete integer variables. The solution methods rely on those that are capable of solving mixed integer programming problems or integer programming problems, or may rely on a heuristic technique. The problem formulation will often contain a number of goals generated from the strategic forest plan as well as spatial constraints imposed by either local forest practices rules, organizational policies, or certification requirements. A generic tactical planning problem formulation related to supply chain management is as follows:

Tactical Planning

Maximize discounted net revenue
  Subject to:
- Area constraints—minimum or maximum limits on productive available land in each time period, including land for purchase, sale, or trade in each compartment or watershed.
- Volume flow constraints—minimum or maximum limits on the volume scheduled for harvest for the overall

(Continued)

(Continued)

property plus those timber sales that are available for purchase in the surrounding area, using guidance from the strategic plan.

- Cash flow constraints—minimum or maximum levels related to the management of the overall property.
- Silvicultural limitations—minimum or maximum limits on areas to be treated with the various available silvicultural prescriptions.
- Habitat requirements—minimum or maximum limits on areas to be maintained in various conditions for habitat maintenance and protection, perhaps by compartment or watershed.
- Volume flow by grade groups—minimum or maximum limits on volume scheduled for particularly important customers.
- Adjacency and green-up constraints—limits related to the timing and size of clearcut harvest areas.
- Road and harvest unit linkage constraints—considerations that allow for issues related to the transportation system to be incorporated into the planning process.

An annual operational plan may be developed to assign the available work crews to forest management activities and to arrange the order in which the management activities will be implemented. This is mainly an integer programming problem, as discrete work crews are assigned to unique management activities. A natural resource manager may want to accommodate constraints in the planning problem that require key contractors to have a continuous work schedule during the year to meet common contractual agreements. At this stage in the supply chain planning process, log yard inventory levels for the jointly-produced products can be tracked to determine if additional sales programs are needed, or if overall wood production needs to be reduced to limit losses in the inventory. A generic annual operational planning problem formulation related to supply chain management might include the following:

Annual Operational Planning

Maximize discounted net revenue
  Subject to:
- Volume flow constraints—minimum or maximum limits on the volume target by wood grade group for each potential customer, using guidance from the tactical plan.
- Work crew utilization rules—considerations for the rules used to assign work crews to management activities to promote high utilization rates, such as a minimum number of weeks worked or continuous work for fixed- or variable-length periods of time.
- Management activities completeness constraints—considerations for the need to ensure that all management

(Continued)

(Continued)

activities are scheduled for implementation during the year, and that they should be completed before moving to the next activity.

- Work crew size constraints—considerations for the rules that ensure that the plan does not exceed the maximum number of crews working in a management activity.

The monthly operational planning process typically places more emphasis on individual sales to an individual customer within the supply chain process. This planning process has the goal of considering how wood volume from each stand can be merchandized to meet different customer orders. This is the first level of planning where the demand of wood has changed from forecasts to actual orders, and the supply of wood must then be consistent with these needs. Through this process, volume tables can be created that are representative of various log merchandizing rules. These are then applied to the stands that currently are being harvested, or will likely be harvested during the 2- to 3-month planning horizon. This problem formulation will also become an integer problem, where discrete cutting schedules are assigned to specific harvest units, and the volume is allocated to specific customers. A generic monthly operational planning problem formulation related to supply chain management might include:

Monthly Operational Planning

Maximize discounted net revenue
  Subject to:
- Volume flow constraint—minimum or maximum limits on the volume target by wood grade group for each potential specific customer, using guidance from the annual operating plan. In addition, minimum or maximum limits on the volume of wood from each harvesting work crew, given specific cutting instructions.
- Wood flow available from short-term supply sales.
- Short-term market potential for specific grades of wood.
- Log yard inventory policy constraints.
- Harvest units are represented by discrete decision variables.
- Harvesting crews are represented by discrete decision variables.

As most domestic mills in the United States (as well as those in many other countries) order wood on a weekly cycle, the weekly operational planning process within the wood supply chain is used to make minor corrections to the monthly orders. This level of planning also is used to adjust production levels in response to changes in demand. The weekly operational plan follows a similar mathematical formulation as the monthly operational plan. If this problem cannot be accommodated, then it is suggested that the monthly operational plan be revisited using the adjustments suggested here, prior to confirming weekly orders. A weekly operational planning problem formulation related to supply chain management might consist of the following:

Weekly Operational Planning

Maximize discounted net revenue
    Subject to:
- Volume flow constraints—minimum or maximum limits on the volume targets by grade group for each potential specific customer, using guidance from the monthly operating plan. This includes the volume scheduled for each customer that has not changed since the development of the monthly schedule, and the available wood flow from short-term supply sales.
- Volume production constraints—minimum or maximum limits on the wood volume to be produced from current harvesting operations using a set of specific cutting instructions.
- Short-term market potential constraints.
- Log yard inventory policy constraints.
- Harvest units are represented by discrete variables.
- Harvesting crews are represented by discrete variables.

One alternative operational plan might involve the transportation logistics of the forestry supply chain. To execute the set of forest plans, some operations may need to be logistically coordinated. For example, to harvest wood in some stands, an adequate transportation system must be developed that considers inventories, truck availability, and customer demands. This mathematical problem formulation might include the following:

Operational Planning—Transportation

Minimize total transportation costs
    Subject to:
- Work intensity constraints—these ensure that the solution to a problem does not suggest that the available hours for a vehicle or a driver exceed some safe limit.
- Customer demand constraints—these ensure that the plan meets the demands of each customer.
- In-woods inventory constraints—these ensure that all inventory stored at a landing are below some maximum stock level.

Throughout the book, we have concentrated on natural resource management planning problems. As you can see in this chapter, when management organizations want to recognize and plan within a forestry supply chain, the concepts we presented earlier are very similar to what we have presented here. We can also extend the conceptual framework of a "supply chain" to the management of other natural resources. For example, we may want to develop a strategic plan for a trail system across the land we manage. The trail users would be considered the customers, and the trail development and maintenance crew could be considered the manufacturer. We may then develop a tactical plan to ensure that the trail system physically connects, and that the costs of construction and maintenance will be within our projected budget levels over the next few years. We may then develop an annual plan that describes the costs and outcomes that are reasonable over the next year, given the availability of trail management personnel and equipment. In addition, we may identify the specific trail construction contractors and the specific list of trail management projects to undertake during the year. As you may have guessed, the outcomes from an annual plan should be consistent with the outcomes suggested at higher levels in the planning process. Finally, we may develop monthly or weekly operational plans to ensure that the appropriate equipment and personnel are assigned to each segment of the trail system, given the specific characteristics of each trail, and given the short-term supply of people and equipment.

V. SOURCES OF VARIATION IN THE FORESTRY SUPPLY CHAIN

Although in this chapter we have discussed many qualitative, value-centric concepts regarding planning and management of natural resources, most of this book has focused on quantitative techniques to apply to various forest management and planning problems. In real-life management problems, costs, prices, supplies, and demands can change dynamically over space and time and relatively quickly as a result of global, regional, and local socioeconomic conditions. As an extreme example, in the aftermath of Hurricane Hugo in 1989, mills in South Carolina were flooded very quickly with storm-damaged wood, and as a result, prices of delivered wood to pulp mills fell sharply due to the increase in supply, and logging costs increased due to the hazardous conditions (Janiskee, 1990). In light of these real-world issues, many land managers, landowners, and students of forest management may find that the deterministic nature inherent in many quantitative planning techniques is a difficult characteristic to accept. It should be obvious that the longer the time frame assumed in a planning process, the more uncertain many planning assumptions become. For example, would assuming a discount rate of 6% for a landowner be acceptable over a 20- or 30-year time horizon, or should the discount rate assumption change as economic conditions change or as the landowner ages? Unfortunately, even though short-term planning processes may better reflect current conditions, in many cases long-term planning is necessary to assure the sustainability of resources and to provide strategic guidance for forest management activities.

When applying quantitative planning techniques to problems related to the forestry supply chain, it is the cost and availability of data that can often prohibit us from fully capturing the essence of a management problem. For strategic planning processes, the supply of data generally is provided from forest growth and yield models, or from strata-based or plot-based averages. Forecasted prices may come from analytical models such as the Timber Assessment Market Model (Adams and Haynes, 1980) or from other projections provided by government agencies or private consultants. Further, logging costs and production rates generally are forecasted using recently obtained local knowledge. As we move down the hierarchy of planning processes from strategic to operational planning, the quality of the information must improve to enable you to utilize the level of detail necessary to obtain improvements in the forestry supply chain. Forest structure information, for example, may shift from strata-based averages (used in strategic planning) to information generated from a mid-rotation inventory of a specific stand that reflects recent silvicultural history (used in tactical or operational planning), or a preharvest estimation of wood quantity and quality. Forecasted prices for products cover much shorter periods of time in operational planning processes, and here you can begin to combine modeled prices with consultation of other supply chain participants to improve price forecasts.

As we move to the operational planning levels (annual, monthly, and weekly), we will be forecasting individual orders and mathematical models may not be able to simulate the actual circumstances that control wood volume desires of individual firms. For example, current logging and hauling rates may be based only on comparable bids made by contractors performing similar work, and not based on the actual production rates for a specific system (since the data is often unavailable). Further, current wood supply information may be generated from a preharvest inventory of each stand, and this information may be useful in supporting log allocation decisions. However, sampling errors are inherent in most estimates of wood volume, and unfortunately these errors can be relatively high for the rare, often more valuable, log grades. Interestingly, there has been limited investigation into the relationship between the quality of these data and the expectations of the customer of the overall efficiency of the supply chain.

Perhaps the most important message is that when you develop plans related to the forestry supply chain, the collection of appropriate data will involve a large commitment of time. The first few attempts to develop a forestry supply chain model may underscore the appropriateness of various types of data for each level of planning. Along the way, high-quality forecasts of prices and volumes may require a significant effort to collect, and you may decide that growth and yield models need to be developed to better represent local conditions. Further, in some areas of the world, the social and economic variables that influence logging system production rates may not be recognizable for some time. These uncertainties are suggestive of reasons why supply chain management systems utilize different information at each level of planning, and they may highlight where improvements in the planning process should be prioritized.

VI. SUMMARY

Evaluating and implementing a specific type of forestry supply chain involves a number of functions that have processes ranging from planning to reporting mechanisms. The strategic, tactical, and operational plans that we have described in the last few chapters of this book all are integral to the successful implementation of forestry supply chains. Whether or not aspects associated with producers or customers outside of the control of a single landowner are included in forest management plans is reflective of whether the supply chain is sufficiently acknowledged. In today's contemporary forest management environment, chain-of-custody assessments can be facilitated with the execution of a forestry supply chain system. Where forest certification is prevalent or necessary, the tracking of wood products may be more efficient when viewed through the supply chain. A number of types of plans that differ in scale and scope can be developed to model the potential supply of resources and efficiency of resource management. However, it is not until the fine-scale operational plans are developed that customer-supplier interactions specifically are acknowledged. The shorter the planning horizon, the better handle we may have on the appropriate prices, costs, supplies, and demands related to wood products. In contrast, variation and uncertainty in prices, costs, supplies, and demands should be greater in longer-term plans. In any event, supply chain plans that are short-term in scope should be consistent with the goals and objectives of longer-term plans.

QUESTIONS

1. *Decentralized forestry supply chain.* You recently have been hired as an analyst for a forestry consultant in Alabama. One client is a large pulp mill, which owns very little land itself. The managers of the pulp mill are interested in understanding the opportunities and challenges associated with the forestry supply chain. Describe in a short memorandum the actors

within a typical forestry supply chain, the roles that they play, and how strategic planning goals of the mill might filter down into the operational plans of their procurement foresters.

2. *Centralized forestry supply chain.* You recently have been hired as an analyst for a forestry company in Texas. The company, in general, operates using a centralized forestry supply chain. The managers of the company are interested in understanding more about the opportunities and challenges associated with other forms of forestry supply chains. Describe in a short memorandum the advantages and disadvantages of a centralized supply chain. How would this differ from a decentralized supply chain?

3. *Recreation supply chain.* Assume that you are a recreation planner with a national forest in Montana. The national forest has developed a strategic forest plan that provides broad-scale direction for the management and development of recreational resources. However, to implement this direction, other plans of action are necessary. Without delving into the intricacies of United States National Forest planning, describe in general how tactical or operational plans could be used to facilitate the management of the recreational supply chain. What type of information should be used to guide decisions in the weekly, monthly, annual, or decadal time frames? Who are the customers, and what products do they desire? How can these be integrated throughout the planning process so that an overall general strategy is achieved?

4. *Chain-of-custody auditing.* You recently have been hired as a chain-of-custody analyst for a nongovernmental organization that is actively involved in the certification of forest resources worldwide. What is a "chain-of-custody" as it relates to wood products? What is the purpose of auditing the wood product chain-of-custody?

REFERENCES

Adams, D.M., Haynes, R.W., 1980. The 1980 softwood timber assessment market model: structure, projections, and policy simulations. For. Sci. Monogr. 22, 64 p.

Boston, K., 2005. The desired future state of primary forested supply chain. In: Naito, K. (Ed.), The Role of Forests for Coming Generations Philosophy and Technology for Forest Research Management. Japan Society of Forest Planning Press, Utsunomiya, pp. 39–46.

Boston, K., Bettinger, P., 2001. The economic impact of green-up constraints in the southeastern United States. For. Ecol. Manage. 145 (3), 191–202.

Boston, K., Bettinger, P., 2006. An economic and landscape evaluation of the green-up rules for California, Oregon, and Washington (USA). For. Policy Econ. 8 (3), 251–266.

Buehlmann, U., Bumgardner, M., Schuler, A., Barford, M., 2007. Assessing the impacts of global competition on the Appalachian hardwood industry. For. Products J. 57 (3), 89–93.

Carlsson, D., Rönnqvist, M., 2005. Supply chain management in forestry-case studies at Södra Cell AB. Eur. J. Operat. Res. 163 (3), 589–616.

Epstein, R., Morales, R., Serón, J., Weintraub, A., 1999. Use of OR systems in the Chilean forest industries. Interfaces. 29 (1), 7–29.

Hall, N.G., Potts, C.N., 2003. Supply chain scheduling: batching and delivery. Operat. Res. 51 (4), 566–584.

Janiskee, R.L., 1990. "Storm of the Century": Hurricane Hugo and its impact on South Carolina. Southeastern Geographer. 30 (1), 63–67.

McTague, J.P., Oppenheimer, M.J., 2015. Rayonier, Inc., Southern United States of America. In: Siry, J.P., Bettinger, P., Merry, K., Grebner, D.L., Boston, K., Cieszewski, C. (Eds.), Forest Plans of North America. Academic Press, New York, NY, pp. 395–402.

Mitchell, C.P., 1992. Biomass supply from conventional forestry. Biomass Bioenergy. 2 (1–6), 97–104.

Murphy, G., Franich, R., 2004. Early experience with aroma tagging and electronic nose technology for log tracking. For. Products J. 54 (2), 28–35.

Perakis, G., Roels, G., 2007. The price of anarchy in supply chains: Quantifying the efficiency of price-only contracts. Manage. Sci. 53 (8), 1249–1268.

Rönnqvist, M., 2003. Optimization in forestry. Math. Prog. B. 97 (1–2), 267–284.

Siry, J.P., Bettinger, P., Merry, K., Grebner, D.L., Boston, K., Cieszewski, C. (Eds.), 2015. Forest Plans of North America. Academic Press, New York, NY, 458 p.

Weintraub, A., Epstein, R., 2002. The supply chain in the forest industry: models and linkages. In: Geunes, J., Pardalos, P.M., Romeijn, H.E. (Eds.), Supply Chain Management: Models, Applications, and Research Directions. Kluwer Academic Publishers, London, pp. 343–362.

Woods, E.J., 2004. Supply-chain management: understanding the concept and its implications in developing countries. In: Johnson, G.I., Hofman, P.J. (Eds.), Agriproduct Supply-Chain Management in Developing Countries. Australian Centre for International Agricultural Research, Canberra, Australia, pp. 18–26. ACIAR Proceedings No. 199e.

Zielke, A., Pohl, M., 1996. Virtual vertical integration: the key to success. McKinsey Quarterly. 3, 160–163.

Chapter 15

Forest Certification and Carbon Sequestration

Objectives

Global forest resources are essential for conservation of biological diversity, water, and soil resources, as well as for meeting our needs for commodities and nontimber forest products. While sustainable forestry had been practiced in some areas for a long time, concerned with progressing deforestation and degradation in many forested regions of the world, in 1992 members of the United Nations Conference on Environment and Development (UNCED)—termed the *Earth Summit* in Rio de Janeiro—developed a nonbinding Statement of Forest Principles that consisted of 17 points outlining guidelines and means for protecting the world's forests (United Nations, 1997), which in essence formed an action plan for the *sustainable forest management* movement. Since then, countries throughout the world have developed regional and international criteria and indicators that can measure and monitor success in achieving sustainable forest management.

Of the criteria and indicator initiatives related to sustainable forest management, the *Montréal Process* is geographically the largest, encompassing most of the world's temperate and boreal forests, which comprise 60% of all the world's forests (The Montréal Process, 2015). The seven criteria identified in the Montréal Process include:

- Conservation of biological diversity
- Maintenance of productive capacity of productive ecosystems
- Maintenance of forest ecosystem health and vitality
- Conservation and maintenance of soil and water resources
- Maintenance of forest contribution to carbon cycles
- Maintenance and enhancement of long-term socioeconomic benefits to meet the needs of societies
- Development of legal, institutional, and economic frameworks for forest conservation and sustainable management

Similar criteria and indicators for measuring and assessing sustainable forest management were developed through FOREST EUROPE (synonym of the Ministerial Conference on the Protection of Forests in Europe, also known as the *Helsinki Process for Europe* until 2009), and the International Tropical Timber Organization for tropical regions.

These sustainability concerns gave raise to two prominent environmental concepts associated with forestry. First, forest certification was prompted by a desire to help achieve sustainable forest management through market forces, since governmental regulations are difficult to create or enforce in many parts of the world. Certification has offered the promise of enhanced forest management and protection, and the promise for generating adequate financial returns from sustainably managed forests to ensure that they are retained in their present use. Second, the growing consensus that global climate change indeed is occurring, and that greenhouse gas (GHG) emissions need to be reduced, has brought attention to the ability of forests to sequester and store large amounts of carbon. As a result, markets for sequestered carbon recently have been developed, and may provide forest landowners with another source of income comparable to an annual hunting lease. Although the issues are complex and new developments occur frequently, at the conclusion of this chapter you should be able to:

1. Understand sustainability concerns driving forest certification and forest carbon sequestration.
2. Compare various forest certification systems and evaluate their applicability.
3. Understand challenges and opportunities associated with forest carbon sequestration.
4. Evaluate management implications of pursuing forest carbon sequestration projects.

I. INTRODUCTION

Forest certification is a method for addressing concerns related to deforestation and degradation of forests, and for promoting the development and maintenance of biological diversity (Rametsteiner and Simula, 2003). Forest certification processes have been called at times *green certification*. Certification programs are basically processes that attempt (1) to identify and promote forest land that is well-managed, and (2) to recognize the products that are produced from these forests as having been sustainably managed. Certification also is a method for verifying a landowner's commitment to sustainable forestry objectives, and has been used as a method for obtaining publicity for voluntary conservation efforts. The underlying idea of

Forest Management and Planning. DOI: http://dx.doi.org/10.1016/B978-0-12-809476-1.00015-1

forest certification is that consumers of wood products will choose to purchase products derived from sustainably managed forests, and pay more for these products (offer a price premium) than for products derived from poorly managed forests. Whether this is true or not is debatable (Hubbard and Bowe, 2005). Forest certification is aimed at recognizing the improved quality of forest management and promoting higher wood product prices, as well as providing better market access for wood products derived from sustainably managed forests. The concept was originally designed for improving tropical forest management, where most forest losses currently take place. Forest products originating from sustainably managed forests can be certified by reliable, independent, third-party auditors. Certified products can then be labeled so that consumers can clearly recognize them and make informed choices at the time of purchase. A wood product with an *eco-label* is meant to help consumers easily identify officially approved *green products* derived from certified forests. The label also allows manufacturers to demonstrate that their finished products are derived from certified forests.

A number of environmental nongovernmental organizations (NGOs) and numerous governmental organizations participating in the Earth Summit (1992) strongly supported binding international agreements and legislation to address deforestation and degradation throughout the world. Despite their efforts, no legally binding commitments were developed, and as a result many NGOs considered the Summit a failure. In an attempt to better protect global forests, NGOs since have devised several forest certification processes that use market-based approaches to address international wood products trade. Forest certification in general consists of four elements: (1) standards (acceptable levels of forest management activity); (2) inspections and audits (objective assessments of whether the standards are fulfilled); (3) chain-of-custody audits (identifying a product's origin and tracking it through the system); and (4) certification (a labeling process that indicates that the previous three elements were met satisfactorily). Depending on the certification program, three types of certificates are awarded: (1) sustainable forest management certificates that verify forests are sustainably managed, (2) chain-of-custody certificates that verify forest products were made with wood harvested from sustainably managed forests, and (3) wood fiber sourcing certificates that verify fiber acquired by wood procurement organizations (which do not own forests) comes from sustainable sources and that sustainable practices are promoted in all forests regardless of their certification status. As a result, forest certification is a market-oriented and consumer-oriented approach that has developed a high level of acceptance among some private and public forest managers (Gallardo Gallardo, 2007).

Forest certification began with the American Tree Farm System in 1941. Of the contemporary processes, certification of forests worldwide began with the Forest Stewardship Council (FSC) in 1993 and has since expanded rapidly. By 2015, certified forest area worldwide amounted to nearly 1.1 billion acres (439 million ha) or nearly 11% of the world's forests (Table 15.1). Most certified forests are in the Northern Hemisphere. North America alone accounts for almost 50% of certified forest area globally, and Western Europe accounts for

TABLE 15.1 Forest Certification by Major Region, 2015

Region	Total Forest Area (million ha)	Certified Forest Area (million ha)	Percentage of Total Forest Area Certified	Certified Industrial Wood Production (million m³)	Percentage of Total Production
North America	614.2	217.3	35.4	245.9	13.9
Western Europe	168.1	109.6	65.2	258.1	14.6
Commonwealth of Independent States (CIS, Russia and others)	836.9	62.9	7.5	12.0	0.7
Oceania	191.4	12.5	6.5	3.6	0.2
Africa	674.4	6.5	1.0	2.0	0.1
Latin America	955.6	17.1	1.8	1.3	0.1
Asia	592.5	13.1	2.2	4.2	0.2
Total	**4,033.1**	**439.0**	**10.9**	**527.1**	**29.8**

Source: United Nations Economics Commission for Europe, 2015. ECE/FAO Forest Products Annual Market Review, 2014−2015. Geneva Timber and Forest Study Paper 29, ECE/TIM/SP/39. United Nations Economic Commission for Europe, Timber Section, Geneva. 120 p.

about 25%. The Commonwealth of Independent States (Russia, etc.) account for about 14% of the certified forest area. Much smaller amounts of certified forest area are located in Oceania (Australia, New Zealand, Micronesia, etc.), Latin America, Africa, and Asia.

The potential wood supply that can arise from certified forests was estimated in 2015 to be about 527 million m^3, or nearly 30% of the total global industrial roundwood supply (Table 15.1). In North America and Europe more than 40% of industrial roundwood production arises from certified forests. In Finland and Austria, 100% of managed forests are certified. However, because of low customer awareness and lack of price premiums, only a small fraction of certified wood products is marketed as such. In other areas of the world (e.g., Commonwealth of Independent States, Latin America), a relatively low percentage of the wood produced arises from certified forests. Compared with the rest of the world, the sustainability of forest resources in the Northern Hemisphere is not threatened, since the quality of forest management is generally satisfactory and since the extent of forest resources is expanding. In most regions of the world, certified forests are usually those forests that have been managed sustainably for many decades. Planted forests managed for wood production account for a substantial share of certified forests in all regions of the world. Although the pace of growth in certified forest area has slowed down in recent years, major expansions in forest certification are expected in Russia, the tropical regions and the Southern Hemisphere.

The major forest certification systems include the FSC, the Programme for the Endorsement of Forest Certification schemes, the Canadian Standards Association sustainable forest management program, the Sustainable Forestry Initiative, and the American Tree Farm System. Both the FSC and the Programme for the Endorsement of Forest Certification serve as umbrella programs endorsing national forest certification systems. Other certification programs gaining importance include, for example, the Australian Forestry Standard, Chile's Certificación Forestal, Brazil's Certificação Florestal, and the Malaysian Timber Certification Council. In the United States, forest certification is dominated by Sustainable Forestry Initiative, which has about 62 million acres (more than 25 million ha) certified. The second most popular system is the FSC certification with about 35 million acres (more than 14 million ha). The American Tree Farm System is third in size with nearly 20 million acres (8 million ha) certified. Since the forested area within the United States is around 750 million acres (about 300 million ha), this suggests that the United States is one of global leaders in forest certification.

About a decade ago, approximately 40 million acres (16 million ha) of the certified forests in the United States were located in the southern states (Cubbage et al., 2005). Thus it is interesting that the southern United States

contains about 29% of the nation's forests and about 46% of the certified forests, and intriguing since federally managed forests are more commonly found in the western portion of the nation, and are not certified. The forest industry owned about 22 million acres (about 9 million ha) of pine plantations in 2005, and almost all this land would have been certified. However, much of the forest industry land has changed ownership in the past few years, mainly through sales from vertically integrated forest products companies to large Timber Investment Management Organizations (TIMOs) or Real Estate Investment Trusts (REITS). What effect this change in ownership may have on the amount of certified forest area remains to be seen.

There are two types of certification processes: (1) performance-based, and (2) systems-based. Performance-based certification processes are ones where a certifying organization determines most, or all, of the performance criteria related to certification, then oversees the assessment process to ensure conformance to the criteria. The systems-based process allows the land management organization seeking certification to identify the environmental concerns related to the property, and to devise an environmental management system that will address these concerns. Verification of adherence to the forest certification standards involves comparing a land management organization's mode of operating to the certification system's set of standards. In general, the verification process is a negotiated agreement, and certification may be awarded on the condition that landowners adopt a certain set of acceptable management practices. The verification process involves preliminary discussions, field verification of activities and conditions, the development of a verification report, and follow-up audits.

The reasons why a landowner would enter into a certification process vary considerably, and are based on environmental, economic, and social considerations. First, there is a belief that certified products may eventually command a price premium among consumers. As we have suggested, this is currently debatable (Hubbard and Bowe, 2005). Second, some manufacturers believe that certified wood products might provide better access to certain markets. This may, in fact, be an important consideration as some of the larger home improvement stores consider this factor in the supply that they offer to their customers. A few retail home-improvement and office supply chains, for example, recognize the value of maintaining a good public image. Whether this leads to higher sales for the store or for the certified wood products is also debatable. However, Tikina et al. (2008) found that interest in certification among landowners in the Pacific Northwest of the United States increases as the proportion of their customers requesting certified goods increases. Some landowners also are concerned about the perception of their land management ability. Tikina et al. (2008)

suggest that in areas where land management may be highly exposed to regulatory requirements (e.g., those related to riparian management zones), landowners may have a higher interest in obtaining forest certification, to receive recognition for the effort required to manage their land within the broader socioeconomic environment. Although all land management organizations claim to manage land consistent with environmental rules and regulations, not all of them do so, thus certification provides a method to prove their case (Ortolan, 2003).

Final consumers of wood products have not been very active in creating the demand for certified wood products, and their role in the market is not expected to change in the short-term. In fact, globally certified wood products represent a small fraction of forest products sales. The primary force behind the drive to certify forest management may be pressure from environmental groups, and the strongest demand drivers are wood products retailers, who are interested in maintaining or increasing their public image, and thus may prefer to avoid direct action protests in their retail outlets (Hansen et al., 2000). Various government policies can also influence the demand for certified wood products, either directly or indirectly. For example, renewable energy policies that induce using wood for energy generation can encourage or require forest certification. Various carbon trading schemes which accept forest carbon offsets also require forest certification. Green building policies and programs in Europe (2010 Energy Performance of Buildings Directive or EED) and the United States (Leadership in Energy and Environmental Design or LEED) also encourage the use of certified wood products in construction (United Nations Economics Commission for Europe, 2015). Also efforts to curb illegal logging worldwide may help encourage forest certification, as it helps to establish the legality of wood harvested through chain of custody documentation requirements.

The promise of price premiums may eventually work better for forests in developed countries that are managed for industrial wood production, and where products are geared toward markets in Europe and North America. There is less promise of a price premium, however, for large forested areas located in developing regions of the world. These forests traditionally have been used to meet the subsistence needs of the local people. In many of these cases, very few resources are available for the active management of forests, and most are threatened by overexploitation, degradation, or conversion to other uses.

II. FOREST CERTIFICATION PROGRAMS

For land managers and landowners engaged in active management of commercial forests, certification of forest plantations is of special importance. In many cases, commercial forests are intensively managed (Fig. 15.1), and

FIGURE 15.1 An intensively managed forest in western Washington state. *Photo courtesy of Kelly A. Bettinger.*

as a result achieve high growth rates. Although very productive, the management environment involves a number of controversial issues, such as the low vegetative diversity and the use of chemicals for site preparation and vegetation control (Cossalter and Pye-Smith, 2003). In light of these and other issues, we might ask whether intensively managed plantations can be certified under the various certification systems. Interestingly, many certified forests are indeed plantations, and the certification process has helped to show how plantations can contribute to broadly understood sustainable forest management concepts, including the attainment of environmental, economic, and social goals. However, the issue of forest certification is contentious, and some environmental movements continue to oppose the certification of forest plantations. The FSC prohibits conversion of natural forests to plantations and places more stringent requirements on the management of plantations. Also, the Sustainable Forestry Initiative prohibits conversion of natural forests to plantations if the natural forests are of high environmental value. Several of the relevant components of each certification program are provided in the sections that follow.

A. Sustainable Forestry Initiative

In 1994, members of the American Forest & Paper Association, primarily composed of forest products companies in the United States, agreed to operate under a set of forestry principles that were suggestive or indicative of sustainable forest management. These principles were designed to demonstrate to the public that the industry could meet the needs of the present generation without compromising the needs of future generations. The *Sustainable Forestry Initiative*, which arose from this effort, called for a land ethic that integrated active management with the conservation of nontimber resources and aesthetics. The objectives of the Sustainable Forestry Initiative program encourage and promote the use of sustainable forestry practices, and promote the efficient use of forest resources. The objectives also call for the protection of certain wildlife habitats and areas that contribute to forest biodiversity, for the protection of water quality, and for the management of potential visual impacts related to harvesting activities.

The Sustainable Forestry Initiative uses in its 2015−2019 Forest Management Standard a hierarchy of 13 principles, 15 objectives, 37 performance measures within those objectives, and 101 indicators of performance toward the standards. Sustainable Forestry Initiative Objectives 1 through 6 provide the means for evaluating a program participant's compliance with the standards on forest lands they own or manage (Table 15.2). Sustainable Forestry Initiative Objective 7

TABLE 15.2 Sustainable Forestry Initiative Objectives, 2015−2019 Edition

Objective	Description
1	Ensure long-term sustainable harvests and measures to avoid forest conversions
2	Ensure long-term forest productivity, carbon storage and conservation of forest resources
3	Protect water quality in streams, lakes, wetlands and other water bodies
4	Manage quality and distribution of wildlife habitats and conserve biological diversity
5	Manage visual impact of harvesting and other forest operations and provide recreational opportunities for the public
6	Protect special geological and cultural sites
7	Minimize waste and ensure efficient use of fiber resources
8	Respect Indigenous People's rights and traditional knowledge
9	Comply with federal, provincial, state, and local laws
10	Invest in forestry research, science and technology
11	Implement appropriate training and education programs
12	Encourage public outreach, education, and involvement
13	Implement sustainable forest management on public lands

Source: Sustainable Forestry Initiative, 2015. SFI 2015−2019 Forest Management Standard. Sustainable Forestry Initiative, Washington, DC. 12 p.

deals with wood procurement systems. Objectives 9 through 13 address research, training, legal compliance, public and landowner involvement, indigenous people's rights and knowledge, management reviews, and continual improvements to the forest management system.

Participants in the Sustainable Forestry Initiative must abide by the standards and substandards relevant to their land ownership or wood-using status. Participants must develop a written policy (or policies) related to:

- Sustainable forestry
- Forest productivity and health
- Protection of water resources
- Protection of biological diversity
- Aesthetics and recreation
- Protection of special sites
- Responsible fiber sourcing
- Legal compliance

- Research
- Training and education
- Community involvement and social responsibility
- Transparency
- Continual improvement

Organizations must demonstrate compliance with these objectives and indicators through a third-person certification audit of their management practices (Sustainable Forestry Initiative, 2015).

As an example of the objectives and indicators relevant to forest plantations, Objective 2 explicitly addresses reforestation and afforestation, stating that:

Performance Measure 2.1. Program Participants shall promptly reforest after final harvest.

Five specific indicators are associated with this measure of performance. They include:

1. A documented reforestation plan.
2. Clear criteria to judge adequate regeneration and respond to problems.
3. Plantings of exotic species that minimize risk to native ecosystems.
4. Protection of desirable or planned advanced natural regeneration during harvest.
5. Afforestation programs that consider potential ecological impacts of the selection and planting of tree species in nonforested landscapes.

In addition to reforestation standards, Sustainable Forestry Initiative objectives address water quality and environmental protection, wildlife habitat and biological diversity, aesthetics, the management of unique sites, economic efficiency, wood procurement systems, forestry research, and the need for public input into the process. To ensure long-term sustainability of forest management, organizations engaged in the Sustainable Forestry Initiative must also develop a management plan and associated analyses conducted at a level appropriate to the size and scale of their operations. The documentation

necessary for the Sustainable Forestry Initiative includes the following:

1. A long-term resource analysis
2. A periodic or on-going forest inventory
3. A land classification system
4. An assessment of biodiversity at the landscape level
5. A soils inventory and associated maps
6. Current maps, or databases associated with a geographic information system (GIS)
7. Recommended sustainable timber harvest levels
8. A review of nontimber issues

Along these lines, there is a need to document current harvest level trends and to determine whether they are within long-term sustainable levels. A forest inventory system and methods to calculate growth and yield of forests are needed, along with processes to periodically update the forest inventory. Finally, an organization needs to document their forest practices.

Through adherence to the standards and objectives of the Sustainable Forestry Initiative, members demonstrate a commitment to sustainability, the environment, and society, while recognizing the importance of maintaining commercially viable forests (Sustainable Forestry Initiative, 2015). As we will see, there are a number of similarities between this program and others (Table 15.3).

B. Forest Stewardship Council

The FSC developed a performance-based forest certification system in 1993. The initial intent of the certification program was to protect tropical forests and assist timber producers with the marketing of their products in the European wood products market. The program has since expanded its scope to include forests in all parts of the world. The goal of the FSC is to promote environmentally and socially responsible, yet economically viable, management of forests. These standards include demonstrating that a forest management plan has been developed, is being

TABLE 15.3 Summary of Forest Certification Systems

	American Tree Farm System	Forest Stewardship Council	Sustainable Forestry Initiative	Canadian Standards Association	ISO 14001	Programme for the Endorsement of Forest Certification
Scope	US	World-wide	US/Canada	Canada	World-wide	Europe
Established	1941	1993	1995	1996	1994	1999
Eco-label?	No	Yes	Yes	Yes	No	Yes
Chain of custody?	Yes	Yes	Yes	Yes	No	Yes

implemented, and is periodically updated. The management plan can include the development of forest plantations as long as they are consistent with areas designated for wildlife habitat and riparian zones, and as long as they have not been recently developed through conversion of naturally regenerated forests.

The FSC certification program has 10 principles and 57 criteria, and associated national indicators in the United States guidelines (Forest Stewardship Council, 2010), representing a national FSC US Forest Management Standard (v1.0). Some regional variations were retained from the previous US-FSC regional standards in indicators pertaining to (1) even-aged systems (Indicator 6.3.g.1), (2) Streamside Management Zones (SMZ) buffers (Indicator 6.5.e.1), and (3) and plantations (Principle 10).

In the national standard, we find language suggesting that the rate of harvest of forest products shall not exceed levels that can be permanently sustained (Criterion 5.6). In addition, a number of standards refer to reforestation and succession and address the context of plantations in the certification process. The United States certification standard (Forest Stewardship Council, 2010) defines plantations as:

Forest areas lacking most of the principal characteristics and key elements of native ecosystems as defined by FSC-approved national and regional standards of forest stewardship, which result from the human activities of either planting, sowing or intensive silvicultural treatments.

The FSC standards require certified forest landowners to maintain or restore forests to natural conditions to the extent possible, and address other means for retaining and managing natural forests. Plantations can be developed, yet should be designed so that pressure on natural forests is not increased, and so that the ecological integrity of natural ecosystems is not diminished or endangered. In addition, forest management practices should be designed to maintain site productivity as well as genetic, species, and community diversity of the forest. More importantly for the United States, the FSC standards (Forest Stewardship Council, 2010) indicate that:

Forest conversion to plantations or non-forest land uses shall not occur, except in circumstances where conversion: a) entails a very limited portion of the forest management unit; and b) does not occur on high conservation value forest areas; and c) will enable clear, substantial, additional, secure, long term conservation benefits across the forest management unit. (Standard 6.10)

Many people have the impression that the forest industry manages most of the land in the southern United States, but in fact most of the forest land is owned by private, nonindustrial landowners. It is likely that a good percentage of industrial plantations were converted from naturally regenerated stands of trees, even though much of the southern United States has been harvested four or five times since the Colonial Era. However, many forest industry plantations today were (and still are) regenerated on old farm fields that are not considered "natural," so this standard may not be as daunting as it appears. Thus in practice, plantations may still be planted in most areas of the southern United States where agricultural practices once occurred. But the language in the standards may still prevent certification of forests where natural stands of trees are converted to plantations. As with the other certification programs, land managers must agree to protect soil resources, minimize forest pests, diseases, and fire, and minimize the use of pesticides. In addition, landowners must assess on- and off-site ecological and social impacts of forest management activities.

C. American Tree Farm System

The American Tree Farm System, perhaps one of the most recognizable programs for forest landowners in the United States, has maintained a Certified Tree Farm program since 1941. Unlike other certification programs, Certified Tree Farms are recognizable by the famous green and white sign posted along the boundary of each forest. And this program, in contrast to the others, was not created in response to various market pressures. The American Tree Farm System requires periodic inspection of the forests of participating "Tree Farms." However, prior to 2002 the rigor of the rules was modest and the inspections were sporadic. To become a more credible forest certification system, new standards and auditing procedures were developed in 2002, and implemented in 2004 (American Forest Foundation, 2014). Audit inspections now are required every 5 years, and are conducted by cooperating foresters with forest industry, private consultants, or state foresters. The program has the same structure as the Sustainable Forestry Initiative, as we might expect, since it also relied in part on the forest industry for development and program support. Wood harvested from American Tree Farm System certified forests can be recognized as sustainably harvested under the Sustainable Forestry Initiative and the Programme for the Endorsement of Forest Certification chain of custody certificates.

This program can be characterized as a performance-based certification system, and membership is limited to those lands that have passed a tree farm inspection, a form of verification process. The American Forest Foundation's program of sustainability currently includes eight broad standards, 14 performance measures, and 22 specific indicators (American Forest Foundation, 2014). Landowners engaged in this program need to demonstrate a commitment to sustainable forestry through the development and implementation of a long-term forest management plan. The McPhail Tree Farm in South Carolina (Straka and Cushing, 2015) is an example of multigenerational tree farm that has been American Tree Farm System certified. On this property, the forest is managed primarily for timber production, with strong support given to forest sustainability objectives. It is a good example of the influences of forest certification on management plan development.

The certification standards are developed by committees within the American Tree Farm System, and the system utilizes its own auditors and auditing process rather than using a third-party auditing mechanism. The certification of forests stresses the protection of environmental benefits and the need for public understanding of managed forestry concepts. As we suggested, landowners enrolled in the tree farm system develop and implement a long-term, written forest management plan that is consistent with the size of the property (Standard 1, Performance Measure 1.1). The standards also suggest that adequate reforestation and stocking are important after final harvests. Standard 3, Performance Measure 3.1 notes:

Reforestation or afforestation shall be achieved by a suitable process that ensures adequate stocking.

and Indicator 3.1.1 for this standard is:

Harvested forest land shall achieve adequate stocking of desired species reflecting the landowner's objectives, within five years after harvest, or within a time interval as specified by applicable regulation.

The American Tree Farm System also emphasizes the protection of soil and water resources (Standard 4). For example, as with other certification programs, landowners must minimize disturbances within riparian areas, and meet or exceed applicable practices prescribed by state forestry best management practices (BMPs) or laws governing forestry practices. Other standards and performance measures relate to the conduct of forest harvesting operations, to compliance with applicable laws and regulations, to conservation of forest biodiversity, to values associated with forest aesthetics, and to the protection of special places. Special places are those recognized for their unique cultural, ecological, archeological, and historical characteristics.

D. Green Tag Forestry System

The Green Tag Certified Forestry program is not as formal as the previously described programs, yet includes 10 criteria and 46 indicators for forest certification. It is administered by the National Forestry Association, and

forestry consultants serve as the inspectors for the program (American Resources, Inc., 2013). Criterion 1 requires forest management planning. Here a 10-year management plan needs to be developed, and a commitment to stewardship needs to be demonstrated. Specifically, this certification program suggests that landowners work under a sustained yield management paradigm, where forest growth exceeds wood harvested over time. Criterion 3 requires postharvest evaluation and reforestation, and the fourth indicator of this criterion requires that a harvested land is regenerated within 3 years. A number of the criteria address soil and water resource protection, as well as the need to maintain forest health and maintain the appropriate aesthetics quality of the landscape. In addition, the program stresses the need to communicate with the public, particularly with neighboring landowners, and suggests that a balance needs to be struck between economic and environmental concerns.

E. Canadian Standards Association

The Canadian Standards Association developed a performance-based and systems-based certification process in 1996. Both the CAN/CSA Z809 and Z804 SFM standards are endorsed by the Programme for the Endorsement of Forest Certification. The process was developed as a voluntary sustainable forestry management standard for lands within Canada, and is consistent with both the Sustainable Forestry Initiative and ISO 14001 programs, yet was meant to be a program that can be influenced by local values and public participation processes. The criteria that are associated with this certification program take into account economic, environmental, and social dimensions of the management of forested areas (Canadian Standards Organization Group, 2016). These include conserving biological diversity, conserving soil and water resources, and maintaining ecosystem productivity. They also address the role of forests in global ecological cycles. In addition, the criteria suggest that land management organizations operating within this framework will provide multiple benefits to society and manage for sustainability in the process. One of the criteria addresses aboriginal relations. The performance measures suggest measuring ecosystem diversity at a broad scale, yet managing for site-specific variation in ecosystems. The conservation of habitat for native species, and the conservation of genetic diversity pervade the objectives of the program. One of the main requirements of landowners engaged in this certification process is the need to adopt planning processes that allow substantial public input, thus allowing citizens to become involved in the decision-making processes of local forests. Interestingly, Criterion 5 suggests that landowners associated with this certification process will contribute to the sustainability of communities by providing diverse opportunities for people to derive benefits from the forest. The level of harvest that is allowed from a certified forest is one that produces, sustainably:

... a mix of both timber and non-timber benefits (Canadian Standards Organization Group, 2016).

In addition to incorporating public input into the planning process, the Canadian Standards Association certification system encourages planning as an adaptive, continuous improvement process that involves monitoring and measuring performance, and one that uses corrective or preventive actions as unplanned variations occur during the normal course of land management (Canadian Standards Organization Group, 2016).

F. International Organization for Standardization, Standard 14001

The International Organization for Standardization (ISO) developed an environmental management standard (ISO 14001) in 1994 in an effort to standardize the environmental management system framework. The ISO 14001 standard is not necessarily specific to forest management activities, but can be used as a forest certification process. ISO 14001 is a systems-based certification process, and the standards for a property are developed through a public input process. This certification process can be applied to forests throughout the world, and the standards reflect a global consensus on management practices that are acceptable within an international context. This process suggests the use of management practices that can reasonably be applied by other land management organizations in other parts of the world operating under similar management situations. Since the program is a systems-based process, the land management organization seeking certification identifies the environmental concerns and develops a management program to address them. The ISO 14001 certification program therefore provides a wide-ranging portfolio of standards to allow land management organizations to take proactive approaches to certification while dealing with specific environmental challenges.

The process for certification involves prioritizing environmental issues, integrating these issues into a management plan, implementing the plan, and communicating the results to the public. Monitoring the performance of the plan in light of the systems-based standards that were adopted, then assessing the actual performance against the expected performance, is an important aspect of the process as well. The ability of a landowner to choose the standards against which performance is evaluated distinguishes this system from some of the others. Two potential downsides to this certification process are

the lack of an eco-label and the lack of chain-of-custody auditing. As a result, some land management organizations may decide to pursue certification under this program concurrently with another.

G. Programme for the Endorsement of Forest Certification

The Programme for the Endorsement of Forest Certification schemes began in 1999 as an alternative to the FSC process. It is an umbrella organization that assesses, endorses, and recognizes national forest certification programs including, among others, the Sustainable Forestry Initiative and the American Tree Farm System. This is a performance-based certification process, and like several of the other processes, includes a chain-of-custody assessment and an eco-label for certified wood products. This certification program emphasizes through its criteria the maintenance and enhancement of forest growing stocks on land that is meant to be available for timber production. In addition, program participants should develop methods to inventory measures of biological diversity, and should assess the impact of management activities on those measures of biological diversity. Although other criteria are associated with this program, one suggests the maintenance of socioeconomic functions and conditions, and suggests that a portion of the revenue generated from a certified forest be reinvested in public awareness about the benefits of forest management. As you can tell, this program has distinct economic, ecological, and social aspects that influence the development of a forest management plan.

III. COST AND BENEFITS OF FOREST CERTIFICATION

One question all landowners and land managers face when considering the certification of forests involves the cost of the program. It is somewhat surprising how little research evaluating the cost and benefits of forest certification has been conducted thus far. Certification has occurred mainly in forests that may have been already sustainably managed (in the Northern Hemisphere) and in areas where only little management adjustments have had to be made. Although many of the contemporary systems are still relatively new, worldwide forest certification has been practiced since 1993 and, given its rapid growth, soon we should have a better idea of the costs and benefits for organizations pursuing these systems. Unfortunately for many land management organizations, the original premise of a price premium for certified forest products largely has been unfulfilled thus far. Forest certification may have improved market access, especially in the United Kingdom, Belgium, and The Netherlands, where demand for certified forest products is more pronounced. In addition, it seems that forest certification has somewhat improved forest management practices, but only modestly, and mainly in developed countries where forests are generally well-managed (Ozinga, 2004). This suggests that certification may overlap existing organizational or regulatory requirements and that the net benefit of certification may be relatively small. Still, much progress is required in tropical regions of the world, where we experience most of the current problems associated with forest decline.

Ultimately, however, the cost of forest certification could be substantial to a landowner or land management organization. The certification process requires a substantial amount of time and energy to document the adherence of management practices to various principles and indicators. Depending on the work required to meet the documentation requirements of a certification system, the direct certification cost may range from a few cents to several dollars per unit area, depending on the size of a forest and the forest conditions (Vidal et al., 2005). Additional costs may be incurred if substantial changes to management activities are required (Mendell and Hamsley Lang, 2013). In some cases, certification may also discriminate against some forest landowners, as was observed in Sweden where private forest owners were found to be disadvantaged in the marketplace by wood produced from FSC certified industrial forests (Elliot and Schlaepfer, 2001). However, some forest certification systems have been developing approaches to accommodate small nonindustrial landowners by providing opportunities for group certification, which should help reduce the direct cost of becoming certified. Forest owners wanting to certify their forestry and manufacturing operations first need to recognize the actions that may be necessary to meet certification requirements, then to estimate the costs associated with these actions. In addition, the choice of a certification program may be important for the marketing of wood products. However, in today's management environment in North America, forest certification is a voluntary endeavor, and in many cases may not be necessary for the efficient or effective management of natural resources.

IV. FOREST CARBON SEQUESTRATION

Although climate change has long been a controversial issue, there is a growing consensus that recently observed climate warming is human-induced through the emission of GHGs. Today, many scientists and politicians believe that global climate change is indeed occurring. There is also a growing consensus that climate change must actively be mitigated to reduce the risk of adverse changes and impacts on human welfare. It is suggested that this goal can be achieved through reductions in the amount of GHGs in the atmosphere, particularly atmospheric carbon dioxide (carbon), the most common GHG. In response to global climate change concerns, representatives from developed countries met in Kyoto, Japan in 1997 and agreed to reduce or limit their GHG emissions. This agreement is a protocol to the United Nations extending its 1992 Framework Convention on Climate Change (UNFCCC), and has been since termed the *Kyoto Protocol*. The Kyoto Protocol also recognized the role that forests and forest management plays in reducing carbon dioxide emissions. Article 3.3 (United Nations, 1998) states that:

The net changes in greenhouse gas emissions by sources and removals by sinks resulting from direct human-induced land-use change and forestry activities, limited to afforestation, reforestation, and deforestation since 1990, measured as verifiable changes in carbon stocks in each commitment period, shall be used to meet the commitments under this Article of each Party included in Annex I.

Annex I includes many developed countries. Article 5 of the *Paris Agreement* (United Nations Framework Convention on Climate Change, 2015) in 2015 also emphasized the role of forests in reducing GHG emissions. It also is important to mention that the United States did not ratify the Kyoto Protocol, yet along with many of the world's countries, did sign the Paris Agreement.

Forests sequester and store large amounts of carbon in their vegetative biomass and surrounding soils. Amounts of carbon sequestered by forest ecosystems can be determined by estimating the forest biomass, because carbon represents about 45−50% of dry biomass of vegetation (Birdsey, 1992). Carbon storage pools include living biomass (trees and understory vegetation), dead biomass (snags, down woody debris), soils, and wood products (off-site). Today, the role that forests and forest management play in reducing carbon emissions is well recognized. Common land management practices, including afforestation, deforestation, reforestation, and harvest, can substantially influence the carbon sequestration potential of the land (Alig, 2003). Forest mitigation strategies may involve eliminating forest land conversions (especially deforestation in tropical regions), postponing harvests, reducing prescribed or controlled burning, or increasing carbon sequestration through intensified forest management of existing resources and conversion of agricultural lands to forests. Although the Kyoto Protocol allows using carbon stored in forests to meet emission targets, it also places several restrictions on how this can be accomplished. First, the Kyoto Protocol forests have to be established after 1990. In addition, they have to be planted on land that historically has not been forested (using afforestation), or on land that was historically forested but which recently has been used for nonforest uses (using reforestation).

V. OPPORTUNITIES AND CHALLENGES IN INCREASING FOREST CARBON STORAGE

Forests have a great potential for sequestering vast amounts of carbon. Forest carbon sequestration is environmentally-friendly, uses known technologies, generates environmental cobenefits, and may be cost effective. It is estimated that forests can account for about 25% of the potential global carbon abatement at a cost of up to 40 Euros (about 45 USD at the time this book was written) per metric ton (Enkvist et al., 2007). At the same time, there are numerous challenges facing effective forest carbon credit markets. They arise from the nature of carbon sequestration projects and from biological and management characteristics of the forestry profession. The four major concerns include the following:

1. *Baseline.* The Kyoto Protocol states that only forests planted after 1990 may be eligible for carbon offsets. Several other carbon trading schemes assume the same base date. This may represent a problem for regions where a large amount of afforestation took place prior to 1990, effectively making these projects ineligible for carbon credits.
2. *Additionality.* The Kyoto Protocol and several other carbon credit schemes require the concept of additionality. A carbon reduction emission project is *additional* only when it was developed solely for the mitigation of climate change. Projects implemented under a *business as usual* or *required by other laws and regulations* set of management actions are not considered additional. Determining what are usual management practices and what are additional management practices may be quite difficult.
3. *Permanence.* Carbon emission reduction schemes implicitly assume that an emission reduction is permanent. However, forest carbon sequestration by its very nature is only temporary. Although trees can store carbon for several decades or more, eventually the trees will die and the carbon will be released. We might logically ask, is there any value in the temporary storage of carbon? There probably is, since if anything, temporary storage will provide more time for the development of alternative approaches to carbon emission reductions. Other questions then arise concerning how carbon should be valued and traded, and how we should treat harvesting and wood product manufacturing. Carbon can be stored in wood products for many years, for example, but some carbon trading schemes

do not consider tree harvests nor do they allow one to recognize and count carbon stored in forest products. A rental payment approach has been suggested for carbon emission reductions that are not permanent (Sedjo and Marland, 2003).

4. *Leakage.* Implementing a forest management project could cause higher carbon emissions outside of the project's scope. For example, a forestry organization may enter part of its forests into a carbon sequestration project, which would impose limits on the size and location of harvested areas. To compensate for the resulting loss of harvest volume, the organization may choose to increase harvest levels in other parts of its ownership. Alternatively, an increase in carbon could be obtained through an application of fertilizer to a forest, yet an accounting of the carbon impact during the manufacture of the fertilizer may be ignored. Well-designed carbon sequestration projects should avoid these types of negative spillover impacts. The shift of the carbon burden to other areas of the world is therefore a major concern (Lauterbach, 2007).

In response to these concerns, many carbon sequestration and reduction programs require that forests are certified using one of the systems described earlier in this chapter.

VI. EMISSIONS TRADING

Three approaches usually are considered for the design of regulations aimed at reducing carbon emissions. The first involves strict regulation of all carbon emitters, which is technically challenging and expensive. The second involves a carbon tax, which would be imposed on most or all activities relying on fossil fuels and generating carbon dioxide and other GHGs. This, however, may imply the imposition of new taxes, which may lack political support. The third, and most popular approach today, involves emissions trading (cap and trade). Emissions trading begins with the establishment, usually by a governmental body, of an emission limit (cap). In many programs, this limit also decreases over time, creating more incentives to reduce emissions. As part of the cap, individual companies are given emission allowances, or credits, which represent their right to emit a specific amount of GHG. In cases where emission allowances are not sufficient to cover their current emissions, companies may choose to invest in cleaner technologies, decrease production, purchase emission allowances (trade) from other companies, or purchase certified emission offsets such as those generated by forest carbon sequestration activities. However, future carbon prices are uncertain for a number of reasons, including varying predictions of the extent of climate change, the number of countries that are pursuing reductions in carbon emissions, the cost of reductions in GHGs, and innovations in energy production (Chen, 2003).

Carbon credits, either noted as emission allowances or as certified emission reductions, represent a right to emit one metric ton of carbon dioxide. As a result, those organizations that can reduce their regular emissions more easily and less expensively, or generate carbon offsets more efficiently, can sell their carbon credits to other organizations who find it harder to limit their emissions. This approach seeks to reduce the overall social costs associated with reducing emissions. An example of such an approach is the trading mechanism provided for by the Kyoto Protocol. The Clean Development Mechanism allows participants to meet part of their emission limits through investment in forest carbon projects that consist only of afforestation or reforestation activities. The Clean Development Mechanism of the Kyoto Protocol set binding constraints on carbon emissions (Chen, 2003). The Mechanism applies to participating countries, and was applicable only through 2012 which marked the end of the Protocol's first commitment period to reduce GHGs emissions. One expected drawback to the process was that reductions in emissions from participating countries would be offset by increases in emissions from non-participating countries (Kallbekken et al., 2007). In practice, developing Clean Development Mechanism forestry projects is quite complex and cumbersome; therefore, only a few projects have been developed to date. Interestingly, an organization in a developed country can undertake a carbon reduction project in an undeveloped country where its costs would usually be lower. As a result, the developed country would receive emission reductions whereas the host country would receive an investment in forestry and perhaps new technologies. Even though the Kyoto Protocol and the Clean Development Mechanism are widely known, there are several other carbon trading schemes, both mandatory and voluntary (Neeff et al., 2007).

In 2012, nearly 40 countries agreed to the second commitment period under the Kyoto Protocol lasting until 2020. This agreement (also known as the *Doha Amendment*), however, has to be accepted by 144 countries before it can come into force. Canada withdrew from the Kyoto Protocol in 2012 citing excessive economic costs as the reason, while Russia and Japan refused to commit to any new GHG emissions reductions targets. Nevertheless, international negotiations have continued efforts, under the framework of the United Nations Framework Convention on Climate Change, to reduce GHGs atmospheric concentrations after the second commitment period expires in 2020. The negotiations resulted in the development of a separate instrument and signing of the Paris Agreement in 2015. The agreement has been adopted by consensus by the representatives of 195 countries. Currently (2016) it is open for signature but has not yet entered into force.

While the Kyoto Protocol may be eventually heading into the sunset, its recognition of the role of forests in climate change mitigation as well as the implementation mechanisms developed, with all their successes and limitations, and in an incremental policy-making environment, continues to inform and influence the development of new solutions. The Paris Agreement (United Nations Framework Convention on Climate Change, 2015) states in its Article 5 that:

1. Parties should take action to conserve and enhance, as appropriate, sinks and reservoirs of GHGs as referred to in Article 4, paragraph 1(d), of the Convention, including forests.

2. Parties are encouraged to take action to implement and support, including through results-based payments, the existing framework as set out in related guidance and decisions already agreed under the Convention for: policy approaches and positive incentives for activities relating to reducing emissions from deforestation and forest degradation, and the role of conservation, sustainable management of forests and enhancement of forest carbon stocks in developing countries; and alternative policy approaches, such as joint mitigation and adaptation approaches for the integral and sustainable management of forests, while reaffirming the importance of incentivizing, as appropriate, noncarbon benefits associated with such approaches.

With this language, the Paris Agreement recognizes the important role of forests in climate change mitigation. Further, the agreement mentions the REDD+ (Reduced Emissions from Deforestation and Degradation) program indicating that tropical and subtropical countries can receive international funding if they succeed in reducing emissions from deforestation and degradation. In addition, several developed countries have promised significant financial resources to support REDD+ projects in developing countries.

VII. SELECTED US CARBON REPORTING AND TRADING SCHEMES

All carbon reporting and trading programs need a carbon registry that is established to record and track carbon emission and storage over time. The registry is used to assess progress in reducing carbon emissions, and is essential for any carbon credit market since it provides quantified and verified carbon offsets. As a result, some of the carbon offset schemes will consist only of a registry and a portfolio of carbon sequestration projects, whereas others will add a trading platform to facilitate buying and selling of carbon offsets. Certification of carbon credits could allow us to differentiate among products and allow the introduction of price premiums for certain types of projects and could increase both the supply and demand of carbon projects as confidence in the quality and quantity increases (Lichtenfeld, 2007).

The European Union Emission Trading Scheme, a Kyoto Protocol derived scheme, is the world's first and only mandatory carbon trading program, however, it has not included forest carbon credits. On the contrary, The Chicago Climate Exchange was the world's first voluntary, legally binding GHG reduction and trading system which included forest carbon credits. The Chicago Climate Exchange relied on independent verification and was active in trading emission reductions from 2003 through 2010, when it ceased its operations due to the lack of carbon market activity. An Exchange member who could not meet their own emission targets could purchase credits from other members who exceeded their emission reductions, or have verifiable offset projects. Forestry projects were eligible under Chicago Climate Exchange rules, and the Exchange accepted forestry emission reductions originating from forests established after 1990. These forests had to be managed sustainably and third-party forest certification was required. Individual projects were audited and reports prepared annually. Recognized carbon pools included above- and below-ground living tree biomass and long-lived wood products. The Chicago Climate Exchange required that 20% of credits registered annually were kept in reserve to cover potential shortages at the conclusion of a forestry project.

The California Climate Action Registry was a non-profit organization formed by the State of California in 2001. The registry was a voluntary GHGs registry that promoted early GHGs emission reduction efforts. The members of this registry voluntarily measured, verified, and reported their GHG emissions. They did so to prepare to participate in market-based solutions (e.g., cap and trade) and future regulatory requirements. Since the registry is purely a reporting mechanism, it does not currently involve trades of carbon offsets. The California Registry closed in 2010. Its activities are continued by the Climate Action Reserve, which was formed to continue the voluntary reporting, and to expand its coverage to all of North America.

In the southern United States, the state of Georgia established the Georgia Carbon Sequestration Registry, the first carbon registry in the region. The purpose of the registry is to encourage voluntary actions to reduce GHG emission, and to ensure that Georgia forest landowners received proper consideration for their management activities in the emerging carbon markets. The ownership of

carbon is not permanently tied to the land or trees, since sequestered carbon is treated as a separate commodity that can be traded independently of wood. Registered forests must be located in the state of Georgia, must be composed of native tree species, and must be managed in a way that is consistent with Georgia's Forestry BMPs.

VIII. FOREST CARBON IMPLICATIONS FOR FOREST MANAGEMENT

At present, the economic value of above-ground or below-ground forest carbon in carbon markets is relatively low, and as a result, the current impact of carbon sequestration on forest management planning is relatively low. One reason for this is that quantifying and verifying carbon offsets is a rather lengthy and expensive process, and low carbon values do not justify very active participation in the carbon offset trade. Further, at this point in time, all carbon trading programs in the United States are voluntary, thus landowners are encouraged, not forced, to participate. Should there eventually arise legally binding regulations that limit emissions, carbon offset values may increase substantially, therefore justifying more attention from forest managers and landowners.

Uncertainty still lingers over whether forest carbon offsets will be eligible as carbon credits under mandatory carbon emission reduction programs. The Kyoto Protocol allows for only a limited use of forest carbon offsets, but the European Union Emission Trading Scheme does not allow any at all. The future of forest carbon offsets under the Paris Agreement still needs to be determined. Interestingly, only voluntary carbon sequestration programs readily admit forest carbon offsets (Hamrick and Goldstein, 2015). The reasons for these differences are not based on biology, but rather on politics, thus the opportunities for (or impacts on) landowners will depend on how policy makers define forestry provisions in potential mandatory regulations.

Another current issue is the organization of the carbon market in the US and the complexity associated with the carbon market protocols. For example, some protocols involve complex accounting and reporting procedures. For landowners, the main questions may involve the type of carbon sequestration that would count as a certified carbon emission offset, and how their current forests can fit into the system. In addition, the level of carbon sequestered may depend on what happens further along in the wood supply chain (see Chapter 14: Forest Supply Chain Management). For example, the amount of carbon sequestered from an intensively managed southern US forest may depend on how the resulting products are treated in carbon accounting schemes. If credit for carbon storage in wood products is severely limited or not permitted at all, then there will be little forest carbon sequestration and trade activity in the southern US, unless, of course, carbon prices rise significantly. Though we have only a short price history of emission allowances and carbon offsets on which to base this, carbons credits eventually could generate income higher than that obtainable from other uses of the land. If this holds true, then we may expect substantial adjustments to forest management practices as landowners respond to rising carbon offset values.

IX. SUMMARY

Forest certification systems and forest carbon accounting systems may eventually become very important issues when planning the management of forests. Many, if not all, of the certification systems require a management plan to be developed. In addition, a management plan must address some form of resource sustainability, either of yields, multiple uses, or of the ecosystem. Certification systems also suggest a number of constraints on forest management planning that relate to the timing and location of management activities. These can be viewed as organizational policy constraints, since an organization generally enters into the certification agreements voluntarily. Forest carbon accounting may, in the future, be an important source of income for forest landowners. It has been suggested that the annual revenue from selling the carbon credits associated with a managed forest may exceed the revenue that could be realized from hunting leases. Assuming the registration process does not cost more than the revenue that might be obtained, this potential additional source of revenue for a landowner may further enhance the attractiveness of forest management investments.

QUESTIONS

1. *Chain of custody certification.* You recently have been hired as a marketing manager for a forest products company in north Florida. The company owns an extensive amount of timberland, and operates several processing facilities, including a paper mill, an oriented strand board (OSB) mill, and several lumber sawmills. The company sells its products in the domestic market (70%) as well as overseas in South East Asia (25%) and Europe (5%). While the company has certified its forests, the CEO recently attended a marketing presentation suggesting that the forest industry worldwide was making a major push to achieve chain of custody certification. Since chain of custody certification is a new concept to the CEO, you are called upon to "bring them up to speed." In a memorandum to the CEO, please address the following:
   a. What is chain of custody certification?
   b. What would chain of custody certification mean to our operations?
   c. What are the advantages and disadvantages of pursuing chain of custody certification?
   d. Should the company choose to engage in chain of custody certification, which certification system should it choose?

2. *Family tree farm certification.* As a county forester in southern Indiana, you assist a number of nonindustrial private landowners with their daily forest management needs. Some of these landowners have certified tree farms through the American Tree Farm System. One of the landowners is interested in understanding how the Tree Farm System differs from the other more recently developed certification systems. Choose one of the other forest certification systems, and develop a short memorandum for the landowner that describes the similarities and differences between it and the American Tree Farm System.

3. *Forest carbon sequestration and management planning.* The forestry organization you work for in California is interested in exploring the market potential for forest carbon offsets. Develop a short memorandum for your supervisor that details the current opportunities for pursuing trade in forest carbon offsets.

4. *Forest carbon trading.* The company that you work for in Georgia is interested in forestry projects that will yield carbon credits that, in turn, other companies can purchase from a carbon exchange. Describe in a memorandum to your supervisor the potential risks associated with entering forestry projects into a carbon trading market.

REFERENCES

Alig, R.J., 2003. US landowner behavior, land use and land cover changes, and climate change mitigation. Silva Fennica. 37 (4), 511−527.

American Forest Foundation, 2014. American Forest Foundation (AFF) 2015−2020 Standards of Sustainability. American Tree Farm System, Washington, DC, 27 p.

American Resources, Inc., 2013. Green Tag Certified Forestry. American Resources, Inc., Vienna, VA. Available from: http://www.greentag.org/default.asp (Accessed 6/14/2016).

Birdsey, R., 1992. Carbon Storage and Accumulation in the United States Forest Ecosystems. US Department of Agriculture, Forest Service, Washington, DC, General Technical Report WO-59. 51 p.

Canadian Standards Organization Group, 2016. CAN/CSA-Z809-16 Sustainable Forest Management. Canadian Standards Association Group, Toronto, ON, 84 p.

Chen, W., 2003. Carbon quota price and CDM potentials for Marrakesh. Energy Policy. 31 (8), 709–719.

Cossalter, C., Pye-Smith, C., 2003. Fast-Wood Forestry. Myths and Realities Center for International Forestry Research, Jakarta, Indonesia, 50 p.

Cubbage, F., Siry, J., Abt, R., 2005. Fast-grown plantations, forest certification, and the US South: Environmental benefits and economic sustainability. N. Z. J. For. Sci. 35 (2/3), 266–289.

Elliot, C., Schlaepfer, R., 2001. The advocacy coalition framework: Application to the policy process for the development of forest certification in Sweden. J. Eur. Public Policy. 8 (4), 642–661.

Enkvist, P., Naucler, T., Rosander, J., 2007. A cost curve for greenhouse gas reduction. McKinsey Quarterly. 2007 (1), 35–45.

Forest Stewardship Council, 2010. FSC-US Forest Management Standard (v1.0). Forest Stewardship Council US, Minneapolis, MN, 109 p.

Gallardo Gallardo, E., 2007. Environmental policy and law influencing forest management practices in Chile. In: Dubé, Y.C., Schmithüsen, F. (Eds.), Cross-Sectoral Policy Developments in Forestry. CABI, Oxfordshire, UK, pp. 231–236.

Hamrick, K., Goldstein, A., 2015. Ahead of the curve, state of the voluntary carbon markets 2015. Forest Trends' Ecosystem Marketplace, Washington, DC, 50 p.

Hansen, E., Forsyth, K., Juslin, H., 2000. Forest Certification Update for the ECE Region: Summer 2000. United Nations Economic Commission for Europe, Timber Section, Geneva, Geneva Timber and Forest Discussion Paper ECE/TIM/DP/20.

Hubbard, S.S., Bowe, S.A., 2005. Environmentally certified wood products: perspectives and experiences of primary wood manufacturers in Wisconsin. For. Products J. 55 (1), 33–40.

Kallbekken, S., Flottorp, L.S., Rive, N., 2007. CDM baseline approaches and carbon leakage. Energy Policy. 35 (8), 4154–4163.

Lauterbach, S., 2007. An assessment of existing demand for carbon sequestration services. J. Sustain. For. 25 (1/2), 75–98.

Lichtenfeld, M., 2007. Improving the supply of carbon sequestration services in Panama. J. Sustain. For. 25 (1/2), 43–73.

Mendell, B., Hamsley Lang, A., 2013. Comparing Forest Certification Standards in the US: Economic Analysis and Practical Considerations. EconoSTATS, George Mason University, Fairfax, VA, 31 p.

Neeff, T., Echler, L., Deecke, I., Fehse, J., 2007. Update on Markets for Forestry Offsets. The Tropical Agricultural Research and Higher Education Center (CATIE), Turrialba, Costa Rica, Manual no. 67. 35 p.

Ortolan, C., 2003. Some thoughts on FSC and environmental certification. Crow's For. Ind. J. 64 (March/April), 24–25.

Ozinga, S., 2004. Time to measure the impacts of certification on sustainable forest management. Unasylva. 219 (55), 33–38.

Rametsteiner, E., Simula, M., 2003. Forest certification—an instrument to promote sustainable forest management? J. Environ. Manage. 67 (1), 87–98.

Sedjo, R., Marland, G., 2003. Inter-trading permanent emissions credits and rented temporary carbon emission offsets: some issues and alternatives. Climate Policy. 3 (4), 435–444.

Straka, T.J., Cushing, T.L., 2015. McPhail Tree Farm, South Carolina, United States of America. In: Siry, J.P., Bettinger, P., Merry, K., Grebner, D.L., Boston, K., Cieszewski, C. (Eds.), Forest Plans of North America. Academic Press, New York, NY, pp. 87–96.

Sustainable Forestry Initiative, 2015. SFI 2015–2019 Forest Management Standard. Sustainable Forestry Initiative, Washington, DC, 12 p.

The Montréal Process, 2015. Criteria and Indicators for the Conservation and Sustainable Management of Temperate and Boreal Forests. 5th ed. Montréal Process Liaison Office, International Forestry Cooperation Office, Tokyo, September 2015. 31 p. Available from: http://www.montrealprocess.org/documents/publications/techreports/MontrealProcessSeptember2015.pdf (Accessed 5/28/2016).

Tikina, A., Kozak, R., Larson, B., 2008. What factors influence obtaining forest certification in the US Pacific Northwest? For. Policy Econ. 10 (4), 240–247.

United Nations, 1997. UN Conference on Environment and Development (1992). United Nations, Department of Public Information. Available from: http://www.un.org/geninfo/bp/enviro.html (Accessed 6/14/2016).

United Nations, 1998. Kyoto Protocol to the United Nations Framework Convention on Climate Change. United Nations, New York, NY, 20 p.

United Nations Economics Commission for Europe, 2015. ECE/FAO Forest Products Annual Market Review, 2014–2015. Geneva Timber and Forest Study Paper 29, ECE/TIM/SP/39. United Nations Economic Commission for Europe, Timber Section, Geneva, 120 p.

United Nations Framework Convention on Climate Change, 2015. Adoption of the Paris Agreement. Conference of the Parties, Twenty-first session, Paris, 30 November to 11 December 2015. Available from: https://unfccc.int/resource/docs/2015/cop21/eng/l09.pdf (Accessed 5/17/2016).

Vidal, N., Kozak, R., Cohen, D., 2005. Chain of custody certification: an assessment of the North American solid wood sector. For. Policy Econ. 7 (3), 345–355.

Chapter 16

Scenario Analysis in Support of Strategic Planning

Objectives

Large uncertain events like fires, hurricanes, and insect outbreaks, or significant social or economic volatility can limit the usefulness of the outcomes of traditional strategic forest planning processes. Strategic planning is necessary, however, to demonstrate that the management of forests can result in sustainable systems and outcomes. Unfortunately, strategic planning models are proxies for real systems, and these models are unable to describe all of the intricacies of real-world systems. In this chapter, we introduce scenario analysis, a set of techniques that can assist with the development of a learning environment within an organization, which in turn can be used to facilitate continual reflection and discussion about possible futures. We also introduce a few processes employed to create scenarios, or alternative plausible futures, that may support strategic plans. Therefore, at the completion of this final chapter, you should be able to:

1. Understand why scenario analysis is important.
2. Understand the basic concepts for creating scenarios.
3. Explain the process for evaluating alternative futures.

I. INTRODUCTION

Throughout this book we have described mathematical models that might be used to develop forest plans. These models required technical coefficients and assumptions that may have seemed to be highly certain and predictable. However, embedded in these assumptions and technical coefficients is a level of error and uncertainty that cannot be explained. Further, in the evaluation of trade-offs among alternative forest plans, both objectively determined technical coefficients and subjective values concerning the future may need to be assessed to acknowledge certain preferences and relationships associated with the world in which we live and work (Nordström et al., 2013). Unfortunately, some of these relationships likely cannot be incorporated into a mathematical representation of the management of land and forests, and thus certain future conditions will be difficult to predict. For example, 30 years ago few people could have predicted that most of the vertically integrated, publicly-owned and traded timber companies in the United States would no longer exist today. Forest products companies such as Willamette Industries, Bohemia, Champion International, Westvaco, Union Camp, Federal Paper Board, and Continental Can (among others) have all been purchased by, or combined into, other companies that then separated the forests owned from the mills managed. Thirty years ago, few could have also predicted that federal forests in the Pacific Northwest (US Forest Service and US Bureau of Land Management), which were providing almost 4 billion board feet of harvested wood in Oregon alone at the time, would provide only about 12% of this harvest level today. The economic, ecological, and social developments of the last three decades were likely too difficult to understand or to believe as plausible 30 years ago.

Much of the purpose of this book has been to describe various mathematical models that can be used to assist in forest planning efforts. These models perform a variety of analytical computations to predict yields of timber, future markets, and wildlife habitat and to present a structured approach to decision making for their management. They are often used in practice to schedule a set of activities that seek to satisfy the constraints associated with managing forests while optimizing the achievement of a goal. These types of models are used in the preparation of strategic or tactical forest plans. Perhaps it was only implied, but these models also often rely on past experiences as representations of future outcomes. For example, market models are often developed using historical data to predict prices or costs. And growth and yield models are developed from tree measurements that represent growth of forests that occurred during past climatic conditions. These models are of value in the sustainable management of forests, as they provide guidance to field foresters and

Forest Management and Planning. DOI: http://dx.doi.org/10.1016/B978-0-12-809476-1.00016-3

landowners. Yet the challenge that some forest planners and managers may face is the development of plans that involve assumptions that are no longer applicable.

Looking forward, forest planners may need to think deeper about the economic, ecological, and social conditions that might be encountered in the future. These revelations may limit the usefulness of traditional planning approaches that rely heavily on historical data and trends. One good example some planners currently face involves modeling the impact of climate change on the sustainable management of forests. A number of aspects of the traditional forest planning process may need to be changed to accommodate assumptions regarding climate change, from the growth of trees to assumptions of the intensity and frequency of natural disasters (Bettinger et al., 2013). Some example questions that planners and managers might ask themselves include:

- Will there be an introduction of an exotic pest that may devastate our forests?
- Will the use of wood products with associated favorable life-cycle assessments increase in commercial buildings?
- Will there be a technological change that can alter the type of wood fibers needed in the future?
- Will there be large social changes that create more or less demand for wood products, or that will increase or decrease the nonmarket services desired from our forests?
- What role does increasing environmental awareness (through the Internet or social media, perhaps) play in the development of forest regulations and practices?

These types of questions are difficult to incorporate into strategic and tactical planning models that rely on fairly predictable knowledge. But you might recall from previous chapters that the goal of a strategic plan is to create a competitive strategy for a forest organization. The strategic plan should therefore allow an organization to distinguish itself from other organizations operating in the same business arena. Some organizations may emphasize the efficiencies associated with management of forests, while other organizations may emphasize the products and habitat that their management system produces. For example, the Newton family in western Oregon has differentiated themselves from other private landowners by growing larger, mature trees that produce logs that ultimately received a market premium (Newton, 2015). Obtaining this type of distinction may have been necessary to obtain higher prices from local mills, but the strategy may have been viewed as a risky venture when it was adopted over 50 years ago. As a result, many organizations have found that formal or informal assessments of alternative scenarios can be a useful for addressing uncertain futures.

II. AN OVERVIEW OF THE ROLE OF SCENARIO ANALYSIS

Although we discussed them earlier in this book in association with alternative constraint sets of forest plans, here *scenarios* are considered synopses of possible alternative courses of action (Klosterman, 2013). They may consist of short narratives or stories concerning future outcomes of forest management, and they may be complemented by maps and other illustrative information. *Scenario analysis* is thus used to translate complex and uncertain relationships into understandable narratives for consumption by broad audiences (Johnson et al., 2016). Scenarios are developed to assist with the examination of potential complex changes in economic, ecological, and social conditions associated with the management of forests. They are therefore developed to further help us understand how the future will unfold, and how it might be different from current or past economic, ecological, and social conditions.

Three approaches are commonly used to develop strategic forest plans: the rational approach, the evolutionary approach, and the process approach (van der Heijden, 2005). The rational approach, as we suggested in Chapter 1, Management of Forests and Other Natural Resources, assumes that there is an ideal forest plan, and the goal is to develop it using the best obtainable information. This approach works sufficiently in a predictable environment where past outcomes are useful indicators of future outcomes. The use of FORPLAN (Iverson and Alston, 1986) for the development of US National Forests was an example of the rational approach to strategic planning. These planning efforts generally had the goal of optimizing net present value over a 100-year strategic planning problem. The solutions to the problems were the basis for land allocation of activities within national forests of the US for the latter half of the twentieth century. Technical coefficients for all outcomes were estimated based on current knowledge and models (e.g., growth and yield models) at the time of the development of each forest plan. However, dramatic changes in the operating environment, due to increases in fire and other natural disturbances, along with additional protections for threatened and endangered species, may have limited the usefulness of the original plans.

The evolutionary approach allows a strategic plan to develop within an organization in response to constraints that are encountered while operating, instead of being guided by goals (van der Heijden, 2005). This approach places a great amount effort on consensus-seeking behavior within an organization. Decisions are made through a serial process involving reactions to the changes in the physical or social environment. One may argue that the current (2012) forest planning rules used by the US Forest

Service are reflective of an evolutionary approach to strategic planning. These rules contain broad guidelines, such as seeking ecosystem resiliency and supporting the economic vitality of rural communities, and use collaborations as key elements in decision making. However, the evolutionary approach often lacks clearly defined goals to influence the organizational performance, and it is often fragmented due to the nature of plan development and the ever-changing sense of consensus among the decision makers. As it related to the management science decision-making process described in Chapter 1, Management of Forests and Other Natural Resources, this approach could reflect more closely the semirational or garbage can models, and is reflective of the adaptive management of forests.

Similarly, the process approach emphasizes that while an organization is following a plan of action, it can intervene in the future, with the goal of improving opportunities for success (van der Heijden, 2005). Thus, one is not simply looking for the right answer in a static world (as in the rational approach), nor reacting to changing environment (as in the evolutionary approach), but creating a *learning loop* (Fig. 16.1). The process approach is one where an organization has a goal of developing the intellectual capacity to continuously react to changes in the operational environment. The process begins by acknowledging the experiences of the group (the planning team) as the basis for this understanding. Then through observations and reflections, changes occur that create the foundation for alternative future states. New theories are created that integrate the knowledge gained through observations and reflections with past experiences. These are then tested, and may create concrete experiences among the planning team. This is the approach that we emphasize in this chapter.

As with linear programming, the process approach originated with military applications. Kahn has been credited with introducing it to business applications in the 1960s, and he developed a procedure called *future-now-thinking* that combined analytical methods with an imagination of the future (Ramírez and Wilkinson, 2016). At the time, similar work was being completed at Stanford Research Institute and Hudson Institute (Ramírez et al., 2010) that allowed elements of large social change to be included in planning processes (van der Heijden, 2005). To illustrate the process approach with a hypothetical example, imagine that a learning loop can be used to develop a strategy for a pulp mill. The concrete experiences of the planning team suggest that paper demand is often a function of many variables, such as per capita income within a geographic region, as shown in the Papyrus model (Gilless and Buongiorno, 1987). However, the rise of digital media continues to erode the demand for printing and writing paper products, as smartphones are rapidly becoming adopted in countries such as China. These represent the team's reflection and observations. The team then creates a new theory that as developing economies grow, there will be an increase in the demand for pulp and paper, but not at a rate as rapid as in the past, due (again) to digital media. This represents the formation of a new theory. If this is observed, a new concrete experience is developed. Thus, the learning loop creates the structure for the organization to intertwine deliberations and actions (van der Heijden, 2005). Scenarios play a critical role in all aspects of the learning loop. They act as a mental device, since their short descriptions are efficient for transmitting information to a large group of people. Diverse opinions can then be incorporated into the discussion, and the scenarios thus allow an organization to reflect on new perspectives of the world outside the organization, which can lead to a more robust decision making environment (van der Heijden, 2005).

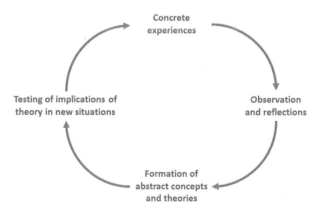

FIGURE 16.1 The learning loop as part of a process approach to strategy. *Adapted from van der Heijden, K., 2005. Scenarios: The Art of Strategic Conversation, 2nd edn. John Wiley & Sons, Ltd., West Sussex.*

Example

Royal Dutch Shell (Shell) was one of the original developers and practitioners of scenario analysis, and has used the process since the mid-1960s. In the early 1970s, oil consumption was very predictable, with growth in worldwide demand of around 6%, a pattern consistent since the late 1940s (van der Heijden, 2005). This represents the concrete experiences shared by most analysts in the learning loop. However, the leader of Shell's scenario analysis team challenged the team to consider deeper questions. They wanted to know more about the organizations who had control over the oil supply, especially those situated in countries of the Middle East. This stage in the learning loop represents the observation and reflections component. Through this analysis they realized that the supply of oil from the Middle East might be tenuous. They found that some oil-producing organizations might not have wanted to continue to meet the demand for oil when it extended beyond their own revenue needs. Some organizations also had a significant amount of control (leverage) on the oil supply, and could affect the price of oil through curtailments of the supply. The formation of a new theory was thus developed, one that limited oil supply to satisfy the revenue needs of oil-producing countries, and several new scenarios were developed, one of which became known as the *crisis scenario*. The scenarios were tested and analyzed, and a new set of experiences were created for a set of plausible futures. When an oil supply crisis actually occurred in 1973, Shell was much better prepared to adapt to the future because it had engaged in this exercise of thinking about the future, instead of relying on forecasts of the future from past experiences. They were able to quickly make changes in their investment policies and refining capacity due to oil shortages. Most of the rest of the oil industry waited for a year to respond the changing oil supply (and price) environment. For example, there was a continued investment in oil tankers by many companies, an assumption aligned with the view of a continually increasing oil supply. This resulted in an over-capacity of the tanker fleet, and shipping rates consequently remained depressed for nearly a decade. As a result of the scenarios developed, Shell was able to rapidly change its strategy and out-perform its competitors by large margins during this period of time (van der Heijden, 2005).

Example

In South Asia, the Republic of India faced tremendous changes in the first decade of the 21st century. They experienced rapid economic growth, yet they faced significant uncertainty with regard to the agricultural sector. The average farm size at the time was less than 3.7 acres (1.5 ha), and employed two-thirds of the country's population under a complex set of production regulations (Ramírez et al., 2010). Yet with greater wealth among the urban populace, significant income inequalities were appearing between the urban and rural sectors. The Indian Ministry of Agriculture, Indian Council of Agricultural Research, and the World Bank were the parties most interested in developing a strategy for the agricultural sector. Over a 2-year period, an iterative approach was employed that used interviews and brainstorming sessions to collect the various concrete experiences, observations, and reflections of stakeholders in the agricultural sector. Significant effort was needed to reach an agreement on the valid experiences and reflections given the large and diverse group involved. However, four scenarios were developed to support new strategies for the agricultural sector (Ramírez et al., 2010). The scenarios represented plausible futures, but the goal of the process approach was to develop a strategy to be able to intervene in the future, if necessary.

The *valley scenario* relied on protectionism and government intervention for the Indian agricultural sector. The outcomes suggested that productivity would remain low and economic growth would not occur in the agricultural sector. The *edge scenario* assumed economic development was the number one priority for the agricultural sector. It represented a free-market approach, and allowed planners to explore the tolerance of Indian society with regard to the inequities between those able to prosper and those struggling to prosper. The *mountain scenario* included a number of crises in the contextual environment such a conflicts with Pakistan and China, and impacts of climate change. The outcomes suggested than an increase in agricultural productivity was one way to react to these crises, but with strong government interaction that included relocation of some rural people. The *hill scenario* attempted to balance the level of government interaction and productivity improvements with the goal of creating a foundation for a market economy. Each of these scenarios were viewed as plausible futures, and new theories regarding how people would respond to these new operating environments were stimulated during the scenario analysis process.

The benefit of the process approach is that it provides a structural framework for analyzing uncertainty, and helps an organization to develop the capacity to adequately adjust their strategic plan. These adjustments may involve intervention in forest management, so that an organization can react more quickly to changes, when they occur. In the Shell example, they hoped to be able to respond more rapidly to dynamic changes in the Middle East oil supplies. With regard to forestry, one could argue that a similar analysis would have been useful in the 1980s for those forest management organizations in the US Pacific Northwest that relied on federal wood supplies. Of course, the scenario analysis process would have needed to occur prior to the northern spotted owl (*Strix occidentalis caurina*) crisis that disrupted and ultimately eliminated a significant portion of the wood supply in the region. In the India example, various levels of government intervention were involved in the economic development of a large sector of the Indian economy.

These approaches to scenario analysis have not been widely applied to forestry problems, but given the long-term nature of forest management and public interest in forestry activities, the potential gain from applying scenario analysis to forest planning problems may be good.

III. DEVELOPING SCENARIOS

Scenario analysis can help planners and managers develop context for important policy decisions, and can facilitate the involvement of people (internal and external to the organization) in a planning process (Johnson et al., 2016). In pursuing a scenario analysis process, planners and managers should recognize what can be controlled within a forest system, and what cannot be controlled. The prediction of long-term forest outcomes is difficult, and forecasting methods need to anticipate the uncertainties that may arise from assumptions concerning the future and the use of incomplete, incorrect, or obsolete data (Klosterman, 2013). That said, there are many ways to develop the scenarios. Schoemaker (1993), for example, suggested a 10-step approach to developing scenarios; other approaches have been offered (Ramírez and Wilkinson, 2016). Our approach uses seven steps:

1. Examine the strategic landscape
2. Look forward and backward in time
3. Identify drivers of change and plot them in a two-dimensional graph as contextual factors
4. Group the drivers of change
5. Position the groups of contextual factors with respect to broad outcomes
6. Develop scenarios
7. Create narratives that illustrate the scenarios

Step 1 begins with an examination of the strategic landscape (Fig. 16.2). At the core of this model is the *scenario planning learner*, or perhaps the *client*. The transactional environment are those actors that interact with the client. One may consider the transactional environment to be the supply chain for a planner's business or organization. The contextual environment is the larger environment for the organization; in forestry terms it represents the broader supply chain, and is uncontrollable by the scenario planner. The learning process focuses on the identification of factors (*drivers of change*) from the contextual

environment that can cause changes that alter the transactional environment. These factors are those that cause disruptions or volatility in the transactions between the scenario planner's organization and others in the transactional environment. The next step (Step 2) is to look forward and backward in time. Backward views help identify potential causes of uncertainty and volatility in the business environment based on historical evidence. Forward views require thinking about what the future may hold. This may involve reflective thought within the group, as there will often be disagreement, but when people share their views about the future, significant learning opportunities in an organization may happen. In Step 3, the drivers of change are identified, and plotted in a two-dimensional graph as *contextual factors* that address the two most important uncertainties (the transactional environment and outcomes of the plan), which are treated as axes of polar opposites (Fig. 16.3).

Step 4 involves grouping the drivers of change; typical groupings include social, technological, economic, environmental, and political aspects of the business environment. Using two axes representing potential outcomes of the strategic plans, the groups are positioned in Step 5 into one of four possible quadrats (Fig. 16.4). From the

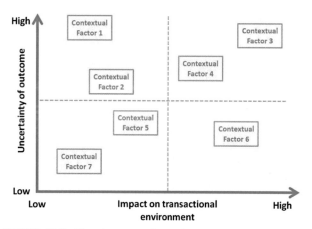

FIGURE 16.3 The placement of groupings contextual factors on a 2×2 matrix.

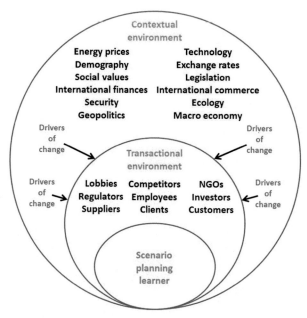

FIGURE 16.2 The strategic landscape. *Adapted from Ramírez, R., Selsky, J.W., van der Heijden, K., 2010. Business Planning for Turbulent Times: New Methods for Applying Scenarios. Earthscan, London. 336 p.*

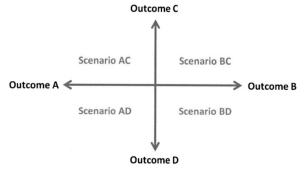

FIGURE 16.4 The grouping of the factors based on the combination of outcomes into scenarios.

groupings, the scenarios are developed (Step 6). The scenarios are then presented as narratives that describe alternative futures. Again, much learning can occur through the grouping as it is the time for discussions among the group, as there is no set formula to perform this task. Thus, scenario AC will use the information from contextual factors 1 and 2, while scenario BC will use contextual factors 3 and 4. Scenario BD will factor 6 and scenario AD will use factors 5 and 7. The last step (Step 7) involves creating the narratives that illustrate the scenarios. The descriptions of scenarios are not extensive, but are designed to place the organization in the context of alternative futures. They provoke conversation about the future and promote learning within an organization. The narratives test and refine the plausibility of scenarios, and they may be able to create greater understanding within an organization. The scenarios are continually refined, which becomes easier as each is developed. With refinement it may also be easier to view the inconsistencies that arise (Johnson et al., 2016).

IV. APPLYING SCENARIO ANALYSIS TO FOREST PLANNING

In applying scenario analysis to strategic forest planning efforts, the models employed may need to be reduced in size and complexity to allow planners and managers to explore potential futures that may result from high levels of uncertainty (Mohren, 2003). These adjustments should be made while maintaining sufficient structure and causality, within the models and among the relationships associated with model variables. In those rare instances where uncertainties within a forest system are recognizable, planners and managers may be able to speculate about future forest management conditions. Since it involves the integration of policy analysis and operations research (soft and hard skills), significant effort may be necessary for understanding how to integrate and use the value-laden objectives of stakeholders and decision makers.

A simple forestry scenario is used to illustrate the process further. Suppose you are a small forest landowner in the western United States, and your land is surrounded by adjacent forests that are managed by a public agency. You have developed a long-term harvest scheduling plan that provides sustained yields for the next 100 years. However, you are concerned about activities on the federal lands, particularly the potential increase in fire frequency and severity on these lands and the potential of fire to spread onto nearby private lands. You realize that your sustained yield calculations do not acknowledge these potential losses, as they are too uncertain. You now also believe that in the future larger and more frequent fires may occur, making modeling then challenging, if not

impossible (Bettinger, 2010). Therefore, you prepare a set of scenarios for the management of the forest, under the assumption of increasing fire frequency and severity on the surrounding public lands. Your first task is to create a strategic landscape (Fig. 16.2) where one driving factor moving from the contextual environment to the transactional environment involves increases in fire threats from surrounding properties. Another driving factor may be the reduction in closed-canopy forests on nearby properties due to increases in fire severity and frequency. Working only in the upper right corner of Fig. 16.3, these factors both contain a high degree of uncertainty and a high impact on the transactional environment. Thus, the scenario, *fire on the mountain*, is created.

The scenario might begin by describing the actions that could be used to prevent and limit the spread of fire onto your property. The deployment of green fuel breaks might be used to mitigate much of the potential damage arising from fires spreading onto the property. The development and maintenance of bare earth fire breaks might also be used in this regard. There are, of course, direct impacts on your forest (tree death) if a fire spreads to your property. Once fires occur in your watershed, sediment generated from burned landscapes may result in limits on activities, which may force a reduction of harvest levels. The loss of closed-canopy forests on nearby properties may result in the movement to your property of many animals that preferred those forests. Some of these species may be considered threatened or endangered, and harming their habitat may be considered an illegal activity. This can further limit harvest opportunities on your property. In the full scenario analysis, you would analyze these potential actions and reactions to determine the potential loss in harvest revenue from these activities and restrictions.

The scenario suggests that increases in fire frequency and severity in the landscape reflects not just the direct loss of trees, but also an increase in regulatory pressure, which may limit other activities from being performed on the property. The question then becomes how you might adjust your strategic plan to lessen the impacts on harvest levels. One course of action may be to address treatable sources of sediment in the forest to reduce the opportunity to restrict future activities. Another course of action may be to consider developing agreements (e.g., Habitat Conservation Plans) with public agencies prior to listing of species as threatened and endangered, so that additional restrictions will not be imposed should the species become listed. Thus, as a small forest landowner, you may need to seek advice from wildlife and planning specialists to determine the impacts of these options and to provide suggestions for the best courses of action to pursue. The scenario analysis allows you, as a small forest landowner, to think more broadly, and plan strategically, for potential losses of trees due to fire.

V. SUMMARY

Scenario analysis involves approaches that support strategic planning efforts. They are useful devices for dealing with uncertain or potentially turbulent futures that are not well supported by outcomes of forecast models developed using historical information. Scenario analysis does not look for the one optimal answer (or forest plan, in our case), but allows an organization to create a set of plausible futures to test strategies against. The potential advantage of using scenario analysis is that it may prepare an organization better for future stochastic events by creating an environment where learning can occur within the organization. There is no standard approach to scenario analysis, but the successful approaches are ones that integrate learning throughout the process of describing the plausible futures.

QUESTIONS

1. *Sources of uncertainty.* What are some sources of uncertainty that can affect the management of a large forest in your region? How can these sources of uncertainty change the outcomes of strategic and tactical plans that are developed for managers of the forest?
2. *Scenario analysis for a timber company.* Select a forest company in your region and describe the strategic operational landscape within which it operates. Describe the likely sources of volatility and uncertainty that surround that business. Finally, describe how these factors can impact the management of the company. Create a scenario, set in the future, that illustrates how these factors may be influential in how the company will operate. Using the previous chapters as guides, discuss the analyses necessary that might support the development of a strategic plan that describes the best course of action for the company.
3. *Scenario analysis for a public forest.* Select a public forest in your region and prepare the strategic operational landscape within which it operates. Describe the likely sources of turbulence and uncertainty that surround the government organization. Further, describe how these factors can impact the management of the public land. Then create a scenario, set in the future, that illustrates how these factors may be influential in how the public land is managed. Finally, using the previous chapters as guides, discuss the analyses necessary that might support the development of a

strategic plan that describes the best course of action for the public forest.
4. *Scenario analysis as a general idea.* You find yourself employed as a planner for a large forestry organization. One day your supervisor suggests that perhaps the organization might need to experiment with scenario analysis, although they admit that they know little about the process. In a short memorandum, describe the general steps that might be used to implement scenario analysis within a forestry organization.

REFERENCES

Bettinger, P., 2010. An overview of methods for incorporating wildfires into forest planning models. Math. Comput. For. Nat. Resour. Sci. 2 (1), 43–52.

Bettinger, P., Siry, J., Merry, K., 2013. Forest management planning technology issues posed by climate change. For. Sci. Technol. 9 (1), 9–19.

Gilless, J.K., Buongiorno, J., 1987. PAPYRUS: a model of the North American pulp and paper industry. For. Sci. Monogr. 28, 37 p.

Iverson, D.C., Alston, R.M., 1986. The Genesis of FORPLAN: A Historical And Analytical Review of Forest Service Planning Models. US Department of Agriculture, Forest Service, Intermountain Research Station, Ogden, UT, General Technical Report INT-214. 31 p.

Johnson, M.L., Bell, K.P., Teisl, M.F., 2016. Does reading scenarios of future land use changes affect willingness to participate in land use planning? Land Use Policy. 57, 44–52.

Klosterman, R.E., 2013. Lessons learned about planning. J. Am. Plan. Assoc. 79 (2), 161–169.

Mohren, G.M.J., 2003. Large-scale scenario analysis in forest ecology and forest management. For. Policy Econ. 5 (2), 103–110.

Newton, M., 2015. Eddyville Tree Farm, Oregon, United States of America. In: Siry, J.P., Bettinger, P., Merry, K., Grebner, D.L., Boston, K., Cieszewski, C. (Eds.), Forest Plans of North America. Academic Press, New York, NY, pp. 11–19.

Nordström, E.-M., Holmström, H., Öhman, K., 2013. Evaluating continuous cover forestry based on the forest owner's objectives by combining scenario analysis and multiple criteria decision analysis. Silva Fennica. 47 (4), article id 1046. 22 p.

Ramírez, R., Wilkinson, A., 2016. Strategic Reframing: The Oxford Scenario Planning Approach. Oxford University Press, Oxford, UK, 272 p.

Ramírez, R., Selsky, J.W., van der Heijden, K. (Eds.), 2010. Business Planning for Turbulent Times: New Methods for Applying Scenarios. Earthscan, London, 336 p.

Schoemaker, P.J.H., 1993. Multiple scenario development: its conceptual and behavioral foundation. Strategic Manage. J. 14 (3), 193–213.

van der Heijden, K., 2005. Scenarios: The Art of Strategic Conversation. 2nd ed. John Wiley & Sons, Ltd, West Sussex, UK.

Appendix A

Databases Used Throughout *Forest Management and Planning*

Many of the examples provided in this book used one of three sets of data: (1) a stand-level projection of a single Douglas-fir (*Pseudotsuga menziesii*) stand located in western Oregon, (2) current and projected conditions of the Lincoln Tract, a western United States coniferous forest, and (3) current and projected conditions of the Putnam Tract, a southern United States pine and hardwood forest. This Appendix provides some of the raw data associated with each of these three databases. In the case of the Douglas-fir stand, the data provided are all that you may need to follow the discussion in the book, and to complete the questions that pertain to the use of the data. In the cases of the Lincoln Tract and the Putnam Tract, you may want to utilize the associated geographic information system databases that can be accessed from the book's website (http://booksite.elsevier.com/9780123743046/).

A.I. A DOUGLAS-FIR STAND FROM WESTERN OREGON

We selected a single Douglas-fir stand to illustrate the development and projection of stand-level conditions through time. The initial inventory data for the stand begins with age 15 (Table A.1), and stand development is tracked through time until stand age 110 (Tables A.2–A.20). The stand was projected through time using the ORGANON growth and yield model (Hann et al., 1997). No intermediate treatments were applied to the stand during the projection period.

A.II. THE LINCOLN TRACT

The Lincoln Tract (Fig. A.1) is a coniferous forest located in the western United States. As we mentioned in Chapter 3, Geographic Information and Land Classification in Support of Forest Planning, the forest is contiguous, and is composed of 87 stands covering 4550.3 acres (1841.5 hectares). At the time that the data was developed for this property, Douglas-fir stands

covered most of the area (about 94%), and these undoubtedly contained a minor percentage of western hemlock (*Tsuga heterophylla*) and other conifers. Some mixed conifer and hardwood stands (about 6% of the area) were also present on the southern side of the tract and along or near the stream system. A ridge crosses the tract from east to west, thus the stream system drains southward and northward from the center of the tract. About 19.9 miles (32.1 km) of intermittent and perennial streams are contained within the tract itself. Some basic stand-level data associated with the Lincoln Tract is provided in Table A.21.

A.III. THE PUTNAM TRACT

The Putnam Tract (Fig. A.2) is a pine and hardwood forest located in the southern United States. As with the Lincoln Tract, we noted in Chapter 3, Geographic Information and Land Classification in Support of Forest Planning, that the Putnam Tract consists of 81 timber stands covering 2602 acres (1053 hectares) in a contiguous block. At the time that the data was developed for this property, pine plantations of various ages covered about 53% of the tract, and natural pine stands comprised about 25% of the forests. Some mixed pine and hardwood forests are present on the tract, but their extent is limited. Hardwood stands occupy most of the lowlands along the streams, and account for about 17% of the area. Numerous streams are intermixed throughout the tract, all draining into a single main stem running from the southwestern portion of the tract through the northeastern portion of the tract. About 11.8 miles (19.1 km) of intermittent and perennial streams can be found within the tract itself. Some of the basic data for the Putnam Tract stands can be found in Table A.22.

REFERENCE

Hann, D.W., Hester, A.S., Olson, C.L., 1997. ORGANON User's Manual. Department of Forest Resources, Oregon State University, Corvallis, OR.

TABLE A.1 Stand Table for a Douglas-Fir Stand, Age 15

DBH Class (inches)	Trees per Acre	Average Height (feet)	Snags per Acre
2	29.0	19.7	0.9
3	75.0	24.5	1.0
4	91.5	27.8	1.1
5	253.5	31.1	0.9
6	16.5	42.0	–
Total	465.5	–	3.9

Average DBH (inches): 4.2
Quadratic mean DBH (inches): 4.4
Basal area per acre (ft^2): 50.1
Volume per acre (Scribner, MBF): 0
Volume per acre (ft^3): 375

TABLE A.3 Stand Table for a Douglas-Fir Stand, Age 25

DBH Class (inches)	Trees per Acre	Average Height (feet)	Snags per Acre
2	–	–	0.6
3	–	–	1.0
4	14.5	41.6	2.2
5	36.9	45.9	2.1
6	67.4	48.8	0.3
7	97.5	52.8	0.2
8	62.3	55.0	–
9	18.7	60.1	–
10	9.2	67.1	–
11	1.8	67.1	–
Total	308.3	–	6.4

Average DBH (inches): 6.8
Quadratic mean DBH (inches): 7.0
Basal area per acre (ft^2): 81.8
Volume per acre (Scribner, MBF): 3.383
Volume per acre (ft^3): 1723

TABLE A.2 Stand Table for a Douglas-Fir Stand, Age 20

DBH Class (inches)	Trees per Acre	Average Height (feet)	Snags per Acre
2	–	–	1.0
3	13.5	30.6	1.1
4	63.8	35.2	1.2
5	138.8	36.3	1.1
6	130.6	41.3	0.1
7	21.2	45.7	–
8	11.0	55.2	–
9	5.5	55.2	–
Total	384.4	–	4.5

Average mean DBH (inches): 5.4
Quadratic mean DBH (inches): 5.5
Basal area per acre (ft^2): 62.7
Volume per acre (Scribner, MBF): 0.653
Volume per acre (ft^3): 867

TABLE A.4 Stand Table for a Douglas-Fir Stand, Age 30

DBH Class (inches)	Trees per Acre	Average Height (feet)	Snags per Acre
2	–	–	0.2
3	–	–	0.9
4	–	–	2.2
5	4.5	51.2	2.2
6	29.1	54.9	0.6
7	47.2	58.7	0.5
8	54.2	62.1	0.1
9	86.1	64.5	0.1
10	56.7	66.5	–
11	14.4	72.8	–
12	9.2	77.9	–
13	1.8	77.9	–
Total	303.2	–	6.8

Average DBH (inches): 8.6
Quadratic mean DBH (inches): 8.7
Basal area per acre (ft^2): 125.5
Volume per acre (Scribner, MBF): 8.531
Volume per acre (ft^3): 3006

TABLE A.5 Stand Table for a Douglas-Fir Stand, Age 35

DBH Class (inches)	Trees per Acre	Average Height (feet)	Snags per Acre
2	–	–	0.1
3	–	–	0.5
4	–	–	1.9
5	–	–	2.0
6	4.4	60.3	0.8
7	23.5	63.8	0.7
8	44.5	68.4	0.3
9	42.0	71.1	0.3
10	74.0	74.0	0.1
11	65.5	75.4	–
12	29.2	78.7	–
13	11.1	84.5	–
14	5.5	87.6	–
15	1.8	87.6	–
Total	301.5	–	6.7

Average DBH (inches): 9.9
Quadratic mean DBH (inches): 10.0
Basal area per acre (ft^2): 166.1
Volume per acre (Scribner, MBF): 15.050
Volume per acre (ft^3): 4534

TABLE A.6 Stand Table for a Douglas-Fir Stand, Age 40

DBH Class (inches)	Trees per Acre	Average Height (feet)	Snags per Acre
3	–	–	0.2
4	–	–	1.7
5	–	–	1.8
6	0.8	65.6	1.0
7	4.5	70.0	0.9
8	29.3	72.8	0.8
9	37.8	77.0	0.7
10	38.1	80.1	0.4
11	70.3	83.1	0.3
12	42.3	83.6	0.1

(Continued)

TABLE A.6 (Continued)

DBH Class (inches)	Trees per Acre	Average Height (feet)	Snags per Acre
13	48.9	86.1	–
14	15.6	91.2	–
15	5.4	96.4	–
16	5.5	96.4	–
Total	298.5	–	7.9

Average DBH (inches): 11.0
Quadratic mean DBH (inches): 11.2
Basal area per acre (ft^2): 204.7
Volume per acre (Scribner, MBF): 22.551
Volume per acre (ft^3): 6194

TABLE A.7 Stand Table for a Douglas-Fir Stand, Age 45

DBH Class (inches)	Trees per Acre	Average Height (feet)	Snags per Acre
3	–	–	0.1
4	–	–	1.5
5	–	–	1.6
6	–	–	1.1
7	1.1	73.8	0.9
8	18.1	77.3	1.8
9	24.2	82.6	1.6
10	31.7	85.1	1.0
11	33.7	88.2	0.8
12	67.2	91.1	0.3
13	36.5	91.9	0.1
14	46.4	93.5	0.1
15	18.3	96.8	–
16	10.9	103.4	–
17	4.2	104.1	–
18	1.2	105.5	–
Total	293.5	–	10.9

Average DBH (inches): 12.0
Quadratic mean DBH (inches): 12.2
Basal area per acre (ft^2): 239.3
Volume per acre (Scribner, MBF): 29.881
Volume per acre (ft^3): 7836

TABLE A.8 Stand Table for a Douglas-Fir Stand, Age 50

DBH Class (inches)	Trees per Acre	Average Height (feet)	Snags per Acre
4	–	–	0.9
5	–	–	1.3
6	–	–	1.0
7	1.1	79.0	0.9
8	3.6	81.9	2.6
9	24.4	85.1	2.4
10	28.4	91.3	1.7
11	28.9	93.8	1.5
12	31.5	95.5	0.6
13	54.4	98.6	0.5
14	35.4	99.1	0.2
15	34.4	100.6	0.1
16	28.2	103.3	–
17	7.4	107.3	–
18	9.0	111.5	–
19	1.8	111.5	–
Total	288.5	–	13.7

Average DBH (inches): 12.9
Quadratic mean DBH (inches): 13.2
Basal area per acre (ft^2): 272.3
Volume per acre (Scribner, MBF): 37.270
Volume per acre (ft^3): 9447

TABLE A.9 Stand Table for a Douglas-Fir Stand, Age 55

DBH Class (inches)	Trees per Acre	Average Height (feet)	Snags per Acre
4	–	–	0.5
5	–	–	1.1
6	–	–	1.0
7	–	–	0.9
8	4.4	85.9	3.1
9	15.7	89.7	2.9
10	21.5	93.9	2.4
11	26.0	97.7	2.2
12	29.6	101.4	0.9
13	26.4	103.4	0.8
14	49.1	105.4	0.4

(Continued)

TABLE A.9 (Continued)

DBH Class (inches)	Trees per Acre	Average Height (feet)	Snags per Acre
15	34.7	105.6	0.4
16	29.9	107.7	0.1
17	25.0	109.8	–
18	10.5	113.4	–
19	9.0	118.1	–
20	1.8	118.2	–
Total	283.6	–	16.7

Average DBH (inches): 13.7
Quadratic mean DBH (inches): 14.0
Basal area per acre (ft^2): 301.4
Volume per acre (Scribner, MBF): 44.858
Volume per acre (ft^3): 10,996

TABLE A.10 Stand Table for a Douglas-Fir Stand, Age 60

DBH Class (inches)	Trees per Acre	Average Height (feet)	Snags per Acre
4	–	–	0.2
5	–	–	0.9
6	–	–	0.7
7	–	–	0.6
8	1.8	87.6	3.5
9	14.3	92.8	3.2
10	18.3	99.7	3.2
11	23.9	101.8	2.9
12	22.3	106.3	1.2
13	25.5	108.0	1.1
14	49.6	110.6	0.5
15	28.1	111.3	0.5
16	20.8	111.5	0.2
17	27.3	113.9	0.1
18	24.4	115.8	–
19	11.1	119.5	–
20	6.6	123.8	–
21	4.2	124.7	–
Total	278.2	–	18.8

Average DBH (inches): 14.3
Quadratic mean DBH (inches): 14.6
Basal area per acre (ft^2): 322.8
Volume per acre (Scribner, MBF): 52.495
Volume per acre (ft^3): 12,464

TABLE A.11 Stand Table for a Douglas-Fir Stand, Age 65

DBH Class (inches)	Trees per Acre	Average Height (feet)	Snags per Acre
4	–	–	0.1
5	–	–	0.6
6	–	–	0.5
7	–	–	0.4
8	0.8	91.1	3.2
9	11.1	95.2	3.0
10	15.7	102.5	3.1
11	20.6	106.9	2.8
12	22.3	109.0	1.1
13	20.0	111.7	1.0
14	23.9	114.6	0.5
15	43.8	116.8	0.4
16	25.7	116.9	0.2
17	28.5	118.6	–
18	16.6	118.9	–
19	22.3	121.6	–
20	10.5	125.2	–
21	6.6	129.3	–
22	4.2	130.2	–
Total	272.6	–	16.9

Average DBH (inches): 14.9
Quadratic mean DBH (inches): 15.3
Basal area per acre (ft^2): 346.5
Volume per acre (Scribner, MBF): 59.555
Volume per acre (ft^3): 13,830

TABLE A.12 Stand Table for a Douglas-Fir Stand, Age 70

DBH Class (inches)	Trees per Acre	Average Height (feet)	Snags per Acre
5	–	–	0.5
6	–	–	0.4
7	–	–	0.3
8	0.7	94.4	3.4
9	9.2	99.0	3.4
10	12.3	105.4	4.1
11	17.3	108.8	3.9
12	19.0	112.6	1.7
13	23.3	117.0	1.6
14	16.7	118.5	0.7
15	47.5	121.1	0.6
16	16.1	121.7	0.3

(Continued)

TABLE A.12 (Continued)

DBH Class (inches)	Trees per Acre	Average Height (feet)	Snags per Acre
17	24.2	122.3	0.1
18	26.7	123.9	0.1
19	20.2	125.5	–
20	14.7	127.3	–
21	7.8	130.7	–
22	6.6	134.5	–
23	4.2	135.4	–
Total	266.5	–	21.1

Average DBH (inches): 15.4
Quadratic mean DBH (inches): 15.8
Basal area per acre (ft^2): 362.9
Volume per acre (Scribner, MBF): 66.092
Volume per acre (ft^3): 15,072

TABLE A.13 Stand Table for a Douglas-Fir Stand, Age 75

DBH Class (inches)	Trees per Acre	Average Height (feet)	Snags per Acre
5	–	–	0.3
6	–	–	0.2
7	–	–	0.2
8	0.6	97.2	3.6
9	4.6	99.2	3.7
10	12.8	107.1	5.1
11	15.2	112.0	4.9
12	18.2	116.2	2.3
13	19.1	119.1	2.2
14	19.1	124.5	1.0
15	43.3	125.2	0.8
16	14.0	126.2	0.4
17	24.3	126.4	0.2
18	15.6	126.6	0.2
19	24.5	129.3	0.1
20	20.0	130.8	–
21	10.6	133.0	–
22	7.4	135.6	–
23	8.4	139.5	–
24	2.4	140.0	–
Total	260.1	–	25.2

Average DBH (inches): 15.9
Quadratic mean DBH (inches): 16.3
Basal area per acre (ft^2): 378.8
Volume per acre (Scribner, MBF): 71.838
Volume per acre (ft^3): 16,189

TABLE A.14 Stand Table for a Douglas-Fir Stand, Age 80

DBH Class (inches)	Trees per Acre	Average Height (feet)	Snags per Acre
6	–	–	0.1
7	–	–	0.1
8	0.5	99.6	3.6
9	1.9	103.8	3.8
10	11.2	107.4	5.9
11	14.7	115.1	5.9
12	16.8	119.8	3.2
13	15.6	125.0	3.1
14	16.7	123.8	1.3
15	19.4	127.7	1.1
16	42.0	130.6	0.4
17	18.8	130.1	0.3
18	15.3	131.8	0.2
19	19.8	133.0	–
20	13.9	132.8	–
21	19.3	135.2	–
22	10.6	137.5	–
23	6.7	142.3	–
24	7.8	144.3	–
25	1.8	144.1	–
Total	252.8	–	29.0

Average DBH (inches): 16.5
Quadratic mean DBH (inches): 16.9
Basal area per acre (ft^2): 395.0
Volume per acre (Scribner, MBF): 78.024
Volume per acre (ft^3): 17,193

TABLE A.15 Stand Table for a Douglas-Fir Stand, Age 85

DBH Class (inches)	Trees per Acre	Average Height (feet)	Snags per Acre
7	–	–	0.1
8	0.4	101.9	3.6
9	1.6	106.3	3.8
10	8.2	110.9	7.0
11	13.1	117.2	7.1
12	15.6	122.8	4.1
13	15.3	127.2	4.0
14	13.8	126.6	1.6
15	15.3	131.2	1.4
16	41.7	134.3	0.5
17	13.5	134.0	0.4
18	18.0	134.6	0.2

(Continued)

TABLE A.15 (Continued)

DBH Class (inches)	Trees per Acre	Average Height (feet)	Snags per Acre
19	15.0	135.1	0.1
20	20.9	138.0	0.1
21	15.4	138.7	–
22	12.3	138.7	–
23	8.7	142.6	–
24	8.5	147.0	–
25	5.4	148.5	–
26	1.8	148.2	–
Total	244.5	–	34.0

Average DBH (inches): 17.0
Quadratic DBH (inches): 17.5
Basal area per acre (ft^2): 406.6
Volume per acre (Scribner, MBF): 83.524
Volume per acre (ft^3): 18,084

TABLE A.16 Stand Table for a Douglas-Fir Stand, Age 90

DBH Class (inches)	Trees per Acre	Average Height (feet)	Snags per Acre
8	0.3	104.3	3.5
9	1.3	108.6	3.8
10	6.9	113.3	8.1
11	9.9	120.2	8.2
12	13.6	125.5	5.2
13	14.9	129.1	5.1
14	13.0	129.7	2.0
15	13.6	135.3	1.8
16	37.5	137.4	0.7
17	13.3	137.9	0.6
18	21.0	138.3	0.3
19	11.2	138.8	0.2
20	18.8	141.3	0.2
21	14.6	139.9	0.1
22	14.3	143.9	–
23	10.5	144.0	–
24	7.4	146.3	–
25	8.5	151.7	–
26	4.2	151.8	–
27	1.2	152.9	–
Total	236.0	–	39.8

Average DBH (inches): 17.5
Quadratic mean DBH (inches): 18.0
Basal area per acre (ft^2): 415.1
Volume per acre (Scribner, MBF): 88.702
Volume per acre (ft^3): 18,871

TABLE A.17 Stand Table for a Douglas-Fir Stand, Age 95

DBH Class (inches)	Trees per Acre	Average Height (feet)	Snags per Acre
8	0.2	105.9	3.4
9	1.0	110.9	3.8
10	5.7	115.8	9.0
11	8.5	122.8	9.2
12	11.4	128.1	6.3
13	12.7	130.9	6.2
14	12.5	134.7	2.5
15	12.9	138.3	2.3
16	38.1	140.3	1.0
17	11.8	141.2	0.8
18	18.1	141.0	0.3
19	10.7	142.5	0.3
20	13.2	142.5	0.2
21	17.7	145.3	0.1
22	7.9	144.7	–
23	16.6	147.0	–
24	9.9	149.4	–
25	7.4	151.8	–
26	5.4	155.6	–
27	5.4	155.7	–
Total	227.1	–	45.4

Average DBH (inches): 17.9
Quadratic mean DBH (inches): 18.5
Basal area per acre (ft^2): 421.8
Volume per acre (Scribner, MBF): 93.108
Volume per acre (ft^3): 19,582

TABLE A.18 Stand Table for a Douglas-Fir Stand, Age 100

DBH Class (inches)	Trees per Acre	Average Height (feet)	Snags per Acre
8	–	–	3.0
9	1.0	111.9	3.8
10	4.7	117.7	9.8
11	6.9	125.8	10.0
12	10.1	129.8	7.4
13	10.7	132.9	7.3
14	12.7	138.1	3.0
15	11.8	141.4	2.8
16	36.0	143.5	1.3
17	11.5	144.1	1.1
18	9.0	144.4	0.4
19	16.7	144.6	0.3
20	13.9	145.3	0.2

(Continued)

TABLE A.18 (Continued)

DBH Class (inches)	Trees per Acre	Average Height (feet)	Snags per Acre
21	13.8	149.2	0.1
22	12.1	147.5	0.1
23	13.4	150.4	–
24	9.4	149.8	–
25	8.1	153.4	–
26	5.5	157.5	–
27	6.6	158.8	–
28	4.2	159.6	–
Total	218.1	–	50.6

Average DBH (inches): 18.4
Quadratic mean DBH (inches): 18.9
Basal area per acre (ft^2): 426.3
Volume per acre (Scribner, MBF): 97.634
Volume per acre (ft^3): 20,237

TABLE A.19 Stand Table for a Douglas-Fir Stand, Age 105

DBH Class (inches)	Trees per Acre	Average Height (feet)	Snags per Acre
8	–	–	2.8
9	0.8	113.6	3.8
10	3.8	119.7	10.2
11	5.4	128.0	10.6
12	7.7	132.2	8.3
13	10.0	134.1	8.2
14	12.4	140.2	3.6
15	10.1	144.8	3.2
16	11.8	142.9	1.6
17	33.2	147.7	1.3
18	9.5	146.9	0.4
19	17.6	148.3	0.3
20	9.0	148.8	0.2
21	14.3	150.7	0.2
22	12.7	150.7	0.2
23	6.5	150.7	–
24	15.8	154.5	–
25	7.6	152.7	–
26	6.8	156.9	–
27	7.3	161.1	–
28	6.0	162.7	–
29	1.8	162.5	–
Total	210.1	–	54.9

Average DBH (inches): 19.0
Quadratic mean DBH (inches): 19.5
Basal area per acre (ft^2): 437.0
Volume per acre (Scribner, MBF): 101.714
Volume per acre (ft^3): 20,850

TABLE A.20 Stand Table for a Douglas-Fir Stand, Age 110

DBH Class (inches)	Trees per Acre	Average Height (feet)	Snags per Acre
8	–	–	2.3
9	0.6	115.5	3.6
10	2.7	121.7	10.5
11	4.1	130.1	10.7
12	7.1	133.7	8.9
13	7.6	135.9	8.7
14	10.4	141.0	4.1
15	10.6	147.0	3.7
16	9.7	145.1	1.9
17	33.8	150.6	1.5
18	8.9	150.5	0.5
19	15.1	150.5	0.4
20	10.0	151.1	0.3
21	10.3	151.2	0.2
22	15.5	155.4	0.2
23	9.0	153.1	0.1
24	11.3	156.9	–
25	9.6	155.7	–
26	7.4	160.5	–
27	7.3	161.5	–
28	5.4	165.4	–
29	4.2	165.2	–
30	1.2	166.3	–
Total	201.8	–	57.6

Average DBH (inches): 19.4
Quadratic mean DBH (inches): 20.0
Basal area per acre (ft^2): 439.7
Volume per acre (Scribner, MBF): 105.705
Volume per acre (ft^3): 21,431

Lincoln Tract

Legend

    lincoln_roads

    lincoln_streams

☐ lincoln_boundary

lincoln_stands

Scale:

1:40,000

FIGURE A.1   A map of the Lincoln Tract.

TABLE A.21 Basic Data Representing the Stands Within the Lincoln Tract

Stand	Acres	Hectares	Age	Species[b]	Volume per Acre (MBF)[a]					
					Period 1	Period 2	Period 3	Period 4	Period 5	Period 6
1	41.913	16.962	18	DF	0.4	2.2	6.1	11.7	18.5	25.3
2	61.130	24.739	10	DF	0.0	0.0	0.7	3.3	8.4	14.9
3	19.911	8.058	7	DF	0.0	0.0	0.3	1.5	4.9	10.0
4	61.228	24.778	100	DF	94.4	98.4	102.2	103.2	104.0	104.8
5	46.946	18.999	70	DF	63.2	68.7	74.6	79.9	84.9	89.1
6	28.837	11.670	55	Mixed	42.1	49.2	55.9	61.9	67.3	73.1
7	97.054	39.276	20	DF	0.7	3.2	7.9	14.0	21.0	27.8
8	55.242	22.356	22	DF	1.5	4.9	10.1	16.4	23.1	29.7
9	105.215	42.579	11	DF	0.0	0.1	1.1	4.2	9.3	15.7
10	79.805	32.296	11	DF	0.0	0.1	1.1	4.0	9.0	15.2
11	119.452	48.341	20	DF	0.6	3.1	7.8	13.8	20.7	27.4
12	118.871	48.106	13	DF	0.0	0.4	2.2	6.3	12.0	19.0
13	119.017	48.165	12	DF	0.0	0.3	1.6	5.2	10.6	17.3
14	71.507	28.938	14	DF	0.0	0.5	2.7	7.3	13.3	20.4
15	73.929	29.918	90	DF	88.1	92.4	96.9	101.0	105.0	106.0
16	46.101	18.656	15	DF	0.0	0.7	3.4	8.4	14.9	22.3
17	26.182	10.596	32	DF	10.4	16.9	23.9	30.7	37.7	44.8
18	65.845	26.646	6	DF	0.0	0.0	0.1	1.1	4.0	8.8
19	21.358	8.643	27	DF	5.2	10.7	17.4	24.6	31.6	38.8
20	78.198	31.646	18	DF	0.4	2.2	6.3	12.0	18.9	25.9
21	31.500	12.748	19	DF	0.5	2.6	6.9	12.6	19.4	26.1
22	45.971	18.604	16	DF	0.1	1.2	4.3	9.6	16.3	23.5
23	63.343	25.634	55	DF	44.9	52.5	59.6	66.0	71.7	77.9
24	65.696	26.586	14	DF	0.0	0.5	2.5	6.8	12.4	19.0
25	83.623	33.841	18	DF	0.4	2.2	6.1	11.7	18.5	25.4
26	35.553	14.388	13	DF	0.0	0.4	2.2	6.3	12.1	19.0
27	42.321	17.127	19	DF	0.5	2.5	6.8	12.4	19.1	25.8
28	30.126	12.191	4	DF	0.0	0.0	0.0	0.5	2.5	6.8
29	25.295	10.236	0	DF	0.0	0.0	0.0	0.0	0.7	3.2
30	31.405	12.709	12	DF	0.0	0.3	1.6	4.9	10.1	16.5
31	5.319	2.153	12	DF	0.0	0.3	1.7	5.2	10.8	17.6
32	70.990	28.729	25	DF	3.2	7.9	14.1	21.0	27.8	34.7
33	3.228	1.306	0	DF	0.0	0.0	0.0	0.0	0.7	3.3
34	63.534	25.711	37	DF	17.6	24.8	32.0	39.3	46.7	53.9
35	46.835	18.954	33	DF	12.0	18.9	26.0	33.1	40.4	47.7
36	41.287	16.709	34	DF	13.5	20.7	27.9	35.2	42.6	50.1
37	21.332	8.633	0	DF	0.0	0.0	0.0	0.0	0.7	3.4

(Continued)

TABLE A.21 (Continued)

Stand	Acres	Hectares	Age	Species[b]	Volume per Acre (MBF)[a]					
					Period 1	Period 2	Period 3	Period 4	Period 5	Period 6
38	16.899	6.839	50	Mixed	33.6	40.5	47.3	53.7	59.6	64.7
39	39.151	15.844	15	DF	0.0	0.7	3.2	8.1	14.4	21.6
40	56.976	23.057	5	DF	0.0	0.0	0.0	0.7	3.4	8.4
41	10.677	4.321	16	DF	0.1	1.1	4.1	9.2	15.6	22.6
42	98.086	39.694	22	DF	1.6	5.0	10.2	16.6	23.4	30.1
43	65.829	26.640	23	DF	2.2	6.2	11.8	18.6	25.6	32.6
44	8.329	3.371	2	DF	0.0	0.0	0.0	0.3	1.6	5.0
45	26.828	10.857	23	DF	2.2	6.2	11.9	18.7	25.7	32.8
46	39.971	16.176	40	DF	20.9	27.6	34.4	41.4	48.4	55.0
47	72.255	29.241	21	DF	1.2	4.3	9.5	16.1	23.2	30.4
48	78.912	31.935	1	DF	0.0	0.0	0.0	0.1	1.2	4.2
49	51.479	20.833	35	DF	13.8	20.7	27.4	34.2	41.2	48.1
50	20.409	8.259	75	DF	67.2	73.0	78.1	83.0	87.1	91.3
51	23.477	9.501	38	DF	18.1	24.9	31.7	38.7	45.7	52.5
52	39.301	15.904	33	DF	12.1	19.0	26.1	33.3	40.6	48.0
53	64.266	26.008	35	DF	13.8	20.7	27.3	34.1	41.1	48.0
54	42.310	17.122	1	DF	0.0	0.0	0.0	0.1	1.1	4.2
55	13.670	5.532	1	DF	0.0	0.0	0.0	0.1	1.1	4.2
56	17.161	6.945	55	Mixed	44.0	51.5	58.4	64.8	70.4	76.5
57	58.149	23.532	38	DF	19.3	26.5	33.8	41.2	48.7	55.9
58	73.249	29.643	40	DF	20.8	27.5	34.3	41.3	48.3	54.8
59	29.787	12.055	65	DF	56.6	62.8	68.2	74.1	79.4	84.3
60	51.636	20.897	31	DF	8.9	15.1	21.8	28.6	35.3	42.2
61	30.762	12.449	9	DF	0.0	0.0	0.5	2.5	6.8	12.5
62	32.205	13.033	50	DF	36.7	44.2	51.6	58.6	65.0	70.6
63	53.579	21.683	62	DF	51.4	57.8	63.6	69.1	74.5	79.6
64	38.905	15.744	54	DF	40.7	48.0	54.6	60.9	66.5	72.2
65	63.824	25.829	53	DF	41.5	49.0	56.3	63.0	69.0	74.9
66	45.018	18.218	0	DF	0.0	0.0	0.0	0.0	0.7	3.4
67	79.428	32.144	22	DF	1.6	5.0	10.4	16.9	23.8	30.7
68	28.334	11.467	49	DF	34.2	41.4	48.8	55.6	62.0	67.6
69	75.572	30.583	46	DF	29.0	35.8	42.8	49.7	56.2	62.0
70	24.380	9.866	16	DF	0.1	1.1	4.1	9.2	15.6	22.5
71	85.393	34.558	55	DF	42.0	49.1	55.7	61.8	67.1	72.9
72	37.455	15.158	24	DF	2.7	7.3	13.3	20.6	27.7	34.9
73	56.148	22.722	5	DF	0.0	0.0	0.0	0.7	3.4	8.5
74	81.661	33.047	40	Mixed	20.7	27.4	34.1	41.1	48.1	54.6
75	62.937	25.470	42	Mixed	25.1	32.3	39.7	47.2	54.5	61.3

(Continued)

TABLE A.21 (Continued)

Stand	Acres	Hectares	Age	Species[b]	Volume per Acre (MBF)[a]					
					Period 1	Period 2	Period 3	Period 4	Period 5	Period 6
76	5.134	2.078	45	DF	28.0	35.0	42.1	49.2	55.9	62.0
77	53.940	21.829	70	DF	61.0	66.3	72.0	77.1	81.9	85.9
78	53.652	21.712	103	DF	95.5	99.3	101.4	102.3	103.1	103.8
79	94.880	38.397	110	DF	103.7	104.7	105.5	106.4	107.1	107.7
80	46.611	18.863	59	Mixed	50.6	57.7	64.3	70.2	76.3	81.8
81	19.701	7.973	37	DF	16.5	23.2	29.9	36.7	43.6	50.4
82	76.216	30.843	50	DF	33.8	40.6	47.5	53.9	59.8	65.0
83	50.515	20.443	90	DF	85.2	89.4	93.7	97.7	101.5	102.5
84	71.207	28.817	50	DF	34.9	42.0	49.1	55.8	61.9	67.2
85	62.715	25.380	101	DF	90.1	93.7	96.8	97.6	98.5	99.2
86	87.598	35.450	65	DF	57.9	64.2	69.7	75.8	81.1	86.1
87	88.657	35.878	55	DF	40.7	47.6	54.0	59.9	65.1	70.7

[a]*Time periods are 5 years long.*
[b]*DF, Douglas-fir.*

Putnam Tract

Legend

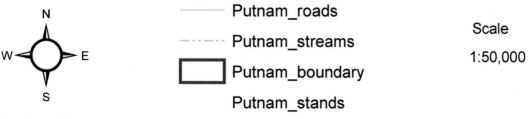

—— Putnam_roads

----- Putnam_streams

▭ Putnam_boundary

Putnam_stands

Scale

1:50,000

FIGURE A.2   A map of the Putnam Tract.

TABLE A.22 Basic Data Representing the Stands Within the Putnam Tract

| Stand | Acres | Hectares | Forest Type | Age | Cords per Acre[a] | | |
					Period 1	Period 2	Period 3
1	73.976	29.937	Pine plantation	2	8.3	12.1	14.9
2	84.525	34.206	Natural pine	45	29.3	32.4	35.8
3	11.882	4.809	Hardwood	5	0.0	2.0	6.0
4	31.615	12.794	Pine plantation	21	13.9	18.1	23.1
5	1.201	0.486	Hardwood	31	20.8	23.6	26.3
6	6.683	2.705	Mixed	25	16.3	20.3	24.1
7	23.277	9.420	Natural pine	44	29.1	33.0	36.2
8	31.157	12.609	Pine plantation	15	8.3	13.5	17.8
9	12.622	5.108	Pine plantation	15	8.2	13.2	17.4
10	48.907	19.792	Pine plantation	19	12.1	16.4	21.5
11	9.067	3.669	Pine plantation	14	7.1	12.2	16.3
12	62.212	25.176	Natural pine	47	31.3	34.1	37.5
13	57.533	23.283	Pine plantation	12	5.4	10.3	15.1
14	56.344	22.802	Pine plantation	11	4.2	9.1	14.5
15	96.191	38.927	Pine plantation	10	2.8	8.4	13.2
16	72.705	29.423	Pine plantation	10	2.4	8.2	12.9
17	85.437	34.575	Pine plantation	10	2.9	8.6	13.4
18	35.667	14.434	Pine plantation	13	6.2	11.2	16.0
19	7.275	2.944	Natural pine	28	20.1	24.9	28.1
20	29.241	11.833	Pine plantation	21	14.2	18.4	23.5
21	29.212	11.822	Natural pine	26	19.0	23.2	26.1
22	10.276	4.159	Pine plantation	20	13.0	17.4	22.4
23	40.580	16.422	Pine plantation	21	13.7	18.0	22.9
24	56.878	23.018	Pine plantation	25	17.6	22.2	26.4
25	26.485	10.718	Natural pine	44	28.4	31.2	34.6
26	4.179	1.691	Natural pine	40	26.8	30.7	33.9
27	41.800	16.916	Pine plantation	26	18.3	23.2	26.2
28	16.948	6.859	Pine plantation	26	18.7	23.4	26.3
29	19.940	8.069	Pine plantation	27	19.4	24.0	26.7
30	109.838	44.450	Pine plantation	28	20.5	25.1	28.0
31	81.232	32.874	Pine plantation	35	26.0	29.2	32.4
32	34.482	13.954	Hardwood	45	27.1	29.8	32.4
33	84.964	34.384	Hardwood	45	24.6	27.3	29.9
34	63.744	25.796	Hardwood	45	26.1	28.7	31.0
35	28.411	11.498	Mixed	35	23.4	27.2	30.1
36	52.379	21.197	Natural pine	25	13.2	17.5	20.2

(Continued)

TABLE A.22 (Continued)

Stand	Acres	Hectares	Forest Type	Age	Cords per Acre[a]		
					Period 1	Period 2	Period 3
37	16.094	6.513	Natural pine	29	21.8	25.3	28.5
38	2.339	0.947	Hardwood	30	19.5	22.3	25.1
39	6.423	2.599	Pine plantation	10	2.8	8.4	13.1
40	10.793	4.368	Pine plantation	20	13.2	17.5	22.6
41	32.247	13.050	Mixed	17	10.5	14.8	18.2
42	7.212	2.919	Hardwood	30	17.4	20.3	23.2
43	9.483	3.838	Natural pine	26	12.8	17.2	21.0
44	11.442	4.631	Mixed	30	20.8	24.9	28.5
45	21.360	8.644	Pine plantation	10	2.6	8.1	13.0
46	37.428	15.146	Pine plantation	10	3.2	8.8	13.6
47	10.968	4.439	Natural pine	25	12.7	17.2	19.9
48	19.064	7.715	Hardwood	30	15.6	18.4	21.1
49	26.269	10.631	Pine plantation	21	14.3	18.3	23.1
50	23.695	9.589	Pine plantation	21	14.1	18.1	23.0
51	25.553	10.341	Natural pine	32	17.2	20.2	22.9
52	23.981	9.705	Hardwood	50	30.8	33.5	35.2
53	10.422	4.218	Pine plantation	14	7.2	12.3	16.5
54	9.027	3.653	Pine plantation	20	13.5	17.6	22.6
55	10.589	4.285	Pine plantation	15	8.0	13.0	17.1
56	5.012	2.028	Mixed	35	22.9	26.8	30.1
57	20.266	8.201	Natural pine	34	17.3	20.3	23.4
58	29.253	11.838	Natural pine	34	18.2	21.5	24.6
59	30.154	12.203	Natural pine	34	16.7	19.6	22.9
60	18.029	7.296	Natural pine	33	16.4	19.5	22.7
61	34.381	13.914	Natural pine	33	15.9	18.9	22.3
62	17.279	6.993	Mixed	35	24.8	28.3	31.2
63	51.807	20.966	Hardwood	60	31.4	34.2	36.3
64	18.988	7.684	Hardwood	62	29.4	32.1	34.4
65	36.426	14.741	Hardwood	65	26.4	29.0	31.2
66	10.634	4.303	Hardwood	67	30.8	33.5	35.1
67	23.976	9.703	Hardwood	70	32.8	35.6	38.0
68	23.035	9.322	Hardwood	68	30.4	33.1	35.0
69	5.315	2.151	Mixed	43	28.9	32.4	35.1
70	63.685	25.772	Pine plantation	26	19.0	23.9	26.6
71	25.719	10.408	Pine plantation	26	18.4	23.3	26.3
72	13.486	5.458	Mixed	25	17.8	21.7	25.4

(Continued)

TABLE A.22 (Continued)

Stand	Acres	Hectares	Forest Type	Age	Cords per Acre[a]		
					Period 1	Period 2	Period 3
73	23.849	9.651	Hardwood	40	25.6	28.4	31.1
74	59.242	23.974	Pine plantation	30	22.3	27.3	30.3
75	0.897	0.363	Pine plantation	4	0.0	1.5	7.1
76	42.720	17.288	Natural pine	40	27.8	31.8	35.2
77	46.265	18.723	Pine plantation	17	10.4	15.1	19.3
78	27.095	10.965	Mixed	45	27.4	31.1	33.8
79	35.157	14.228	Natural pine	40	27.5	31.6	35.0
80	0.881	0.357	Hardwood	30	20.5	23.1	25.6
81	85.555	34.623	Natural pine	45	29.2	32.1	35.4

[a]*Time periods are 5 years long.*

Appendix B

The Simplex Method for Solving Linear Planning Problems

In several chapters of this book, we introduced the notion that many forestry and natural resource management problems could be arranged as a set of linear equations. In two-variable problems, these equations can be graphed to illustrate the solution space and to determine the optimal decision, although the graphing scale may lead to imprecisions. In larger problems, a solution procedure is necessary and more precise. The Simplex Method is one process that can be used to develop a solution for a set of linear equations that represent a management problem such that an optimal allocation to the decision variables will be made. Admittedly, the process appears to be complex. However, in this Appendix we attempt to make the solution procedure clear. The emphasis of this discussion is on placing a set of linear equations into a matrix, and transforming the matrix through a series of manipulations, leading to the identification of the optimal solution for a management problem. A small, two-variable, two-constraint problem is used to illustrate the process. Linear programming continues to be widely used in natural resource management, therefore after reading this Appendix, you should be able to:

1. Understand how a set of linear equations are converted to a detached coefficient matrix, or tableau
2. Identify the optimal solution to a linear programming model based on mathematical relationships
3. Transform a matrix from the initial representation to the optimal solution

B.I. INTRODUCTION

As you may have learned, there are a variety of methods that can be used to solve natural resource management problems, from traditional optimization techniques, to simulation, to heuristics. The Simplex Method is a traditional optimization technique, and one of the most basic ways we can solve a problem that is composed of a set of linear equations. It is considered an *exact technique*, since when completed, we are assured that the optimal solution (if one exists) to a problem has been located.

The Simplex Method was first introduced in 1947 by George Dantzig, who was a planner for the United States Air Force. As we noted in Chapter 7, Linear Programming, the term *linear programming* was suggested to Dantzig as the type of solution technique he developed because linear equations are used, and because the term *programming* was synonymous with *planning* in the military. At the time of the development of the Simplex Method, the process was viewed as a mechanization of the traditional military planning process. For an interesting overview of the life and contributions of George Dantzig, students should read the summary provided by O'Connor and Robertson (2003).

Essentially, the Simplex Method searches the edges of an *n*-dimensional solution space, seeking improvements to a solution for a set of linear equations. When graphing linear equations, these large problems may contain numerous corners. The search process stops when the method determines that no other corners can be reached (from the current corner being assessed) that will improve the objective function value. Other types of search methods designed to handle linear equations have been developed since Dantzig's development of the Simplex Method. Some of these include searching through the middle of an *n*-dimensional solution space. Although we concentrate here on the basic linear programming search process, interested students should read Morgan (1997), who describes several variations of the Simplex Method.

Linear programming assumes that all the variables will be assigned a continuous real number (e.g., 0, 0.5689, 24, 156.3) that is either zero or positive. This is the assumption of divisibility being applied. When we assume that variables can be assigned integers only (e.g., 0, 1, 2, 3 ...), the search technique is considered either mixed-integer programming (some variables will be integers, some will be continuous numbers), or integer programming (all variables will be integers). Within linear programming, variables also are considered basic (i.e., have a positive value) and nonbasic (i.e., have a zero value).

As we mentioned in earlier chapters, a linear problem formulation consists of a set of linear equations. Some of these equations contain inequalities. These equations need to be converted to equalities, by introducing new variables (slack variables). For example, the equation

$$X1 + X2 \leq 250$$

is an inequality, and is converted to an equality by introducing a slack variable ($S1$),

$$X1 + X2 + S1 = 250$$

to represent the fact that both $X1$ and $X2$ can be 0, thus $S1$ has to be 250. We could say that this equation, if it were a constraint, has 250 units of slack in it (i.e., 250 units of the right-hand side (RHS) of the equation are unused). For any solution in the feasible region of the solution space, the slack variables are also either zero or a positive continuous number.

B.II. TEN STEPS THAT REPRESENT THE SIMPLEX METHOD

The Simplex Method can be summarized as a set of ten steps (Fig. B.1). The full set of steps is repeated (iterated) numerous times until the method has determined that the optimal solution has been located.

Step 1: Develop the Detached Coefficient Matrix

At the start of the Simplex Method, the equations that comprise the linear problem are placed into a detached coefficient matrix, or tableau. In addition to this tableau, we will note the variables that are currently in the solution, and their contribution to the objective function value. For example, in Fig. B.2 we have illustrated a simple problem where we want to maximize $2.5X + 8Y$. The constraints are:

$$15X + 20Y \leq 60$$

$$X + 5Y \leq 10$$

These constraints were converted to equalities with the addition of slack variables:

$$15X + 20Y + S1 = 60$$

$$X + 5Y + S2 = 10$$

They can then be inserted into a detached coefficient matrix (Fig. B.2). The initial feasible solution in this case is the null set, where all variables except the slack have a value of 0 in the solution. If an initial feasible solution

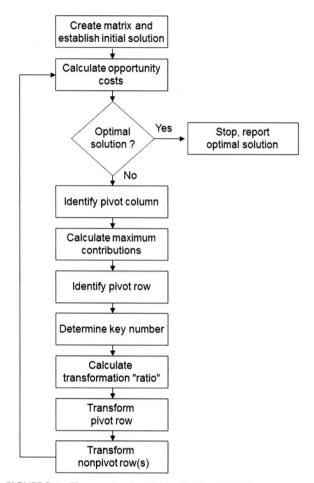

FIGURE B.1   The steps involved in the Simplex Method.

X	Y	S1	S2	Q	Variables in the solution	Their value
2.5	8					
15	20	1		60	S1	0
1	5		1	10	S2	0

FIGURE B.2   An example detached coefficient matrix.

cannot be located, then the solution process terminates, and a new set of constraints need to be determined. To avoid confusion at this stage in the process, remember that in Fig. B.2 the "their value" column represents the value of each variable to the objective or solution. While the slack variables do in fact have values of 60 ($S1$) and 10 ($S2$) that cannot be seen in Fig. B.2, their value to the solution is zero.

The top row of the matrix simply recognizes the variable that is being represented by each column. The Q column initially is represented by the RHS of each constraint. Row 2 includes the coefficients of each variable that are recognized in the objective function. The objective function we are using (maximize $2.5X + 8Y$) does not include any slack. Row 3 represents the first constraint ($15X + 20Y + S1 = 60$), and row 4 represents the second constraint ($X + 5Y + S2 = 10$).

To the right of the matrix, we provide an indication of the variables that are currently in the solution (the slack variables $S1$ and $S2$), and their contribution ("their value") to the objective function value (nothing).

Step 2: Calculate the Opportunity Costs

The *opportunity costs* are calculated for each of the columns (variables) in the management problem. This is performed to determine whether an optimal solution has been located (when all opportunity costs are less than or equal to zero), and to locate the pivot column. All variables currently in a solution have opportunity costs of zero or less. A pivot column and row identifies the place around which the matrix will be transformed.

The opportunity costs are calculated for each column (variable) using the equation:

OC = (Objective function value of variable in the column)
− (Constraint Row 1 value) × ("Their value" on Row 1)
− (Constraint Row 2 value) × ("Their value" on Row 2)
− (Constraint Row 3 value) × ("Their value" on Row 3)
...

For example, the opportunity cost of column 1 in Fig. B.2 is:

$$OC = (2.5) - (15 \times 0) - (1 \times 0)$$
$$OC = 2.5$$

Step 3: Make a Decision Regarding Optimality

If all opportunity costs are less than or equal to zero, then the process stops and the optimal solution is reported. If one or more opportunity costs are greater than zero, then the process continues, as further improvements can still be made to the objective function value.

Step 4: Identify the Pivot Column

The column of the matrix that is associated with the highest positive opportunity cost is the pivot column.

Step 5: Calculate the Maximum Contributions of Variables to the Current Solution

These maximum contributions of variables indicate the maximum amount of the variable represented by the pivot column that can be brought into the solution, as affected by each constraint. The equation for computing the maximum contributions is:

$$MC = (Q_{row}/\text{Pivot column value}_{row})$$

For example, assuming that the pivot column is Y, the maximum contribution for the first constraint in our simple model is:

$$MC = (60/20)$$

The value of 60 is found in column Q, row 1 (the first full row after the objective function values). The value of 20 is the row 1 value associated with the pivot column.

The smallest of these maximum contributions indicates how much (at most) the variable represented by the pivot column can enter the solution without violating any constraints.

Step 6: Identify the Pivot Row

The constraint row of the matrix that was associated with the lowest maximum contribution is the pivot row.

Step 7: Determine the Key Number

The "key number" is the value of the cell in the matrix that is located at the intersection of the pivot column and pivot row.

Step 8: Calculate the Transformation Ratio

Each constraint that is not identified as the pivot row needs to be transformed. The transformation ratio assists us with these computations. It is simply the value of the cell in the pivot column of the nonpivot row divided by the key number.

TR = (Value of cell in pivot column of non-pivot row/
Key number)

Step 9: Transform the Pivot Row

In this step we transform the matrix, starting with the pivot row, to represent that the Simplex Method is moving to another corner of the solution space by allowing a variable currently not in the solution, to enter the solution.

New pivot row value = (Old pivot row value/Key number)

Step 10: Transform the Nonpivot Row(s)

For all the cells in the nonpivot rows, we need to remember one thing: the value of the cell in the pivot row for each column in question (the value before we transformed the pivot row in Step 9). Given this, along with the current value of each cell in the nonpivot rows, and the transformation ratio associated with each nonpivot row, we can transform the cells of the nonpivot rows using the equation:

New nonpivot row value = (Old nonpivot row cell value)
  − (Old pivot row cell value for the column in question)
  × (TR)

Assuming that TR = 4, and the old pivot row cell value for the column is 1, the value for column 1, constraint 1, will change from 15 to 11:

New nonpivot row value = (15) − (1 × 4) = 11

B.III. A TWO-VARIABLE, TWO-CONSTRAINT PROBLEM SOLVED WITH THE SIMPLEX METHOD

Assume that recently you have been hired as a forester for a medium-sized company in southern Mississippi. One of your responsibilities is to manage a site preparation and planting budget of $500,000 per year. Based on some preliminary discussions with the company's preferred site preparation contractor, there are two main reforestation options. You have determined that it will cost the company about $178 per acre to use a 3-in-1 plow in conjunction with machine planting to reforest a cutover area. It will cost about $203 per acre to shear the residual stems and apply a chemical site preparation treatment, then hand plant cutover areas. Although estimates of these types of costs vary over time and with sampling method, this data seems consistent with recent cost data for the United States southern Coastal Plain (Barlow and Dubois, 2011; Barlow and Levendis, 2015).

The contractor has indicated that given their current machinery and personnel, on average they can site prepare 12 acres per day using the 3-in-1 plow. On average, they can site prepare 15 acres per day using the other system. These figures are yearly averages, and in any given year there are about 200 work days. You have decided that you want to maximize the number of acres site prepared over the next year.

The set of linear equations that are needed to solve this problem are:

$$\text{Maximize } SPM1 + SPM2$$

Subject to:

$$178\ SPM1 + 203\ SPM2 \le 500{,}000$$

$$0.0833\ SPM1 + 0.0667\ SPM2 \le 200$$

The objective function $(SPM1 + SPM2)$ simply implies that you need to maximize the number of acres assigned to site preparation method 1 $(SPM1)$ or site preparation method 2 $(SPM2)$. The first constraint is the budget constraint, and the second constraint is an operational constraint. The values in the second constraint are days per acre, or the inverse of the acres per day suggested earlier for each type of treatment. First, we convert the constraints from inequalities to equalities by adding slack variables to each.

$$178\ SPM1 + 203\ SPM2 + S1 = 500{,}000$$

$$0.0833\ SPM1 + 0.0667 SPM2 + S2 = 200$$

Next, we begin one or more iterations of the Simplex Method.

Iteration 1 of the Simplex Method

Iteration 1, Step 1. The initial detached coefficient matrix, or tableau, is created (Fig. B.3). As you can see, since this is the initial formulation of the problem, the slack variables are "in the solution," and the variables related to the site preparation methods are not. This is an initial feasible solution. In other words, no acres have been assigned to each of the site preparation methods yet.

Iteration 1, Step 2. The opportunity costs are calculated for each column (variable).

$$OC_{SPM1} = (1) - (178 \times 0) - (0.0833 \times 0) = 1$$
$$OC_{SPM2} = (1) - (203 \times 0) - (0.0667 \times 0) = 1$$
$$OC_{S1} = (0) - (1 \times 0) - (0 \times 0) = 0$$
$$OC_{S2} = (0) - (0 \times 0) - (0 \times 1) = 0$$

Iteration 1, Step 3. Since one or more opportunity costs are greater than zero, the process continues, and we note that the optimal solution has not yet been located.

Iteration 1, Step 4. The column of the matrix that is associated with the highest positive opportunity cost is either the $SPM1$ or $SPM2$ columns, since each have opportunity costs of 1. We will choose $SPM1$ as the pivot column, but we could have chosen $SPM2$ as well.

Iteration 1, Step 5. Calculate the maximum contributions, or the maximum amount of the variable represented in the pivot column that can be brought into the solution, as affected by each constraint.

SPM1	SPM2	S1	S2	Q	Variables in the solution	Their value
1	1					
178	203	1		500,000	S1	0
0.0833	0.0667		1	200	S2	0

FIGURE B.3 The initial detached coefficient matrix.

$$\text{MC}_1 = (500,000/178) = 2808.989$$
$$\text{MC}_2 = (200/0.0833) = 2400.960$$

The smallest of these values tells us that at most, 2400.96 acres of $SPM1$ can enter the solution without violating any of the constraints. This is related to constraint row 2.

Iteration 1, Step 6. The constraint row of the matrix that was associated with the lowest maximum contribution is the pivot row (constraint row 2).

Iteration 1, Step 7. The key number is the value of the cell in the matrix that is located at the intersection of the pivot column and pivot row, or 0.0833.

Iteration 1, Step 8. Since each constraint that is not identified as the pivot row needs to be transformed, and there is only one other constraint, other than the one identified as the pivot row, we simply need one transformation ratio.

$$\text{TR}_1 = (178/0.0833) = 2136.8547$$

Iteration 1, Step 9. Transform the pivot row values by dividing each by the key number (0.0833). As you can see (Fig. B.4), the value of Q in constraint row 2 has changed to the maximum contribution value we calculated in Step 5. In addition, we note to the right of the matrix that the variable $SPM1$ has entered the solution, and that it has a contribution of 1 to the objective function value for each acre assigned to $SPM1$.

Iteration 1, Step 10. Transform the nonpivot row (constraint 1).

New nonpivot row value for Column $SPM1$
$= (178) - (0.0833) \times (2136.8547) = 0$
New nonpivot row value for Column $SPM2$
$= (203) - (0.0667) \times (2136.8547) = 60.4718$
New nonpivot row value for Column $S1$
$= (1) - (0) \times (2136.8547) = 1$
New nonpivot row value for Column $S2$
$= (0) - (1) \times (2136.8547) = -2136.8547$
New nonpivot row value for Column Q
$= (500,000) - (200) \times (2136.8547) = 72,629.06$

Fig. B.5 illustrates the condition of the detached coefficient matrix after one complete transformation (one iteration). When viewed as a graph, in two-dimensional space, we find that the Simplex Method moved from an initial null solution ($SPM1 = 0$, $SPM2 = 0$) to a solution where $SPM1 = 2400.96$ and $SPM2 = 0$ (Fig. B.6).

Iteration 2 of the Simplex Method

Since we have not determined whether the optimal decision has been located, we must begin Iteration 2 with the assessment of the opportunity costs. Step 1 of Iteration 1

$SPM1$	$SPM2$	$S1$	$S2$	Q	Variables in the solution	Their value
1	1					
178	203	1		500,000	$S1$	0
1	0.80072		12.0048	2400.96	$SPM1$	1

FIGURE B.4 The detached coefficient matrix after transforming the pivot row.

$SPM1$	$SPM2$	$S1$	$S2$	Q	Variables in the solution	Their value
1	1					
0	60.4718	1	-2136.85	72629.06	$S1$	0
1	0.80072		12.0048	2400.96	$SPM1$	1

FIGURE B.5 The detached coefficient matrix after transforming the nonpivot row.

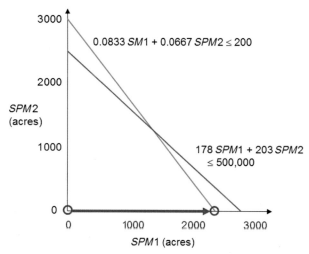

FIGURE B.6 Movement of the search (in graphical space) after Iteration 1.

(develop initial detached coefficient matrix) is not repeated in subsequent iterations of the Simplex Method.

Iteration 2, Step 2. The opportunity costs are calculated for each column (variable).

$$\text{OC}_{SPM1} = (1) - (0 \times 0) - (1 \times 1) = 0$$
$$\text{OC}_{SPM2} = (1) - (60.472 \times 0) - (0.8007 \times 1) = 0.1993$$
$$\text{OC}_{S1} = (0) - (1 \times 0) - (0 \times 1) = 0$$
$$\text{OC}_{S2} = (0) - (-2136.855 \times 0) - (12.005 \times 1) = -12.005$$

Iteration 2, Step 3. Since one or more opportunity costs are greater than zero, the process continues, and we note that the optimal solution has not yet been located.

Iteration 2, Step 4. The column of the matrix that is associated with the highest positive opportunity cost is the $SPM2$ column, since it has an opportunity cost of 0.1993, and the others have zero or negative opportunity costs.

Iteration 2, Step 5. Calculate the maximum contributions, or the maximum amount of the variable represented in the pivot column that can be brought into the solution, as affected by each constraint.

$$MC_1 = (72,629/60.4718) = 1201.0403$$
$$MC_2 = (2400.96/0.80072) = 2998.5007$$

The smallest of these values tells us that at most, 1201.04 acres of *SPM2* can enter the solution without violating any of the constraints. This is related to constraint row 1.

Iteration 2, Step 6. The constraint row of the matrix that was associated with the lowest maximum contribution is the pivot row (constraint row 1).

Iteration 2, Step 7. The key number is the value of the cell in the matrix that is located at the intersection of the pivot column and pivot row, or 60.4718.

Iteration 2, Step 8. Since each constraint that is not identified as the pivot row needs to be transformed, and there is only one other constraint, other than the one identified as the pivot row, we simply need one transformation ratio.

$$TR_1 = (0.80072/60.4718) = 0.01324122$$

Iteration 2, Step 9. Transform the pivot row values by dividing each by the key number (60.4718). As you can see (Fig. B.7), the value of Q in constraint row 1 has changed to the maximum contribution value we calculated in Iteration 2, Step 5. In addition, we note to the right of the matrix that the variable *SPM2* has entered the solution, and that it has a contribution of 1 to the objective function value for each acre assigned to *SPM2*.

Iteration 2, Step 10. Transform the nonpivot row (constraint 1).

New nonpivot row value for Column *SPM1*
= $(1) - (0) \times (0.01324122) = 1$
New nonpivot row value for Column *SPM2*
= $(0.80072) - (60.4718) \times (0.01324122) = 0$
New nonpivot row value for Column *S1*
= $(0) - (1) \times (0.01324122) = -0.01324122$
New nonpivot row value for Column *S2*
= $(12.005) - (-2136.86) \times (0.01324122)$
= 40.29937
New nonpivot row value for Column Q
= $(2400.96) - (72629.06) \times (0.01324122)$
= 1439.263

Fig. B.8 illustrates the condition of the detached coefficient matrix after two complete transformations (two iterations). When viewed as a graph, in two-dimensional space, we find that the Simplex Method moved from the previous solution (*SPM1* = 2400.96, *SPM2* = 0) to a solution where *SPM1* = 1439.26 and *SPM2* = 1201.04 (Fig. B.9).

SPM1	SPM2	S1	S2	Q	Variables in the solution	Their value
1	1					
0	1	0.01654	-35.3364	1201.04	SPM2	1
1	0.80072		12.0048	2400.96	SPM1	1

FIGURE B.7 The detached coefficient matrix after transforming the pivot row.

SPM1	SPM2	S1	S2	Q	Variables in the solution	Their value
1	1					
0	1	0.01654	-35.3364	1201.04	SPM2	1
1	0	-0.01324	40.2994	1439.26	SPM1	1

FIGURE B.8 The detached coefficient matrix after transforming the nonpivot row.

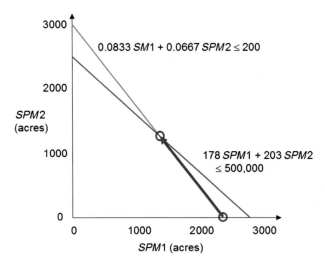

FIGURE B.9 Movement of the search (in graphical space) after Iteration 2.

Iteration 3 of the Simplex Method

As with iteration 2, since we have not determined whether the optimal decision has been located, we must first assess the opportunity costs. Again, Step 1 of Iteration 1 (develop initial detached coefficient matrix) is not repeated in subsequent iterations of the Simplex Method.

Iteration 3, Step 2. The opportunity costs are calculated for each column (variable).

$$OC_{SPM1} = (1) - (0 \times 1) - (1 \times 1) = 0$$
$$OC_{SPM2} = (1) - (1 \times 1) - (0 \times 1) = 0$$
$$OC_{S1} = (0) - (0.01654 \times 1) - (-0.01324 \times 1) = -0.0033$$
$$OC_{S2} = (0) - (-35.3364 \times 1) - (40.29937 \times 1) = -4.9630$$

Iteration 3, Step 3. Since all of the opportunity costs are zero or less, we learn the optimal solution has been located. The optimal solution is $SPM2 = 1201.04$ acres, and $SPM1 = 1439.26$ acres. As a check of the solution against the constraints, we can input each of the optimal solution values into the two constraints:

Budget constraint:

($178 per acre) × (1439.26 acres) + ($203 per acre)
   × (1201.04 acres) = $499,999.40

Operational constraint:

(0.0833 days acre) × (1439.26 acres) + (0.0667 days
   per acre) × (1201.04 acres) = 200 days

Both of the constraints are binding. That is, all the resources (budget and allowable work days) are needed for one of the two site preparation methods. Even though the budget is not exactly at $500,000 in the preceding example, some rounding led to our result.

Verification of the Simplex Method Results

As a check to the solution developed by hand earlier, we solved the problem using LINGO (LINDO Systems, Inc., 2016), and found the following:

Problem formulation:

```
max SPM1 + SPM2
subject to
2) 178 SPM1 + 203 SPM2 < = 500000
3) 0.0833 SPM1 + 0.0667 SPM2 < = 200
end
```

Result:

```
Global optimal solution found.
Objective value:              2640.303
Infeasibilities:              0.000000
Total solver iterations:             2
Elapsed runtime seconds:          1.60

Variable    Value       Reduced Cost
SPM1        1439.263    0.000000
SPM2        1201.040    0.000000

Row    Slack or Surplus    Dual Price
2)     0.000000            0.003295
3)     0.000000            4.962975
```

As we had hoped, the optimal solution provided by LINGO is consistent with the solution we generated by hand using the Simplex Method.

QUESTIONS

1. You work for the United States Forest Service in California, and manage a budget to develop fish structures (to facilitate the development of pools). Your budget is $100,000 per year. Based on the availability of a local contractor, you determine that there are two options: place logs in the stream system, or place boulders in the stream system. You would like to develop as much fish habitat in as many miles of stream as possible each year. It will cost $10,000 per stream mile to create fish structure using logs, and $21,000 per stream mile to create structures using boulders, due to the availability of logs and boulders. You would like the project to be completed during a 30-day time period, and the contractor indicates that they will require 5 days per mile of stream to install logs, and 3 days per mile of stream to install boulders. Develop the detached coefficient matrix, and solve this problem using the Simplex Method.

2. You work as a procurement forester in south Alabama, and manage a cruising budget, which is $2000 per month. You would like to cruise as many areas as possible to develop bids for timber purchases. Based on recent experience, you think it will cost $8 per acre to sample with 1/10 acre fixed radius circular plots, and $4 per acre to sample using a BAF 10 prism. You have some local help (a forest technician), but they prefer to sample with plots. After discussing your needs with the technician, you determine, given the other work available to the technician, that they can measure 500 plots per month, and 400 points. Develop the detached coefficient matrix, and solve this problem using the Simplex Method.

REFERENCES

Barlow, R., Levendis, W., 2015. Special report: 2014 cost and cost trends for forestry practices in the South. For. Landown. 74 (5), 22–31.

Barlow, R.J., Dubois, M.R., 2011. Cost & cost trends for forestry practices in the South. For. Landown. 70 (6), 14–23.

LINDO Systems, Inc., 2016. Extended Lingo/Win64. LINDO Systems, Inc., Chicago, IL.

Morgan, S.S., 1997. A Comparison of Simplex Method Algorithms. M.S. Thesis. University of Florida, Gainesville, FL, 110 p.

O'Connor, J.J., Robertson, E.F., 2003. George Dantzig. University of St. Andrews, St. Andrews, Scotland, UK. Available from: http://www-groups.dcs.st-and.ac.uk/~history/Biographies/Dantzig_George.html (Accessed 5/11/2016).

Appendix C

Writing a Memorandum or Report

As a professional in your field, you will be expected to communicate effectively with your colleagues, supervisors, and the public. E-mail is a standard method for communicating ideas, information, and results. However, written memorandums and reports are still quite commonly used in natural resource organizations. This Appendix focuses on two types of written communication (memorandums and reports), although we would hope that you will apply some of the same concepts described here to e-mail communications as well. This Appendix is designed to illustrate standard memorandum and report formats as well as the numerous pitfalls students may encounter when developing them. Upon reading this Appendix, you should be able to:

1. Understand the elements and characteristics of a well-written memorandum.
2. Understand the elements and characteristics of a well-written report.
3. Understand the difference between the two reporting options.

C.I. MEMORANDUMS

A memorandum is a relatively brief form of written communication typically used within office settings for communicating specific information. A basic memorandum contains five distinct structural elements:

- The date that the memorandum was created
- The name of the person who created the memorandum
- The name of the person to whom the memorandum is addressed
- The subject of the memorandum
- The content of the memorandum

It is not imperative, but at the top of the memo, the word *Memorandum* might also be placed. The most common errors related to the date of the memorandum include using the incorrect year, and using a date that is prior to the date that the memorandum actually was written. The name of the person writing the memorandum should be obvious (you). The name of the person to whom the memorandum is being addressed ideally should be represented by the name of an actual person or group of persons, as opposed to "Instructor," "Lab TA," or some other set of words. If you find an entry for "cc:" ("courtesy copy", or the older term "carbon copy") in a memorandum template, then you would place here the name of a person to whom the memorandum was copied.

The subject line of a memorandum should adequately describe the content, and use descriptive terms such as "Sampling results from stand 35," rather than more vague terms such as "Lab 5." Some memorandum templates use "Re:" to represent the subject line ("Re" is an abbreviation for "Regarding"). A memorandum typically refers only to one subject, so that the results or discussion that it contains can be filed in the most appropriate place.

The content of a memorandum needs to present information in a logical manner, and in a manner that conveys results clearly, correctly, and concisely. Some important considerations for the content of a memorandum include the following:

- The objective of the work
- The approach that was used to collect and process information
- A summary

A number of items are probably not appropriate for a memorandum, such as a description of the standard equipment that might have been used (diameter tape, etc.), and other basic information that natural resource management professionals should take for granted, such as how many paces were needed to locate sample plots. However, other appropriate topics for memoranda may include changes in organizational procedures, changes in strategy with a supporting justification, or recommendations for new capital purchases like buying a truck. If figures or tables are referenced within a memorandum, then they must be attached to the memorandum. The converse holds as well: if figures or tables are attached to a memorandum, then they must have been referenced within the memorandum.

To ensure that your memorandum has a professional appearance, try to answer the following questions:

1. Is the font size consistent throughout the memorandum?
2. Is the font type (Times New Roman, Arial, etc.) consistent throughout the memorandum?
3. Has the memorandum been checked for typographical and grammatical errors?
4. Are units such as ft^3 appropriately superscripted?
5. Have the main results been provided in the text of the memorandum?
6. Have the main results been interpreted?

To ensure that tables are high quality, consider answers to the following questions:

1. Are descriptive column headers placed over the top of the data that they were meant to represent?
2. Is consistent precision used within each column of data (i.e., are the decimal places consistent within a column)?
3. Is the precision used within each column of data (i.e., the decimal places) appropriate for the data being described?
4. Does the table have a title?
5. Has the table been spell-checked?
6. Is the data right-justified within in a column? Center- or left-justification of data is less intuitive to many people.
7. Can the table stand alone so that anyone could understand it should it become separated from the memo?

Figures are graphs, maps, and other items that are not considered tables of data. To ensure that figures are high quality, consider answers to the following questions:

1. Are the axes of graphs labeled?
2. Does the figure have a title?
3. Is a consistent font used throughout each figure?
4. Has the figure been spell-checked?
5. Can your figure stand alone so that anyone could understand it should it become separated from the memo?

An example of a brief memorandum is presented in Fig. C.1. Of course, Fig. C.1 represents only the text of a memorandum, and its associated table (Fig. C.2) and figure (Fig. C.3) should be attached (preferably stapled to the memo).

C.II. REPORTS

In general, a report is a written document that is meant to inform other people of an idea or situation that you have studied. The content of a report is much different than that of a memorandum. In a report the content is designed to be much more informative of the background, purpose, importance, and influence of a topic or issue. In contrast, the content of a memorandum is brief and generally to the point. Reports generally contain the six basic structural elements:

- A summary or abstract
- An introduction to the report
- A description of the methods employed
- A summary of the results that were found
- A discussion of the importance of the results
- The important conclusions

These sections represent the *summary*, *introduction*, *methods*, *results*, *discussion*, and *conclusions* sections that commonly are used in reports. The summary or abstract can also be called the *executive summary*, depending on the situation. The summary presents the main findings or arguments that are presented within the report. The summary also helps readers quickly determine what may be found within the report. The summary may be one of the most difficult parts of a report to write, since you must describe the content of the report in a condensed form, and since it may be the only part of the report that is read thoroughly by busy managers. The summary, in effect, is used to convince people to read further. The introduction, by comparison, is a section of the report that presents the purpose, scope, and objectives of the written document. The introduction can include hypotheses as well as a brief guide to the material that will follow. The introduction to a report can also often include a summary of prior studies or reports on the same (or similar) subject. This section of a report prepares readers for what follows.

The methods section of a report provides readers with an understanding of the types of materials and data that were used to develop results from which conclusions or decisions are drawn. In addition to a description of the equipment and analytical techniques used to arrive at results, the methods section may also include a critique of other previously used methods, and may also include the hypotheses. If the report concerned an experiment, then the methods section would describe how the experiment was designed. If the report described a survey, then the methods section would describe the sequence of events that were followed to effectively design and deliver a survey. The methods should be thoroughly described so that the reader can understand how the results were generated and whether the analysis performed on the data collected is valid. Ultimately, the methods section would contain enough information so that someone else could repeat the study or analysis.

The results section of a report provides the outcomes of the inquiry, either quantitatively or qualitatively. Comparisons of outcomes are also provided here, although interpretation of the reasons for different (or similar) outcomes is usually left for the discussion

Memorandum

To: Dr. Boston
From: Steven Johnson
Date: 5/10/2016
Re: Sample cruise of stand 45 on the Miller Tract

On March 24, 2016, I performed a prism sample of stand 45 on the Miller Tract. The stand type can be considered natural pine / hardwood, and the trees are approximately 25 years old. My BAF 10 prism points were spaced apart on a 4 chain by 5 chain grid (Figure 1), and I was able to sample approximately 13 points within the stand.

Upon developing the stand-level summaries from this sample, I found that the basal area is, on average, about 120 ft^2 per acre. In addition, the timber volume within the stand is around 2,652 ft^3 per acre (Table 1). Given the large number of dispersed oaks throughout the stand, deer habitat suitability averaged is estimated to be about 0.515 (good quality). This is a well-stocked, natural pine / hardwood stand that has significant timber and wildlife value.

FIGURE C.1   An example of a brief memorandum.

Table 1. Stand and stock table for stand 45.

Diameter class (inches)	Trees per acre	Basal area (ft^2/ac)	Volume (f^3/ac)
4	5	0.4	6.3
5	12	1.6	26.1
6	17	3.3	58.8
7	33	8.8	168.8
8	49	17.1	350.5
9	56	24.7	536.5
10	51	27.8	632.3
11	29	19.1	452.6
12	12	9.4	230.5
13	7	6.5	162.4
14	2	2.1	55.2
15	1	1.2	30.1
Total	274	122.3	2710.1

FIGURE C.2   An example table from a memorandum.

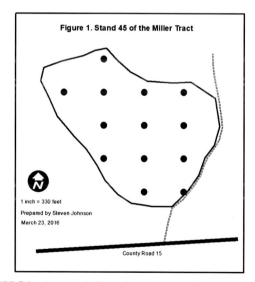

Figure 1. Stand 45 of the Miller Tract

1 inch = 330 feet
Prepared by Steven Johnson
March 23, 2016

County Road 15

FIGURE C.3   An example figure from a memorandum.

section of a report. It is in the results section that we find the highest quantity of tables and figures. As a result, the suggestions we provided earlier regarding tables and figures for memorandums also should be followed. The results should be presented in a relatively straightforward manner, and their presentation should be organized in a logical manner as well. The discussion section of a report represents an interpretation of the results provided. As with the presentation of results, the discussion should flow in a logical manner: if a reader needs to understand one specific set of information prior to considering a broader set of information, then the discussion should be organized accordingly. The discussion section is where you would compare the outcomes of the current study with those from previous studies or reports. In addition, it is here where the hypotheses that may have been proposed earlier are addressed (e.g., did you reject or fail to reject the hypothesis you proposed?).

As with the summary section, the conclusions section needs to be brief and concise, and it needs to present the main thoughts of the report clearly. This section reiterates the important concepts presented in the report, and it may also present areas for future work, make recommendations for decisions, or raise questions that continue to linger. The conclusions section is your last chance to convince the reader of the value of your work.

Index

Note: Page numbers followed by "*f*," "*t*," and "*b*" refer to figures, tables, and boxes, respectively.

A

Accounting rows, for linear programming
    models, 156−160
  habitat-related accounting rows, 158−160
  land areas scheduled for treatment, 156−157
  wood flow-related accounting rows,
    157−158
Accretion, 88, 98
Acid rain, 57
Adaptive management planning process,
    273−274, 274*f*
Adjacency and green-up restrictions
  adjacency relationship types, 250*f*
  area restriction model (ARM), 249−250,
    253*f*
  group selection patch harvests, 254, 254*f*
  Lincoln Tract planning, 251, 252*f*, 253*f*,
    260*f*, 262−263, 263*f*, 263*t*
  unit restriction model (URM), 249−250,
    252*f*
Adjacency relationships, three types of, 250*f*
Advanced planning techniques, 177
  binary search, 184−188, 185*t*, 187*t*, 189*f*
  forest planning software, 194−196
    Habplan, 195
    Multiple-resource Analysis and
      Geographic Information System
      (MAGIS), 195
    Remsoft spatial planning system (RSPS),
      195−196
    Spectrum model, 194−195
    Tigermoth, 196
  heuristic methods, 189−194
    ant colony optimization, 194
    genetic algorithms, 192−194, 193*f*
    HERO algorithm, 194
    Monte Carlo simulation, 190, 190*f*
    raindrop method, 194
    simulated annealing, 190−191, 191*f*
    tabu search, 192, 192*f*
    threshold accepting, 191−192
  linear programming, extensions to, 177−184
    goal programming, 182−184
    integer programming, 179−182
    mixed integer programming, 178−179
Aerial photogrammetry, GIS data collection,
    67−68
African Participatory Management plan, 12−13
After-tax cost, calculation of, 53

Age class distribution of the Lincoln Tract, 224*f*
Age classifications, 28−29
Age of stand, evaluation of, 28−29
Age of trees, 28−29
Air quality, calculation of, 56−58
Allowable cut effect, 243−245
Alternative rate of return, 51
American Tree Farm System, 9−10, 17*b*,
    292−293, 297−298
Annual operational planning, 287*b*
Ant colony optimization, 194
Aquatic habitat values, calculation of, 56
Arapaho National Forest plan, 80*b*, 81*t*
Area control, 229−230
  application of, to Putnam Tract, 240−242
Area restriction model (ARM), 249−250, 253*f*
Area to harvest, control, 230, 241
Area−volume check, 242
ARM. *See* Area restriction model (ARM)
Asian Private and Communal Forest Area
    Management plan, 11−12
ASPEN model, 107
Australian Forestry Standard, 293
Australian State Forest plan, 10−11
Austrian formula for volume control, 236−238,
    241
Average height, 24

B

Backward recursion, dynamic programming,
    125
Bands, 67
Bare land value (BLV), 49
Basal area, evaluation of, 23−24, 24*b*
BCR. *See* Benefit/cost ratio (BCR)
BDq method, volume control, 240
Benefit/cost ratio (BCR), calculation of, 48
Bimodal forest age class structure, 223*f*
Binary search, 184−188, 185*t*, 187*t*, 189*f*, 233
Biological maturity, 25
Biological rotation age, 117
Biomass and carbon, 30
Bird species, spatial restrictions in forest
    planning and habitat quality
    considerations, 255−256
BLV. *See* Bare land value (BLV)
Brazil's Certificação Florestal, 293
Breast height age, 28
Broader-scale volume estimates, 98−99

Broad-scale habitat estimates, 99−101
Brule River State Forest (Wisconsin), 16*b*
Buffering, 73−74

C

CACTOS. *See* California Conifer Timber
    Output Simulator (CACTOS)
CAI. *See* Current annual increment (CAI)
California Climate Action Registry, 303
California Conifer Timber Output Simulator
    (CACTOS), 106−107
California Forest Practices Rules, 210
Canadian Standards Association, forest
    certification, 293, 298−299
Canopy cover, 28, 92
Capital gain, 53
Carbon credits, 302
Carbon dioxide, air quality, 57
Carbon fraction of dry wood, 30
Carbon measurement, in trees, 30
Carbon sequestration, 300−301
  emissions trading, 301−302
  forest management
    implications, 303−304
    opportunities and challenges, 301
  Kyoto Protocol, 300−303
  reporting and trading schemes, 302−303
Carbon storage, 300−301
Cartography, 76
Certification. *See* Forest certification
Chain-of-custody verification process, 282
Chattahoochee-Oconee National Forest, 10,
    80*b*, 81*t*
Chicago Climate Exchange, 303
Chico Mendes Extractive Reserve, 12
Chile's Certificación Forestal, 293
City of San Francisco urban forest plan,
    269−270
Clean Development Mechanism, 302
Clearcut harvesting
  adjacency and green-up rules as they relate
    to, 249−254
Climate change. *See* Carbon sequestration
Collaborative forest management, 8−9
Community/cooperative forest plans, 8−9
Compound interest, 35
Conceptual model of synergies related to forest
    planning processes, 274*f*
Conservation management plan, 12

Constraints, for linear programming problems, 160–162
  on habitat availability, 162
  on harvested areas, 160–161
  on harvested volume, 161–162
  policy constraints, 160–162
  resource constraints, 160
Control techniques. *See* Harvesting
Cooperative Forest Plans, 8–9
Cost of capital, 51
Cross-ownership planning, 9
Crown cover. *See* Canopy cover
Crown Forest Sustainability Act of 1994, 208
Culmination of mean annual increment, 216
Current annual increment (CAI), 26, 27*f*
Curse of dimensionality, management problems, 114, 125

D
Databases used throughout forest management and planning, 315
  Douglas-fir stand from western Oregon, 315
DBH. *See* Diameter at breast height (DBH)
Decision-making, 14–15
  irrational model of, 14
  characterizing, 13–18
    hierarchy of planning within natural resource management organizations, 17–18
    planning within natural resource management organizations, 15–17
    view from management sciences, 13–15
Decision tree analysis, 122–124, 123*f*
Demand management, 281
DEMs. *See* Digital elevation models (DEMs)
Density, evaluation of, 33–34
Depletion account, 53
Desired forest structure models, 215
  irregular forest structures, 222–224, 222*f*
  normal forest, 215–221, 216*f*
    normal yields, 227
  regulated forest, 221–222
  structures guided by historical range of variability, 224–225
  structures not easily classified, 225–226
Detached coefficient matrix, 162–163
  Simplex Method, 332–333, 332*f*
Diameter. *See* Quadratic mean diameter (QMD); Tree average diameter; Tree diameter distribution
Diameter at breast height (DBH), 22–23, 23*b*
Digital elevation models (DEMs), 68
Digital orthophotographs, 68–69
Digital raster graphics (DRGs), 68–69
Diminution quotient, 93
Discount rates, calculation of, 51–52
Discounted cash flow model, 52
Doha Amendment, 302
Douglas-fir simulator (DFSIM) model, 107
Douglas-fir stand from western Oregon, 315, 316*t*, 322*t*
  dynamic programming of thinning with fixed rotation length, 129–131, 130*f*, 130*t*
  stand tables by age, 316*t*, 317*t*, 318*t*, 319*t*, 320*t*, 321*t*, 322*t*

Down woody debris
  classes, 28*t*
  volume evaluation, 27–28
DRGs. *See* Digital raster graphics (DRGs)
Dry weight, evaluation, 30
Duel prices, linear programming models, 166–167
Dynamic programming, 124–135
  caveats, 126
  disadvantages, 126
  examples
    cost of evening out, 126–129
    southern stand thinning, varying rotation lengths, 132–135
    western stand thinning, fixed rotation length, 129–131
  nodes, 125
  recursive relationships, 125
  stages, 124–125
  states, 125

E
EAE. *See* Equal annual equivalent (EAE)
Ecological approaches to plans, 9
Economic challenges, 6–7
Economic evaluation
  benefit/cost ratio (BCR), 48
  discount rate, 51–52
  equal annual equivalent (EAE), 48
  forest taxation, 52–53
  future value
    equation, 36
    non-terminating annual cost or revenue, 41
    non-terminating periodic cost or revenue, 42
    one revenue or cost, 40–41
    terminating annual cost or revenue, 41–42
    terminating periodic cost or revenue, 42–43
  income and employment, 58
  internal rate of return (IRR), 47–48
  mixed-method economic assessments, 50–51
  net present value (NPV), 46–47
  present value, 35
    equation, 35–36, 40*t*
    non-terminating annual revenue or cost, 36–37
    non-terminating periodic revenue or cost, 38–39
    periodic revenue or cost, 39–40
    single revenue or cost, 36
    terminating annual revenue or cost, 37–38
  prices and costs, 43–46
  soil expectation value (SEV), 49–50
Economic rotation age, 119
Ecosystem management, 9
Ecosystems and social values, sustainability of, 206–208
Ecosystem services, 202
Elk, spatial restrictions in forest planning and habitat quality considerations, 255
Elliott State Forest plan (Oregon) (2011), 16

E-mail communications, 339
Empirical yield tables, 218–220
Employment, evaluation, 58
Environmental and social evaluation, 53–58
  air quality, 56–58
  aquatic habitat values, 56
  habitat suitability, 53–54
  income and employment, 58
  recreation values, 54–55
  water resources, 55–56
Equal annual equivalent (EAE), calculation of, 48
Equal annual income, 48
Equivalent annual cash flow, 48
European Estate plan, 11
European Union Emission Trading Scheme, 303
Even-aged management, 51, 121, 125
Expansion factor, 103–104

F
Federal Clean Water Act, 55
Field data, GIS data collection, 67
Fish habitat development plan, graphical solutions for two-variable linear problems, 146–147
Forest age, 28–29
Forest certification, 291, 296*t*
  cost and benefits of, 299–300
  by major region, 292*t*
  programs, 294–299
    American Tree Farm System, 297–298
    Canadian Standards Association, 298–299
    Forest Stewardship Council (FSC), 296–297
    Green Tag Certified Forestry program, 298
    International Organization for Standardization (ISO) 14001, 299
    Programme for the Endorsement of Forest Certification, 299
    Sustainable Forestry Initiative, 295–296
Forest conditions, valuing and characterizing, 21
  economic evaluation, 34–53
    benefit/cost ratio (BCR), 48
    equal annual equivalent (EAE), 48
    forest taxation, 52–53
    internal rate of return (IRR), 47–48
    mixed-method economic assessments, 50–51
    net present value (NPV), 46–47
    present and future values, 35–43
    prices and costs, 43–46
    selecting discount rates, 51–52
    soil expectation value (SEV), 49–50
  environmental and social evaluation, 53–58
    air quality, 56–58
    aquatic habitat values, 56
    habitat suitability, 53–54
    income and employment, 58
    recreation values, 54–55
    water resources, 55–56

structural evaluation, 21–34
  average diameter of trees, 22–23
  average height, 24
  basal area, 23–24
  biomass and carbon, 30
  crown or canopy cover, 28
  diameter distribution of trees, 23
  down woody debris, 27–28
  mean annual increment (MAI), 25–27,
    27*f*
  nontimber forest products, 30–31
  periodic annual increment, 26
  pine straw, 30
  quadratic mean diameter (QMD), 24
  site quality, 31–33
  snags per unit area, 27
  stocking and density, 33–34
  timber volume, 24–25
  tree, stand, or forest age, 28–29
  trees per unit area, 21–22
Forest growth. *See* Growth
Forest planning situations, 1–2
Forest planning software, 194–196
  Habplan, 195
  Multiple-resource Analysis and Geographic
    Information System (MAGIS), 195
  Remsoft spatial planning system (RSPS),
    195–196
  Spectrum model, 194–195
  Tigermoth, 196
Forest rent. *See* Income generation rotation age
Forest-level planning. *See* Binary search; Goal
    programming; Graphical solution
    techniques, for two-variable linear
    problems; Heuristic methods; Integer
    programming; Linear programming
    models; Mixed integer programming
Forestry supply chain. *See* Supply chain
Forest Stewardship Council (FSC), forest
    certification, 292–293, 296–297
Forest structure. *See* Desired forest structure
    models; Structural evaluation
Forest succession, 3
Forest taxation, 52–53
Forest vegetation simulator (FVS), 106
FORPLAN, 194–195, 308
Forward recursion, dynamic programming,
    125
FSC. *See* Forest Stewardship Council (FSC)
Future-now-thinking procedure, 309–310
Future value
  equation, 36
  non-terminating annual cost or revenue, 41
  non-terminating periodic cost or revenue, 42
  one revenue or cost, 40–41
  terminating annual cost or revenue, 41–42
  terminating periodic cost or revenue, 42–43
FVS. *See* Forest vegetation simulator (FVS)

G

Gap simulators, 105
Garbage can model, decision making, 13–14
Genetic algorithms, 192–194, 193*f*

Geographic information system (GIS), 6,
    65–76
  data collection, 66–68
    aerial photogrammetry, 67–68
    field data collection, 67
    map digitizing, 67
    remote sensing, 67
  data structures, 68–70
    raster data, 68–69
    topology, 70
    vector data, 69–70
  GIS databases, 7, 70, 74–75, 178–179
  Lincoln Tract database, 70–71
  processes, 71–76
    buffering, 73–74
    clipping and erasing, 73
    combining and splitting, 74–75
    joining, 75
    mapping, 76
    overlaying, 75–76
    proximity analysis, 74
    querying, 72–73
  Putnam Tract database, 70
Georgia Carbon Sequestration Registry,
    303
GIS. *See* Geographic information system (GIS)
Goal programming, 182–184
Graphical solution techniques, for two-variable
    linear problems, 139
  examples, 140–148
    fish habitat development, 146–147
    hurricane clean-up plan, 147–148, 148*f*
    road construction plan, 140–144
    snags development to enhance wildlife
      habitat, 144–146, 145*f*
  production possibility frontier, 150
  solution, 148–149
    efficiency, 149–150
    feasible solution, 148–149
    infeasible solution, 148–149
    optimal solution, 149
    suboptimal solution, 149
  steps used to solve a problem, 139–140
Green certification, 291–292
Green Tag Certified Forestry program, 298
Green-up rules. *See* Adjacency and green-up
    restrictions
Group selection patch harvests
  adjacency and green-up of, 254, 254*f*
Growth, 87–101
  accretion, 88, 98
  even-aged stands, 89–91
  ingrowth, 88
  model evaluation, 108–109
  mortality, 88
  output, 108
  self-thinning, 88
  simulators, yield, 103–108
    ASPEN model, 107
    California Conifer Timber Output
      Simulator, 106–107
    diameter class models, 105
    Douglas-fir simulator (DFSIM) model,
      107

Forest Vegetation Simulator, 106
    gap simulators, 105
    individual tree, distance-dependent
      models, 104
    individual tree, distance-independent
      models, 103–104
    landscape management system (LMS), 108
    ORGANON model, 107
    PTAEDA 4.0, 107–108
    Simulator for Intensively Managed Stands,
      108
    snag and coarse woody debris models,
      105–106
    Tree and Stand Simulator, 108
    whole-stand models, 104–105
    Zelig, 107
  survivor growth, 94–95
  transition through time, 94–101
    broader-scale volume estimates, 98–99
    broad-scale habitat estimates, 99–101
    stand-level volume estimates, 95–98
  two-aged forests, 94
  uneven-aged forests, 91–94
  value growth rate, 114–115, 119
  volume table, 101–102
  and yield data, 7
  and yield tables, 101–103

H

Habitat
  accounting rows, 156–160
  broad-scale estimates, 99–101
  fish habitat development plan, 146–147
  requirements, 256
  snag development, 145, 145*f*
  spatial restrictions in forest planning and
    quality considerations, 255–260
    bird species habitat (case), 255–256
    elk habitat (case), 255
    Lincoln Tract, 260*f*
    red-cockaded woodpecker habitat (case),
      256–258
    spotted owl habitat (case), 258–260
  wildlife habitat control, 242–243
Habitat availability, constraints on, 162
Habitat capability index (HCI) model, 258
Habitat suitability index (HSI), 53–54, 76*f*,
    158, 159*t*
Habplan, forest planning software, 195
Hanzlik formula, for volume control, 233–235,
    241
Hardwood-dominated forest in West Virginia,
    202*f*
Harvested areas, constraints on, 160–161
Harvested volume, constraints on, 161–162
Harvesting, 229
  allowable cut effect, 243–245
  area control method, 229–230
  area–volume check, 242
  Putnam Tract, application of area control to,
    241
  Putnam Tract, application of volume control
    to, 242*f*
  Austrian formula, 241

Harvesting (*Continued*)
  BDq method, 240
  Hanzlik formula, 241
  Hundeshagen formula, 241
  Meyer formula, 242
  Von Mantel formula, 241
  volume control, 231–240
    Austrian formula for, 236–238
    Hanzlik formula for, 233–235
    Heyer method for, 239–240
    Hundeshagen formula for, 238
    Meyer amortization method for, 238–239
    structural methods for, 240
    Von Mantel formula for, 235–236
  wildlife habitat control, 242–243
HERO algorithm, 194
Heuristic methods, 189–194
  ant colony optimization, 194
  genetic algorithms, 192–194, 193*f*
  HERO algorithm, 194
  Monte Carlo simulation, 190, 190*f*
  raindrop method, 194
  simulated annealing, 190–191, 191*f*
  tabu search, 192, 192*f*
  threshold accepting, 191–192
Heyer method for volume control, 239–240
Hierarchical system for planning and
      scheduling management activities, 269
  blended, combined, and adaptive approaches,
      273–275
  flow of information, 275*f*
  operational planning, 271–272
  role of planners and managers, 275
  strategic planning, 269–270
  tactical planning, 270–271
  vertical integration of planning processes,
      272–273
Hierarchy of forest planning, 282–285
  within natural resources management
      organizations, 17–18, 17*f*
Historical range of variability, structures
      guided by, 224–225
HSI. *See* Habitat suitability index (HSI)
Hufnagl method, 232–233
Humboldt-Toiyabe National Forest (Nevada),
      15–16
Hundeshagen formula for volume control, 238,
      241
Hurdle rate, 36
Hurricane clean-up plan, graphical solutions for
      two-variable linear problems, 147–148,
      148*f*
Hurricane Katrina, 147

I

Immediate forest, sustainability beyond, 211
Income and employment, 58
Income generation rotation age, 117–118, 119*f*
Ingrowth, 88, 94–95
Integer programming, 179–182, 181*f*
Intensity of forest management, 204
Intensively managed forest, 294–295
  in western Washington state, 294*f*

Internal rate of return (IRR), calculation of,
      47–48
International Organization for Standardization
      (ISO) 14001, 299
Inventories of resources, 4
IRR. *See* Internal rate of return (IRR)
Irrational model of decision-making, 14
Irregular data structure, 69–70
Irregular forest structures, 222–224, 222*f*, 223*f*

K

Kakamega Forest Ecosystem management plan,
      12–13
Karl's method, 236–237
Key number, Simplex Method, 333
Knowledge collection and reporting, 282
Kyoto Protocol, 300–303

L

Land classification, 76–84
  Putnam Tract example, 79*t*
  spatial position-based classifications, 82–84
  strata-based classifications, 80–82
  units of land-based classifications, 82, 82*t*
Land expectation values (LEVs), 35
Landscape management system (LMS), 108
Landscape plans, 8
Law of de Liocourt, 93
Leadership in Energy and Environmental
      Design (LEED), 294
Learning loop, 309–310, 309*f*
LEED. *See* Leadership in Energy and
      Environmental Design (LEED)
LEVs. *See* Land expectation values (LEVs)
LiDAR. *See* Light Detection and Ranging
      (LiDAR)
Light Detection and Ranging (LiDAR), 67
Lincoln Tract, 251, 315
  age class distribution of, 224*f*
  area restriction model of adjacency applied
      to the first-period harvests on, 253*f*
  costs associated with the development of a
      trail system, 263*t*
  geographic information system, 70–71
  potential trails system for, 263*f*
  unit restriction model of adjacency applied
      to the first-period harvests on, 252*f*
  wildlife habitat restrictions applied to, 260*f*
Linear programming models, 153, 182, 270
  accounting rows, 156–160
    habitat-related accounting rows, 158–160
    land areas scheduled for treatment,
      156–157
    wood flow-related accounting rows,
      157–158
  alternative management scenarios
      assessment, 167–168
  assumptions, 153–154
    of additivity, 154
    of certainty, 154
    of divisibility, 154
    of proportionality, 154

case study
    Northern United States hardwood forest,
      170–174, 172*t*
    Western United States forest, 168–170,
      169*t*
  constraints, 160–162
    on habitat availability, 162
    on harvested areas, 160–161
    on harvested volume, 161–162
    policy constraints, 160–162
    resource constraints, 160
  detached coefficient matrix, 162–163
  extensions, 177–184
    goal programming, 182–184
    integer programming, 179–182
    mixed integer programming, 178–179
  interpretation of results, 164–167
    duel prices, 166–167
    objective function value, 165–166
    reduced costs, 165–166
    slack, 166–167
    variable values, 165–166
  Model I problems, 163–164
  Model II problems, 163–164
  Model III problems, 163–164
  objective functions, 154–156
  Simplex Method. *See* Simplex Method
LMS. *See* Landscape management system
      (LMS)

M

MAGIS. *See* Multiple-resource Analysis and
      Geographic Information System
      (MAGIS)
MAI. *See* Mean annual increment (MAI)
Malaysian Timber Certification Council, 293
Management activities, types of, 2*t*
Management of forested lands, 1
Management problem
  check the solution to, 140
  solving the problem using mathematical/
      graphical methods, 140
  translation into mathematical equations.,
      139–140
  understanding, 139
Management recommendation, 4–5
Map digitizing, GIS data collection, 67
Mapping, 76
Market fluctuations, 204
Mathematical formulations associated with
      forestry supply chain components,
      286–288
Mathematical programming, defined, 153
McPhail Tree Farm, 10, 269–270, 297–298
Mean annual increment (MAI), evaluation of,
      25–27, 27*f*
Memorandum, writing, 339–340, 341*f*
Meyer amortization method for volume control,
      238–239, 242
Mixed integer programming, 178–179, 180*f*
Mixed-method economic assessments, 50–51
Molpus Timberlands Management, LLC, 16*b*
Monte Carlo simulation, 190, 190*f*, 253

Monthly operational planning, 287, 287*b*
Mortality, 88
Movement of information during a planning cycle, 6*f*
Moving window approach, 258, 258*f*
Multiple uses, sustainability of, 205–206, 206*f*
Multiple Use Sustained Yield Act of 1960, 207
Multiple-resource Analysis and Geographic Information System (MAGIS), 195
Mushroom production, evaluation of, 31*b*

N

National Forest Management Act, 15, 203, 207
Natural forces, forest depletions from, 204
Natural succession, 3
Need for forest management plans, 3–7
    information necessary to develop forest management plan, 4–6
    necessity of plans, planners, and planning processes, 3–4
    plan development challenges, 6–7
Net present value (NPV), 155
    calculation of, 46–47
NGOs. *See* Nongovernmental organizations (NGOs)
Nodes, dynamic programming, 125
Nominal discount rate, calculation of, 51
Nongovernmental organizations (NGOs), 292
Nonlinear programming, 124
Nonpivot row(s), Simplex Method, 333–334
Nonpoint source (NPS) pollution, 55
Nontimber forest products, 30–31
Normal forest, 34, 215–221, 216*f*
Normal yield tables, 104, 218–220
North American Industrial Forest plan, 13
North American National Forest plan, 10
North American Small Private Landowner plan, 10
North American Urban Forest plan, 13
Northern United States hardwood forest (case study), 170–174
NPV. *See* Net present value (NPV)

O

Objective function value, linear programming models, 154–156, 165–166
Operational planning, 18, 284–285, 287–288
    hierarchical system, 271–272
    transportation, 288*b*
Opportunity costs, Simplex Method, 333
Optimality decision, Simplex Method, 333
Optimal rotation age, 117
Optimization, of tree- and stand-level objectives, 113
    dynamic programming, 124–135
        caveats, 126
        disadvantages, 126
        example, 126–135
    mathematical models for, 124
    nodes, 125
    recursive relationships, 125–126
    stages, 124–125
    stand-level optimization, 116–122

optimum stand density/stocking, 121–122
optimum thinning timing, 120–121
optimum timber rotation, 116–119
recent developments in scientific literature, 122
states, 125
tree-level optimization, 114–116
Organic Act of 1897, 203
Organization-specific plans, 8
ORGANON, 107, 315

P

PAI. *See* Periodic annual increment (PAI)
Paris Agreement, 300, 302
Periodic annual increment (PAI), evaluation of, 26, 26*f*
Physical rotation age, 116
Pine straw, evaluation of, 30, 31*f*
Pivot column, Simplex Method, 333
Pivot row, Simplex Method, 333
Plan development challenges, 6–7
Policy constraints, 160–162
    habitat availability, constraints on, 162
    harvested areas, constraints on, 160–161
    harvested volume, constraints on, 161–162
Potential trails system, network flow diagram for, 263*f*
Prescribed Burning Act (1993), 58
Present value, 35
    equation, 35–36, 40*t*
    non-terminating annual revenue or cost, 36–37
    non-terminating periodic revenue or cost, 38–39
    single revenue or cost, 36
    terminating annual revenue or cost, 37–38
    terminating periodic revenue or cost, 39–40
Primary succession, 3
Production, sustainability of, 202–205
Production possibility frontiers, 150
Programme for the Endorsement of Forest Certification, 293, 299
Projecting stand conditions, 101–108
Proximity analysis, 74
PTAEDA 4.0, 107–108
Pulpwood Conservation Act of 1929, 203
Putnam Tract, 70, 179–180, 180*f*, 181*f*, 182, 183*f*, 186–188, 187*t*, 189*f*, 240–242, 315, 327*f*, 328*t*
    application of area control to, 241
    application of volume control to, 242*f*
        Austrian formula, 241
        Hanzlik formula, 241
        Hundeshagen formula, 241
        Meyer formula, 242
        Von Mantel formula, 241
    geographic information system, 70

Q

QMD. *See* Quadratic mean diameter (QMD)
Quadratic mean diameter (QMD), 24

R

RADR. *See* Risk adjusted discount rate (RADR)
Raindrop method, 194
Raster data, GIS, 68–69
Rate of growth, calculation of, 114–115
Rational model of decision-making, 13–14
Real discount rate, calculation of, 51
Real Estate Investment Trusts (REITS), 293
Recreation values, evaluation of, 54–55
Recursive relationships, 125–126
Red-cockaded woodpecker, spatial restrictions in forest planning and habitat quality considerations, 256–258
Red-cockaded woodpecker habitat considerations (case), 256–258
REDD + (Reduced Emissions from Deforestation and Degradation) program, 302
Reduced costs, linear programming models, 165–166
Regular data structure, 68
Regulated forests, 221–222, 230, 233, 236–238
Reineke's stand density index, 87
REITS. *See* Real Estate Investment Trusts (REITS)
Remote sensing, GIS data collection, 67
Remsoft spatial planning system (RSPS), 195–196
Reports, writing, 340–342
Resource constraints, 160
Risk adjusted discount rate (RADR), 52
Road construction plan, 140–144
Road construction problem, 143, 261*f*, 262
Road management problem (case), 261–262
RSPS. *See* Remsoft spatial planning system (RSPS)

S

Scenario analysis, in support of strategic planning, 307 310
    application to forest planning, 312
    developing scenarios, 310–312
Scenario planning learner, 311
Scenic beauty index, calculation of, 54
Secondary succession, 3
Self-thinning, 88
Semirational model, 14
SEV. *See* Soil expectation value (SEV)
Silvicultural rotation age, 117
Simplex Method, 331–332
    steps, 332–334, 332*f*
        detached coefficient matrix development, 332–333, 332*f*
        key number determination, 333
        maximum contributions of variables, 333
        nonpivot row(s) transformation, 333–334
        opportunity costs calculation, 333
        optimality decision, 333
        pivot column identification, 333
        pivot row identification, 333
        pivot row transformation, 333
        transformation ratio calculation, 333
        two-variable, two-constraint problem solution, 334–337

SiMS. *See* Simulator for Intensively Managed Stands (SiMS)

Simulated annealing, 190–191, 191*f*, 253

Simulator for Intensively Managed Stands (SiMS), 108

Site index, 32–33, 33*f*, 89–90

Site quality, evaluation of, 31–33

Slack, linear programming models, 166–167

Snags
development to enhance wildlife habitat, 144–146, 145*f*
evaluation of, 27
models, 105–106

Social values, sustainability of, 206–208

Software, forest planning, 194–196
Habplan, 195
Multiple-resource Analysis and Geographic Information System (MAGIS), 195
Remsoft spatial planning system (RSPS), 195–196
Spectrum model, 194–195
Tigermoth, 196

Soil expectation value (SEV), calculation of, 49–50

South American Community Forest plan, 12

Spatial position-based classifications, 82–84

Spatial restrictions and considerations in forest planning, 249
adjacency and green-up rules
of group selection patch harvests, 254, 254*f*
as they relate to clearcut harvesting, 249–254
habitat quality considerations, 255–260
bird species habitat considerations (case), 255–256
elk habitat quality (case), 255
red-cockaded woodpecker habitat considerations (case), 256–258
spotted owl habitat quality (case), 258–260
road management problem (case), 261–262
trail development problem (case), 262–264

Spatial Woodstock, 195–196

Spectrum model, 194–195, 270

Spotted owl, spatial restrictions in forest planning and habitat quality considerations, 256–258

Spotted owl habitat quality (case), 258–260

Stages, dynamic programming, 124–125

Stand-level volume estimates, 95–98

Stanley, forest planning software, 195–196

States, dynamic programming, 125

Stocking, evaluation of, 33–34

Strata-based land classifications, 80–82

Strata-based planning, 65

Strategic forest planning, hierarchical system, 269–270

Strategic landscape, 311*f*

Strategic planning, 283, 286*b*

Structural evaluation, 21–34
age of stand, 28–29
average diameter of trees, 22–23

average height, 24
basal area, 23–24
biomass and carbon, 30
canopy cover, 28
current annual increment (CAI), 26, 26*f*
diameter distribution of trees, 23
down woody debris, 27–28
mean annual increment (MAI), 25–27, 26*f*
mushroom production, 31*b*
nontimber forest products, 30–31
periodic annual increment, 25–27
pine straw, 30
quadratic mean diameter (QMD), 24
site quality, 31–33
snags per unit area, 27
stocking and density, 33–34
timber volume, 24–25
tree, stand, or forest age, 28–29
trees per unit area, 21–22

Structural methods for volume control, 240

Sulfur, air quality, 57

Supply chain, 279
components of, 281–282
demand management, 281
execution, 281
knowledge collection and reporting, 282
management, 281
planning and scheduling, 281
hierarchy of forest planning, 282–285
mathematical formulations, 286–288
process for forest products, 281*f*
variation sources, 288–289

Survivor growth, 94–95

Sustainability, 1, 3, 13, 201
concepts beyond immediate forest, 211
of ecosystems and social values, 206–208
incorporating measures of, into forest plans, 208–211, 209*f*, 210*f*, 211*f*
of multiple uses, 205–206, 206*f*
of production, 202–205

Sustainable annual harvest level, 233

Sustainable forest management, 3, 294–295

Sustainable Forestry Initiative, 293, 295–296

Sustainable harvest, 230, 232–235

Sustained Yield Forest Management Act of 1944, 205

T

Tabu search, 192, 192*f*, 253

Tactical planning, hierarchical system, 18, 270–271, 283, 286*b*

TASS. *See* Tree and Stand Simulator (TASS)

Taxation, 51–53

Technical rotation age, 116–117

Technological challenges, 7

Thinning timing, optimum, 120–121

Threshold accepting, 191–192

Tigermoth, 196

Timber Investment Management Organizations (TIMOs), 17, 293

Timber price, 43–46

Timber production, 201–203, 210–212

Timber rotation, optimization, 116–119, 117*f*, 118*f*, 119*f*, 120*f*

Timber volume, evaluation of, 24–25

Time value of money, 35

TIMOs. *See* Timber Investment Management Organizations (TIMOs)

Topology, GIS, 70

TPA. *See* Trees per acre (TPA)

TPH. *See* Trees per hectare (TPH)

Trail development problem (case), 262–264

Transformation ratio, Simplex Method, 333

Tree and Stand Simulator (TASS), 108

Tree average diameter, 22–23

Tree diameter distribution, 23

Tree list, 103–104

Tree quadratic mean diameter, evaluation of, 24

Tree-level optimization, 114–116

Trees per acre (TPA), 21–22

Trees per hectare (TPH), 21–22

Trees per unit area, evaluation of, 21–22

Two-aged forests, 94

U

Uncertainty, 307, 311–312

Uneven-aged forests
growth, 90*t*, 91–94
management, 50–51, 121–122, 125

Unit restriction model (URM), 249–250, 252*f*

United States Geologic Service (USGS), 66

Urban forest plans, 9

URM. *See* Unit restriction model (URM)

USGS. *See* United States Geologic Service (USGS)

V

Value growth percent rotation age, 119

Variability, structures guided by a historical range of, 224–225

Variable values, linear programming models, 165–166

Vector data, geographic information system, 69–70

Vertical integration of planning processes, 272–273

VicForests, 10–11

Volume control, 231–240
Austrian formula for, 236–238, 241
Hanzlik formula for, 233–235, 241
Heyer method for, 239–240
Hundeshagen formula for, 238, 241
Meyer amortization method for, 238–239, 241
structural methods for, 240
Von Mantel formula for, 235–236, 241

Volume table, 101–102

Von Mantel formula for volume control, 235–236, 241

W

Water resources, evaluation of, 55–56

Weekly operational planning, 288*b*

Western United States forest (case study), 168–170
Wildlife habitat control, 242–243
Wood flow-related accounting rows, 157–158
Woodstock, 195–196, 270

Y

Yield
  normal yields, 227
  simulators, 103–108
    ASPEN model, 107
    California Conifer Timber Output
      Simulator (CACTOS), 106–107

diameter class models, 105
Douglas-fir simulator (DFSIM) model, 107
forest vegetation simulator (FVS), 106
gap simulators, 105
individual tree, distance-dependent models, 104
individual tree, distance-independent models, 103–104
landscape management system (LMS), 108
ORGANON model, 107

PTAEDA 4.0, 107–108
Simulator for Intensively Managed Stands (SiMS), 108
snag and coarse woody debris models, 105–106
Tree and Stand Simulator (TASS), 108
whole-stand models, 104–105
Zelig, 107
tables, 101–103

Z

Zelig, 107

Printed in the United States
By Bookmasters